Results and Problems in Cell Differentiation

55

Series Editors

Dietmar Richter, Henri Tiedge

Jacek Z. Kubiak
Editor

Mouse Development

From Oocyte to Stem Cells

Springer

Editor
Jacek Z. Kubiak
Institute of Genetics and Development of Rennes (IGDR)
UMR 6290, CNRS/University Rennes 1
Faculty of Medicine, UEB, IFR 140 6290
Rennes Cedex
France

Series Editors
Dietmar Richter
Center for Molecular Neurobiology
University Medical Center
Hamburg-Eppendorf (UKE)
University of Hamburg
Hamburg Germany

Henri Tiedge
The Robert F. Furchgott Center for Neural and Behavioral Science
Department of Physiology and Pharmacology
Department of Neurology
SUNY Health Science Center at Brooklyn
Brooklyn, NY, USA

ISSN 0080-1844 ISSN 1861-0412 (electronic)
ISBN 978-3-642-30405-7 ISBN 978-3-642-30406-4 (eBook)
DOI 10.1007/978-3-642-30406-4
Springer Heidelberg New York Dordrecht London

Library of Congress Control Number: 2012945633

Printed on acid-free paper

Springer is part of Springer Science+Business Media (www.springer.com)

Preface

My purpose in preparing "Mouse Development. From Oocyte to Stem Cells" for the Springer series "Results and Problems in Cell Differentiation" was to create a comprehensive review of the current knowledge on the most popular mammalian model in developmental biology. Most scientific achievements that have had an important impact on the understanding of basic mechanisms governing embryo development in humans originated from mouse experimental embryology. Stem cell research, which now offers the promise of regenerative medicine, began with the derivation of mouse embryonic stem cells. One of the authors of this pioneering research—Martin Evans—was awarded the Nobel Prize in Medicine or Physiology in 2007. Attracting the experts to create this book I was hoping to provide an overview of mouse development, spanning from oocytes and early embryos to the state-of-the-art description of embryonic and adult stem cells. The book begins with different aspects of mouse oocyte biology, mostly gene expression control at various levels of complexity, and the oocyte-to-embryo transition. Special attention is focused on the currently most vigorously discussed and controversial issues concerning mechanisms involved in the earliest steps of cells diversification occurring just before and during blastocyst formation. As the topic is very hot and highly discussed, five independent views are presented. The authors of these chapters used different methods in their research and emphasized different aspects of processes leading to the blastocyst formation. However, despite their conclusions diverge, they share several important common points which, I hope, will allow to develop a cohesive vision. Further chapters are devoted to the control of development of selected organs or tissues with emphasis at transdifferentiation, transdetermination, and involvement of stem cells. Examples of ecto-, meso-, and endoderm-derived stem cells are presented. Finally, induced pluripotent stem cells are discussed. The book, thus, speak on mouse development starting from oogenesis and finishing on aging.

At the end, I would like to thank to Rafal Piprek (Krakow), Jarosław Czyż (Krakow), and Malathesha Ganachari (Houston) for their help in creating the final version of this book.

Rennes, France Jacek Z. Kubiak

Contents

Chapter 1
Post-transcriptional Control of Gene Expression During Mouse Oogenesis

Hugh J. Clarke

Abstract Post-transcriptional mechanisms play a central role in regulating gene expression during oogenesis and early embryogenesis. Growing oocytes accumulate an enormous quantity of messenger RNAs (mRNAs), but transcription decreases dramatically near the end of growth and is undetectable during meiotic maturation. Following fertilization, the embryo is initially transcriptionally inactive and then becomes active at a species-specific stage of early cleavage. Meanwhile, beginning during maturation and continuing after fertilization, the oocyte mRNAs are eliminated, allowing the embryonic genome to assume control of development. How the mammalian oocyte manages the storage, translation, and degradation of the huge quantity and diversity of mRNAs that it harbours has been the focus of enormous research effort and is the subject of this review. We discuss the roles of sequences within the 3′-untranslated region of certain mRNAs and the proteins that bind to them, sequence-non-specific RNA-binding proteins, and recent studies implicating ribonucleoprotein processing (P-) bodies and cytoplasmic lattices. We also discuss mechanisms that may control the temporally regulated translational activation of different mRNAs during meiotic maturation, as well as the signals that trigger silencing and degradation of the oocyte mRNAs. We close by highlighting areas for future research including the potential key role of small RNAs in regulating gene expression in oocytes.

H.J. Clarke (✉)
Departments of Obstetrics and Gynecology, Biology, and Experimental Medicine, McGill University and the Research Institute of the McGill University Health Centre, F3.50, Women's Pavilion, Royal Victoria Hospital, 687 av. des Pins O, Montréal, QC, Canada H3A1A1
e-mail: hugh.clarke@mcgill.ca

J.Z. Kubiak (ed.), *Mouse Development*, Results and Problems in Cell Differentiation 55,
DOI 10.1007/978-3-642-30406-4_1, © Springer-Verlag Berlin Heidelberg 2012

1.1 Introduction

Translational control of mRNA plays a central role in regulating gene expression during oogenesis and early embryogenesis. This reflects the developmentally regulated nature of transcriptional activity in oocytes and after fertilization. During their growth phase, mammalian oocytes are transcriptionally active (Fig. 1.1). Near the time that oocytes reach full size, transcription drops to very low or undetectable levels, and it remains so throughout the terminal stage of oogenesis, termed meiotic maturation. Transcription remains low after fertilization until it resumes at a species-specific stage of early embryogenesis—for example, at the late 2-cell stage in mice, 4-cell stage in humans, and 8- to 16-cell stage in the cow. This means that the period of development encompassing late growth of the oocyte until early/mid-cleavage of the embryo depends on mRNAs synthesized during oocyte growth. This poses an enormous regulatory challenge to the growing oocyte, which at full size has accumulated almost twice as much mRNA as is present in a blastocyst. Some newly synthesized mRNAs must be translated to support growth of the oocyte, whereas others must be safely stored and then translationally activated at the appropriate stage of late oogenesis or early embryogenesis. In addition, these oocyte (maternal) mRNAs must subsequently be degraded to permit the newly transcribed embryonic mRNAs to assume control of development. How the mammalian oocyte manages the storage, translation, and degradation of the huge quantity and diversity of mRNAs that it harbours has been the focus of enormous research effort and is the subject of this review.

1.2 A Brief Primer on Translational Control

Newly transcribed mRNAs typically undergo three major processing events within the nucleus—splicing, capping, and polyadenylation (some mRNAs, notably those encoding the replication-dependent histones, are neither spliced nor polyadenylated). The latter two are particularly relevant to translation. Capping describes the addition of a 7-methylguanosine structure (m_7GpppN) to the 5′-end of an mRNA during transcription (Cowling 2009). Polyadenylation, by contrast, occurs at the 3′-end. A polyadenylation signal (AAUAAA) in the 3′-untranslated region of the primary mRNA signals recruitment of a protein termed cleavage and polyadenylation specificity factor (CPSF) to the 3′-end. CPSF in turn both triggers cleavage of the transcript at a site 3′ to the polyadenylation sequence and recruits polyA polymerases (PAPs) that catalyse the addition of ~250 adenosine nucleotides to the 3′-end of the mRNA (Millevoi and Vagner 2010). Six mammalian PAPs have been identified: PAPα (*Papola*) and PAPγ (*Papolg*) in the nucleus, GLD2 (*Papd4*) and GLD4 (*Papd5*) in the cytoplasm at least in some cell types, and two that are restricted to round spermatids and mitochondria, respectively (Burns et al. 2011;

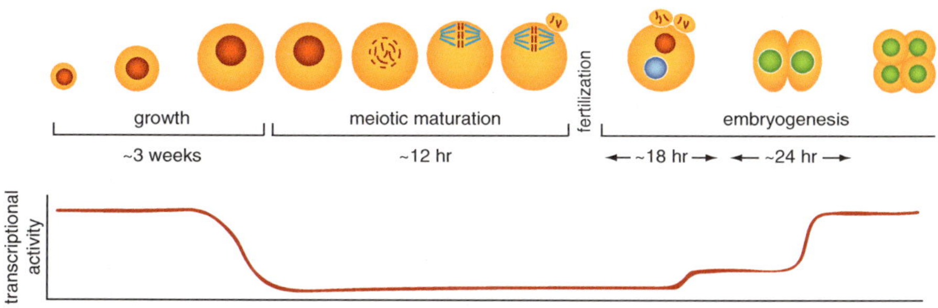

Fig. 1.1 Ooogenesis, early embryogenesis, and transcriptional activity. Time periods indicated correspond to the mouse. During oocyte growth, which requires about 3 weeks, the oocyte is at late prophase I of the cell cycle and is transcriptionally active. Transcription drops substantially near or at the end of the growth phase. During meiotic maturation, which spans about 12 h, the oocyte enters M-phase, completes the first meiotic division, and then arrests at metaphase II. Transcription is not detectable during this time. Following fertilization, weak transcription can be detected during the first cell cycle, which spans about 18 h. The major activation of transcription occurs at G2 of the 2-cell stage. Hence, late oogenesis and early embryogenesis depend on mRNAs that have been transcribed during oocyte growth

Nakanishi et al. 2006; Schmid et al. 2009). Capped and polyadenylated mRNA is exported to the cytoplasm.

In the cytoplasm, the 5′-end of the RNA interacts with the cap-binding protein, eIF4E (Abaza and Gebauer 2008; Groppo and Richter 2009; Jackson et al. 2010; Richter and Sonenberg 2005). At the other end, polyA-binding protein (PABP) associates both with the poly(A) tail and, through its N-terminal region, with eIF4G. eIF4G and eIF4E together with eIF4A form the cap-binding complex, eIF4F, which in turn binds via eIF3 to the 40S ribosomal subunit. Through the association of eIF4G with PABP, the mRNA becomes pseudo-circularized, which renders its translation substantially more efficient (Fig. 1.2). Thus, long or short poly(A) tails are generally correlated with high or low translational activity, respectively. Nonetheless, certain mRNAs have been identified whose translation is associated with deadenylation or is independent of polyadenylation (Kleene et al. 1984; Tadros et al. 2007; Vardy and Orr-Weaver 2007a).

Translational inactivation and subsequent degradation of an mRNA comprises three steps. First, deadenylases shorten the poly(A) tail. Three families of deadenylases have been identified: the CNOT (also known as CCR4-NOT) family, the PAN2-PAN3 heterodimer, and the poly(A)-specific ribonuclease (PARN) (Goldstrohm and Wickens 2008). It has been proposed that PAN2–PAN3 shortens the tail from several hundred to about 100 nt, after which CCR4-NOT removes most or all of the remaining tail (Yamashita et al. 2005). PARN is mainly nuclear, but could also contribute to cytoplasmic deadenylation. Numerous regulatory proteins as well as short RNAs can influence the activity of these deadenylases (Goldstrohm and Wickens 2008; Standart and Jackson 2007; von Roretz and Gallouzi 2008). Deadenylation is followed by, and may be required for, removal of the 5′-cap by decapping enzymes (Franks and Lykke-Andersen 2008; Garneau

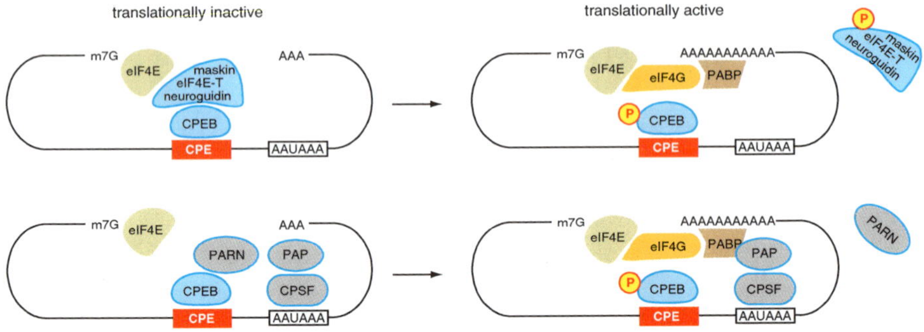

Fig. 1.2 Two models of CPE-dependent translational regulation. *Upper*: In translationally inactive mRNA, an adaptor protein such as maskin, eIF4-T, or neuroguidin associates with both CPEB and eIF4E, thus preventing interaction between eIF4E and eIF4G. During meiotic maturation, CPEB becomes phosphorylated and the adaptor protein no longer inhibits eIF4E–eIF4G interaction, permitting translational activation of the mRNA. *Lower*: In translationally inactive mRNA, both deadenylases (PARN) and adenylases (PAP) are associated directly or indirectly with CPEB and the mRNA carries a short poly(A) tail. During maturation, the deadenylase is 'ejected' from the complex, permitting elongation of the poly(A) tail and translational activation of the mRNA. Other factors in the complexes are not shown for clarity. These models are not necessarily mutually exclusive and other models of CPE-dependent regulation exist. Modified from Radford et al. (2008)

et al. 2007). The deadenylated and decapped RNA is then degraded by nucleases that digest it from both the 5′- and 3′-ends. Although the signals that target a translationally active mRNA for degradation are not fully understood, PABP can attract PAN deadenylases to the mRNA. This suggests two points. First, the length of the poly(A) tail of an mRNA, and hence its translational efficiency, may be determined by the relative strength of competing PAPs and deadenylases that act simultaneously on it. Second, translation of an mRNA may promote its deadenylation and subsequent degradation, thus providing a translation-coupled pathway of mRNA turnover.

Complementing these studies of the biochemical basis of mRNA metabolism, recent work has addressed where in the cell these events occur and specifically the potential role of cytoplasmic ribonucleoprotein particles (RNPs) (Balagopal and Parker 2009; Eulalio et al. 2007; Kulkarni et al. 2010; Parker and Sheth 2007). Microscopically visible RNPs have been observed in a variety of cell types and given different names; here, they will be termed (processing) P-bodies. Although the molecular composition of these large RNPs likely varies depending on cell type, they typically harbour many of the enzymes involved in mRNA metabolism, including deadenylases, decapping factors, and exonucleases, as well as factors that modulate these activities. miRNAs and proteins involved in miRNA-mediated RNA silencing, such as Argonaute proteins and GW182 (TNRC6A), have also been detected in P-bodies and recently discovered to interact with PABPs and with deadenylases (Tritschler et al. 2010) and with both PAN2-PAN3 and CCR4-NOT (Braun et al. 2011; Chekulaeva et al. 2011; Fabian et al. 2011). Based on their composition and several lines of experimental evidence, it has been proposed that

P-bodies can act as sites of mRNA degradation. mRNAs localized in P-bodies can also return to the translationally active pool, however, suggesting a more general function as sites to which (some) mRNAs are transported for subsequent storage or degradation. Nevertheless, it remains unresolved whether P-bodies, by providing locally high concentrations of enzymes and co-factors, are the sites of processes such as deadenylation and decapping or, alternatively, are aggregates that serve no regulatory function but simply become detectable when a large quantity of non-polysomal mRNA is present.

1.3 Translational Control During Oocyte Growth

Although non-growing oocytes are transcriptionally active, entry into the growth phase is accompanied by dramatic changes in the pattern of gene expression (Pan et al. 2005). The growing oocyte remains transcriptionally active until near the end of growth when activity drops considerably. Transcriptional arrest is correlated with a change in the nuclear distribution of the chromatin, termed the non-surrounded nucleolus (NSN) to surrounded nucleolus (SN) configuration (De La Fuente 2006). Many of the mRNAs synthesized in growing oocytes are immediately translated; however, a significant and developmentally crucial fraction—estimates run as high as 30 % (Piqué et al. 2008)—are translationally repressed until meiotic maturation or after fertilization. Notably, these repressed mRNAs are not degraded but instead are stored in a stable form; bulk oocyte mRNA in mouse oocytes has a remarkably long half-life estimated at 8–12 days (Brower et al. 1981). Considerable work, both in mammals and non-mammalian organisms, has focused on the mechanisms by which these silent mRNAs are translationally repressed and stored.

The best-known mechanism of translational repression operates through the cytoplasmic polyadenylation element (CPE, also known as the adenylation control element [ACE]) (Brook et al. 2009; Kang and Han 2011; Radford et al. 2008). This U-rich element (consensus: UUUUA(A)U) is located in the 3′-utr within about 200 nt of the polyadenylation signal and was first identified in mRNAs known to be translationally repressed in growing oocytes and activated during meiotic maturation (Fox et al. 1989; Huarte et al. 1992; McGrew et al. 1989). Its role has been clearly demonstrated using reporter mRNAs carrying the 3′-utr of the mRNA of interest—insertion of a CPE represses translation whereas mutation of existing CPE activates translation (Huarte et al. 1992; Tay et al. 2000). Although these experiments were carried out using fully grown immature oocytes, it is believed that the CPE functions in the same way in growing oocytes. The CPE is bound by the CPE-binding protein (CPEB), a 62-kDa protein first identified in oocytes of *Xenopus* and having homologues in a wide range of species. CPEB is highly conserved, particularly in the C-terminal portion that contains two RNA recognition motifs and a Zn-finger domain (Gebauer and Richter 1996; Welk et al. 2001). Vertebrates also contain related proteins, CPEB-2, -3, and -4 that are expressed in different cell types including oocytes and show different RNA-binding preferences (Standart and Minshall 2008). The crucial role of CPEB-dependent translational repression in maintaining the normal programme of oocyte development is illustrated by the severe abnormalities in growing oocytes that lack CPEB (Racki and Richter 2006).

Many lines of evidence support the link between a CPE-dependent short poly(A) tail and translational repression. First, although mRNAs containing a CPE are initially provided with a long poly(A) tail in the nucleus, this is rapidly removed following their export to the cytoplasm leaving a short tail of 20–40 bases (Huarte et al. 1992; Kim and Richter 2006). Second, mRNAs provided with a long poly(A) tail in vitro, but not a short tail, are translated following injection into immature oocytes. Third, overexpression of *Gld2* in immature oocytes of *Xenopus* lengthens the poly(A) tail of mRNAs including *Ccnb1* and *Mos*, both of which contain CPE sequences (Nakanishi et al. 2006), consistent with the model that competing polymerase and deadenylase activities determine the steady-state length of the poly(A) tail. Intriguingly, when a competitor RNA containing a CPE was injected into immature oocytes, endogenous mRNAs containing a CPE became translationally activated even though their mRNA did not become polyadenylated (Stutz et al. 1998). Although the large excess of competitor that was required means that the results should be interpreted with caution, this suggests that translational repression may depend on more than shortening of the poly(A) tail.

While the role of CPEB in mediating translational repression is firmly established, its mechanism remains to be fully elucidated (Fig. 1.2). *Xenopus* oocytes contain a protein termed maskin that is able to bind both CPEB and eIF4E. This prevents eIF4E from binding to eIF4G, and translation is consequently repressed. Although the mammalian maskin homologue, TACC4, is expressed in oocytes (Yang et al. 2010), the site to which eIF4E binds is not conserved, implying that this mechanism may not operate in mammals. Neuroguidin, originally identified in cells of the nervous system, is also able to bind CPEB and eIF4E and to repress translation (Jung et al. 2006), and *Ngd* is expressed in mouse oocytes (http://www.ncbi.nlm.nih.gov/geoprofiles), suggesting that it could serve the same function as maskin. In addition, at least two other mechanisms through which CPEB may repress translation have been identified. CPEB in amphibians can bind to the PARN deadenylase, suggesting a simple model by which it represses translation. Although most PARN appears to be nuclear in amphibian oocytes (Kim and Richter 2006; Standart and Minshall 2008) and in mammalian somatic cells and oocytes (K.-F. Vieux & H.C., unpublished) it is possible that a quantitatively small fraction in the cytoplasm is sufficient to serve this function. CPEB can also interact with 4E-T (eIF4E-transporter), a protein that in turn binds to eIF4E (eIF4E1b in oocytes (Evsikov et al. 2006)) and prevents the latter from interacting with eIF4G (Richter and Sonenberg 2005; Standart and Minshall 2008). This would inhibit pseudo-circularization of the mRNA thus repressing translation, although it does not seem to account for the shortening of the poly(A) tail. These interactions have been established biochemically in *Xenopus* and *Drosophila*, whose 4E-T homologue is termed Cup, and 4E-T is expressed in mouse oocytes (Villaescusa et al. 2006).

The central role of CPEB in translational repression in immature oocytes does not preclude a role for other mechanisms. The PUmilio-Fem-3-binding factor (PUF) family of RNA-binding proteins are found throughout the plant and animal kingdoms (Wickens et al. 2002). They are characterized by the presence of eight repeats of a ~40-amino acid sequence, termed the Pumilio-homology domain, that are located in the C-terminal region of the proteins. Each of the eight repeats contacts a different RNA base in the PUF-binding element (PBE; consensus UGUANUAU) that is found in many translationally repressed mRNAs. A clue to their mechanism of repression may lie in the observation that they can bind directly to the CNOT8 deadenylase in vitro (Goldstrohm et al. 2006; Morris et al. 2008). Intriguingly, through their conserved C-terminal region, PUF proteins can also bind to CPEB (Nakahata et al. 2001). Moreover, many mRNAs that contain CPEs also contain putative PBEs (Piqué et al. 2008; Radford et al. 2008). Thus, PUF proteins might repress translation of PBE-containing mRNAs either independently or in co-operation with CPEB (Piqué et al. 2008). Indeed this cooperativity has been demonstrated in *Xenopus* for cyclin B1 mRNA (Nakahata et al. 2003).

Two PUF proteins exist in mammals, PUM1 and PUM2. PUM1 is expressed in a wide variety of cell types, whereas PUM2 is expressed only in ES cells, testis, and the ovary where it is restricted to the oocyte (Moore et al. 2003). Although mice bearing a mutation in *Pum2* are fertile (Xu et al. 2007), these observations are consistent with a potential role for PUM proteins in translational control in mammalian oocytes.

The factors discussed so far repress translation through binding to specific sequences in the 3′-utr of specific target mRNAs. A different paradigm of translational control may be illustrated by YBX2 (formerly MSY2). This protein is very abundant in oocytes but rapidly degraded during early embryogenesis, implying that its function is restricted to oogenesis (Yu et al. 2001). Mice lacking YBX2 are hypo-fertile, and their oocytes manifest a variety of abnormalities during growth (Medvedev et al. 2011; Yu et al. 2004). Although the function of YBX2 is not yet fully understood, its Y-box contains a conserved cold-shock domain that includes RNA-binding motifs and Y-box proteins in amphibian oocytes associate with and repress translation of numerous mRNAs. Mammalian YBX2 displays some sequence preference in RNA binding, but also binds to a wide range of mRNAs in vitro, suggesting that it may act as a sequence-independent RNA-binding protein (Yu et al. 2002). The observations that oocytes lacking YBX2 have a reduced quantity of mRNA and that YBX2 is associated with the detergent-insoluble fraction of oocytes suggest that it may be part of a complex that sequesters and stabilizes translationally repressed mRNAs. It would be interesting to test whether mRNAs that are normally repressed in growing oocytes become translationally activated in the absence of YBX2.

The potential role of P-bodies in translational control in somatic cells raises the possibility that they play important roles in oocytes as well. Supporting this idea is the presence of RNP complexes in the oocytes of a variety of non-mammalian species. Recent work in *C. elegans* has identified structures termed storage bodies that may be derived from the well-known germinal (P-) granules and are defined by their position in the oocyte, mechanism of assembly, and presence of an RNA helicase, CGH-1 (Boag et al. 2008). CGH-1 associates with a subset of oocyte mRNAs and, in *cgh-1* mutants, these mRNAs are depleted. These observations suggest that CGH-1 in the storage bodies may protect this subset of mRNAs—perhaps those that will be translationally activated later during development—from degradation. Oocytes of *Xenopus* contain RNPs whose protein composition shows a remarkable resemblance to P-bodies. Intriguingly, these oocyte P-bodies also contain CPEB, providing circumstantial evidence that they also may store translationally silent mRNA.

Two recent studies have addressed P-bodies in mammalian oocytes. In the first, GFP-tagged DCP1A was expressed in fully grown immature oocytes, where it became localized in many small and a few large cytoplasmic foci (Swetloff et al. 2009). Staining using anti-DCP1A revealed only the large foci. A subset of the GFP-DCP1A foci were also stained by antibodies recognizing the P-body components DDX6, an RNA helicase that is orthologous to CGH-1, and RAP55, an RNA-binding protein. CPEB, however, could not be immunologically detected. In the second study (Flemr et al. 2010), several P-body components (AGO2, DCP1A, DDX6, GW182) were found to co-localize in cytoplasmic foci in growing oocytes. YBX2, an RNA-binding protein discussed earlier, was also present in these foci. In oocytes larger than 50 μm diameter, however, these foci were not detected. Instead, at least for DDX6, GW182, and YBX2, staining was enriched in the periphery of the oocyte, in a narrow band that the authors term the subcortical RNP domain (SCRD). Interestingly, c-*mos* mRNA, which is translationally repressed in immature oocytes, also localized at the oocyte periphery, although technical limitations mean that co-localization with specific SCRD components could not be tested.

Both studies indicate that mammalian oocytes contain structures that biochemically and functionally resemble P-bodies, but obviously differ with respect to the dynamics of their assembly and localization during oocyte growth. It is interesting to note that in *Xenopus*, maskin and PARN are not detectable until late in oogenesis, whereas P-bodies may be more abundant at earlier stages (Minshall et al. 2007). Hence, there is precedent for the notion that oocytes at different stages of growth may employ different mechanisms to silence and store mRNAs. Nonetheless, especially in view of the similarities between P-bodies and stress granules and the crucial supporting role of the granulosa cells in oogenesis, it will be important to verify that oocytes growing in vivo also contain P-bodies.

Mouse oocytes also contain RNA- and protein-containing fibrillar matrices, termed cytoplasmic lattices (CPLs) (Capco et al. 1993; Lehtonen et al. 1983). When PADI6, a peptidyl arginine deiminase, was found to co-localize with the lattices (Wright et al. 2003), this opened the way to experimental analysis of their function. In oocytes of *Padi6−/−* mice, the lattices cannot be detected and ribosomal components—specifically rRNA and protein S6—display increased solubility (Yurttas et al. 2008). This has led to the proposal that the CPLs may be storage sites for the large fraction (~70%) of oocyte ribosomes that are not in polysomes (Bachvarova and De Leon 1977). Very recent studies have identified PADI6 within a large protein complex that includes FLOPED, MATER, TLE6, and Filia (Li et al. 2008). Oocytes lacking any of these genes give rise to embryos that arrest during early cleavage, consistent with the proposal that the CPL-associated ribosomes are required for protein synthesis in the early embryo. Although protein synthesis in the oocyte is not quantitatively affected in the absence of PADI6 (Yurttas et al. 2008), it is conceivable that the CPLs play a role in storing silent mRNAs that will be translationally activated following fertilization. Finally, it should be noted that a sub-population of mRNAs is preferentially associated with the spindle region in mature eggs, validating the concept that mRNAs can be differentially distributed within the cytoplasm of mammalian germ cells (VerMilyea et al. 2011) as also demonstrated in amphibians (Eliscovich et al. 2008).

What about the very large population of oocyte mRNAs that are actively translated in growing oocytes? It is generally assumed that this represents a default pathway—an mRNA will be translated unless some structural element such as a CPE acts to repress translation—and mRNAs known to be translated typically do not possess such sequences in the region of the 3′-utr close to the polyadenylation signal. Despite its repressive activity, however, not all mRNAs that contain a CPE are completely silent in immature oocytes. For example, mouse immature oocytes contain a small quantity of cyclin B1, indicating that even though *Ccnb1* is activated during maturation, it is nonetheless weakly translated in immature oocytes (Holt et al. 2010; Marangos and Carroll 2008). Since keeping cyclin B1 at a low level is essential to prevent precocious initiation of maturation, this weak translation likely reflects incomplete repression. A more telling example is the stem-loop-binding protein (SLBP). Growing oocytes absolutely require SLBP to accumulate mRNAs encoding the histones that will be used to assemble chromatin during the embryonic cell cycles (Arnold et al. 2008). Yet, *Slbp* contains a functional CPE that represses its translation in growing oocytes (Yang et al. 2010). How are immature oocytes able to overcome the repressive activity of the CPE and produce the SLBP that they need?

One explanation may be that *Slbp* contains other sequences that enable a sufficient level of expression to be achieved in spite of the repressive effects of the CPE. We have found that the 3′-utr of *Slbp* confers a higher translational activity to a reporter RNA than does the 3′-utr of *Ccnb1* (Q. Yang and H.J.C., unpublished). Within the *Slbp* 3′-utr is a highly conserved U-rich sequence that is not present in the *Ccnb1* 3′-utr. Mutation of the sequence reduces the activity conferred by the *Slbp* 3′-utr, whereas its insertion in to the *Ccnb1* 3′-utr increases activity. This suggests that the translation of an mRNA is determined by net activity of positive and negative regulatory elements. This could provide a mechanism for the oocyte to regulate the supply of proteins whose steady-state quantity needs to be restricted within a narrow range.

1.4 Translational Control During Oocyte Maturation

Meiotic maturation is the final stage of oocyte development before fertilization (Fig. 1.2). During maturation the nuclear envelope becomes disassembled (germinal vesicle breakdown, GVBD), releasing the contents of the nucleus into the cytoplasm.

The chromosomes condense, become aligned on the first meiotic spindle, and the oocyte undergoes the first meiotic division. The oocyte chromosomes then become aligned on the second meiotic spindle, at which point meiosis becomes arrested until the egg is activated by fertilization. In contrast to oocyte growth, which is a protracted process requiring several weeks or months, depending on the species, oocyte maturation is relatively rapid, spanning ~12 h in mouse and ~36 h in humans. Importantly for this discussion, it is characterized by significant changes in translational activity.

About 25 years ago, a series of papers appeared demonstrating that, during oocyte maturation in mice, the mRNA encoding tissue plasminogen activator (tPA) becomes elongated and translationally activated (Huarte et al. 1987; Strickland et al. 1988). Further work demonstrated that the lengthening is due to an increase in the length of the poly(A) tail and that this increase depends on a sequence in the 3′-utr subsequently identified as the CPE (Stutz et al. 1997, 1998; Vassalli et al. 1989). Since then, numerous mRNAs have been identified that become translationally activated during maturation, and these consistently carry a CPE that is necessary for activation (Dai et al. 2005; Gebauer et al. 1994; Gershon et al. 2006; Murai et al. 2010; Oh et al. 2000; Salles et al. 1992; Tay et al. 2000; Tremblay et al. 2005; Yang et al. 2010). In addition to the CPE, polyadenylation also requires the nuclear polyadenylation signal. Conversely, mRNAs such as *Actb* that do not carry a CPE near the polyadenylation signal become deadenylated and translationally silenced (Bachvarova et al. 1985; Yang et al. 2010; Chen et al. 2011). These results suggested a simple 'binary-switch' model in which mRNAs bearing a CPE are silent in growing and fully grown immature oocytes and then activated in maturing oocytes, whereas mRNAs lacking a CPE show the opposite behaviour.

Recent work comparing the mRNAs associated with polysomes, as a marker of translational activity, in immature (GV-stage) and mature oocytes has extended and refined this model (Chen et al. 2011). Of the ~7,600 mRNAs analysed, about two-thirds showed a less than twofold change and were defined as stable. The remaining one-third were about equally divided between those becoming more abundant (translationally activated) or less abundant (translationally repressed) during maturation. Comparison of the 3′-utr of the entire mRNA population revealed that several motifs, including the CPE and a sequence resembling the consensus DAZL-binding sequence (see below), were enriched in those recruited to polysomes. Conversely, mRNAs whose 3′-utr lacked either of these sequences became less abundant on the polysomes. These studies extend the role of the CPE in translational activation to a population-wide basis and implicate other sequences in the 3′-utr, but also highlight that many mRNAs are associated with polysomes in both immature and mature oocytes.

The mechanism by which the CPE switches from repressing to activating translation has been carefully studied in amphibians (Radford et al. 2008). In one model, maskin or an analogous protein becomes phosphorylated, weakening its interaction with eIF4E and thus permitting assembly of a translationally active complex (Fig. 1.2). A non-exclusive second model focuses on a shift in the balance of power within the protein complex bound via CPEB to the CPE from the deadenylase PARN to the cytoplasmic PAP, GLD-2. The first event is phosphorylation of CPEB on Ser-174 (Thr-171 in mice). Although there is not yet consensus regarding the responsible kinase—both Aurora A (Hodgman et al. 2001; Mendez et al. 2000a) and calmodulin kinase II (Radford et al. 2008) have been proposed—this event occurs well before the activation of cyclin-dependent kinase-1 (CDK1) and thus precedes GVBD. Phosphorylation of CPEB increases the strength of its binding to CPSF and to GLD-2, which leads to 'ejection' of PARN from the protein complex at the 3′-end of the mRNA, thus permitting GLD-2-mediated elongation of the poly(A) tail (Barnard et al. 2005; Mendez et al. 2000a, b). The longer poly(A) tail attracts the embryonic poly(A)-binding protein (ePAB) that interacts with eIF4G and thereby increases the affinity of the eIF4F complex for the cap structure and hence the efficiency of translation. CPEB phosphorylation may serve a second function through its effect to decrease its affinity for Pumilio proteins, which would be predicted to relieve translational repression. Expression of a mutant CPEB lacking Ser-174 acts in a dominant-negative manner to inhibit polyadenylation (and maturation), confirming the essential role of CPEB phosphorylation in translational activation. After CDK1 has become activated and GVBD has occurred, CPEB becomes phosphorylated by a CDK-dependent mechanism. This leads to its ubiquitination and degradation (Reverte et al. 2001; Setoyama et al. 2007), which is required for certain mRNAs to become polyadenylated during the later stages of maturation (Mendez et al. 2002).

Although less well studied in mammals owing to the small amount of material that can be obtained, the available evidence (references cited earlier) suggests that CPEB plays a similar role during the translational activation of silent mRNAs in maturing oocytes. Despite the broad similarity, certain differences between mammals and amphibians are worth noting. Unlike amphibians, where the crucial Ser-174 phosphorylation precedes activation of CDK1, there is no detectable phosphorylation of CPEB in mammalian oocytes until after GVBD (Chen et al. 2011; Uzbekova et al. 2008; Yang et al. 2010). This relatively late phosphorylation of CPEB does not, however, challenge the importance of this event in driving polyadenylation and translational activation, as no mRNAs have yet been demonstrated to become translationally activated before or independently of CPEB phosphorylation. In addition, although CPEB becomes degraded during maturation in mammals as in amphibians (Chen et al. 2011; Uzbekova et al. 2008; Yang et al. 2010), it is unknown whether this is required to activate translation of specific mRNAs. Finally, although *Gld2* mRNA is present in oocytes, it is translationally repressed until maturation (Nakanishi et al. 2006) and, more importantly, deletion of the gene does not inhibit polyadenylation during maturation (Nakanishi et al. 2007). These results seem to rule out GLD2 as the PAP responsible for mRNA polyadenylation during maturation, at least in the mouse. Although canonical PAPα, which would be released from the nucleus at GVBD, might provide the activity, in vitro studies indicate that it is inactivated by phosphorylation during maturation (Colgan et al. 1996) and experiments in *Xenopus* show that prior removal of the nucleus does not prevent polyadenylation during maturation (Fox et al. 1989). The recently identified GLD4 (see above) is an alternative candidate.

Comparative analysis of different mRNAs has revealed that they do not become polyadenylated at the same time during maturation. In *Xenopus*, for example, the mRNA encoding Mos becomes polyadenylated before GVBD but those encoding different cyclin B proteins not until metaphase I (Piqué et al. 2008). What determines the difference in timing? Piqué et al (2008) synthesized 3′-utr sequences that contained different combinations and spacings of CPE, PBE, and polyadenylation signals and injected mRNAs carrying each 3′-utr into *Xenopus* oocytes. They then determined when each became polyadenylated and translated. Based on the results, they derived a code according to which the specific arrangement of these three elements determines at what stage of maturation an mRNA will become translationally activated.

Analysis of mRNAs recruited to polysomes during oocyte maturation in the mouse identified several motifs including the CPE that were enriched in these mRNAs, but did not reveal a specific code analogous to that in the frog (Chen et al. 2011). Instead, another mechanism by which different mRNA species are sequentially activated was uncovered. Among the enriched motifs was the consensus binding site (UUU[C/G]UUU) for the DAZ family of proteins. *Dazl* contains a putative CPE and is translationally activated during maturation, leading to an increase in the amount of DAZL (see also (Liu et al. 2009)). Next, it was shown that DAZL immunoprecipitates contained numerous mRNAs (including *Dazl* itself) possessing putative DAZL-binding elements, and in the case of a known DAZL target, *Tex19.1*, these were confirmed to be functional. Finally, when translational activation of *Dazl* was inhibited, oocyte maturation was disrupted as illustrated by an absent or grossly abnormal metaphase II spindle. The authors propose that *Dazl* becomes translationally activated during early maturation by a classical CPEB-regulated mechanism, leading to the accumulation of DAZL, which subsequently activates ‘late’ translation of DAZL-responsive mRNAs. This provides an elegant mechanism by which activation of different mRNAs can be temporally regulated during maturation.

1.5 Translational Silencing and Degradation of Oocyte mRNAs

Although the continuing transcriptional silence following fertilization means that early embryogenesis relies on mRNAs synthesized in growing oocytes, and some stored mRNAs are not translationally activated until after fertilization, the elimination of this maternal legacy in preparation for the transition to embryonic control of development begins during meiotic maturation. Early studies showed that as much as half of the mRNA in fully grown oocytes becomes either deadenylated or degraded during maturation (Bachvarova et al. 1985). More recent microarray and polysomal analyses (Chen et al. 2011; Su et al. 2007) suggest that when the number of mRNA species rather than bulk quantity of mRNA is assessed, approximately ¼ are affected. These results clearly establish that a large subset of mRNAs, which presumably have remained stable and polyadenylated throughout the extended period of oocyte growth, become rapidly targeted for deadenylation and/or degradation when maturation begins. What triggers these processes and why some mRNAs are targeted and others spared remain largely mysterious.

Some recent studies, however, have begun to shed light on the details and the underlying molecular control.

YBX2 (MSY2) was discussed earlier in the context of its potential role in stabilizing mRNA in growing oocytes. During maturation, YBX2 becomes phosphorylated by a CDK1-dependent mechanism and this is associated with a decrease in the amount of YBX2 and mRNA present in the detergent-insoluble cytoplasmic fraction (Medvedev et al. 2008). Expression of a non-phosphorylatable YBX2 in maturing oocytes prevents the decrease in some mRNAs that normally occurs during maturation. Conversely, insertion of acidic residues into putative phosphorylation sites triggers a decrease in some mRNAs in immature oocytes. These results suggest that upon phosphorylation, YBX2 might lose its ability to protect mRNAs from degrading activities present in the cytoplasm. It would be interesting to learn whether mRNAs that are translationally active during growth or maturation, respectively, are differentially sensitive to the loss of YBX2 activity.

As discussed earlier, mRNA degradation in somatic cells is preceded by and tightly linked to deadenylation. A simple model is that mRNAs lacking a CPE become deadenylated and degraded during oocyte maturation, whereas those bearing a CPE are polyadenylated and stable (at least during maturation). We have observed three patterns of polyadenylation during maturation (Yang et al. 2010; Q.Y. and H.J.C., unpublished). As expected, mRNAs that become translationally silenced during maturation, such as *Actb*, become deadenylated. Some mRNAs that become translationally activated during maturation, such as *Ccnb1*, become continuously polyadenylated during maturation. Some of these will be deadenylated after fertilization, a process mediated in amphibians by the embryonic deadenylation element (EDEN) (Paillard et al. 1998) and its associated protein, EDEN-BP, in a cell cycle-dependent manner (Detivaud et al. 2003). In contrast, others such as *Slbp* and *Orc6l* become polyadenylated and activated during early maturation and then deadenylated during late maturation. Importantly, those mRNAs that become deadenylated are not degraded to a greater extent than those that remain polyadenylated. Consistent with these results, deadenylated actin mRNA was previously reported to remain stable in mature oocytes (Bachvarova et al. 1985; Paynton et al. 1988) and a substantial fraction of the mRNAs removed from polysomes during maturation are not degraded (Chen et al. 2011). Thus, deadenylation of mRNAs during maturation does not necessarily target them for immediate degradation as also observed in *Xenopus* (Audic et al. 1997).

Work in amphibians indicates that PARN, released into the cytoplasm at GVBD, is responsible for most of the mRNA deadenylation that occurs during maturation (Copeland and Wormington 2001; Korner et al. 1998). In mammals, analysis of a small set of mRNAs revealed that these were not deadenylated until after metaphase I of maturation (Yang et al. 2010). Similarly, analysis of mRNAs released from polysomes indicates that although some are degraded before metaphase I, most appear to remain stable until after metaphase I (Chen et al. 2011). This suggests that the activities responsible for deadenylation and degradation do not appear until late maturation. Recent work in amphibians offers a potential clue to understanding this temporal control. The tristetraprolins are Zn-finger proteins that can bind AU-rich elements (ARE) found in many short-lived mRNAs. C3H-4, a tristetraprolin in *Xenopus* that is encoded by a CPE-containing mRNA, is synthesized during oocyte maturation and is required, likely by recruiting deadenylases, for deadenylation of specific oocyte mRNAs (Belloc and Mendez 2008). By analogy, perhaps a functionally similar protein that is newly synthesized during early maturation leads to deadenylation and or degradation of specific mRNAs during late maturation. According to this model, mRNAs carrying a binding site for the protein (analogous to the ARE site recognized by C3H-4) would be selectively targeted through this mechanism. It should be noted, however, that ARE sites have not been linked to mRNA degradation during maturation in mammals (Thelie et al. 2007). Another candidate is the decapping protein, DCP1A, which accumulates during maturation owing to translational activation of its mRNA (Flemr et al. 2010) and could be a limiting factor that controls mRNA degradation during late maturation.

1.6 What's Next for Translational Control?

What does the future hold for studies of translational control in oocytes? An immediate challenge is to determine whether silent mRNAs are stored in specific cytoplasmic structures and, if so, whether their stability depends on these structures. The localization of *mos* mRNA in a sub-cortical domain in mouse oocytes is consistent with this possibility, and the co-localization of CPEB with P-bodies in somatic cells and amphibian oocytes suggests a mechanism by which the mRNAs could be recruited. The possibility that the mechanisms regulating mRNA storage and activity differ in oocytes at different stages of growth, evoked both by the P-body studies described earlier and by the observation that neither maskin nor PARN is present in mid-growth stage amphibian oocytes (Flemr et al. 2010; Minshall et al. 2007), needs to be explored. Although the role of the CPE and CPEB is long established, much remains to be learned about how they work both in immature and maturing oocytes. For example, KHDC1B is a newly identified RNA-binding protein that can interact with CPEB and whose oocyte-specific expression suggests a function in translational control (Cai et al. 2010). The expression of PUM1 and PUM2 in oocytes suggests a role for these proteins also. In addition, numerous other CPE-independent mechanisms of translational control that have been identified in non-mammalian oocytes may play key roles in regulating translation in growing, fully grown, or maturing mammalian oocytes (Charlesworth et al. 2006; Sugimura and Lilly 2006; Vardy and Orr-Weaver 2007b).

Finally, might there be a role for small RNAs? In the zebrafish, a specific micro (mi) RNA regulates deadenylation and degradation in embryos of a specific subset of maternal mRNAs (Giraldez et al. 2006). Similarly, embryonically expressed Piwi-interacting (pi) RNA is required in *Drosophila* for deadenylation and degradation of maternal *nos* mRNA (Rouget et al. 2010). It appears to act through a complex that includes an Argonaute protein, the CCR4 (CNOT 6) deadenylase, and

Smaug, a protein previously implicated in the degradation of maternal mRNA in early embryos (Tadros and Lipshitz 2009). Moreover, the recent discovery that GW182 can interact with PABP (Tritschler et al. 2010) and with both PAN2-PAN3 and CCR4-NOT (Braun et al. 2011; Chekulaeva et al. 2011; Fabian et al. 2011) also suggests a link between miRNAs, P-bodies, and mRNA metabolism. Different classes of small RNAs, including miRNAs and endogenous small interfering (endo-si) RNAs, are present in mammalian oocytes (Suh and Blelloch 2011; Tam et al. 2008; Tang 2010; Watanabe et al. 2008). Loss of either AGO2 or Dicer in oocytes triggers changes in the abundance of a large number of mRNAs and prevents their normal development (Kaneda et al. 2009; Murchison et al. 2007; Tang et al. 2007). In contrast, oocytes lacking *Dgcr*, which is required for the production of miRNAs but not of endo-siRNAs, develop normally (Ma et al. 2010; Suh et al. 2010). This suggests that endo-siRNAs regulate mRNA levels in oocytes. The availability particularly in the mouse of genetic tools and RNAi technologies will enable us to further dissect this and other mechanisms of translational control.

Acknowledgements Work in the author's laboratory is supported by the Canadian Institutes of Health Research and the Natural Sciences and Engineering Research Council, as well as by funds from the Royal Victoria Hospital and McGill University Health Centre Foundations. The MUHC Research Institute is supported by the Fonds de Recherche du Quebec. Thanks to Professor Marco Conti (University of California at San Francisco) for comments and suggestions on the manuscript. I apologize to those whose work could not be cited owing to space limitations.

References

Abaza I, Gebauer F (2008) Trading translation with RNA-binding proteins. RNA 14:404–409

Arnold DR, Francon P, Zhang J, Martin K, Clarke HJ (2008) Stem-loop binding protein expressed in growing oocytes is required for accumulation of mRNAs encoding histones H3 and H4 and for early embryonic development in the mouse. Dev Biol 313:347–358

Audic Y, Omilli F, Osborne HB (1997) Postfertilization deadenylation of mRNAs in *Xenopus laevis* embryos is sufficient to cause their degradation at the blastula stage. Mol Cell Biol 17:209–218

Bachvarova R, De Leon V (1977) Stored and polysomal ribosomes of mouse ova. Dev Biol 58:248–254

Bachvarova R, De Leon V, Johnson A, Kaplan G, Paynton BV (1985) Changes in total RNA, polyadenylated RNA, and actin mRNA during meiotic maturation of mouse oocytes. Dev Biol 108:325–331

Balagopal V, Parker R (2009) Polysomes, P bodies and stress granules: states and fates of eukaryotic mRNAs. Curr Opin Cell Biol 21:403–408

Barnard DC, Cao Q, Richter JD (2005) Differential phosphorylation controls Maskin association with eukaryotic translation initiation factor 4E and localization on the mitotic apparatus. Mol Cell Biol 25:7605–7615

Belloc E, Mendez R (2008) A deadenylation negative feedback mechanism governs meiotic metaphase arrest. Nature 452:1017–1021

Boag PR, Atalay A, Robida S, Reinke V, Blackwell TK (2008) Protection of specific maternal messenger RNAs by the P body protein CGH-1 (Dhh1/RCK) during *Caenorhabditis elegans* oogenesis. J Cell Biol 182:543–557

Braun JE, Huntzinger E, Fauser M, Izaurralde E (2011) GW182 proteins directly recruit cytoplasmic deadenylase complexes to miRNA targets. Mol Cell 44:120–133

Brook M, Smith JWS, Gray NK (2009) The DAZL and PABP families: RNA-binding proteins with interrelated roles in translational control in oocytes. Reproduction 137:595–617

Brower PT, Gizang E, Boreen SM, Schultz RM (1981) Biochemical studies of mammalian oogenesis: synthesis and stability of various classes of RNA during growth of the mouse oocyte in vitro. Dev Biol 86:373–383

Burns DM, D'Ambrogio A, Nottrott S, Richter JD (2011) CPEB and two poly(A) polymerases control miR-122 stability and p53 mRNA translation. Nature 473:105–108

Cai C, Tamai K, Molyneaux K (2010) KHDC1B is a novel CPEB binding partner specifically expressed in mouse oocytes and early embryos. Mol Biol Cell 21:3137–3148

Capco DG, Gallicano GI, McGaughey RW, Downing KH, Larabell CA (1993) Cytoskeletal sheets of mammalian eggs and embryos: a lattice-like network of intermediate filaments. Cell Motil Cytoskeleton 24:85–99

Charlesworth A, Wilczynska A, Thampi P, Cox LL, MacNicol AM (2006) Musashi regulates the temporal order of mRNA translation during Xenopus oocyte maturation. EMBO J 25:2792–2801

Chekulaeva M, Mathys H, Zipprich JT, Attig J, Colic M, Parker R, Filipowicz W (2011) miRNA repression involves GW182-mediated recruitment of CCR4-NOT through conserved W-containing motifs. Nat Struct Mol Biol 18:1218–1226

Chen J, Melton C, Suh N, Oh JS, Horner K, Xie F, Sette C, Blelloch R, Conti M (2011) Genome-wide analysis of translation reveals a critical role for deleted in azoospermia-like (Dazl) at the oocyte-to-zygote transition. Genes Dev 25:755–766

Colgan DF, Murthy KG, Prives C, Manley JL (1996) Cell-cycle related regulation of poly(A) polymerase by phosphorylation. Nature 384:282–285

Copeland PR, Wormington M (2001) The mechanism and regulation of deadenylation: identification and characterization of *Xenopus* PARN. RNA 7:875–886

Cowling VH (2009) Regulation of mRNA cap methylation. Biochem J 425:295–302

Dai Y, Newman B, Moor R (2005) Translational regulation of MOS messenger RNA in pig oocytes. Biol Reprod 73:997–1003

De La Fuente R (2006) Chromatin modifications in the germinal vesicle (GV) of mammalian oocytes. Dev Biol 292:1–12

Detivaud L, Pascreau G, Karaiskou A, Osborne HB, Kubiak JZ (2003) Regulation of EDEN-dependent deadenylation of Aurora A/Eg2-derived mRNA via phosphorylation and dephosphorylation in *Xenopus laevis* egg extracts. J Cell Sci 116:2697–2705

Eliscovich C, Peset I, Vernos I, Mendez R (2008) Spindle-localized CPE-mediated translation controls meiotic chromosome segregation. Nat Cell Biol 10:858–865

Eulalio A, Behm-Ansmant I, Izaurralde E (2007) P bodies: at the crossroads of post-transcriptional pathways. Nat Rev Mol Cell Biol 8:9–22

Evsikov AV, Graber JH, Brockman JM, Hampl A, Holbrook AE, Singh P, Eppig JJ, Solter D, Knowles BB (2006) Cracking the egg: molecular dynamics and evolutionary aspects of the transition from the fully grown oocyte to embryo. Genes Dev 20:2713–2727

Fabian MR, Cieplak MK, Frank F, Morita M, Green J, Srikumar T, Nagar B, Yamamoto T, Raught B, Duchaine TF, Sonenberg N (2011) miRNA-mediated deadenylation is orchestrated by GW182 through two conserved motifs that interact with CCR4-NOT. Nat Struct Mol Biol 18:1211–1217

Flemr M, Ma J, Schultz RM, Svoboda P (2010) P-body loss is concomitant with formation of a messenger RNA storage domain in mouse oocytes. Biol Reprod 82:1008–1017

Fox CA, Sheets MD, Wickens MP (1989) Poly(A) addition during maturation of frog oocytes: distinct nuclear and cytoplasmic activities and regulation by the sequence UUUUUAU. Genes Dev 3:2151–2162

Franks TM, Lykke-Andersen J (2008) The control of mRNA decapping and P-body formation. Mol Cell 32:605–615

Garneau NL, Wilusz J, Wilusz CJ (2007) The highways and byways of mRNA decay. Nat Rev Mol Cell Biol 8:113–126

Gebauer F, Richter JD (1996) Mouse cytoplasmic polyadenylylation element binding protein: an evolutionarily conserved protein that interacts with the cytoplasmic polyadenylylation elements of c-mos mRNA. Proc Natl Acad Sci USA 93:14602–14607

Gebauer F, Xu WH, Cooper GM, Richter JD (1994) Translational control by cytoplasmic polyadenylation of c-mos messenger-RNA is necessary for oocyte maturation in the mouse. EMBO J 13:5712–5720

Gershon E, Galiani D, Dekel N (2006) Cytoplasmic polyadenylation controls cdc25B mRNA translation in rat oocytes resuming meiosis. Reproduction 132:21–31

Giraldez AJ, Mishima Y, Rihel J, Grocock RJ, Van Dongen S, Inoue K, Enright AJ, Schier AF (2006) Zebrafish MiR-430 promotes deadenylation and clearance of maternal mRNAs. Science 312:75–79

Goldstrohm AC, Wickens M (2008) Multifunctional deadenylase complexes diversify mRNA control. Nat Rev Mol Cell Biol 9:337–344

Goldstrohm AC, Hook BA, Seay DJ, Wickens M (2006) PUF proteins bind Pop2p to regulate messenger RNAs. Nat Struct Mol Biol 13:533–539

Groppo R, Richter JD (2009) Translational control from head to tail. Curr Opin Cell Biol 21:444–451

Hodgman R, Tay J, Mendez R, Richter JD (2001) CPEB phosphorylation and cytoplasmic polyadenylation are catalyzed by the kinase IAK1/Eg2 in maturing mouse oocytes. Development 128:2815–2822

Holt JE, Weaver J, Jones KT (2010) Spatial regulation of APCCdh1-induced cyclin B1 degradation maintains G2 arrest in mouse oocytes. Development 137:1297–1304

Huarte J, Belin D, Vassalli A, Strickland S, Vassalli JD (1987) Meiotic maturation of mouse oocytes triggers the translation and polyadenylation of dormant tissue-type plasminogen activator mRNA. Genes Dev 1:1201–1211

Huarte J, Stutz A, Oconnell ML, Gubler P, Belin D, Darrow AL, Strickland S, Vassalli JD (1992) Transient translational silencing by reversible mRNA deadenylation. Cell 69:1021–1030

Jackson RJ, Hellen CU, Pestova TV (2010) The mechanism of eukaryotic translation initiation and principles of its regulation. Nat Rev Mol Cell Biol 11:113–127

Jung MY, Lorenz L, Richter JD (2006) Translational control by neuroguidin, a eukaryotic initiation factor 4E and CPEB binding protein. Mol Cell Biol 26:4277–4287

Kaneda M, Tang F, O'Carroll D, Lao K, Surani MA (2009) Essential role for Argonaute2 protein in mouse oogenesis. Epigenetics Chromatin 2:9

Kang MK, Han SJ (2011) Post-transcriptional and post-translational regulation during mouse oocyte maturation. BMB Rep 44:147–157

Kim JH, Richter JD (2006) Opposing polymerase-deadenylase activities regulate cytoplasmic polyadenylation. Mol Cell 24:173–183

Kleene KC, Distel RJ, Hecht NB (1984) Translational regulation and deadenylation of a protamine mRNA during spermiogenesis in the mouse. Dev Biol 105:71–79

Korner CG, Wormington M, Muckenthaler M, Schneider S, Dehlin E, Wahle E (1998) The deadenylating nuclease (DAN) is involved in poly(A) tail removal during the meiotic maturation of Xenopus oocytes. EMBO J 17:5427–5437

Kulkarni M, Ozgur S, Stoecklin G (2010) On track with P-bodies. Biochem Soc Trans 38:242–251

Lehtonen E, Lehto VP, Vartio T, Badley RA, Virtanen I (1983) Expression of cytokeratin polypeptides in mouse oocytes and preimplantation embryos. Dev Biol 100:158–165

Li L, Baibakov B, Dean J (2008) A subcortical maternal complex essential for preimplantation mouse embryogenesis. Dev Cell 15:416–425

Liu J, Linher K, Li J (2009) Porcine DAZL messenger RNA: its expression and regulation during oocyte maturation. Mol Cell Endocrinol 311:101–108

Ma J, Flemr M, Stein P, Berninger P, Malik R, Zavolan M, Svoboda P, Schultz RM (2010) MicroRNA activity is suppressed in mouse oocytes. Curr Biol 20:265–270

Marangos P, Carroll J (2008) Securin regulates entry into M-phase by modulating the stability of cyclin B. Nat Cell Biol 10:445–451

McGrew LL, Dworkin-Rastl E, Dworkin MB, Richter JD (1989) Poly(A) elongation during *Xenopus* oocyte maturation is required for translational recruitment and is mediated by a short sequence element. Genes Dev 3:803–815

Medvedev S, Yang J, Hecht NB, Schultz RM (2008) CDC2A (CDK1)-mediated phosphorylation of MSY2 triggers maternal mRNA degradation during mouse oocyte maturation. Dev Biol 321:205–215

Medvedev S, Pan H, Schultz RM (2011) Absence of MSY2 in mouse oocytes perturbs oocyte growth and maturation, RNA stability, and the transcriptome. Biol Reprod 85:575–583

Mendez R, Hake LE, Andresson T, Littlepage LE, Ruderman JV, Richter JD (2000a) Phosphorylation of CPE binding factor by Eg2 regulates translation of c-mos mRNA. Nature 404:302–307

Mendez R, Murthy KG, Ryan K, Manley JL, Richter JD (2000b) Phosphorylation of CPEB by Eg2 mediates the recruitment of CPSF into an active cytoplasmic polyadenylation complex. Mol Cell 6:1253–1259

Mendez R, Barnard D, Richter JD (2002) Differential mRNA translation and meiotic progression require Cdc2-mediated CPEB destruction. EMBO J 21:1833–1844

Millevoi S, Vagner S (2010) Molecular mechanisms of eukaryotic pre-mRNA 3′ end processing regulation. Nucleic Acids Res 38:2757–2774

Minshall N, Reiter MH, Weil D, Standart N (2007) CPEB interacts with an ovary-specific eIF4E and 4E-T in early *Xenopus* oocytes. J Biol Chem 282:37389–37401

Moore FL, Jaruzelska J, Fox MS, Urano J, Firpo MT, Turek PJ, Dorfman DM, Pera RAR (2003) Human Pumilio-2 is expressed in embryonic stem cells and germ cells and interacts with DAZ (Deleted in AZoospermia) and DAZ-Like proteins. Proc Natl Acad Sci USA 100:538–543

Morris AR, Mukherjee N, Keene JD (2008) Ribonomic analysis of human Pum1 reveals cis-trans conservation across species despite evolution of diverse mRNA target sets. Mol Cell Biol 28:4093–4103

Murai S, Stein P, Buffone MG, Yamashita S, Schultz RM (2010) Recruitment of Orc6l, a dormant maternal mRNA in mouse oocytes, is essential for DNA replication in 1-cell embryos. Dev Biol 341:205–212

Murchison EP, Stein P, Xuan Z, Pan H, Zhang MQ, Schultz RM, Hannon GJ (2007) Critical roles for Dicer in the female germline. Genes Dev 21:682–693

Nakahata S, Katsu Y, Mita K, Inoue K, Nagahama Y, Yamashita M (2001) Biochemical identification of *Xenopus pumilio* as a sequence-specific cyclin B1 mRNA-binding protein that physically interacts with a Nanos homolog, Xcat-2, and a cytoplasmic polyadenylation element-binding protein. J Biol Chem 276:20945–20953

Nakahata S, Kotani T, Mita K, Kawasaki T, Katsu Y, Nagahama Y, Yamashita M (2003) Involvement of *Xenopus pumilio* in the translational regulation that is specific to cyclin B1 mRNA during oocyte maturation. Mech Dev 120:865–880

Nakanishi T, Kubota H, Ishibashi N, Kumagai S, Watanabe H, Yamashita M, S-i K, Miyado K, Baba T (2006) Possible role of mouse poly(A) polymerase mGLD-2 during oocyte maturation. Dev Biol 289:115–126

Nakanishi T, Kumagai S, Kimura M, Watanabe H, Sakurai T, Kashiwabara S, Baba T (2007) Disruption of mouse poly(A) polymerase mGLD-2 does not alter polyadenylation status in oocytes and somatic cells. Biochem Biophys Res Commun 364:14–19

Oh B, Hwang SY, McLaughlin J, Solter D, Knowles BB (2000) Timely translation during the mouse oocyte-to-embryo transition. Development 127:3795–3803

Paillard L, Omilli F, Legagneux V, Bassez T, Maniey D, Osborne HB (1998) EDEN and EDEN-BP, a cis element and an associated factor that mediate sequence-specific mRNA deadenylation in *Xenopus* embryos. EMBO J 17:278–287

Pan H, O'Brien MJ, Wigglesworth K, Eppig JJ, Schultz RM (2005) Transcript profiling during mouse oocyte development and the effect of gonadotropin priming and development in vitro. Dev Biol 286:493–506

Parker R, Sheth U (2007) P bodies and the control of mRNA translation and degradation. Mol Cell 25:635–646

Paynton BV, Rempel R, Bachvarova R (1988) Changes in state of adenylation and time course of degradation of maternal mRNAs during oocyte maturation and early embryonic development in the mouse. Dev Biol 129:304–314

Piqué M, López JM, Foissac S, Guigó R, Méndez R (2008) A combinatorial code for CPE-mediated translational control. Cell 132:434–448

Racki WJ, Richter JD (2006) CPEB controls oocyte growth and follicle development in the mouse. Development 133:4527–4537

Radford HE, Meijer HA, de Moor CH (2008) Translational control by cytoplasmic polyadenylation in *Xenopus* oocytes. Biochim Biophys Acta 1779:217–229

Reverte CG, Ahearn MD, Hake LE (2001) CPEB degradation during *Xenopus* oocyte maturation requires a PEST domain and the 26S proteasome. Dev Biol 231:447–458

Richter JD, Sonenberg N (2005) Regulation of cap-dependent translation by eIF4E inhibitory proteins. Nature 433:477–480

Rouget C, Papin C, Boureux A, Meunier AC, Franco B, Robine N, Lai EC, Pelisson A, Simonelig M (2010) Maternal mRNA deadenylation and decay by the piRNA pathway in the early Drosophila embryo. Nature 467:1128–1132

Salles FJ, Darrow AL, O'Connell ML, Strickland S (1992) Isolation of novel murine maternal mRNAs regulated by cytoplasmic polyadenylation. Genes Dev 6:1202–1212

Schmid M, Kuchler B, Eckmann CR (2009) Two conserved regulatory cytoplasmic poly(A) polymerases, GLD-4 and GLD-2, regulate meiotic progression in *C. elegans*. Genes Dev 23:824–836

Setoyama D, Yamashita M, Sagata N (2007) Mechanism of degradation of CPEB during *Xenopus* oocyte maturation. Proc Natl Acad Sci USA 104:18001–18006

Standart N, Jackson RJ (2007) MicroRNAs repress translation of m7G ppp-capped target mRNAs in vitro by inhibiting initiation and promoting deadenylation. Genes Dev 21:1975–1982

Standart N, Minshall N (2008) Translational control in early development: CPEB, P-bodies and germinal granules. Biochem Soc Trans 36:671–676

Strickland S, Huarte J, Belin D, Vassalli A, Rickles RJ, Vassalli JD (1988) Antisense RNA directed against the 3′ noncoding region prevents dormant mRNA activation in mouse oocytes. Science 241:680–684

Stutz A, Huarte J, Gubler P, Conne B, Belin D, Vassalli JD (1997) In vivo antisense oligodeoxynucleotide mapping reveals masked regulatory elements in an mRNA dormant in mouse oocytes. Mol Cell Biol 17:1759–1767

Stutz A, Conne B, Huarte J, Gubler P, Volkel V, Flandin P, Vassalli JD (1998) Masking, unmasking, and regulated polyadenylation cooperate in the translational control of a dormant mRNA in mouse oocytes. Genes Dev 12:2535–2548

Su Y-Q, Sugiura K, Woo Y, Wigglesworth K, Kamdar S, Affourtit J, Eppig JJ (2007) Selective degradation of transcripts during meiotic maturation of mouse oocytes. Dev Biol 302:104–117

Sugimura I, Lilly MA (2006) Bruno inhibits the expression of mitotic cyclins during the prophase I meiotic arrest of *Drosophila* oocytes. Dev Cell 10:127–135

Suh N, Blelloch R (2011) Small RNAs in early mammalian development: from gametes to gastrulation. Development 138:1653–1661

Suh N, Baehner L, Moltzahn F, Melton C, Shenoy A, Chen J, Blelloch R (2010) MicroRNA function Is globally suppressed in mouse oocytes and early embryos. Curr Biol 20:271–277

Swetloff A, Conne B, Huarte J, Pitetti J-L, Nef S, Vassalli J-D (2009) Dcp1-bodies in mouse oocytes. Mol Biol Cell 20:4951–4961

Tadros W, Lipshitz HD (2009) The maternal-to-zygotic transition: a play in two acts. Development 136:3033–3042

Tadros W, Goldman AL, Babak T, Menzies F, Vardy L, Orr-Weaver T, Hughes TR, Westwood JT, Smibert CA, Lipshitz HD (2007) SMAUG is a major regulator of maternal mRNA destabilization in Drosophila and its translation is activated by the PAN GU kinase. Dev Cell 12:143–155

Tam OH, Aravin AA, Stein P, Girard A, Murchison EP, Cheloufi S, Hodges E, Anger M, Sachidanandam R, Schultz RM, Hannon GJ (2008) Pseudogene-derived small interfering RNAs regulate gene expression in mouse oocytes. Nature 453:534–538

Tang F (2010) Small RNAs in mammalian germline: tiny for immortal. Differentiation 79:141–146

Tang F, Kaneda M, O'Carroll D, Hajkova P, Barton SC, Sun YA, Lee C, Tarakhovsky A, Lao K, Surani MA (2007) Maternal microRNAs are essential for mouse zygotic development. Genes Dev 21:644–648

Tay J, Hodgman R, Richter JD (2000) The control of cyclin B1 mRNA translation during mouse oocyte maturation. Dev Biol 221:1–9

Thelie A, Papillier P, Pennetier S, Perreau C, Traverso JM, Uzbekova S, Mermillod P, Joly C, Humblot P, Dalbies-Tran R (2007) Differential regulation of abundance and deadenylation of maternal transcripts during bovine oocyte maturation in vitro and in vivo. BMC Dev Biol 7:125

Tremblay K, Vigneault C, McGraw S, Sirard MA (2005) Expression of cyclin B1 messenger RNA isoforms and initiation of cytoplasmic polyadenylation in the bovine oocyte. Biol Reprod 72:1037–1044

Tritschler F, Huntzinger E, Izaurralde E (2010) Role of GW182 proteins and PABPC1 in the miRNA pathway: a sense of deja vu. Nat Rev Mol Cell Biol 11:379–384

Uzbekova S, Arlot-Bonnemains Y, Dupont J, Dalbies-Tran R, Papillier P, Pennetier S, Thelie A, Perreau C, Mermillod P, Prigent C, Uzbekov R (2008) Spatio-temporal expression patterns of Aurora kinases A, B, and C and cytoplasmic polyadenylation-element-binding protein in bovine oocytes during meiotic maturation. Biol Reprod 78:218–233

Vardy L, Orr-Weaver TL (2007a) The Drosophila PNG kinase complex regulates the translation of cyclin B. Dev Cell 12:157–166

Vardy L, Orr-Weaver TL (2007b) Regulating translation of maternal messages: multiple repression mechanisms. Trends Cell Biol 17:547–554

Vassalli JD, Huarte J, Belin D, Gubler P, Vassalli A, O'Connell ML, Parton LA, Rickles RJ, Strickland S (1989) Regulated polyadenylation controls mRNA translation during meiotic maturation of mouse oocytes. Genes Dev 3:2163–2171

VerMilyea MD, Maneck M, Yoshida N, Blochberger I, Suzuki E, Suzuki T, Spang R, Klein CA, Perry ACF (2011) Transcriptome asymmetry within mouse zygotes but not between early embryonic sister blastomeres. EMBO J 30:1841–1851

Villaescusa JC, Allard P, Carminati E, Kontogiannea M, Talarico D, Blasi F, Farookhi R, Verrotti AC (2006) Clast4, the murine homologue of human eIF4E-Transporter, is highly expressed in developing oocytes and post-translationally modified at meiotic maturation. Gene 367:101–109

von Roretz C, Gallouzi IE (2008) Decoding ARE-mediated decay: is microRNA part of the equation? J Cell Biol 181:189–194

Watanabe T, Totoki Y, Toyoda A, Kaneda M, Kuramochi-Miyagawa S, Obata Y, Chiba H, Kohara Y, Kono T, Nakano T, Surani MA, Sakaki Y, Sasaki H (2008) Endogenous siRNAs from naturally formed dsRNAs regulate transcripts in mouse oocytes. Nature 453:539–543

Welk JF, Charlesworth A, Smith GD, MacNicol AM (2001) Identification and characterization of the gene encoding human cytoplasmic polyadenylation element binding protein. Gene 263:113–120

Wickens M, Bernstein DS, Kimble J, Parker R (2002) A PUF family portrait: 3′UTR regulation as a way of life. Trends Genet 18:150–157

Wright PW, Bolling LC, Calvert ME, Sarmento OF, Berkeley EV, Shea MC, Hao Z, Jayes FC, Bush LA, Shetty J, Shore AN, Reddi PP, Tung KS, Samy E, Allietta MM, Sherman NE, Herr

JC, Coonrod SA (2003) ePAD, an oocyte and early embryo-abundant peptidylarginine deiminase-like protein that localizes to egg cytoplasmic sheets. Dev Biol 256:74–89

Xu EY, Chang R, Salmon NA, Reijo Pera RA (2007) A gene trap mutation of a murine homolog of the Drosophila stem cell factor Pumilio results in smaller testes but does not affect litter size or fertility. Mol Reprod Dev 74:912–921

Yamashita A, Chang TC, Yamashita Y, Zhu W, Zhong Z, Chen CY, Shyu AB (2005) Concerted action of poly(A) nucleases and decapping enzyme in mammalian mRNA turnover. Nat Struct Mol Biol 12:1054–1063

Yang Q, Allard P, Huang M, Zhang W, Clarke HJ (2010) Proteasomal activity is required to initiate and to sustain translational activation of messenger RNA encoding the stem-loop-binding protein during meiotic maturation in mice. Biol Reprod 82:123–131

Yu J, Hecht NB, Schultz RM (2001) Expression of MSY2 in mouse oocytes and preimplantation embryos. Biol Reprod 65:1260–1270

Yu JY, Hecht NB, Schultz RM (2002) RNA-binding properties and translation repression in vitro by germ cell-specific MSY2 protein. Biol Reprod 67:1093–1098

Yu J, Deng M, Medvedev S, Yang J, Hecht NB, Schultz RM (2004) Transgenic RNAi-mediated reduction of MSY2 in mouse oocytes results in reduced fertility. Dev Biol 268:195–206

Yurttas P, Vitale AM, Fitzhenry RJ, Cohen-Gould L, Wu W, Gossen JA, Coonrod SA (2008) Role for PADI6 and the cytoplasmic lattices in ribosomal storage in oocytes and translational control in the early mouse embryo. Development 135:2627–2636

Chapter 2
Polarity and Asymmetry During Mouse Oogenesis and Oocyte Maturation

Malgorzata Kloc, Rafik M. Ghobrial, Ewa Borsuk, and Jacek Z. Kubiak

Abstract Cell polarity and asymmetry play a fundamental role in embryo development. The unequal segregation of determinants, cues, and activities is the major event in the differentiation of cell fate and function in all multicellular organisms. In oocytes, polarity and asymmetry in the distribution of different molecules are prerequisites for the progression and proper outcome of embryonic development. The mouse oocyte, like the oocytes of other mammals, seems to apply a less stringent strategy of polarization than other vertebrates. The mouse embryo undergoes a regulative type of development, which permits the full rectification of development even if the embryo loses up to half of its cells or its size is experimentally doubled during the early stages of embryogenesis. Such pliability is strongly related to the proper oocyte polarization before fertilization. Thus, the molecular mechanisms leading to the development and maintenance of oocyte polarity must be included in any fundamental understanding of the principles of embryo development. In this chapter, we provide an overview of current knowledge regarding the development and maintenance of polarity and asymmetry in the distribution of organelles and molecules in the mouse oocyte. Curiously, the

M. Kloc (✉) • R.M. Ghobrial
The Methodist Hospital, Department of Surgery, and The Methodist Hospital Research Institute, Houston, TX, USA
e-mail: mkloc@tmhs.org; rmghobrial@tmhs.org

E. Borsuk
Department of Embryology, Institute of Zoology, Faculty of Biology, University of Warsaw, Miecznikowa 1, 02-096 Warsaw, Poland
e-mail: borsuk@biol.uw.edu.pl

J.Z. Kubiak
Institut de Génétique et Développement de Rennes, CNRS, UMR 6290, Rennes F-35043, France

Université Rennes 1, IFR 140, Faculté de Médecine, UEB, Rennes F-35043, France
e-mail: Jacek.Kubiak@univ-rennes1.fr

J.Z. Kubiak (ed.), *Mouse Development*, Results and Problems in Cell Differentiation 55,
DOI 10.1007/978-3-642-30406-4_2, © Springer-Verlag Berlin Heidelberg 2012

mouse oocyte becomes polarized at least twice during ontogenesis; the question of how this phenomenon is achieved and what role it might play is addressed in this chapter.

2.1 Two Keywords: Polarity and Asymmetry

Throughout the last century, extensive studies performed on the formation, structure, and physiology of animal oocytes and eggs indicated that the majority of invertebrate and vertebrate (including mammalian) oocytes and eggs are asymmetrical and/or polar (Albertini and Barrett 2004; Brunet and Verlhac 2011; De Smedt et al. 2000; Kloc et al. 2004a, b, 2008). Surprisingly, for many decades, the issues of mouse oocyte asymmetry and polarity and their effects on future embryo development have been, and still are, vigorously debated (Hiiragi et al. 2006; Zernicka-Goetz and Huang 2010; Johnson 2009; Kloc et al. 2008; VerMilyea et al. 2011). In dealing with this subject, it is important to recognize the fundamental conceptual difference between asymmetry and polarity, with the first being an asymmetrical distribution of organelles, molecules, or functions within individual cells of the same type, and the second being a situation in which the asymmetrically distributed entities have an invariable, noninterchangeable, and irreplaceable position in all cells of the same type. Additional confusion arises from the fact that asymmetry/polarity can be either permanent or temporary, i.e., limited to a specific stage/time frame of oocyte development (Brunet and Verlhac 2011; Kloc et al. 2008). Thus, only comprehensive studies of all stages of oogenesis or oocyte maturation, and not arbitrarily selected stages, can validate generalized conclusions. Because the natural or experimental disruption of the existing asymmetry/polarity of the oocyte can lead to disastrous developmental consequences in the resulting embryo (Anifandis et al. 2010; Eichenlaub-Ritter et al. 2011; Evans et al. 2000), knowledge of the asymmetry/polarity of the mature mammalian oocyte is of the utmost importance in this era of the ever-expanding usage of experimental manipulations and in vitro fertilization (Anifandis et al. 2010; Edwards 2000, 2001; Edwards and Ludwig 2003; Eichenlaub-Ritter et al. 2011).

2.2 The Balbiani Body, Mitochondria, and Centrioles

Over a century ago, a peculiar structure, subsequently called the Balbiani body (Bb) after its discoverer, was described in the oocytes of spiders and myriapods (reviewed in Kloc et al. 2004a, b). Since then, the term Balbiani body (also called a mitochondrial cloud) has been ubiquitously used to describe a variety of spherical structures with distinct morphologies and ultrastructures and probably unrelated functions, which are asymmetrically located within the oocyte cytoplasm in the majority of animals. These structures are not surrounded by a membrane and usually consist of an aggregate of mitochondria (often interspersed with the electron-dense nuage), ER cisternae, and Golgi complexes (Kloc et al. 2004a, b). The function(s) of Bb in animal oocytes, with the exception of the oocytes of the frog *Xenopus laevis*, is completely elusive. In *Xenopus*, the Balbiani body (mitochondrial cloud) is a vehicle for the delivery of various developmentally relevant localized RNAs, proteins, germ cell determinants (nuage and germinal granules), and maternal mitochondria to the vegetal pole of the oocyte, and it is a prerequisite for the normal development of the germ cell line (Kloc et al. 1996, 1998, 2001; Kloc and Etkin 1995, 1998). Detailed studies of Bb formation during consecutive stages of oogenesis showed that *Xenopus* female germ cells are not only asymmetrical but also polar—in all oogonia (cystocytes), the Bb, from the time of its conception, is located at the vicinity of the centriole and the cytoplasmic bridges connecting the cystocytes, and in all oocytes, the Bb is located at the vegetal pole (Kloc et al. 2004a, b).

Although the presence of the Bb in the oocytes of many mammals, including humans, has been well documented, until last decade it was believed that mouse oocytes are an exception among other mammalian species and that they do not contain the Bb (De Smedt et al. 2000; Kloc et al. 2004a, b). In 2007, Pepling and colleagues were first to describe the presence of the Bb in neonatal mouse oogonia and in the oocytes of primordial follicles (Pepling et al. 2007). This study also indicated that mouse oogonia and early oocytes are asymmetrical (the Bb was always located at one side of the nucleus) and that the Bb is a transient structure that disperses during late oogenesis. Subsequent three-dimensional reconstruction studies performed by Kloc et al. (2008) showed that mouse neonatal oocytes are not only asymmetrical but also polar and that the aspects of polarity in mouse oocytes are highly reminiscent of those in *Xenopus* early oocytes, i.e., in both species the Bb forms around the centriole in the vicinity of the cytoplasmic bridges connecting the oogonia (Fig. 2.1; Kloc et al. 2004a, b, 2008). In spite of this similarity, the structure of the mouse oocyte Bb differs from that of *Xenopus*. Whereas in *Xenopus*, the bulk of the Bb is composed of hundreds of thousands of mitochondria interspersed with some Golgi and ER cisternae, the mouse Bb contains an elaborate aggregate of Golgi cisternae surrounded at its periphery by mitochondria (Figs. 2.1 and 2.2; Kloc et al. 2008; Pepling et al. 2007). Although we do not know whether, in the mouse, the Bb Golgi apparatus delivers any molecules or organelles to the oocyte surface in a polarized manner, it is possible that the Golgi vesicles secrete certain components of the extracellular matrix in this way, such as the *zona pellucida* (ZP), which is involved in fertilization and early development (El-Mestrah et al. 2002; Hoodbhoy et al. 2006). If this supposition were true, then the transient polarity of the mouse oocyte leading to polar secretions from the Bb Golgi would translate into the polarity of the *zona pellucida*, which would have possible developmental consequences (Kloc et al. 2008). Another possibility is that the polar distribution of the Golgi in the mouse oocyte is related to the polar/asymmetric division during oocyte maturation. The most recent study by Zhang and collaborators (2011) showed that GM130, a Golgi resident protein, associates with the meiotic spindle and plays a key role (possibly via its cooperation with the MAPK pathway) in the

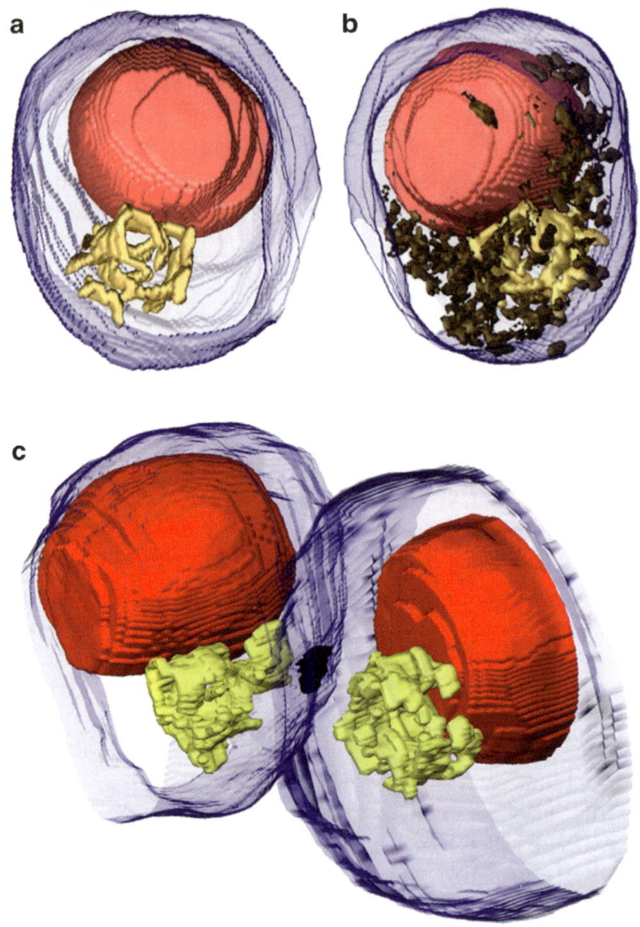

Fig. 2.1 Asymmetry and polarity of the mouse early oocyte. The three-dimensional reconstruction of mouse P0 oocytes from 16 semithin sections. (**a**, **b**) An image of a single P0 oocyte showing the asymmetrical distribution of the Balbiani body's Golgi complex (*yellow*) and mitochondria (*black*) at one side of the nucleus (*red*). (**c**) An image of two P0 oocytes connected by a cytoplasmic bridge (*black*), showing the polarity of the oocytes: in both oocytes, the Balbiani body faces the cytoplasmic bridge. For clarity, images (**a**) and (**c**) show only the nucleus and the Golgi complex. For further details and methods used for 3D reconstruction, see Kloc et al. (2008)

organization and polar migration of the spindle during the extrusion of the polar body in the maturing mouse oocyte.

Another exciting possibility is that, in the mouse, the Bb Golgi apparatus plays a role in the transduction of stress/apoptotic signals to the Bb mitochondria. Recent studies on the role of the Golgi in neurodegenerative diseases indicate the existence of cross talk among the Golgi, the mitochondria, and the ER and suggest that the Golgi (via Golgi cisternal stacking protein Grasp65) plays a role of a common stress sensor, downstream effector, and transducer of cell death signals

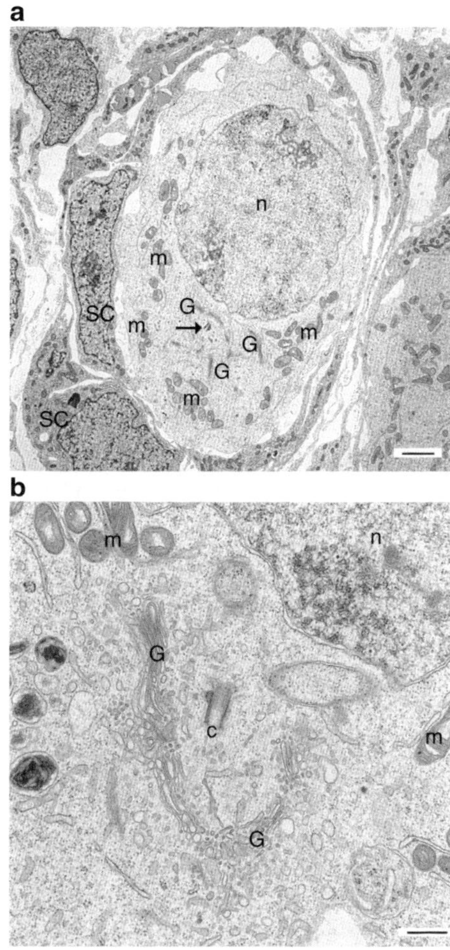

Fig. 2.2 Asymmetric distribution of the Balbiani body in the mouse early oocyte. (**a**) The electron microscopy image of a P0 oocyte surrounded by somatic cells (sc). The asymmetrically positioned Balbiani body composed of Golgi cisternae (G) and mitochondria (m) is organized around the centrally located centriole (*arrow*). (**b**) The electron microscopy image of the oocyte fragment showing structural details of the Balbiani body. The Golgi cisternae (G) and the mitochondria (m) are concentrically arranged around the centriole (C). Oocyte nucleus (n). The scale bar is equal to 2 μm in (**a**) and 500 nm in (**b**). For further details and methods, see Kloc et al. (2008)

(Nakagomi et al. 2008). Applying this scenario to the mouse oocyte, it is possible that the aggregation of mitochondria in the vicinity of the Bb Golgi facilitates perception of the stress signal by the mitochondria, eventually leading to the elimination of damaged or substandard mitochondria and preventing their inheritance by the offspring.

In the majority of animals, including the mouse, the mitochondria are one of the invariable components of the Bb (mitochondrial cloud). The importance of the health of maternally transmitted mitochondria for the quality of the future embryo and the role of mitochondrial DNA disorders related to aging and common diseases with maternal inheritance have been well documented in mammals, including humans (Bouchet et al. 2006; Dean et al. 2003; Eichenlaub-Ritter et al. 2011; Monnot et al. 2011). Studies in cattle and mice showed a dramatic shift in the variant/wild-type mtDNA ratio between the mother and offspring, indicating the presence of a tight bottleneck that insures that new mutations are fixed rapidly or lost (Laipis et al. 1988; Bergstrom and Pritchard 1998; Roze et al. 2005; Jenuth et al. 1996; Cree et al. 2008; Cao et al. 2007, 2009; Wai et al. 2008). Recently, Zhang and colleagues (2008) showed that, in the zebrafish oocyte, the mitochondrial cloud attracts high-functional (i.e., high-inner membrane potential) mitochondria. These and other authors suggest that the mitochondrial cloud may function as a place for the selection of healthy mitochondria and the transmission of the fittest mitochondria to the offspring. Thus, by preventing deleterious mtDNA mutations from being passed to offspring, the mitochondrial cloud could play the role of a mitochondrial genetic bottleneck (Kloc et al. 2004a, b; Pepling and Spradling 1998; Cox and Spradling 2003, 2006; Zhou et al. 2010).

The polarity of mouse oogonia at the cytocyst stage may be related to the organization of the Balbiani body around the classical centriolar centrosome present at this stage (Fig. 2.2). Later in mouse oocyte development, the centrioles are lost. There is no information available describing at what stage this loss occurs. One may only speculate that the disappearance of the centrioles occurs during the long-lasting growth phase of the oocytes. If this supposition is true, then the transformation of centriolar into acentriolar centrosomes may have an important impact on the apparent loss of polarity of fully grown mouse oocytes.

2.3 Asymmetry of Maturation Division and Germinal Vesicle (GV) Positioning

The meiotic maturation of the oocyte is asymmetric and generates two unequal cells: the large mature germinal cell and the small polar body. The asymmetry of the meiotic division depends on the proper migration of the spindle containing the bivalents to the cortical region of the oocyte during the first meiotic M-phase. Because the starting point for spindle migration is the position of the GV (central or eccentric) within the oocyte, the proper positioning of the GV is paramount for the successful outcome of meiotic division, and the position of the GV has been used as a marker of oocyte quality (Gönczy 2002; Brunet and Maro 2007; Brunet and Verlhac 2011).

The positioning of the GV, and thus the route of mitotic spindle migration, varies between different animal species. In *Drosophila,* the eccentric migration of the GV in the vicinity of oocyte cortex employs microtubules, Lissencephaly1 (DLis1), Bicaudal (BicD), and dynein-dependent mechanisms (Lei and Warrior 2000; Swan et al. 1999). Similarly, in sea cucumber oocytes, the off-center positioning of the GV and the future spindle are both mediated by microtubule/centrosome-dependent mechanisms (Miyazaki et al. 2005). In some organisms, the eccentric migration of the GV is followed by the subsequent migration of the spindle to the oocyte cortex, and these two processes are mechanistically independent and mediated by different mechanisms and molecules. In *C. elegans,* the eccentric migration of the GV depends on microtubules and zygote defective protein 9, whereas the migration of the spindle involves microtubules, microtubule-severing enzyme katanin, kinesin-1, and kinesin-cargo adaptor protein (Yang et al. 2003, 2005).

In mammalian oocytes, the GV is located in the oocyte center or slightly off-center, which is where the meiotic spindle forms before subsequently migrating towards the oocyte cortex. Studies on mice showed that the ability of the oocyte to complete maturation depends on the location of the GV within the oocyte and that the GV is positioned centrally in maturation-competent oocytes and at the periphery in maturation-incompetent oocytes (Brunet and Maro 2007). In addition, this study showed a positive correlation between mouse age, reduced ability to progress through meiosis, and decreased efficiency in GV centering. Although, at present, little is known about the molecular processes involved in the positioning of the GV in mammalian oocytes, recent studies on mice indicate that a complex meshwork of actin filaments physically connects the GV to the cortex and may lead to its centering by exerting counterpoising forces (Azoury et al. 2011). Interestingly, recent studies also show that the proper positioning of the GV, as well as the positioning and structure of the spindle, depends on the maintenance of the contact between the somatic (cumulus) cells and the oocyte (Barrett and Albertini 2010). It is known that in vitro culture reduces the integrity of the contact between the oocyte and its somatic cells, resulting in weakened developmental competence (Barrett and Albertini 2010; Sanfins et al. 2003, 2004). This observation underscores the importance of the somatic cell component and indicates that the bidirectional signaling between the oocyte and its somatic partners regulates intrinsic oocyte processes and has a lasting (and often undervalued) influence on oocyte developmental potential. The importance of the interaction between oocytes and somatic cells in the developing mouse ovary was recently underscored by the results of studies by Lechowska et al. (2011). These authors showed that, in the ovaries of Nobox-deficient mice, the somatic cells are unable to surround and separate individual primordial follicles, leading to the formation of syncytial follicles instead of primordial follicles. Because the NOBOX is a key regulator of oocyte-specific genes in mice and its expression is misregulated in women with premature ovarian failure (POF) syndrome, this study may shed light on the mechanism of POF in humans (Albertini 2011; Lechowska et al. 2011; Rajkovic et al. 2004).

The more or less central (with emphasis on "less") position of the GV in fully grown, prophase-arrested mouse oocytes may be the key factor determining the future direction of the spindle migration to the nearest subcortical region that therefore determines the axis of the future egg polarity. There is ample information on the molecular mechanisms responsible for the positioning and migration of the meiotic spindle from the center of the oocyte to its periphery. In mouse oocytes, the migration of the spindle from the slightly off-center position (where the GV was located) toward the nearest region of the cortex occurs along its long axis and involves actin filaments, actin nucleator factor formin-2 (Fmn2; Azoury et al. 2009, 2011; Dumont et al. 2007; Leader et al. 2002; Longo and Chen 1985; Verlhac et al. 2000; Terada et al. 2000), and possibly the function of Golgi resident protein GM130 (Zhang et al. 2011). It must be stressed in this discussion that the MI spindle in mouse oocytes is acentriolar and that only the electron-dense pericentriolar material is found at the spindle poles (Szollosi et al. 1972). This description also applies to the MII spindle and, as mentioned earlier, the stage of oogenesis at which the centrioles disappear is unknown. In recent years, mounting evidence has pointed to the critical role of cytoplasmic sheets or lattices (CPLs) in the regulation of various stages of oocyte maturation, spindle movement, and emission of the polar body. The cytoplasmic lattices or sheets are functionally puzzling structures, unique to the oocytes and early embryos of all mammals studied so far, including mouse and humans. CPLs are composed of cylindrical bundles of keratin-containing intermediate filaments and undergo dramatic reorganizations at critical stages of oocyte maturation and early development (Capco et al. 1993; Esposito et al. 2007; Gallicano et al. 1994; Kan et al. 2011; Wright et al. 2003). One of the possible functions of CPLs is the regulation of microtubule dynamics through the storage of soluble tubulin (Kan et al. 2011). CPLs are enriched in peptidylarginine deiminase 6 (PADI6), which is necessary for their formation, and it seems that the PADI6/CPL superstructure plays a key role in regulating microtubule-mediated organelle positioning and movement (Esposito et al. 2007; Wright et al. 2003; Kan et al. 2011). Interestingly, the subcortical positioning of the acentriolar spindle and chromosomes results in subsequent local remodeling of the oocyte cortex.

2.4 Remodeling of the Oocyte Cortex During Maturation

It is well established that the oocyte cortex and its remodeling (reorganization and polarization) during maturation play pivotal roles in polar body emission and fertilization, as well as the subsequent proper spatiotemporal development of the embryo (Sardet et al. 2002; Kloc and Etkin 1995; Kloc et al. 1996, 1998). Although the usage of the term "cortex" varies between different authors, when defined in the broadest sense, the oocyte/egg cortex contains three major components: the plasma membrane, the submembrane cytoskeletal elements, which anchor different organelles (such as mitochondria and cortical granules) and molecules, and the extracellular matrix. After GVBD (GV breakdown) and the translocation of the spindle, the region of the cortex overlying the spindle becomes free of the cortical granules (CGs) and microvilli and forms an actin filament-enriched cap that is also enriched in the mammalian homologs of the evolutionarily conserved polarity proteins Par3 and Par6 (Vinot et al. 2004; Duncan et al. 2005) and Rac-GTP (Halet and Carroll 2007). In many animal species, the size and the position of this domain regulate the size of the polar body and mark the animal pole of the future embryo (Azory et al. 2008, 2009; Deng et al. 2003; Longo and Chen 1985; Van Blerkom and Bell 1986). The formation of a CG- and microvilli-free cortical domain requires the migration of the spindle (Longo and Chen 1985) but occurs also in the absence of intact microtubule components of the spindle, thus demonstrating that it is chromosome/chromatin dependent (Maro et al. 1986;

Deng et al. 2003, 2005). Other studies show that this remodeling of the cortex depends on the balanced spatial segregation of gamma-tubulin between the oocyte cortex and the meiotic acentriolar spindle (Barrett and Albertini 2007). Studies in mouse oocytes indicate that the formation of the actin cap and the cortical granule-free domain also depend on the Arp2/3 (actin-related protein 2/3) complex and its activator, the actin nucleation factor/p53-cofactor, JMY (Sun et al. 2011a, b; Zuchero et al. 2009). Thus, coordinated cytoskeleton rearrangements seem to be key phenomena leading to the development of oocyte polarity.

To achieve the highly asymmetric division required during the first meiosis of the mouse oocyte, two steps are necessary: (1) the meiotic spindle must be relocated from the position where it was formed to the oocyte cortex; and (2) the polar body must be extruded. Because it was shown that, in maturing mouse oocytes, the disassembly of microtubules does not affect the movement of the chromatin to the oocyte cortex (Longo and Chen 1985; Verlhac et al. 2000), but the destruction or stabilization of actin filaments does (Terada et al. 2000), the role of actin in this process was extensively studied. The nucleation of actin filaments requires the activity of several proteins: the Arp 2/3 complex (Mullins et al. 1998) and the members of two different protein families, Formin (Evangelista et al. 2003; Harris and Higgs 2004; Zigmond 2004) and Spire (Quinlan et al. 2005; Ducka et al. 2010). At the GV stage and during the early stages of maturation, the cortex of the mouse oocyte possesses numerous actin filaments that emanate from the plasma membrane, forming a uniform layer of filamentous actin (F-actin). The distribution of cortical actin changes during the progression of meiotic maturation, finally forming a thick cap over the spindle (Longo 1987). Except for this cortical actin, a network of cytoplasmic actin filaments surrounds the GV and the oocyte chromatin after GVBD. The relocation of the meiosis I spindle requires the interaction between the chromosomes, microtubules, and actin filaments. The studies on Fmn2 knockout mouse oocytes, in which the spindle does not relocate to the cortex because of the lack of cytoplasmic actin filaments nucleated by Formin2, have led to the creation of a mechanistic model of asymmetric spindle positioning (Dumont et al. 2007; Schuh and Ellenberg 2008). According to this model, Fmn2 nucleates the continuous remodeling of the actin network, which bridges the gap between the cortex and the spindle, which is located slightly off-center in the oocyte. The dynamic connection of F-actin to the spindle is mediated by myosin II, which accumulates at the poles (Simerly et al. 1998). The spindle poles pull on the actin network via activated myosin. The pulling force of the pole that was initially positioned closer to the cortex is stronger than the force of the opposite pole. This inequality generates a higher number of stable actin filaments and results in the relocation of the spindle toward the cortex (Schuh and Ellenberg 2008).

Formin2 is not the only actin nucleator involved in the generation of the cytoplasmic actin network and the regulation of spindle relocation in maturing mouse oocytes. Recently, it was demonstrated that two other proteins, members of the Spire family, are key regulators of this process and in first polar body extrusion (Pfender et al. 2011). In oocytes co-depleted of both proteins, not only the asymmetric spindle position but also the efficiency of the first polar body extrusion

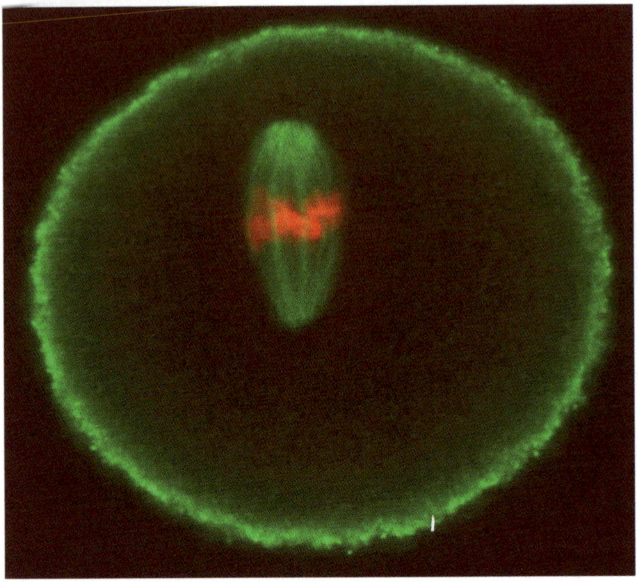

Fig. 2.3 Cdc42 localization in the MI mouse oocyte (Cdc42 *green*, DNA *red*)

becomes severely affected. In such oocytes, karyokinesis was completed, but cytokinesis failed because the cleavage furrow was not assembled. Groups of chromosomes separated in telophase I eventually fused together and formed one metaphase II spindle. Spire1 and Spire2 cooperate with themselves and associate with Formin2 in a functional unit to assemble the cytoplasmic actin network, which mediates the relocation of the meiotic spindle and is indispensable for the formation of the cleavage furrow during the first meiotic cytokinesis.

During cytokinesis, the separation of daughter cells occurs due to a properly functioning contractile ring, composed of actin and myosin (Guertin et al. 2002; Glotzer 2005). In mitotic cells, the formation of the contractile ring is regulated by proteins belonging to the Rho family of small guanosine triphosphatases (Rho GTPase) and Formins. The Cdc42 (cell division cycle 42) protein is, besides Rac1 and RhoA, one of the best-characterized members of the Rho GTPases (Kozma et al. 1995; Nobes and Hall 1995; Etienne-Manneville and Hall 2002). Cdc42 is involved in establishing the polarity of many cells, controlling the spindle positioning, the actin cytoskeleton, and the asymmetric distribution of determinants (Cau and Hall 2005). Its role in the polarization and asymmetrical cell division in vertebrate oocytes was demonstrated for *Xenopus laevis* (Ma et al. 2006) and mouse (Na and Zernicka-Goetz 2006; Cui et al. 2007). In MI and MII mouse oocytes, Cdc42 is localized on the microtubules of the spindle (Fig. 2.3), and it moves to the mid-body region at telophase of the first and second meiosis (Bielak-Zmijewska et al. 2008). Disruption of Cdc42 function causes the formation of abnormally elongated spindles, which fail to migrate to the oocyte cortex. Cdc42 disruption results in a reduction in the number of oocytes that are able to finish the first meiotic division and extrude the first polar body (Na and Zernicka-Goetz 2006; Cui et al. 2007). Cdc42 is not the only Rho GTPase involved in the regulation of the meiotic spindle formation and its relocation. The effects of the inhibition of Rac1 in maturing mouse oocytes mimic the effects of Cdc42 inactivation. Additionally, in MII oocytes, the Rac1 protein is involved in the anchoring of the spindle in the cortical region (Halet and Carroll 2007). RhoA is engaged in the organization of actin filaments and, like Cdc42, accumulates in the region of the mid-body during telophase I and II (Zhong et al. 2005).

All of these data indicate that Rho GTPases and actin nucleators regulate the movement of the meiotic spindle to the oocyte cortex and the asymmetric cytokinesis of the first and second meiosis. In polarized mitotic cells, the regulation of asymmetric division requires coordination of the activity of Rho GTPases and their effector proteins, including IQGAP1 (IQ-domain GTPase-activating protein 1). This conserved protein is engaged in the organization of microtubules and microfilaments. Through its binding to selected partners, such as E-cadherin, beta-catenin (Fukata et al. 1999; Kuroda et al. 1998; Shannon and Li 1999), or F-actin (Bashour et al. 1997; Fukata et al. 1997), IQGAP1 participates in several signaling pathways. IQGAP1 interacts with Rho GTPases and is involved in the polarization of mitotic cells (Noritake et al. 2005; Yasuda et al. 2006). It binds Cdc42 and Rac1 in their active, GTP-bound forms (Kuroda et al. 1998), but it does not bind RhoA. In the GV of intact mouse oocytes, IQGAP1 is present in the cytoplasm and forms a continuous layer in the cortex in a pattern that resembles the localization of Cdc42 (Bielak-Zmijewska et al. 2008). The initiation of maturation causes a dramatic reorganization of IQGAP1, including the loss of its co-localization with its partner protein, Cdc42. IQGAP1 loses its cortical localization and remains dispersed in the cytoplasm until telophase I and II, when it accumulates in the contractile ring, a pattern that suggests its involvement in cytokinesis. This function of IQGAP1 seems to depend on its direct relationship with the Rho GTPases. Inhibition of this interaction by toxin B, a glucosyltransferase that keeps the Rho GTPases in their inactive, GDP-bound forms, blocks the extrusion of the first polar bodies. In toxin B-treated oocytes, the formation of contractile rings is abolished (Bielak-Zmijewska et al. 2008). This change is accompanied by a massive depolymerization of cortical actin and rearrangement of IQGAP1 localization. These data suggest that, in mouse oocytes, IQGAP1 acts downstream of Cdc42 and that the activity of both proteins could be necessary for the proper arrangement of actin filaments in the contractile ring. Although it was postulated that Cdc42 regulates the relocation of the meiotic spindle in mouse oocytes (Na and Zernicka-Goetz 2006), it seems that the step of asymmetric division is achieved mainly due to the activity of actin nucleators and motor proteins (Schuh and Ellenberg 2008; Pfender et al. 2011). Rho GTPases and their regulators or effectors (such as IQGAP1) are involved in the process of polar body extrusion, arranging the assembly of the cleavage furrow and the contractile ring.

There are also recent studies indicating the paramount importance of strictly mechanical properties of the oocyte/egg, such as cortical tension and stiffness, on the subsequent molecular events that lead to the remodeling of the maturing oocyte.

Larson et al. (2010) showed that the meiotic maturation of the mouse oocyte is accompanied by a sixfold drop in tension. After fertilization, the tension increases by approximately 1.6-fold. In addition, they showed that there is a 2.5-fold mechanical tension differential between the actin- and myosin II-enriched microvilli-free cortical domain overlaying the spindle and the opposite microvillar cortex, which is enriched in radixin. Experimental perturbation of the expression of radixin, actin, and myosin II resulted in the reduction of tension during maturation and in spindle abnormalities (Larson et al. 2010).

In summary, all of these studies indicate that the extremely complex and multifaceted signaling between the multitude of mechanical, structural, and molecular components of the oocyte and the somatic cells leading to egg polarization still remains a worthwhile challenge.

2.5 Polarity of Cell Cycle Regulators

Besides the polarity of the cytoskeleton elements and the cortex, the more "fluid" elements are also clearly polarized in the mouse oocyte. The notion of the potential "fluidity" of certain enzymes may be controversial because they are potentially linked to the cytoskeletal elements. Some examples are the M-phase regulators (oocytes spend the vast majority of the time during meiotic maturation in M-phases), such as the CDK1 (Cyclin-Dependent Kinase 1) and MAP (Mitogen-Activated Protein) kinases from the ERK (Extracellular-Regulated Kinases) family (including ERK1, 2, 3, and 5), which are major serine/threonine kinases.

CDK1, also called MPF for M-phase Promoting Factor, is a major kinase active during M-phase that was first identified in amphibian oocytes (Masui and Markert 1971) and later in mouse oocytes (Balakier and Czolowska 1977). It is activated by binding to the regulatory subunit cyclin B (Labbé et al. 1988; Gautier et al. 1988, 1990), which is followed by a series of phosphorylation reactions at the activating sites as well as coordinated dephosphorylations at the inhibitory sites. CDK1/cyclin B is activated upon M-phase entry (during GVBD in oocytes) and inactivated at the end of each M-phase (upon anaphase I during oocyte maturation and anaphase II during fertilization). Inactivation of this kinase requires the separation of cyclin B from CDK1 and its proteolytic degradation by the proteasome (Glotzer et al. 1991; Ledan et al. 2001). Cyclin B separation from CDK1 and its further degradation is mediated by polyubiquitination via the ubiquitin ligase, APC/C (Anaphase Promotin Complex/ Cyclosome) (Nishiyama et al. 2000; Chesnel et al. 2006). CDK1/cyclin B exhibits typical histone H1 kinase activity in oocytes of different species (e.g., sea urchin: Arion and Meijer 1989, starfish: Labbé et al. 1989, and mouse: Kubiak et al. 1991; Kubiak et al. 1992; Ciemerych et al. 1998; Kubiak 2012). In mouse MII-arrested oocytes, H1 kinase activity is clearly more concentrated within the spindle-containing oocyte half (Kubiak et al. 1994). Accordingly, the CDK1 and cyclin B proteins concentrate within the spindle in mouse oocytes (Mitra and Schultz 1996; Huo et al. 2005), as well as in somatic cells (Bailly et al. 1989; Pines and Hunter 1991;

Ookata et al. 1995). Thus, active CDK1/cyclin B concentration with the spindle is not a unique characteristic of oocytes. However, in contrast to somatic cells, which are small and have spindles positioned in the center, in large cells like oocytes, the high concentration of active CDK1 within the asymmetrically positioned meiotic spindle creates a highly polarized gradient of this kinase in the cytoplasm (Kubiak et al. 1994). Cyclin B undergoes turnover during MII, which requires the presence of the intact meiotic spindle (Kubiak et al. 1994). Interestingly, cyclin B is stable in MI oocytes (Winston 1997). It is therefore likely that the components of the molecular machinery involved in cyclin B degradation (e.g., the APC/C components and regulators) are also asymmetrically distributed, at least within MII oocytes. Indeed, ubiquitin concentrates around the meiotic spindle in mouse oocytes, especially during anaphase and telophase, when cyclin B is ubiquitinated and degraded (Huo et al. 2004). Additionally, a regulatory subunit of the proteasome, the high molecular weight protease complex that degrade cyclin B, was shown to concentrate within the spindle of the mouse oocyte (Tan et al. 2005). Altogether, the distribution of CDK1, its regulator cyclin B, and the inactivating molecular machinery is highly asymmetric and polarized within the mouse oocyte, with the meiotic spindle acting as a concentration and/or diffusion point. Recent study by the group of Rong Li has shown the presence of a cytoplasmic stream in MII-arrested oocytes, which is involved in localization of the second meiotic spindle to the restricted subcortical area (Yi et al. 2011). This indicates that the seemingly stagnant MII state is much more dynamic than previously thought. Not only the dynamic state of cyclin B regulating stable CDK1 activity (Kubiak et al., 1993), but also the physical movement of the cytoplasm fixing the meiotic spindle in a stable position argues for the dynamic equilibrium preserving the polarity of mouse MII oocyte.

MAP kinases from the ERK family show a very similar asymmetry in their localization in the oocyte. ERK1 and 2, ERK3, and phosphorylated ERK5 MAP kinases are localized to the mouse oocyte spindle and form a gradient in the cytoplasm (ERK1/2: Verlhac et al. 1993; ERK3: Li et al. 2010; ERK5: Maciejewska et al. 2011). Additionally, MEK1, the ERK1/2 MAP kinase activating kinase, localizes to the spindle and especially to the pericentriolar material foci at the spindle poles (Yu et al. 2007). Most importantly, the phosphorylated, and thus active, form of MEK1 has a very similar localization within the specific region of the spindle. This observation suggests that the activation of ERK1 and 2 occurs in this specific region of the spindle and that it is followed by the distribution of the active MAP kinases throughout the whole spindle and in the ooplasm (Xiong et al. 2007). The fact that the ERK1/2 MAP kinase pathway (including the ERK 1/2 MAP kinase kinase kinase Mos) has an essential role in the meiotic maturation and spindle positioning in MI mouse oocytes, as well as in the sizing of the first polar body, indicates its functional relevance to oocyte polarity (Weber et al. 1991; Colledge et al. 1994; Verlhac et al. 1996; Araki et al. 1996). Thus, the polarized localizations of these MAP kinases and the components of their activation pathways seem to be essential for the asymmetric division of mouse oocytes. These major kinases are not the only spindle-associated kinases in mouse oocytes. Fyn kinase, a member of the Src family of tyrosine kinases, also has a similar distribution (see Levi et al. 2012 in this issue).

Protein kinase action is counterbalanced by specific phosphatases, which dephosphorylate their substrates and form the equilibrium between the turnover of phosphorylated and dephosphorylated substrates. The role of the phosphatases PP2A and PP1 in the regulation of the cell cycle was described in detail in both mouse oocytes (Rime and Ozon 1990; de Pennart et al. 1993; Schindler 2011) and rat oocytes (Zernicka-Goetz et al. 1997). So far, the localization pattern of these two essential cell cycle-regulating phosphatases in mouse oocytes remains unknown, but another cell cycle-regulating phosphatase, CDC14, has been found to be concentrated within the meiotic spindle (Schindler and Schultz 2009). This observation suggests that other phosphatases could also accumulate in the meiotic spindle, where they would counterbalance kinase-dependent phosphorylation. This arrangement, in turn, would imply the polarized distribution of a number of phosphorylated substrates. Such complex organization of various molecules in the oocyte cytoplasm would play a key role in the formation and temporal maintenance of oocyte polarity.

Other examples of the asymmetric distribution of proteins guided by the meiotic spindle are the mouse PAR polarity proteins, such as the PAR6-related proteins mPARD6a and mPARD6b (Vinot et al. 2004). During the first meiotic M-phase, the two proteins are located within the MI acentriolar spindle as it forms and migrates to the cortex. The similar amounts of mPARD6a and b found in each spindle half demonstrate an equal distribution of these proteins within the spindle. However, by the end of the first meiosis, mPARD6a concentrates at this spindle pole, which is closer to cortex, i.e., its localization within the single spindle becomes polarized. During the MII arrest of oocytes, mPARD6b becomes rapidly relocated to the neighboring cortex. Interestingly, when MTs are experimentally disassembled during MI, the MT-free bivalents still migrate towards the oocyte surface and mPARD6a accumulates preferentially on their side facing the cortex, suggesting that this protein may play an active role in migration (Vinot et al. 2004). This behavior of mPARD6a and b seems to be limited to the oocytes, as it was not observed in early embryos, where this protein is equally distributed in the mitotic spindles (Vinot et al. 2005). The asymmetric distribution of mPARD6a within the meiotic spindle and the delocalization of mPAR6b from the spindle to the MII oocyte cortex suggest that the meiotic spindle not only is the central structure involved in polarization of the oocyte but also becomes an active vector bringing the cortical proteins necessary for asymmetric division to the right place. It is absolutely unknown whether the evolution of mPARD6a and b proteins described earlier is causally related to the acentriolar character of the meiotic spindle.

2.6 Conclusions

Oocyte polarity is a prerequisite for the proper completion of embryogenesis. In mouse oocyte ontogenesis, polarity develops twice. The final polarity of the mature egg involves spindle migration and successful anchoring at the cortex.

Once harbored at the cortex, the acentriolar spindle induces modifications of the cell cortex and the cytoplasm, which maintain the polarity until fertilization and permit asymmetric division to occur properly. In addition, the meiotic spindle is a vector that brings the proteins necessary for unequal meiotic division (such as CDK1, ERK1/2, and mPARD6a and b) to the right place. The complete understanding of mouse oocyte polarity will enable us to understand many important aspects of mammalian (including human) embryo development and answer such fundamental questions as when and how the embryo axes and body plan become determined.

Acknowledgements We are grateful to Guillaume Halet for reading the manuscript and valuable discussions. While writing this article, MK was supported by NSF grant 0904186 and JZK by ARC.

References

Albertini DF (2011) NOBOX does right for the follicle reserve: insights into premature ovarian failure. J Assist Reprod Genet 28:567–568

Albertini DF, Barrett SL (2004) The developmental origins of mammalian oocyte polarity. Semin Cell Dev Biol 15:599–606

Anifandis G, Dafopoulos K, Messini CI, Chalvatzas N, Messinis IE (2010) Effect of the position of the polar body during ICSI on fertilization rate and embryo development. Reprod Sci 17:849–853

Araki K, Naito K, Haraguchi S, Suzuki R, Yokoyama M, Inoue M, Aizawa S, Toyoda Y, Sato E (1996) Meiotic abnormalities of c-mos knockout mouse oocytes: activation after first meiosis or entrance into third meiotic metaphase. Biol Reprod 55:1315–1324

Arion D, Meijer L (1989) M-phase-specific protein kinase from mitotic sea urchin eggs: cyclic activation depends on protein synthesis and phosphorylation but does not require DNA or RNA synthesis. Exp Cell Res 183:361–375

Azoury J, Lee KW, Georget V, Rassinier P, Leader B, Verlhac MH (2008) Spindle positioning in mouse oocytes relies on a dynamic meshwork of actin filaments. Curr Biol 18:1514–1519

Azoury J, Verlhac MH, Dumont J (2009) Actin filaments: key players in the control of asymmetric divisions in mouse oocytes. Biol Cell 101:69–76

Azoury J, Lee KW, Georget V, Hikal P, Verlhac MH (2011) Symmetry breaking in mouse oocytes requires transient F-actin meshwork destabilization. Development 138:2903–2908

Bailly E, Dorée M, Nurse P, Bornens M (1989) p34cdc2 is located in both nucleus and cytoplasm; part is centrosomally associated at G2/M and enters vesicles at anaphase. EMBO J 8:3985–3995

Balakier H, Czolowska R (1977) Cytoplasmic control of nuclear maturation in mouse oocytes. Exp Cell Res 110:466–469

Barrett SL, Albertini DF (2007) Allocation of gamma-tubulin between oocyte cortex and meiotic spindle influences asymmetric cytokinesis in the mouse oocyte. Biol Reprod 6:949–957

Barrett SL, Albertini DF (2010) Cumulus cell contact during oocyte maturation in mice regulates meiotic spindle positioning and enhances developmental competence. J Assist Reprod Genet 27:29–39

Bashour AM, Fullerton AT, Hart MJ, Bloom GS (1997) IQGAP1, a Rca- and Cdc42-binding protein, directly binds and cross-links microfilaments. J Cell Biol 137:1555–1566

Bergstrom CT, Pritchard J (1998) Germline bottlenecks and the evolutionary maintenance of mitochondrial genomes. Genetics 149:2135–2146

Bielak-Zmijewska A, Kolano A, Szczepanska K, Maleszewski M, Borsuk E (2008) Cdc42 protein acts upstream of IQGAP1 and regulates cytokinesis in mouse oocytes and embryos. Dev Biol 322:21–32

Bouchet C, Steffann J, Corcos J, Monnot S, Paquis V, Rotig A, Lebon S, Levy P, Royer G, Giurgea I, Gigarel N, Benachi A, Dumez Y, Munnich A, Bonnefont JP (2006) Prenatal diagnosis of myopathy, encephalopathy, lactic acidosis, and stroke-like syndrome: contribution to understanding mitochondrial DNA segregation during human embryofetal development. J Med Genet 43:788–792

Brunet S, Maro B (2007) Germinal vesicle position and meiotic maturation in mouse oocyte. Reproduction 133:1069–1072

Brunet S, Verlhac MH (2011) Positioning to get out of meiosis: the asymmetry of division. Hum Reprod Update 17:68–75

Cao L, Shitara H, Horii T, Nagao Y, Imai H, Abe K, Hara T, Hayashi J, Yonekawa H (2007) The mitochondrial bottleneck occurs without reduction of mtDNA content in female mouse germ cells. Nat Genet 39:386–390

Cao L, Shitara H, Sugimoto M, Hayashi J, Abe K, Yonekawa H (2009) New evidence confirms that the mitochondrial bottleneck is generated without reduction of mitochondrial DNA content in early primordial germ cells of mice. PLoS Genet 5:e1000756

Capco DG, Gallicano GI, McGaughey RW, Downing KH, Larabell CA (1993) Cytoskeletal sheets of mammalian eggs and embryos: a lattice-like network of intermediate filaments. Cell Motil Cytoskeleton 24:85–99

Cau J, Hall A (2005) Cdc42 controls the polarity of the actin and microtubule cytoskeletons through two distinct signal transduction pathways. J Cell Sci 118:2579–2587

Chesnel F, Bazile F, Pascal A, Kubiak JZ (2006) Cyclin B dissociation from CDK1 precedes its degradation upon MPF inactivation in mitotic extracts of *Xenopus laevis* embryos. Cell Cycle 5:1687–1698

Ciemerych MA, Tarkowski AK, Kubiak JZ (1998) Autonomous activation of histone H1 kinase, cortical activity and microtubule organization in one- and two-cell mouse embryos. Biol Cell 90:557–564

Colledge WH, Carlton MB, Udy GB, Evans MJ (1994) Disruption of c-mos causes parthenogenetic development of unfertilized mouse eggs. Nature 370:65–68

Cox RT, Spradling AC (2003) A Balbiani body and the fusome mediate mitochondrial inheritance during Drosophila oogenesis. Development 130:1579–1590

Cox RT, Spradling AC (2006) Milton controls the early acquisition of mitochondria by Drosophila oocytes. Development 133:3371–3377

Cree LM, Samuels DC, de Sousa Lopes SC, Rajasimha HK, Wonnapinij P, Mann JR, Dahl HH, Chinnery PF (2008) A reduction of mitochondrial DNA molecules during embryogenesis explains the rapid segregation of genotypes. Nat Genet 40:249–254

Cui XS, Li XY, Kim NH (2007) Cdc42 is implicated in polarity during meiotic resumption and blastocyst formation in the mouse. Mol Reprod Dev 74:785–794

de Pennart H, Verlhac MH, Cibert C, Santa Maria A, Maro B (1993) Okadaic acid induces spindle lengthening and disrupts the interaction of microtubules with the kinetochores in metaphase II-arrested mouse oocytes. Dev Biol 157:170–181

De Smedt V, Szollosi D, Kloc M (2000) The Balbiani body: asymmetry in the mammalian oocyte. Genesis 26:208–212

Dean NL, Battersby BJ, Ao A, Gosden RG, Tan SL, Shoubridge EA, Molnar MJ (2003) Prospect of preimplantation genetic diagnosis for heritable mitochondrial DNA diseases. Mol Hum Reprod 9:631–638

Deng M, Kishikawa H, Yanagimachi R, Kopf GS, Schultz RM et al (2003) Chromatin-mediated cortical granule redistribution is responsible for the formation of the cortical granule-free domain in mouse eggs. Dev Biol 257:166–176

Deng M, Williams CJ, Schultz RM (2005) Role of MAP kinase and myosin light chain kinase in chromosome-induced development of mouse egg polarity. Dev Biol 278:358–366

Ducka AM, Joel P, Popwicz GM, Trybus KM, Schleicher M, Noegel AA, Huber R, Holak TA, Sitar T (2010) Structures of actin-bound Wiskott-Aldrich syndrome protein homology 2 (WH2) domains of Spire and the implication for filament nucleation. Proc Natl Acad Sci USA 107:11757–11762

Dumont J, Million K, Sunderland K, Rassinier P, Hyunjung L, Leader B, Verlhac M-H (2007) Formin-2 is required for spindle migration and for late steps of cytokinesis in mouse oocytes. Dev Biol 301:254–265

Duncan FE, Moss SB, Schultz RM, Williams CJ (2005) PAR-3 defines a central subdomain of the cortical actin cap in mouse eggs. Dev Biol 280:38–47

Edwards RG (2000) The role of embryonic polarities in preimplantation growth and implantation of mammalian embryos. Hum Reprod Suppl 6:1–8

Edwards RG (2001) Ovarian differentiation and human embryo quality. 1. Molecular and morphogenetic homologies between oocytes and embryos in Drosophila, *C. elegans*, Xenopus and mammals. Reprod Biomed Online 3:138–160

Edwards RG, Ludwig M (2003) Are major defects in children conceived in vitro due to innate problems in patients or to induced genetic damage? Reprod Biomed Online 7:131–138

Eichenlaub-Ritter U, Wieczorek M, Lüke S, Seidel T (2011) Age related changes in mitochondrial function and new approaches to study redox regulation in mammalian oocytes in response to age or maturation conditions. Mitochondrion 11(5):783–796

El-Mestrah M, Castle PE, Borossa G, Kan FW (2002) Subcellular distribution of ZP1, ZP2, and ZP3 glycoproteins during folliculogenesis and demonstration of their topographical disposition within the zona matrix of mouse ovarian oocytes. Biol Reprod 66:866–876

Esposito G, Vitale AM, Leijten FP, Strik AM, Koonen-Reemst AM, Yurttas P, Robben TJ, Coonrod S, Gossen JA (2007) Peptidylarginine deiminase (PAD) 6 is essential for oocyte cytoskeletal sheet formation and female fertility. Mol Cell Endocrinol 273:25–31

Etienne-Manneville S, Hall A (2002) Rho GTPases in cell biology. Nature 420:629–635

Evangelista M, Zigmond S, Boone C (2003) Formins: signaling effectors for assembly and polarization of actin filaments. J Cell Sci 116:2603–2611

Evans JP, Foster JA, McAvey BA, Gerton GL, Kopf GS, Schultz RM (2000) The effects of perturbation of cell polarity on molecular markers of sperm-egg binding sites on mouse eggs. Biol Reprod 62:76–84

Fukata M, Kuroda S, Fujii K, Nakamura T, Shoji I, Matsuura Y, Okawa K, Iwamatsu A, Kikuchi A, Kaibuchi K (1997) Regulation of cross-linking of actin filament by IQGAP1, a target for Cdc42. J Biol Chem 272:29579–29583

Fukata M, Kuroda S, Nakagawa M, Kawajiri A, Itoh N, Shoji I, Matsuura Y, Yonehara S, Fujisawa H, Kikuchi A, Kaibuchi K (1999) Cdc42 and Rac1 regulate the interaction of IQGAP1 with beta-catenin. J Biol Chem 274:26044–26050

Gallicano GI, Larabell CA, McGaughey RW, Capco DG (1994) Novel cytoskeletal elements in mammalian eggs are composed of a unique arrangement of intermediate filaments. Mech Dev 45:211–226

Gautier J, Norbury C, Lohka M, Nurse P, Maller J (1988) Purified maturation-promoting factor contains the product of a Xenopus homolog of the fission yeast cell cycle control gene cdc2+. Cell 54:433–439

Gautier J, Minshull J, Lohka M, Glotzer M, Hunt T, Maller JL (1990) Cyclin is a component of maturation-promoting factor from Xenopus. Cell 60:487–494

Glotzer M (2005) The molecular requirements for cytokinesis. Science 307:1735–1739

Glotzer M, Murray AW, Kirschner MW (1991) Cyclin is degraded by the ubiquitin pathway. Nature 349:132–138

Gönczy P (2002) Mechanisms of spindle positioning: focus on flies and worms. Trends Cell Biol 12:332–339

Guertin DA, Trautmann S, McCollum D (2002) Cytokinesis in eukaryotes. Microbiol Mol Biol Rev 66:155–178

Halet G, Carroll J (2007) Rac activity is polarized and regulates meiotic spindle stability and anchoring in mammalian oocytes. Dev Cell 12:309–317

Harris ES, Higgs HN (2004) Actin cytoskeleton: formins lead the way. Curr Biol 14:R520–R522

Hiiragi T, Alarcon VB, Fujimori T, Louvet-Vallee S, Maleszewski M, Marikawa Y, Maro B, Solter D (2006) Where do we stand now? Mouse early embryo patterning meeting in Freiburg, Germany (2005). Int J Dev Biol 50:581–586; discussion 586–587

Hoodbhoy T, Avilés M, Baibakov B, Epifano O, Jiménez-Movilla M, Gauthier L, Dean J (2006) ZP2 and ZP3 traffic independently within oocytes prior to assembly into the extracellular zona pellucida. Mol Cell Biol 26:7991–7998

Huo LJ, Fan HY, Zhong ZS, Chen DY, Schatten H, Sun QY (2004) Ubiquitin-proteasome pathway modulates mouse oocyte meiotic maturation and fertilization via regulation of MAPK cascade and cyclin B1 degradation. Mech Dev 121:1275–1287

Huo LJ, Yu LZ, Liang CG, Fan HY, Chen DY, Sun QY (2005) Cell-cycle-dependent subcellular localization of cyclin B1, phosphorylated cyclin B1 and p34cdc2 during oocyte meiotic maturation and fertilization in mouse. Zygote 13:45–53

Jenuth JP, Peterson AC, Fu K, Shoubridge EA (1996) Random genetic drift in the female germline explains the rapid segregation of mammalian mitochondrial DNA. Nat Genet 14:146–151

Johnson MH (2009) From mouse egg to mouse embryo: polarities, axes, and tissues. Annu Rev Cell Dev Biol 25:483–512

Kan R, Yurttas P, Kim B, Jin M, Wo L, Lee B, Gosden R, Coonrod SA (2011) Regulation of mouse oocyte microtubule and organelle dynamics by PADI6 and the cytoplasmic lattices. Dev Biol 350:311–322

Kloc M, Etkin LD (1995) Two distinct pathways for the localization of RNAs at the vegetal cortex in Xenopus oocytes. Development 121:287–297

Kloc M, Etkin LD (1998) Apparent continuity between the METRO and late RNA localization pathways during oogenesis in Xenopus. Mech Dev 73:95–106

Kloc M, Larabell C, Etkin LD (1996) Elaboration of the messenger transport organizer pathway (METRO) for localization of RNA to the vegetal cortex of Xenopus oocytes. Dev Biol 180:119–130

Kloc M, Larabell C, Chan P-YA, Etkin L (1998) Contribution of METRO pathway localized molecules to the organization of the germ cell lineage. Mech Dev 75:81–93

Kloc M, Bilinski S, Chan AP, Allen LH, Zearfoss NR, Etkin LD (2001) RNA localization and germ cell determination in Xenopus. Int Rev Cytol 203:63–91

Kloc M, Bilinski S, Dougherty MT, Brey EM, Etkin LD (2004a) Formation, architecture and polarity of female germline cyst in Xenopus. Dev Biol 266:43–61

Kloc M, Bilinski S, Etkin LD (2004b) The Balbiani body and germ cell determinants: 150 years later. Curr Top Dev Biol 59:1–36

Kloc M, Jaglarz M, Dougherty M, Stewart MD, Nel-Themaat L, Bilinski S (2008) Mouse early oocytes are transiently polar: three-dimensional and ultrastructural analysis. Exp Cell Res 314:3245–3254

Kozma R, Ahmed S, Best A, Lim L (1995) The Ras-related protein Cdc42Hs and bradykinine promote formation of peripheral actin microspikes and filopodia in Swiss 3T3 fibroblasts. Mol Cell Biol 15:1942–1952

Kubiak JZ (2012) Protein kinase assays for measuring MPF and MAPK activities in mouse and rat oocytes and early embryos. In: Homer H (ed) Methods in molecular biology. Springer, Heidelberg (in press)

Kubiak J, Paldi A, Weber M, Maro B (1991) Genetically identical parthenogenetic mouse embryos produced by inhibition of the first meiotic cleavage with cytochalasin D. Development 111:763–769

Kubiak JZ, Weber M, Géraud G, Maro B (1992) Cell cycle modification during the transitions between meiotic M-phases in mouse oocytes. J Cell Sci 102:457–467

Kubiak JZ, Weber M, de Pennart H, Winston NJ, Maro B (1994) The metaphase II arrest in mouse oocytes is controlled through microtubule-dependent destruction of cyclin B in the presence of CSF. EMBO J 12:3773–3778

Kuroda S, Fukata M, Nakagawa M, Fujii K, Nakamura T, Ookubo T, Izawa I, Nagase T, Nomura N, Tani H, Shoji I, Matsuura Y, Yonehara S, Kaibuchi K (1998) Role of IQGAP1, a target of the small GTPases Cdc42 and Rac1, in regulation of E-cadherin-mediated cell-cell adhesion. Science 281:832–835

Labbé JC, Picard A, Karsenti E, Dorée M (1988) An M-phase-specific protein kinase of Xenopus oocytes: partial purification and possible mechanism of its periodic activation. Dev Biol 127:157–169

Labbé JC, Picard A, Peaucellier G, Cavadore JC, Nurse P, Doree M (1989) Purification of MPF from starfish: identification as the H1 histone kinase p34cdc2 and a possible mechanism for its periodic activation. Cell 57:253–263

Laipis PJ, Van de Walle MJ, Hauswirth WW (1988) Unequal partitioning of bovine mitochondrial genotypes among siblings. Proc Natl Acad Sci USA 85:8107–8110

Larson SM, Lee HJ, Hung PH, Matthews LM, Robinson DN, Evans JP (2010) Cortical mechanics and meiosis II completion in mammalian oocytes are mediated by myosin-II and Ezrin-Radixin-Moesin (ERM) proteins. Mol Biol Cell 21:3182–3192

Leader B, Lim H, Carabatsos MJ, Harrington A, Ecsedy J, Pellman D, Maas R, Leder P (2002) Formin-2, polyploidy, hypofertility and positioning of the meiotic spindle in mouseoocytes. Nat Cell Biol 4:921–929

Lechowska A, Bilinski S, Choi Y, Shin Y, Kloc M, Rajkovic A (2011) Premature ovarian failure in nobox-deficient mice is caused by defects in somatic cell invasion and germ cell cyst breakdown. J Assist Reprod Genet 28:583–589

Ledan E, Polanski Z, Terret ME, Maro B (2001) Meiotic maturation of the mouse oocyte requires an equilibrium between cyclin B synthesis and degradation. Dev Biol 232:400–413

Lei Y, Warrior R (2000) The Drosophila Lissencephaly1 (DLis1) is required for nuclear migration. Dev Biol 226:57–72

Levi M, Ninio-Mani L, Shalgi R (2012). Src Protein Kinases in Mouse and Rat Oocytes and Embryos. Results Probl Cell Differ. 55., 93–106

Li S, Ou XH, Wang ZB et al (2010) ERK3 is required for metaphase-anaphase transition in mouse oocyte meiosis. PLoS One 5(9):pii: e13074

Longo FJ (1987) Actin-plasma membrane associations in mouse eggs and oocytes. J Exp Zool 243:299–309

Longo FJ, Chen DY (1985) Development of cortical polarity in mouse eggs: involvement of the meiotic apparatus. Dev Biol 107:382–394

Ma C, Benink HA, Cheng D, Montplaisir V, Wang L, Xi Y, Zheng PP, Bement WM, Liu XJ (2006) Cdc42 activation couples spindle positioning to first polar body formation in oocyte maturation. Curr Biol 16:214–220

Maciejewska Z, Pascal A, Kubiak JZ, Ciemerych MA (2011) Phosphorylated ERK5/BMK1 transiently accumulates within division spindles in mouse oocytes and preimplantation embryos. Folia Histochem Cytobiol 49(3):528–534

Maro B, Johnson MH, Webb M, Flach G (1986) Mechanism of polar body formation in the mouse oocyte: an interaction between the chromosomes, the cytoskeleton and the plasma membrane. J Embryol Exp Morphol 92:11–32

Masui Y, Markert CL (1971) Cytoplasmic control of nuclear behavior during meiotic maturation of frog oocytes. J Exp Zool 177:129–145

Mitra J, Schultz RM (1996) Regulation of the acquisition of meiotic competence in the mouse: changes in the subcellular localization of cdc2, cyclin B1, cdc25C and wee1, and in the concentration of these proteins and their transcripts. J Cell Sci 109:2407–2415

Miyazaki A, Kato KH, Nemoto S (2005) Role of microtubules and centrosomes in the eccentric relocation of the germinal vesicle upon meiosis reinitiation in sea-cucumber oocytes. Dev Biol 280:237–247

Monnot S, Gigarel N, Samuels DC, Burlet P, Hesters L, Frydman N, Frydman R, Kerbrat V, Funalot B, Martinovic J, Benachi A, Feingold J, Munnich A, Bonnefont JP, Steffann J (2011) Segregation of mtDNA throughout human embryofetal development: m.3243A $>$ G as a model system. Hum Mutat 32:116–125

Mullins RD, Heuser JA, Pollard TD (1998) The interaction of Arp2/3 complex with actin: nucleation, high affinity pointed end capping, and formation of branching networks of filaments. Proc Natl Acad Sci USA 95:6181–6186

Na J, Zernicka-Goetz M (2006) Asymmetric positioning and organization of the meiotic spindle of mouse oocytes requires CDC42 function. Curr Biol 16:1249–1254

Nakagomi S, Barsoum MJ, Bossy-Wetzel E, Sütterlin C, Malhotra V, Lipton SA (2008) A golgi fragmentation pathway in neurodegeneration. Neurobiol Dis 29:221–231

Nishiyama A, Tachibana K, Igarashi Y, Yasuda H, Tanahashi N, Tanaka K, Ohsumi K, Kishimoto T (2000) A nonproteolytic function of the proteasome is required for the dissociation of Cdc2 and cyclin B at the end of M phase. Genes Dev 14:2344–2357

Nobes CD, Hall A (1995) Rho, Rac and Cdc42 GTPases: regulators of actin structures, cell adhesion and motility. Biochem Soc Trans 23:456–459

Noritake J, Watanabe T, Sato K, Wang S, Kaibuchi K (2005) IQGAP1: a key regulator of adhesion and migration. J Cell Sci 118:2085–2092

Ookata K, Hisanaga S, Bulinski JC, Murofushi H, Aizawa H, Itoh TJ, Hotani H, Okumura E, Tachibana K, Kishimoto T (1995) Cyclin B interaction with microtubule-associated protein 4 (MAP4) targets p34cdc2 kinase to microtubules and is a potential regulator of M-phase microtubule dynamics. J Cell Biol 128:849–862

Pepling ME, Spradling AC (1998) Female mouse germ cells form synchronously dividing cysts. Development 125:3323–3328

Pepling ME, Wilhelm JE, O'Hara AL, Gephardt GW, Spradling AC (2007) Mouse oocytes within germ cell cysts and primordial follicles contain a Balbiani body. Proc Natl Acad Sci USA 104:187–192

Pfender S, Kuznetsov V, Pleiser S, Kerkhoff E, Schuh M (2011) Spire-type actin nucleators cooperate with Formin-2 to drive asymmetric oocyte division. Curr Biol 21:955–960

Pines J, Hunter T (1991) Human cyclins A and B1 are differentially located in the cell and undergo cell cycle-dependent nuclear transport. J Cell Biol 115:1–17

Quinlan ME, Heuser JE, Kerkhoff E, Mullins RD (2005) Drosophila Spire is an actin nucleation factor. Nature 433:382–388

Rajkovic A, Pangas SA, Ballow D, Suzumori N, Matzuk MM (2004) NOBOX deficiency disrupts early folliculogenesis and oocytespecific gene expression. Science 305:1157–1159

Rime H, Ozon R (1990) Protein phosphatases are involved in the in vivo activation of histone H1 kinase in mouse oocyte. Dev Biol 141:115–122

Roze D, Rousset F, Michalakis Y (2005) Germline bottlenecks, biparental inheritance and selection on mitochondrial variants: a two-level selection model. Genetics 170:1385–1399

Sanfins A, Lee GY, Plancha CE, Overstrom EW, Albertini DF (2003) Distinctions in meiotic spindle structure and assembly during in vitro and in vivo maturation of mouse oocytes. Biol Reprod 69:2059–2067

Sanfins A, Plancha CE, Overstrom EW, Albertini DF (2004) Meiotic spindle morphogenesis in in vivo and in vitro matured mouse oocytes: insights into the relationship between nuclear and cytoplasmic quality. Hum Reprod 19:2889–2899

Sardet C, Prodon F, Dumollard R, Chang P, Chênevert J (2002) Structure and function of the egg cortex from oogenesis through fertilization. Dev Biol 241:1–23

Schindler K (2011) Protein kinases and protein phosphatases that regulate meiotic maturation in mouse oocytes. Results Probl Cell Differ 53:309–341

Schindler K, Schultz RM (2009) The CDC14A phosphatase regulates oocyte maturation in mouse. Cell Cycle 8:1090–1098

Schuh M, Ellenberg J (2008) A new model for asymmetric spindle positioning in mouse oocytes. Curr Biol 18:1986–1992

Shannon KB, Li R (1999) The multiple roles of Cyk1p in the assembly and function of the actomyosin ring in budding yeast. Mol Biol Cell 10:283–296

Simerly C, Nowak G, de Lanerolle P, Schatten G (1998) Differential expression and functions of cortical myosin IIA and IIB isotypes during meiotic maturation, fertilization and mitosis in mouse oocytes and embryos. Mol Biol Cell 9:2509–2525

Sun SC, Sun QY, Kim NH (2011a) JMY is required for asymmetric division and cytokinesis in mouse oocytes. Mol Hum Reprod 17:286–295

Sun SC, Wang ZB, Xu YN, Lee SE, Cui XS, Kim NH (2011b) Arp2/3 complex regulates asymmetric division and cytokinesis in mouse oocytes. PLoS One 6:e18392

Swan A, Nguyen T, Suter B (1999) Drosophila Lissencephaly-1 functions with Bic-D and dynein in oocyte determination and nuclear positioning. Nat Cell Biol 1:444–449

Szollosi D, Calarco P, Donahue RP (1972) Absence of centrioles in the first and second meiotic spindles of mouse oocytes. J Cell Sci 11:521–541

Tan X, Peng A, Wang Y, Tang Z (2005) The effects of proteasome inhibitor lactacystin on mouse oocyte meiosis and first cleavage. Sci China C Life Sci 48:287–294

Terada Y, Simerly C, Schatten G (2000) Microfilament stabilization by jasplakinolide arrests oocyte maturation, cortical granule exocytosis, sperm incorporation cone resorption, and cell-cycle progression, but not DNA replication, during fertilization in mice. Mol Reprod Dev 56:89–98

Van Blerkom J, Bell H (1986) Regulation of development in the fully grown mouse oocyte: chromosome-mediated temporal and spatial differentiation of the cytoplasm and plasma membrane. J Embryol Exp Morphol 93:213–238

Verlhac MH, de Pennart H, Maro B, Cobb MH, Clarke HJ (1993) MAP kinase becomes stably activated at metaphase and is associated with microtubule-organizing centers during meiotic maturation of mouse oocytes. Dev Biol 158, 330–340

Verlhac MH, Kubiak JZ, Weber M, Géraud G, Colledge WH, Evans MJ, Maro B (1996) Mos is required for MAP kinase activation and is involved in microtubule organization during meiotic maturation in the mouse. Development 122:815–822

Verlhac M-H, Lefebvre C, Guillaud P, Rassinier P, Maro B (2000) Asymmetric division in mouse oocytes: with or without Mos. Curr Biol 10:1303–1306

VerMilyea MD, Maneck M, Yoshida N, Blochberger I, Suzuki E, Suzuki T, Spang R, Klein CA, Perry AC (2011) Transcriptome asymmetry within mouse zygotes but not between early embryonic sister blastomeres. EMBO J 30:1841–1851

Vinot S, Le T, Maro B, Louvet-Vallee S (2004) Two Par6 proteins become asymmetrically localized during establishment of polarity in mouse oocytes. Curr Biol 14:520–525

Vinot S, Le T, Ohno S, Pawson T, Maro B, Louvet-Vallée S (2005) Asymmetric distribution of PAR proteins in the mouse embryo begins at the 8-cell stage during compaction. Dev Biol 282:307–319

Wai T, Teoli D, Shoubridge EA (2008) The mitochondrial DNA genetic bottleneck results from replication of a subpopulation of genomes. Nat Genet 40:1484–1488

Weber M, Kubiak JZ, Arlinghaus RB, Pines J, Maro B (1991) c-mos proto-oncogene product is partly degraded after release from meiotic arrest and persists during interphase in mouse zygotes. Dev Biol 148:393–397

Winston NJ (1997) Stability of cyclin B protein during meiotic maturation and the first mitotic cell division in mouse oocytes. Biol Cell 89:211–219

Wright PW, Bolling LC, Calvert ME, Sarmento OF, Berkeley EV, Shea MC, Hao Z, Jayes FC, Bush LA, Shetty J, Shore AN, Reddi PP, Tung KS, Samy E, Allietta MM, Sherman NE, Herr JC, Coonrod SA (2003) ePAD, an oocyte and early embryo-abundant peptidylarginine deiminase-like protein that localizes to egg cytoplasmic sheets. Dev Biol 256:73–88

Xiong B, Yu LZ, Wang Q, Ai JS, Yin S, Liu JH, OuYang YC, Hou Y, Chen DY, Zou H, Sun QY (2007) Regulation of intracellular MEK1/2 translocation in mouse oocytes: cytoplasmic dynein/dynactin-mediated poleward transport and cyclin B degradation-dependent release from spindle poles. Cell Cycle 6:1521–1527

Yang HY, McNally K, McNally FJ (2003) MEI-1/katanin is required for translocation of the meiosis I spindle to the oocyte cortex in C. elegans. Dev Biol 260:245–259

Yang HY, Mains PE, McNally FJ (2005) Kinesin-1 mediates translocation of the meiotic spindle to the oocyte cortex through KCA-1, a novel cargo adapter. J Cell Biol 169:447–457

Yasuda S, Taniguchi H, Oceguera-Yanez F, Ando Y, Watanabe S, Monypenny J, Narumiya S (2006) An essential role of Cdc42-like GTPases in mitosis of HeLa cells. FEBS Lett 580:3375–3380

Yi K, Unruh JR, Deng M, Slaughter BD, Rubinstein B, Li R (2011) Dynamic maintenance of asymmetric meiotic spindle position through Arp2/3-complex-driven cytoplasmic streaming in mouse oocytes. Nat Cell Biol 13:1252–1258

Yu LZ, Xiong B, Gao WX, Wang CM, Zhong ZS, Huo LJ, Wang Q, Hou Y, Liu K, Liu XJ, Schatten H, Chen DY, Sun QY (2007) MEK1/2 regulates microtubule organization, spindle pole tethering and asymmetric division during mouse oocyte meiotic maturation. Cell Cycle 6:330–338

Zernicka-Goetz M, Huang S (2010) Stochasticity versus determinism in development: a false dichotomy? Nat Rev Genet 11:743–744

Zernicka-Goetz M, Verlhac M-H, Géraud G, Kubiak JZ (1997) Protein phosphatases control MAP kinase activation and microtubule organization during rat oocyte maturation. Eur J Cell Biol 72:30–38

Zhang YZ, Ouyang YC, Hou Y, Schatten H, Chen DY, Sun QY (2008) Mitochondrial behavior during oogenesis in zebrafish: a confocal microscopy analysis. Dev Growth Differ 50:189–201

Zhang CH, Wang ZB, Quan S, Huang X, Tong JS, Ma JY, Guo L, Wei YC, Ouyang YC, Hou Y, Xing FQ, Sun QY (2011) GM130, a cis-Golgi protein, regulates meiotic spindle assembly and asymmetric division in mouse oocyte. Cell Cycle 10:1861–1870

Zhong ZS, Huo LJ, Liang CC, Chen DY, Sun QY (2005) Small GTPase RhoA is required for ooplasmic segregation and spindle rotation, but not for spindle organization and chromosome separation during mouse oocyte maturation, fertilization, and early cleavage. Mol Reprod Dev 41:256–261

Zhou RR, Wang B, Wang J, Schatten H, Zhang YZ (2010) Is the mitochondrial cloud the selection machinery for preferentially transmitting wild-type mtDNA between generations? Rewinding Müller's ratchet efficiently. Curr Genet 56:101–107

Zigmond SH (2004) Formin-induced nucleation of actin filaments. Curr Opin Cell Biol 16:99–105

Zuchero JB, Coutts AS, Quinlan ME, Thangue NB, Mullins RD (2009) p53-cofactor JMY is a multifunctional actin nucleation factor. Nat Cell Biol 11:451–459

Chapter 3
Chromatin Structure and ATRX Function in Mouse Oocytes

Rabindranath De La Fuente, Claudia Baumann, and Maria M. Viveiros

Abstract Differentiation of chromatin structure and function during oogenesis is essential to confer the mammalian oocyte with meiotic and developmental potential. Errors in chromosome segregation during female meiosis and subsequent transmission of an abnormal chromosome complement (aneuploidy) to the early conceptus are one of the leading causes of pregnancy loss in women. The chromatin remodeling protein ATRX (α-thalassemia mental retardation X-linked) has recently emerged as a critical factor involved in heterochromatin formation at mammalian centromeres during meiosis. In mammalian oocytes, ATRX binds to centromeric heterochromatin domains where it is required for accurate chromosome segregation. Loss of ATRX function induces abnormal meiotic chromosome morphology, reduces histone H3 phosphorylation, and promotes a high incidence of aneuploidy associated with severely reduced fertility. The presence of centromeric breaks during the transition to the first mitosis in the early embryo indicates that the role of ATRX in chromosome segregation is mediated through an epigenetic mechanism involving the maintenance of chromatin modifications associated with pericentric heterochromatin (PCH) formation and chromosome condensation. This is consistent with the existence of a potential molecular link between centromeric and PCH in the epigenetic control of centromere function and maintenance of chromosome stability in mammalian oocytes. Dissecting the molecular mechanisms of ATRX function during meiosis will have important clinical implications towards uncovering the epigenetic factors contributing to the onset of aneuploidy in the human oocyte.

R. De La Fuente (✉) • C. Baumann • M.M. Viveiros
Department of Physiology and Pharmacology, College of Veterinary Medicine, University of Georgia, 501 D.W. Brooks Drive, Athens, GA 30602, USA
e-mail: rfuente@uga.edu

J.Z. Kubiak (ed.), *Mouse Development*, Results and Problems in Cell Differentiation 55,
DOI 10.1007/978-3-642-30406-4_3,

3.1 Regulation of Meiotic Division in Oocytes: The Role of the Centromere

Meiosis is a unique division, which occurs exclusively in gametes, to reduce the number of chromosomes to a haploid complement prior to fertilization. Chromosome reduction necessitates the progression through two successive divisions without an intermediate S-phase. Meiosis-I entails the segregation of homologous chromosomes (also referred to as bivalents)-paired copies of each maternal and paternal chromosome, held together at sites of recombination (crossovers). In contrast, meiosis-II is more analogous to mitosis as it involves the separation of sister chromatids. This second division is initiated only upon fertilization and oocyte activation. Both the segregation of homologues (meiosis-I) and sister chromatids (meiosis-II) are dependent on the (1) correct attachment and pulling force of spindle microtubules as well as (2) the dissolution of cohesin complexes that link sister chromatids (Fig. 3.1).

Errors in meiotic division disrupt genomic stability in gametes and are the leading genetic cause of congenital birth defects and pregnancy loss in women (Hassold et al. 2007; Hunt and Hassold 2008). Hence, faithful chromosome segregation is of fundamental importance for embryonic development. This process is critically dependent on stable and correct chromosome–microtubule attachments. Microtubules bind to chromosomes specifically at the kinetochore, a specialized region of the centromere (Cheeseman and Desai 2008). Kinetochores are transient and only assemble upon entry into M-phase, then disassemble with mitotic exit (Oegema et al. 2001). It is essential that a kinetochore assemble only at one location on each chromosome, and various properties of the centromere mark it as the exclusive site for kinetochore formation. This includes a combination of epigenetic marks such as nucleosomes that contain Centromere Associated Protein-A (CENP-A) as well as histone modifications that affect chromatin structure (Black et al. 2007; Howman et al. 2000; Oegema et al. 2001). While the inner kinetochore plate contact scentromeric heterochromatin, the outer kinetochore plate and corona layer contain proteins required for microtubule interaction. Identified proteins with microtubule binding activity include the dimeric kinesin CENP-E (Duesbery et al. 1997; Wood et al. 1997), Ska1 complex (Welburn et al. 2009), as well as the KMN network composed of KNL1, Mis12, and Ndc80 (Cheeseman et al. 2002). Elegant studies in mitotic cells demonstrate that the NDC80 microtubule interacting complex directly binds to microtubules and is essential for the formation of kinetochore–microtubule attachments (Cheeseman et al. 2002). While it is not known if these mechanisms are fully conserved in gametes, studies have shown that mouse oocytes express CENP-E at the kinetochores and microinjection of specific antibodies to inhibit CENP-E function lead to MI arrest (Duesbery et al. 1997).

For chromosomes to segregate properly, the kinetochore from each sister chromatid must attach to microtubules from opposing spindle poles, referred to as bi-oriented or amphitelic attachment. Different types of attachment errors can occur. For example, syntelic errors denote when kinetochores from sister chromatids both attach to the same spindle pole. In contrast, merotelic errors occur when a single kinetochore is attached to

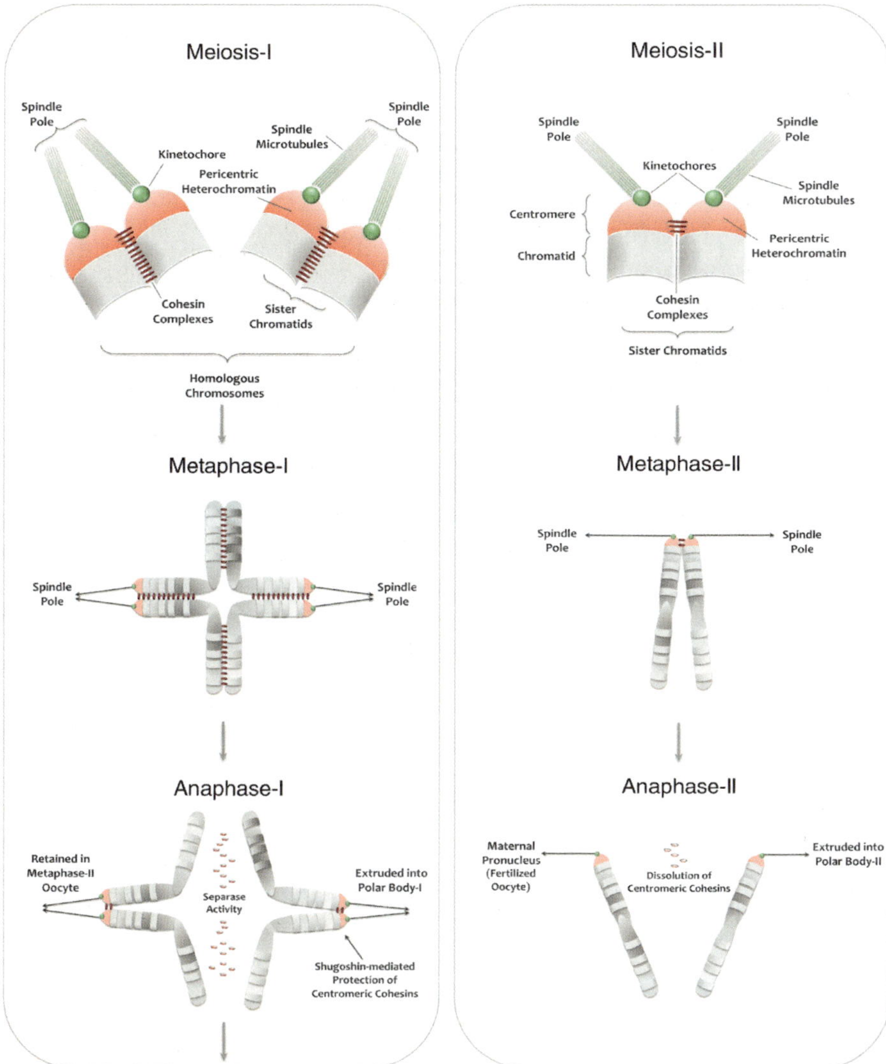

Fig. 3.1 Mechanisms of spindle-microtubule attachments and chromosome cohesion during oocyte meiosis. Accurate segregation of homologous chromosomes (Meiosis-I) and sister chromatids (Meiosis-II) depends on the establishment of bi-oriented (amphitelic) spindle-microtubule attachments as well as the timely dissolution of cohesin complexes. Separation of homologous chromosomes during Meiosis-I requires the attachment of both sister chromatids of each homologue to the same spindle pole (co-orientation), while Meiosis-II involves the association of each sister chromatid with microtubules originating from opposing spindle poles. In addition, faithful chromosome segregation requires key mechanisms that prevent premature sister chromatid separation prior to initiation of meiosis-II, including (1) sustained attachments of homologous chromosomes at sites of crossovers and (2) chromosome cohesion along chromatids and at centromeres during Meiosis-I. In oocytes, separase activity promotes dissolution of the cohesin complex along the chromosome arms during anaphase-I. However, centromeric cohesion remains protected from degradation by shugoshin proteins preserving sister centromere cohesion until anaphase-II onset

both spindle poles simultaneously. Chromosome–microtubule interactions during meiosis-I are more complex as it involves the separation of homologous chromosomes, rather than sister chromatids. This requires the attachment and segregation of sister chromatids of each homologue to the same spindle pole (co-orientation). Hence, sister kinetochores must function as a unit. The fidelity of chromosome segregation is promoted by cellular mechanisms that stabilize appropriate bi-oriented (amphitelic) kinetochore–microtubule attachments, while inappropriate (e.g., syntelic or merotelic) attachments are eliminated (Nicklas and Ward 1994; Nicklas et al. 2001). In mitotic cells, the phosphorylation of kinetochore substrates by the conserved Ipl1/Aurora B kinase selectively eliminates incorrect attachments by promoting the turnover of microtubule attachments in the absence of tension. Aurora B reportedly controls kinetochore–microtubule attachments by modulating the microtubule binding activity of key kinetochore proteins such as Hec1/Ndc8 (Cheeseman et al. 2002; DeLuca et al. 2006; Welburn et al. 2010). Importantly, studies in mouse oocytes have identified Aurora C kinase, not Aurora B, as playing a pivotal role in regulating chromosome–microtubules interaction during meiotic division (Yang et al. 2010). Inhibition of Aurora C activity in oocytes results in significant defects, including abnormal kinetochore–microtubule attachments as well as premature chromosome segregation and cytokinesis failure during meiosis-I (Yang et al. 2010). This points to potentially unique regulatory mechanisms that govern chromosome–microtubule interactions during meiotic division in mammalian oocytes.

Errors in chromosome attachment activate the spindle assembly checkpoint (SAC), which inhibits the activity of the anaphase-promoting complex APC/C^{cdc20} to prevent anaphase onset. Hence, SAC functions as a critical surveillance mechanism and delays anaphase, allowing for error correction mechanisms to promote bi-orientation. Key component proteins of SAC include Mad2, BubR1, and Bub3, which localize to the outer kinetochore until stable chromosome–microtubule interactions are established. Studies have confirmed the expression and function of key checkpoint proteins in mammalian oocytes (Homer et al. 2005; Leland et al. 2009; McGuinness et al. 2009; Niault et al. 2007), which activate the SAC to prevent anaphase onset until all chromosomes are attached to the meiotic spindle apparatus. Interestingly, elegant high-resolution live cell imaging analyses indicate that homologous chromosome co-orientation during meiosis-I is potentially error prone in mammalian oocytes, as the majority of chromosomes undergo error correction of kinetochore–microtubule attachments prior to anaphase onset and completion of the first meiotic division (Kitajima et al. 2011). Moreover, studies have demonstrated that oocytes can still progress to metaphase-II, despite activation of the SAC in response to meiotic spindle defects during meiosis-I, leading to a high incidence of aneuploidy (Ma et al. 2010; Nagaoka et al. 2011).

Accurate chromosome segregation is also critically dependent on the maintenance of sister chromatid pairing until meiosis-II is initiated. Hence, key mechanisms operate during meiosis-I to prevent premature sister chromatid separation. For example, crossovers allow homologues to remain attached when kinetochore–microtubule pulling forces are exerted. In addition, cohesin proteins hold sister chromatids together along the chromosome arms as well as the centromere prior to chromosome segregation (Fig. 3.1). Known proteins that form the cohesin complex in mitotic cells, include Rad21/Scc1, Smc1a, Smc3, Stag1/Sa1, and Stag2/Sa2 (Nasmyth and Haering 2009) for review. Notably, meiosis-specific cohesin components such as Rec8 and Smc1b have also been identified, which are loaded onto the chromosomes in oocytes during fetal development (Prieto et al. 2004). Upon anaphase onset cdc20-activated APC/C (APC/C^{cdc20}) promotes destruction of the anaphase inhibitor (Pds1/securin) that normally maintains the protease separase inactive. In oocytes, separase activity promotes Rec8 degradation during anaphase-I onset, which leads to dissolution of the cohesin complex along the chromosome arms (Kudo et al. 2006). Importantly, centromeric Rec8 is specifically protected from degradation during meiosis-I and, therefore, maintains the cohesin complex and sister chromatid attachment exclusively at the centromeres during homologue separation. Studies have identified the Shugoshin protein family (Sgo1 and Sgo2) as being essential for protection of centromeric cohesin. Targeted deletion of *Sgo2* leads to Rec8 degradation and, therefore, premature sister chromatid separation in mouse oocytes prior to meiosis-II (Lee et al. 2008; Llano et al. 2008).

Studies in yeast and human somatic cell lines have shown that centromeric heterochromatin functions in recruiting Shugoshin (Yamagishi et al. 2008). Moreover, cohesin complexes, enriched at the pericentric region, play a key role in the organization and packaging of mitotic chromosomes (Wood et al. 2010). Understanding whether similar mechanisms operate in mammalian oocytes is essential as the centromere plays a critical role in the timing and accuracy of chromosome segregation during meiosis. Specific chromatin remodeling proteins act at the centromere to maintain its unique chromatin structure, and changes in higher order chromatin structure and/or function critically impact chromosome segregation. In this chapter we focus on the chromatin-remodeling protein ATRX and its function in heterochromatin formation during meiosis.

3.2 ATRX Structure and Function in Mammalian Cells

The alpha-thalassemia/mental retardation X-linked protein (ATRX) is a switch/sucrose non-fermenting (Swi2/Snf2) helicase/ATPase that exhibits chromatin-remodeling activity in mammalian cells (Gibbons et al. 1997; Picketts et al. 1996; Tang et al. 2004; Villard et al. 1997; Xue et al. 2003). The Atrx gene resides within the long arm of the submetacentric human X chromosome and localizes to the acrocentric mouse X chromosome where it encodes an evolutionarily conserved protein of approximately 280 kDa containing several characteristic protein domains. Among these is an N-terminal cysteine-rich ADD domain reminiscent of sequences within the DNMT3 family of de novo DNA methyltransferases that comprises a zinc finger and a plant homeodomain (PHD) as well as an alpha-helical sequence motif (Aapola et al. 2000; Argentaro et al. 2007; Picketts et al. 1998). The carboxyl terminus exhibits several helicase domains containing ATPase activity required for protein–protein interactions (Mitson et al. 2011). A number of specific sequence motifs facilitate binding with various protein partners and hence enable the interaction of ATRX with large protein complexes. For example, through its C terminus, ATRX associates with the methyl-CpG-binding protein MeCP2 (Kernohan et al. 2010; Nan et al. 2007) as well as protein components of promyelocytic leukemia (PML) nuclear bodies (Bérubé et al. 2007; Ishov et al. 2004; Luciani et al. 2006; Xue et al. 2003). In addition, a primary DAXX interacting domain (DID) within the intermediate section of the protein confers binding to the transcriptional regulator death domain associated protein (DAXX) via paired amphipathic alpha helices domains (Ishov et al. 2004; Tang et al. 2004; Xue et al. 2003). Motifs adjacent to the ADD domain of the N terminus enable binding of ATRX protein to HP1-alpha (Kourmouli et al. 2005), the polycomb group protein EZH2 (Cardoso et al. 1998), and histone H3 residues tri-methylated at lysine 9 (Dhayalan et al. 2011; Eustermann et al. 2011; Iwase et al. 2011). Therefore, ATRX has the potential to assemble into large protein complexes that confer chromatin-remodeling activity and/or exert transcriptional regulation according to the cell type or stage of differentiation (Ishov et al. 2004; Kernohan et al. 2010; Nan et al. 2007; Tang et al. 2004; Xue et al. 2003).

Analysis of cells from human patients with hypomorph mutations in the *Atrx* coding sequence supports a role of this chromatin-remodeling factor in multiple cellular processes. For example, patients with ATRX mutations exhibit varying degrees of a neurodevelopmental syndrome (ATRX syndrome) that include symptoms such as mental retardation, facial anomalies, and hematological pathologies such as alpha-thalassemia secondary to deregulated alpha-globin gene transcription (Gibbons et al. 1995, 2003) as well as evidence for urogenital abnormalities such as gonadal dysgenesis (Gibbons et al. 2003). Importantly, abnormal ATRX expression patterns have been described in patients suffering from various types of cancers, such as acute myeloid leukemia (AML) (Serrano et al. 2006), myelodysplastic syndrome (Gibbons et al. 2003), and pancreatic neuroendocrine tumors (Elsasser et al. 2011; Heaphy et al. 2011; Jiao et al. 2011). This suggests that loss of ATRX function may be involved in the onset of malignant neoplastic transformation, potentially through mechanisms involving compromised genome and chromosome stability (Elsasser et al. 2011; Heaphy et al. 2011; Jiao et al. 2011). Several studies have shown that ATRX associates with heterochromatin at the centromere, whereby it can impact chromosome stability (Baumann et al. 2010; McDowell et al. 1999).

3.3 Role of ATRX in Heterochromatin Formation at the Centromere

The centromere is composed of unique DNA sequences and chromatin structure, including centric and pericentric heterochromatin (PCH). In mice, the 120 bp minor satellite repeats of centric heterochromatin are characterized by the specific incorporation of CENP-B protein which is required for the formation of a functional kinetochore on each sister chromatid of condensed chromosomes (Henikoff et al. 2001; Karpen and Allshire 1997; Maison and Almouzni 2004; Murphy and Karpen 1998; Perpelescu and Fukagawa 2011; Pluta et al. 1995; Polo and Almouzni 2006; Sarma and Reinberg 2005; Wiens and Sorger 1998). In contrast, PCH is comprised of repetitive major satellite elements as well as transcriptionally repressive histone modifications. PCH has recently emerged as an essential factor for the timely coordination of sister centromere cohesion and chromosome segregation during mitosis (Guenatri et al. 2004; Maison and Almouzni 2004). Notably, ATRX is a prominent mark at PCH domains in human and mouse cells (Baumann et al. 2010; Bérubé et al. 2000, 2002; De La Fuente et al. 2004b; Garrick et al. 2006; Kernohan et al. 2010; Law et al. 2010; McDowell et al. 1999; Ritchie et al. 2008). The *Drosophila* homologue of ATRX (dATRX/XNP) is required for heterochromatin formation as well as transcriptional silencing of this nuclear domain (Bassett et al. 2008; Emelyanov et al. 2010; Schneiderman et al. 2009). Moreover, in human and mouse embryonic stem (ES) cells, ATRX exhibits a direct interaction with telomeric DNA sequences where it is required for deposition of the histone variant Histone H3.3 and telomere heterochromatin structural integrity (Drane et al. 2010; Goldberg et al. 2010). ATRX also marks constitutive heterochromatin at the Y chromosome in mouse cells (Baumann et al. 2008) where it might be involved in the maintenance of facultative heterochromatin at the inactive X chromosome suggesting a critical role for ATRX in heterochromatin formation (Baumann and De La Fuente 2009; De La Fuente et al. 2011). Consistent with this hypothesis, spontaneous mutations in patients with ATRX syndrome are associated with abnormal DNA methylation patterns at repetitive sequences of ribosomal DNA, at a Y chromosome-specific repeat as well as telomeric sequences (Gibbons et al. 2000).

Heterochromatin formation during gametogenesis is essential, yet the regulatory mechanisms that govern this key process are poorly understood. Studies in somatic cells indicate that transcriptionally repressive epigenetic marks, such as tri-methylation of histone H3 at lysine 9 (H3K9me3), are a hallmark of PCH and essential for the maintenance of a transcriptionally quiescent chromatin environment (Guenatri et al. 2004; Peters et al. 2001; Puschendorf et al. 2008). In turn, H3K9me3 is proposed to act as a docking site for additional chromatin binding proteins such as heterochromatin protein 1 (HP1) and ATRX (Bannister et al. 2001; Dhayalan et al. 2011; Kourmouli et al. 2005; Lachner et al. 2001; Maison and Almouzni 2004; Rea et al. 2000; Schotta et al. 2004). Peptide binding assays demonstrate that the association of ATRX with PCH requires of a combinatorial readout of several histone modifications in which the ADD domain of ATRX interacts with H3K9me3 only in the absence of transcriptionally permissive chromatin modifications such as H3K4me3 and H3K4me2 (Dhayalan et al. 2011; Eustermann et al. 2011).

3.4 ATRX is a Critical Epigenetic Regulator of Chromatin Structure and Function During Mitotic and Meiotic Division

ATRX has recently emerged as a major chromatin-remodeling factor capable of modulating chromatin structure at both the nucleosomal level as well as over large chromosomal domains including PCH. At the nucleosomal level, ATRX regulates gene expression of a subset of imprinted genes in brain tissue (Kernohan et al. 2010). ATRX is also known to regulate the expression of several ancestral genes that have translocated from their original position at the pseudo-autosomal region (PAR) on the sex chromosomes to specific autosomal chromosomes in mice (Levy et al. 2008). Importantly, ATRX modulates key aspects of large-scale chromatin modifications that are essential for the maintenance of genome and chromosome stability during mitosis as well as meiosis (Bérubé 2011; De La Fuente et al. 2011). The dual role of ATRX function in the control of gene expression as well as large-scale chromatin structure has important clinical implications for understanding the role of genome and chromosome stability during development, the onset of malignant neoplastic transformation, and the pathogenesis of several human conditions including neurodevelopmental syndromes (Bérubé 2011; De La Fuente et al. 2011). In the postnatal mouse brain, ATRX regulates the expression patterns at several imprinted genes by establishing a functional interaction with the cohesin proteins SMC1/SMC3, the chromatin insulator CTCF, as well as the methyl cytosine binding protein (MeCP2) where ATRX might be directly involved in regulating histone modifications and chromatin loop formation at the imprinted control region of the H19 locus (Kernohan et al. 2010).

Several lines of evidence underscore a critical role for ATRX in the regulation of higher order chromatin structure and chromosome segregation. For example, functional ablation of ATRX in mammalian oocytes through antibody microinjection or short interfering RNAi sequences induces chromosome misalignment on the meiotic spindle (De La Fuente et al. 2004b). In addition, lack of ATRX function in HeLa cells following RNAi microinjection promotes severe chromosome cohesion and segregation defects with micronuclei formation (Ritchie et al. 2008). The evidence to date indicates that ATRX might be essential for the epigenetic control of centromere function and, hence, maintenance of chromosome stability during mitotic and meiotic division. There is an essential need for a solid mechanistic understanding of ATRX function in the female germ line. Although ATRX is required for proper chromosome segregation during mitosis and meiosis, the rates of chromosomal defects observed following functional ablation of ATRX in mouse oocytes are twice as high compared to those observed in somatic cells (De La Fuente et al. 2004b; Ritchie et al. 2008; Baumann et al. 2010). The type of chromosome segregation defects observed in ATRX deficient oocytes, i.e., premature centromere separation and chromosome lagging closely resemble those observed in oocytes from women of advanced reproductive age (Vialard et al. 2006). Importantly, although more than 70% of ATRX deficient oocytes exhibit some kind of chromosomal defect, meiotic progression and the transition to the first mitosis following fertilization ensue uninterrupted with subsequent transmission of severe aneuploidy to the early embryo (Baumann et al. 2010; De La Fuente et al. 2004b).

The mechanisms involved in the differential rates of chromosome instability due to the loss of ATRX function during mitosis and meiosis are not clear at present. However, the high rates of chromosomal non-disjunction in ATRX deficient oocytes might reflect important differences in ATRX function during meiosis. This might be attributed to the ability of ATRX to regulate gene expression and/or protein–protein interactions on a cell type-specific basis (Tang et al. 2004; Xue et al. 2003). In support of this premise, recent studies show key differences in the temporal association of ATRX with the transcriptional regulator DAXX in mammalian oocytes (Baumann et al. 2010). In somatic cells, ATRX establishes a physical interaction with DAXX at PCH only during a brief period restricted to the final stages of the S phase of the cell cycle (Ishov et al. 2004). By contrast, in mammalian oocytes, the association of DAXX with ATRX at PCH is maintained beyond S-phase extending into the G2/M transition in fully grown oocytes (Baumann et al. 2010). This process might be of functional significance for centromeric heterochromatin formation during oogenesis as well as for the establishment of the unique chromatin remodeling events that characterize the onset of meiotic resumption in mammalian oocytes (Baumann et al. 2010; De La Fuente 2006). Heterochromatin formation during oogenesis is essential to confer the mammalian oocyte with both meiotic and developmental potential, as well as to establish the maternal contribution of chromatin remodeling factors to the epigenetic landscape of the embryonic genome following fertilization. Therefore, both the ability of ATRX to interact with different protein partners according to the cell type, stage of the cell cycle, and/or stage of differentiation as well as the germ line-specific mechanisms regulating chromatin remodeling in the oocyte genome underscore the importance to determine the germ line-specific functions of ATRX.

3.5 Chromatin Remodeling During Meiosis in the Female Germ Line

To cope with the unique functional requirements and properties of meiotic chromosomes, specialized mechanisms are set in place for the control of chromatin remodeling in the germ line (De La Fuente 2006; Kimmins and Sassone-Corsi 2005; Revenkova et al. 2001, 2004). Importantly, differentiation of chromatin structure and function during oogenesis is critical for remodeling the oocyte genome into a transcriptionally quiescent environment in preparation for accurate chromosome condensation and segregation following meiotic resumption (De La Fuente 2006; De La Fuente and Eppig 2001; De La Fuente et al. 2004b). Beginning on day 16 of postnatal growth, the mouse oocyte genome undergoes striking changes in large-scale chromatin structure in which a decondensed, transcriptionally active nucleus exhibiting a configuration termed non-surrounded nucleolus (NSN), progressively acquires a condensed and transcriptionally quiescent chromatin configuration called surrounded nucleolus (SN). At this stage, major centromeric satellite DNA sequences in the germinal vesicle of preovulatory oocytes undergo a striking nuclear redistribution forming a prominent heterochromatic rim or karyosphere around the nucleolus (De La Fuente 2006; De La Fuente et al. 2004b). While the underlying mechanisms are not fully understood, remodeling chromatin into the SN configuration may confer centromeric domains with a functional configuration to recruit heterochromatin-binding proteins such as ATRX that are required to mediate proper chromosome alignment on the meiotic spindle (De La Fuente et al. 2004a, b).

It is postulated that regulation on global transcriptional repression and chromatin remodeling in the germinal vesicle are under the control of different pathways (De La Fuente et al. 2004a, b). Although chromatin remodeling into the SN configuration and global transcriptional silencing take place concomitantly in preovulatory oocytes, timely repression of transcription is observed in oocytes lacking the nuclear chaperone nucleoplasmin 2 (Npm2), which fail to remodel chromatin into the SN configuration (De La Fuente et al. 2004a, b). Moreover, oocytes deficient for the histone H3K4 methyltransferase (MLL2) fail to undergo global transcriptional silencing despite the presence of a SN configuration (Andreu-Vieyra et al. 2010). Consistent with these observations, recent studies in mouse oocytes indicate that global transcriptional silencing may be attributed to dissociation of the largest subunit of RNA polymerase II (RPB1) from the chromatin template independent from global chromatin condensation in the germinal vesicle in preovulatory oocytes (Abe et al. 2010). Therefore, although temporally linked, global chromatin remodeling and transcriptional silencing in the oocyte genome are autonomous developmental processes in the mammalian female gamete.

What then is the functional significance of the large-scale chromatin remodeling events taking place in the germinal vesicle? In *Drosophila* oocytes, chromosomes acquire a meiosis-specific nuclear organization or karyosome with functional similarities to the karyosphere observed in preovulatory oocytes from several mammalian species including human, mouse, and primates (Ivanovska et al. 2005; Ivanovska and Orr-Weaver 2006). These structures are thought to be essential to maintain a specific chromosome arrangement in order to promote chromosome–nucleolar interactions and perhaps provide a transcriptionally repressive chromatin environment to facilitate the association of major satellite DNA sequences closely bound to the nucleolus with heterochromatin binding proteins critical for proper chromosome segregation (Gruzova and Parfenov 1993). Importantly, histone posttranslational modifications play a vital role in the formation of these nuclear structures in both *Drosophila* and mouse oocytes. For example, in *Drosophila* oocytes the histone H2A kinase (NHK-1) induces H2A phosphorylation at specific tyrosine residues and its functional deletion results in lack of karyosome formation, chromosome segregation defects, and complete sterility (Ivanovska et al. 2005; Ivanovska and Orr-Weaver 2006). Moreover, functional deletion of the vaccinia-related kinase (VRK1), the mammalian homologue of *Drosophila* NHK-1, results in delayed meiotic progression, abnormal chromosome segregation, and fertilization failure with female sterility (Schober et al. 2011). The TATA-binding protein (TBP2) (Gazdag et al. 2009) is also required for the formation of the karyosphere in mammalian oocytes. Whether these factors play a direct structural role in large-scale chromatin remodeling, or in mediating the expression of additional chromatin associated factors that induce the SN configuration in mammalian oocytes warrants further investigation (Gazdag et al. 2009; Schober et al. 2011).

Global chromatin remodeling in mammalian oocytes is regulated through a (1) series of hierarchical histone posttranslational modifications such as histone acetylation or histone methylation at different lysine residues (De La Fuente et al. 2004a; Kouzarides 2007), the incorporation of histone variants such as histone H2A.Z, H3.3, or CENP-A (Polo and Almouzni 2006; Torres-Padilla et al. 2006), or (2) through the action of ATP-dependent chromatin remodeling proteins such as ATRX (Baumann et al. 2010; De La Fuente et al. 2004b). Importantly, covalent histone modifications provide the chromatin context that confers structural and functional identity to specialized chromosomal subdomains such as centromeres or telomeres that are essential for heterochromatin formation and proper chromosome segregation (Durand-Dubief and Ekwall 2008; Ekwall 2007). Initial studies in mouse oocytes revealed the presence of high levels of histone H3 and histone H4 acetylation in the germinal vesicle of preovulatory oocytes. However, coincident with the onset of GV breakdown and the resumption of meiosis, a wave of global histone deacetylation removes the majority of acetylated histone marks throughout the entire length of condensing chromatids in a process that is essential for proper chromosome condensation and the establishment of specialized chromosome–-microtubule interactions in mammalian oocytes (De La Fuente 2006; De La Fuente et al. 2004a; Kim et al. 2003). Although the molecular mechanisms involved in the removal of specific histone residues from meiotic chromosomes are not known, exposure of mammalian oocytes to the histone deacetylase inhibitor trichostatin A (TSA) prevents the onset of global deacetylation and induces the formation of hyperacetylated chromosomes that exhibit abnormal chromosome condensation and improper attachment to the microtubular network (De La Fuente et al. 2004a). These data indicate that histone deacetylases (HDACs) play a critical role in the onset of global histone deacetylation during female meiosis. The nature of the specific HDAC isoforms involved in this process remains to be established. Identification of the HDAC isoforms responsible for deacetylation of histone H4 at lysine 12, which is specific to meiotic chromosome condensation (Kim et al. 2003; Kruhlak et al. 2001) is of particular importance.

Developmental transitions in chromatin modifications during meiotic resumption are critical for the maintenance of chromosome stability in the female gamete as demonstrated by the high levels of chromosome segregation defects observed in mammalian oocytes following transient exposure to TSA (De La Fuente 2006; De La Fuente et al. 2004a) as well as the transmission of an aneuploid chromosome complement to the early conceptus following fertilization in the presence of TSA (Akiyama et al. 2006). Interestingly, $>80\%$ of metaphase-I and metaphase-II stage human oocytes obtained from women of advanced reproductive age exhibited persistence of H4K12 acetylation associated with misaligned chromosomes.

Thus, abnormal regulation of histone deacetylases and the subsequent lack of global histone acetylation during oocyte aging might contribute to the high rates of aneuploidy observed in human oocytes during advanced maternal age (van den Berg et al. 2011). Global histone deacetylation during meiosis might be an essential mechanism to recruit large protein complexes to centromeric domains essential for heterochromatin formation, regulation of centromeric cohesion, and the establishment of centromere–microtubule interactions (Akiyama et al. 2006; De La Fuente 2006; De La Fuente et al. 2004a). This mechanism seems to be particularly susceptible to environmental disruption in the mammalian oocyte. For example, inhibition of HDACs in somatic cells with TSA interferes with recruitment of heterochromatin protein 1 (HP1β) to centromeric domains and results in abnormal chromosome segregation defects after 5 days of treatment (Taddei et al. 2001, 2005). In contrast, TSA induced chromosome-wide hyperacetylation and interfered with recruitment of ATRX to centromeric domains within 8 h of exposure resulting in the presence of chromosomal non-disjunction in >60% of metaphase-II stage oocytes and leading to the transmission of aneuploidy and subsequent post-implantation embryo loss (Akiyama et al. 2006; De La Fuente 2006; De La Fuente et al. 2004a).

3.6 Role of ATRX in Chromosome Segregation During Meiotic Division in Oocytes

Centromeres are embedded within large PCH blocks in both mitotic as well as meiotic cells. Accumulating evidence indicates the presence of a potential functional interaction or "crosstalk" between PCH protein complexes and components of the kinetochore, the specialized centromeric domain that mediates chromosome–microtubule interactions and chromosome segregation (Malik and Henikoff 2009). PCH is not only required for chromosome structural integrity, but is also essential to recruit cohesin protein complexes that maintain centromeric cohesion between sister chromatids during metaphase-I and subsequently ensure an amphitelic microtubular attachment to the metaphase-II spindle in order to segregate sister chromatids (Bernard et al. 2001; Petronczki et al. 2003). PCH is particularly important during meiosis to prevent illegitimate recombination between highly repetitive sequences at meiotic centromeres that would otherwise result in compromised centromere integrity and severely disrupt chromosome segregation (Lamb et al. 2005; Malik and Henikoff 2009; Talbert and Henikoff 2010). Moreover, chromatin remodeling proteins as well as transcriptionally repressive histone marks associated with PCH during female meiosis constitute an essential contribution to the epigenetic inheritance of the zygotic genome necessary to coordinate the striking nuclear reprogramming events essential for the activation of the embryonic genome following fertilization (Probst and Almouzni 2011; Santenard and Torres-Padilla 2009; Santenard et al. 2010).

The mouse centromere consists of several hundred kilobases of minor satellite repeat units as well as the centromere-specific histone H3 variant CENP-A, which confers the epigenetic information required for the formation of a functional kinetochore (Karpen and Allshire 1997; Maison and Almouzni 2004; Polo and Almouzni 2006; Sarma and Reinberg 2005). Epigenetic inheritance of the centromeric modifications through mitotic cell division depends on the assembly of newly synthesized CENPA and nucleosomes at centromeric domains in daughter cells during the G1 phase of the cell cycle and this process is mediated by histone chaperones such as histone H3.1 and H3.3 (Black et al. 2010).

PCH is formed by repeats of major satellite sequences and contributes to the formation of large chromatin remodeling complexes through the recruitment of DNA-binding proteins such as HP1 and helicases of the switch/sucrose non-fermenting (SWI/SNF2) family such as ATRX (De La Fuente et al. 2004b; Maison and Almouzni 2004; McDowell et al. 1999). Interestingly, PCH is a key determinant of CENP-A localization during neocentromere formation in human chromosomes and its functional integrity is essential to coordinate sister centromere cohesion during mitosis (Durand-Dubief and Ekwall 2008; Guenatri et al. 2004; Maison and Almouzni 2004). Heterochromatin-binding complexes such as SNF2h are essential to load the cohesin subunit Rad21 onto the centromeres in human mitotic cells (Hakimi et al. 2002). Moreover, loss of HP1 from PCH in mouse somatic cells deficient for the SUV39 histone methyltransferase severely affects cohesion between sister chromatids (Maison et al. 2002; Peters et al. 2001). Collectively, these studies indicate that PCH formation plays a major role in centromere cohesion in mitotic cells and hence has a direct impact in mediating accurate chromosome segregation.

In the mammalian germ line, PCH regulates important aspects of homologous chromosome interactions during prophase-I of meiosis (Takada et al. 2011), as well as during the formation and nuclear reorganization of chromocenters in the preimplantation embryo (Houlard et al. 2006; Peters et al. 2001; Probst and Almouzni 2011). Importantly, functional ablation of the ATRX protein in mammalian oocytes points to an essential role for PCH in the epigenetic control of centromere function during female meiosis and the transition to the first mitosis at the onset of preimplantation embryo development (Baumann et al. 2010; De La Fuente et al. 2004b).

ATRX is co-localized with histone H3 trimethylated at lysine 9 (H3K9me3) at large heterochromatin domains in the germinal vesicle of preovulatory oocytes where some of these foci are associated with the perinucleolar heterochromatin rim at the karyosphere (De La Fuente et al. 2004b). Interestingly, upon germinal vesicle breakdown, ATRX remains predominantly associated with centromeric domains of condensing chromosomes where it is found partially co-localized with kinetochore proteins detected by the CREST antiserum whereas H3K9me3 is redistributed throughout the entire length of condensing meiotic chromosomes. The mechanisms or functional significance of chromosome-wide H3K9me3 redistribution during meiotic resumption are not known, although this process might reflect the importance of global chromatin remodeling events in chromosome condensation. Global histone deacetylation during meiosis onset is required for the association of ATRX with centromeric domains as treatment of maturing oocytes with the histone deacetylase inhibitor TSA induced a dramatic redistribution of ATRX throughout the chromosomes. Notably, lack of ATRX at pericentric domains induced severe abnormalities in chromosome segregation (De La Fuente et al. 2004a). Two independent experimental approaches provided initial evidence indicating a potential role for ATRX in mediating chromosome–microtubule interactions during female meiosis. For example, both microinjection of preovulatory oocytes with a neutralizing antibody against ATRX as well as selective ablation of the ATRX protein following microinjection with short interfering RNAi sequences had no effect on meiotic progression to the metaphase-II stage. However, lack of ATRX function induced severe abnormalities on the alignment of metaphase-II stage chromosomes with the meiotic spindle in >60% of oocytes which exhibited chromosome lagging and in extreme cases loss of the characteristic bipolar structure of the metaphase-II spindle (De La Fuente et al. 2004b). These studies indicate that ATRX is involved in mediating proper chromosome alignment to the meiotic spindle and strongly suggested that lack of ATRX function might result in the transmission of aneuploidy to the early conceptus.

To address the potential molecular mechanisms of ATRX function during meiosis, a mouse model with an oocyte-specific knockdown of ATRX was generated (Baumann et al. 2010) using a transgenic RNAi approach, whereby the expression of an *Atrx* hairpin double-stranded RNA is under the control of the zona pellucida 3 (*Zp3*) gene promoter. Expression of the *Atrx* hairpin induced an efficient degradation of ATRX transcripts resulting in a significant decrease in the levels of ATRX protein in >95% of preovulatory oocytes. Consistent with previous studies (De La Fuente et al. 2004b), functional ablation of ATRX had no effect on meiotic progression to the metaphase-II stage. However, lack of ATRX function induced severe defects in chromosome–microtubule interactions as well as abnormal chromosome segregation in >70% of transgenic oocytes. Importantly, chromosome segregation defects were observed at similar levels in transgenic oocytes matured in vivo following gonadotropin stimulation or in mutant oocytes matured in vitro (Baumann et al. 2010). In vivo matured mutant oocytes exhibited a spectrum of chromosome segregation defects at the metaphase-II stage including severely misaligned chromosomes, chromosome lagging, presence of single chromatids loosely attached to the meiotic spindle, and in some instances premature anaphase-II onset demonstrating that ATRX function is required for the establishment of proper chromosome–microtubule interactions during meiosis in vivo. Analysis of surface spread chromosomes under high resolution revealed that the majority of mutant oocytes exhibit normal centromere cohesion at the metaphase-I stage. However, lack of ATRX function induced abnormal chromosome morphology, premature centromere separation, as well as chromatid breaks in the majority of oocytes at the metaphase-II stage. These chromosomal defects were also associated with a high incidence of aneuploidy in mutant oocytes and severe infertility. The high incidence of chromosome non-disjunction and presence of single chromatids together with the chromosome alignment defects indicate that ATRX plays a critical role in chromosome segregation during female meiosis and is required for female fertility.

This model provided the first insight into the molecular mechanisms of ATRX function in the mammalian oocyte. For example, ATRX mutant oocytes exhibit similar levels of H3K9me3 and CREST signals at centromeric domains suggesting that H3K9me3 binding to centromeric domains is upstream from that of ATRX. Yet, ATRX deficient oocytes at the germinal vesicle stage fail to recruit the transcriptional repressor DAXX to centromeric domains suggesting that ATRX is essential for the binding of DAXX to this nuclear domain. This observation is of particular significance since it is well established that in somatic cells it is DAXX that recruits ATRX to PML nuclear bodies. However, the mechanisms responsible for the association of DAXX to heterochromatin domains remained to be determined (Ishov et al. 2004). The unique functional interactions of ATRX with DAXX in the female gamete are currently under investigation as studies revealed that lack of ATRX function has a significant impact on the molecular composition of centromeric heterochromatin (Baumann et al. 2010). Consistent with this hypothesis, ATRX deficiency at PCH domains results in abnormal individualization of chromosomes and is associated with reduced levels of histone H3 phosphorylation (H3S10ph), known to be an epigenetic hallmark of chromosome condensation. Notably, the establishment of this histone modification is known to begin at centromeric heterochromatin domains and subsequently spread throughout the chromatids of condensing chromosomes during mitosis (Hendzel et al. 1997). Therefore, lack of ATRX function during female meiosis might interfere with the establishment, or chromosomal distribution, of critical chromatin modifications that are required for proper chromosome condensation and centromere cohesion during the metaphase-II stage. Collectively these studies indicate some degree of evolutionary conservation of the functional interactions between PCH protein composition, centromeric cohesion, and chromosome segregation in yeast and mammals (Bernard and Allshire 2002; Bernard et al. 2001; Hakimi et al. 2002; Nonaka et al. 2002). The meiotic phenotype observed in ATRX deficient oocytes is consistent with the potential interaction (i.e., "crosstalk") between centromeric and PCH in the epigenetic control of centromere function in mammalian oocytes (Baumann et al. 2010).

3.7 ATRX Deficient Oocytes as an experimental Model: Towards Uncovering the Epigenetic Origins of Human Aneuploidy

Several independent lines of evidence underscore the importance of ATRX deficient oocytes as a unique model to gain a better understanding of the molecular basis of aneuploidy in the human oocyte: (1) The type of chromosome segregation defects observed in ATRX deficient oocytes (i.e., whole chromosome non-disjunction and premature separation of sister chromatids) are the two primary mechanisms leading to aneuploidy in oocytes from women of advanced reproductive age (Fragouli et al. 2011; Vialard et al. 2006). (2) Analysis of global gene expression patterns by microarrays revealed that the expression of multiple chromatin remodeling proteins is severely compromised in oocytes from older females (Fragouli et al. 2010; Hamatani et al. 2004). There was a notable reduction in transcript levels as well as protein expression patterns of chromosome-associated proteins such as BRCA1 as well as ATRX, in which lower expression has been associated with high rates of aneuploidy in aging mouse oocytes (Pan et al. 2008). (3) Both cytogenetic analysis and live cell imaging studies indicate that chromosome segregation defects in ATRX deficient oocytes arise predominantly during the metaphase-II stage of meiosis and the transition to the first mitosis in the zygote (Baumann et al. 2010). Therefore, this model is highly relevant for understanding human aneuploidy since chromosome segregation errors that do not arrest the mammalian oocyte at metaphase-I pose a significant risk for the transmission of aneuploidy to the early conceptus. Importantly, ATRX deficient oocytes afford the direct visualization of the critical transition from metaphase-II to the first mitosis to better assess the molecular mechanisms involved in the transmission of aneuploidy to the zygote stage embryo. Ethical guidelines restrict direct investigation of this process in human oocytes, such that most data obtained so far are based mainly on the analysis of second polar bodies (Fragouli et al. 2011; Kuliev et al. 2011).

Studies of genetic polymorphisms initially indicated that most human oocyte aneuploidies are due to defects in metaphase-I of meiosis (Hassold et al. 2007; Hassold and Hunt 2001). However, a series of recent comprehensive analyses of first and second polar bodies by whole genome amplification (WGA) and comparative genomic hybridization (CGH) revealed similar rates of aneuploidy arising in human oocytes at the metaphase-I and metaphase-II stage (Fragouli et al. 2011). These results were confirmed by two independent research groups following analysis of human oocytes from women of varying ages, and support that not only metaphase-I, but also the metaphase-II stage is prone to segregation errors. In oocytes from older women, metaphase-II aneuploidies frequently involved chromosomes 16 and 22, a condition resulting in a high incidence of embryonic demise before implantation (Fragouli et al. 2011; Kuliev et al. 2011).

Loss of maternal ATRX function in mouse oocytes results in pericentromeric DNA breaks leading to severe meiotic centromere instability, the formation of aneuploid embryos, and female subfertility (Baumann et al. 2010; De La Fuente et al. 2004b). Taken together, these findings underscore a crucial role for ATRX in the formation and molecular composition of PCH. Notably, centromeric heterochromatin formation is susceptible to the aging process (Adams 2007) and its role in mediating chromosome–microtubule interactions is potentially disrupted by hormonal (Roberts et al. 2005) as well as environmental fluctuations (Yang et al. 2009). Therefore, it is important to determine whether there is an association between PCH content of individual human chromosomes and their frequency of non-disjunction in metaphase-II stage oocytes. PCH is also dynamically regulated by epigenetic modifications and, contrary to cohesins (Chiang et al. 2010; Lister et al. 2010; Revenkova et al. 2010), might be amenable to potential dietary (Delage and Dashwood 2008) and/or pharmacological intervention through "epigenetic therapy" to prevent the onset of aneuploidy.

Acknowledgments Research in the author's laboratory is supported by grants from the National Institutes of Health (NICHD): 2RO1-HDO42740 and the Georgia Cancer Coalition. We apologize to colleagues whose work could not be cited due to space limitations.

References

Aapola U, Shibuya K, Scott HS, Ollila J, Vihinen M, Heino M, Shintani A, Kawasaki K, Minoshima S, Krohn K, Antonarakis SE, Shimizu N, Kudoh J, Peterson P (2000) Isolation and initial characterization of a novel zinc finger gene, DNMT3L, on 21q22.3, related to the cytosine-5-methyltransferase 3 gene family. Genomics 65:293–298

Abe K, Inoue A, Suzuki MG, Aoki F (2010) Global gene silencing is caused by the dissociation of RNA polymerase II from DNA in mouse oocytes. J Reprod Dev 56:502–507

Adams PD (2007) Remodeling of chromatin structure in senescent cells and its potential impact on tumor suppression and aging. Gene 397:84–93

Akiyama T, Nagata M, Aoki F (2006) Inadequate histone deacetylation during oocyte meiosis causes aneuploidy and embryo death in mice. Proc Natl Acad Sci USA 103:7339–7344

Andreu-Vieyra CV, Chen R, Agno JE, Glaser S, Anastassiadis K, Stewart AF, Matzuk MM (2010) MLL2 is required in oocytes for bulk histone 3 lysine 4 trimethylation and transcriptional silencing. PLoS Biol 8:e1000453

Argentaro A, Yang J, Chapman L, Kowalczyk M, Gibbons R, Higgs D, Neuhaus D, Rhodes D (2007) Structural consequences of disease-causing mutations in the ATRX-DNMT3-DNMT3L (ADD) domain of the chromatin-associated protein ATRX. Proc Natl Acad Sci USA 104:11939–11944

Bannister AJ, Zegerman P, Partridge JF, Miska EA, Thomas JO, Allshire RC, Kouzarides T (2001) Selective recognition of methylated lysine 9 on histone H3 by the HP1 chromo domain. Nature 410:120–124

Bassett AR, Cooper SE, Ragab A, Travers AA (2008) The chromatin remodelling factor dATRX is involved in heterochromatin formation. PLoS One 3:e2099

Baumann C, De La Fuente R (2009) ATRX marks the inactive X chromosome (Xi) in somatic cells and during imprinted X chromosome inactivation in trophoblast stem cells. Chromosoma 118:209–222

Baumann C, Schmidtmann A, Muegge K, De La Fuente R (2008) Association of ATRX with pericentric heterochromatin and the Y chromosome of neonatal mouse spermatogonia. BMC Mol Biol 9:29

Baumann C, Viveiros MM, De La Fuente R (2010) Loss of maternal ATRX results in centromere instability and aneuploidy in the mammalian oocyte and pre-implantation embryo. PLoS Genet 6:e1001137

Bernard P, Allshire R (2002) Centromeres become unstuck without heterochromatin. Trends Cell Biol 12:419–424

Bernard P, Maure J, Partridge J, Genier S, Javerzat J, Allshire R (2001) Requirement of heterochromatin for cohesion at centromeres. Science 294:2539–2542

Bérubé NG (2011) ATRX in chromatin assembly and genome architecture during development and disease. Biochem Cell Biol 89:435–444

Bérubé N, Smeenk C, Picketts D (2000) Cell cycle-dependent phosphorylation of the ATRX protein correlates with changes in nuclear matrix and chromatin association. Hum Mol Genet 9:539–547

Bérubé NG, Jagla M, Smeenk CA, De Repentigny Y, Kothary R, Picketts DJ (2002) Neurodevelopmental defects resulting from ATRX overexpression in transgenic mice. Hum Mol Genet 11:253–261

Bérubé NG, Healy J, Medina CF, Wu S, Hodgson T, Jagla M, Picketts DJ (2007) Patient mutations alter ATRX targeting to PML nuclear bodies. Eur J Hum Genet 16:192–201

Black BE, Brock MA, Bedard S, Woods VL, Cleveland DW (2007) An epigenetic mark generated by the incorporation of CENP-A into centromeric nucleosomes. Proc Natl Acad Sci USA 104:5008–5013

Black BE, Jansen LET, Foltz DR, Cleveland DW (2010) Centromere identity, function, and epigenetic propagation across cell divisions. Cold Spring Harb Symp Quant Biol 75:403–418

Cardoso C, Timsit S, Villard L, Khrestchatsy M, Fontés M, Colleaux L (1998) Specific interaction between the XNP/ATR-X gene product and the SET domain of the human EZH2 protein. Hum Mol Genet 7:679–684

Cheeseman IM, Desai A (2008) Molecular architecture of the kinetochore-microtubule interface. Nat Rev Mol Cell Biol 9:33–46

Cheeseman IM, Anderson S, Jwa M, Green EM, Kang J-S, Yates Iii JR, Chan CSM, Drubin DG, Barnes G (2002) Phospho-regulation of kinetochore-microtubule attachments by the Aurora Kinase Ipl1p. Cell 111:163–172

Chiang T, Duncan FE, Schindler K, Schultz RM, Lampson MA (2010) Evidence that weakened centromere cohesion is a leading cause of age-related aneuploidy in oocytes. Curr Biol 20:1522–1528

De La Fuente R (2006) Chromatin modifications in the germinal vesicle (GV) of mammalian oocytes. Dev Biol 292:1–12

De La Fuente R, Eppig JJ (2001) Transcriptional activity of the mouse oocyte genome: companion granulosa cells modulate transcription and chromatin remodeling. Dev Biol 229:224–236

De La Fuente R, Viveiros M, Burns K, Adashi E, Matzuk M, Eppig J (2004a) Major chromatin remodeling in the germinal vesicle (GV) of mammalian oocytes is dispensable for global transcriptional silencing but required for centromeric heterochromatin function. Dev Biol 275:447–458

De La Fuente R, Viveiros M, Wigglesworth K, Eppig J (2004b) ATRX, a member of the SNF2 family of helicase /ATPases, is required for chromosome alignment and meiotic spindle organization in metaphase II stage mouse oocytes. Dev Biol 272:1–14

De La Fuente R, Baumann C, Viveiros MM (2011) Role of ATRX in chromatin structure and function: implications for chromosome instability and human disease. Reproduction 142:221–234

Delage B, Dashwood RH (2008) Dietary manipulation of histone structure and function. Annu Rev Nutr 28:347–366

DeLuca JG, Gall WE, Ciferri C, Cimini D, Musacchio A, Salmon ED (2006) Kinetochore microtubule dynamics and attachment stability are regulated by Hec1. Cell 127:969–982

Dhayalan A, Tamas R, Bock I, Tattermusch A, Dimitrova E, Kudithipudi S, Ragozin S, Jeltsch A (2011) The ATRX-ADD domain binds to H3 tail peptides and reads the combined methylation state of K4 and K9. Hum Mol Genet 20(11):2195–2203

Drane P, Ouararhni K, Depaux A, Shuaib M, Hamiche A (2010) The death-associated protein DAXX is a novel histone chaperone involved in the replication-independent deposition of H3.3. Genes Dev 24:1253–1265

Duesbery NS, Choi T, Brown KD, Wood KW, Resau J, Fukasawa K, Cleveland DW, Vande Woude GF (1997) CENP-E is an essential kinetochore motor in maturing oocytes and is masked during Mos-dependent, cell cycle arrest at metaphase II. Proc Natl Acad Sci USA 94:9165–9170

Durand-Dubief M, Ekwall K (2008) Heterochromatin tells CENP-A where to go. Bioessays 30:526–529

Ekwall K (2007) Epigenetic control of centromere behavior. Annu Rev Genet 41:63–81

Elsasser SJ, Allis CD, Lewis PW (2011) New epigenetic drivers of cancers. Science 331:1145–1146

Emelyanov AV, Konev AY, Vershilova E, Fyodorov DV (2010) Protein complex of Drosophila ATRX/XNP and HP1A is required for the formation of pericentric beta-heterochromatin in vivo. J Biol Chem 285:15027–15037

Eustermann S, Yang J-C, Law MJ, Amos R, Chapman LM, Jelinska C, Garrick D, Clynes D, Gibbons RJ, Rhodes D, Higgs DR, Neuhaus D (2011) Combinatorial readout of histone H3 modifications specifies localization of ATRX to heterochromatin. Nat Struct Mol Biol 18:777–782

Fragouli E, Bianchi V, Patrizio P, Obradors A, Huang Z, Borini A, Delhanty JDA, Wells D (2010) Transcriptomic profiling of human oocytes: association of meiotic aneuploidy and altered oocyte gene expression. Mol Hum Reprod 16:570–582

Fragouli E, Alfarawati S, Goodall N, Sanchez-Garcia JF, Colls P, Wells D (2011) The cytogenetics of polar bodies: insights into female meiosis and the diagnosis of aneuploidy. Mol Hum Reprod 17:286–295

Garrick D, Sharpe JA, Arkell R, Dobbie L, Smith AJ, Wood WG, Higgs DR, Gibbons RJ (2006) Loss of Atrx affects trophoblast development and the pattern of X-inactivation in extraembryonic tissues. PLoS Genet 2:e58

Gazdag E, Santenard A, Ziegler-Birling C, Altobelli G, Poch O, Tora L, Torres-Padilla M-E (2009) TBP2 is essential for germ cell development by regulating transcription and chromatin condensation in the oocyte. Genes Dev 23:2210–2223

Gibbons RJ, Picketts DJ, Villard L, Higgs DR (1995) Mutations in a putative global transcriptional regulator cause X-linked mental retardation with alpha-thalassemia (ATR-X syndrome). Cell 80:837–845

Gibbons R, Bachoo S, Picketts D, Aftimos S, Asenbauer B, Bergoffen J, Berry S, Dahl N, Fryer A, Keppler K, Kurosawa K, Levin M, Masuno M, Neri G, Pierpont M, Slaney S, Higgs D (1997) Mutations in transcriptional regulator ATRX establish the functional significance of a PHD-like domain. Nat Genet 17:146–148

Gibbons RJ, McDowell TL, Raman S, O'Rourke DM, Garrick D, Ayyub H, Higgs DR (2000) Mutations in ATRX, encoding a SWI/SNF-like protein, cause diverse changes in the pattern of DNA methylation. Nat Genet 24:368–371

Gibbons RJ, Pellagatti A, Garrick D, Wood WG, Malik N, Ayyub H, Langford C, Boultwood J, Wainscoat JS, Higgs DR (2003) Identification of acquired somatic mutations in the gene encoding chromatin-remodeling factor ATRX in the [alpha]-thalassemia myelodysplasia syndrome (ATMDS). Nat Genet 34:446–449

Goldberg AD, Banaszynski LA, Noh K-M, Lewis PW, Elsaesser SJ, Stadler S, Dewell S, Law M, Guo X, Li X, Wen D, Chapgier A, DeKelver RC, Miller JC, Lee Y-L, Boydston EA, Holmes MC, Gregory PD, Greally JM, Rafii S, Yang C, Scambler PJ, Garrick D, Gibbons RJ, Higgs DR, Cristea IM, Urnov FD, Zheng D, Allis CD (2010) Distinct factors control histone variant H3.3 localization at specific genomic regions. Cell 140:678–691

Gruzova MN, Parfenov VN (1993) Karyosphere in oogenesis and intranuclear morphogenesis. Int Rev Cytol 144:1–52

Guenatri M, Bailly D, Maison C, Almouzni G (2004) Mouse centric and pericentric satellite repeats form distinct functional heterochromatin. J Cell Biol 166:493–505

Hakimi M, Bochar D, Schmiesing J, Dong Y, Barak O, Speicher D, Yokomori K, Shiekhattar R (2002) A chromatin remodelling complex that loads cohesin onto human chromosomes. Nature 418:994–998

Hamatani T, Falco G, Carter MG, Akutsu H, Stagg CA, Sharov AA, Dudekula DB, VanBuren V, Ko MS (2004) Age-associated alteration of gene expression patterns in mouse oocytes. Hum Mol Genet 13:2263–2278

Hassold T, Hunt P (2001) To err (meiotically) is human: the genesis of human aneuploidy. Nat Rev Genet 2:280–291

Hassold T, Hall H, Hunt P (2007) The origin of human aneuploidy: where we have been, where we are going. Hum Mol Genet 16:203–208

Heaphy CM, de Wilde RF, Jiao Y, Klein AP, Edil BH, Shi C, Bettegowda C, Rodriguez FJ, Eberhart CG, Hebbar S, Offerhaus GJ, McLendon R, Rasheed BA, He Y, Yan H, Bigner DD, Oba-Shinjo SM, Marie SKN, Riggins GJ, Kinzler KW, Vogelstein B, Hruban RH, Maitra A, Papadopoulos N, Meeker AK (2011) Altered telomeres in tumors with ATRX and DAXX mutations. Science 333:425

Hendzel MJ, Wei Y, Mancini MA, Van Hooser A, Ranalli T, Brinkley BR, Bazett-Jones DP, Allis CD (1997) Mitosis-specific phosphorylation of histone H3 initiates primarily within pericentromeric heterochromatin during G2 and spreads in an ordered fashion coincident with mitotic chromosome condensation. Chromosoma 106:348–360

Henikoff S, Ahmad K, Malik HS (2001) The centromere paradox: stable inheritance with rapidly evolving DNA. Science 293:1098–1102

Homer HA, McDougall A, Levasseur M, Yallop K, Murdoch AP, Herbert M (2005) Mad2 prevents aneuploidy and premature proteolysis of cyclin B and securin during meiosis I in mouse oocytes. Genes Dev 19:202–207

Houlard M, Berlivet S, Probst AV, Quivy JP, Hery P, Almouzni G, Gerard M (2006) CAF-1 is essential for heterochromatin organization in pluripotent embryonic cells. PLoS Genet 2(11):e181

Howman EV, Fowler KJ, Newson AJ, Redward S, MacDonald AC, Kalitsis P, Choo KHA (2000) Early disruption of centromeric chromatin organization in centromere protein A (Cenpa) null mice. Proc Natl Acad Sci USA 97:1148–1153

Hunt PA, Hassold TJ (2008) Human female meiosis: what makes a good egg go bad? Trends Genet 24:86–93

Ishov AM, Vladimirova OV, Maul GG (2004) Heterochromatin and ND10 are cell-cycle regulated and phosphorylation-dependent alternate nuclear sites of the transcription repressor Daxx and SWI/SNF protein ATRX. J Cell Sci 117:3807–3820

Ivanovska I, Orr-Weaver TL (2006) Histone modifications and the chromatin scaffold for meiotic chromosome architecture. Cell Cycle 5:2064–2071

Ivanovska I, Khandan T, Ito T, Orr-Weaver TL (2005) A histone code in meiosis: the histone kinase, NHK-1, is required for proper chromosomal architecture in Drosophila oocytes. Genes Dev 19:2571–2582

Iwase S, Xiang B, Ghosh S, Ren T, Lewis PW, Cochrane JC, Allis CD, Picketts DJ, Patel DJ, Li H, Shi Y (2011) ATRX ADD domain links an atypical histone methylation recognition mechanism to human mental-retardation syndrome. Nat Struct Mol Biol 18:769–776

Jiao Y, Shi C, Edil BH, de Wilde RF, Klimstra DS, Maitra A, Schulick RD, Tang LH, Wolfgang CL, Choti MA, Velculescu VE, Diaz LA, Vogelstein B, Kinzler KW, Hruban RH, Papadopoulos N (2011) DAXX/ATRX, MEN1, and mTOR pathway genes are frequently altered in pancreatic neuroendocrine tumors. Science 331:1199–1203

Karpen GH, Allshire RC (1997) The case for epigenetic effects on centromere identity and function. Trends Genet 13:489–496

Kernohan KD, Jiang Y, Tremblay DC, Bonvissuto AC, Eubanks JH, Mann MRW, Berube NG (2010) ATRX partners with cohesin and MeCP2 and contributes to developmental silencing of imprinted genes in the brain. Dev Cell 18:191–202

Kim J, Liu H, Tazaki M, Nagata M, Aoki F (2003) Changes in histone acetylation during mouse oocyte meiosis. J Cell Biol 162:37–46

Kimmins S, Sassone-Corsi P (2005) Chromatin remodelling and epigenetic features of germ cells. Nature 434:583–589

Kitajima TS, Ohsugi M, Ellenberg J (2011) Complete kinetochore tracking reveals error-prone homologous chromosome biorientation in mammalian oocytes. Cell 146:568–581

Kourmouli N, Sun YM, van der Sar S, Singh PB, Brown JP (2005) Epigenetic regulation of mammalian pericentric heterochromatin in vivo by HP1. Biochem Biophys Res Commun 337:901–907

Kouzarides T (2007) Chromatin modifications and their function. Cell 128:693–705

Kruhlak MJ, Hendzel MJ, Fischle W, Bertos NR, Hameed S, Yang XJ, Verdin E, Bazett-Jones DP (2001) Regulation of global acetylation in mitosis through loss of histone acetyltransferases and deacetylases from chromatin. J Biol Chem 276:38307–38319

Kudo NR, Wassmann K, Anger M, Schuh M, Wirth KG, Xu H, Helmhart W, Kudo H, McKay M, Maro B, Ellenberg J, de Boer P, Nasmyth K (2006) Resolution of chiasmata in oocytes requires separase-mediated proteolysis. Cell 126:135–146

Kuliev A, Zlatopolsky Z, Kirillova I, Spivakova J, Cieslak J (2011) Meiosis errors in over 20,000oocytes studied in the practice of preimplantation aneuploidy testing. Reprod Biomed Online 22:2–8

Lachner M, O'Carroll D, Rea S, Mechtler K, Jenuwein T (2001) Methylation of histone H3 lysine 9 creates a binding site for HP1 proteins. Nature 410:116–120

Lamb NE, Sherman SL, Hassold TJ (2005) Effect of meiotic recombination on the production of aneuploid gametes in humans. Cytogenet Genome Res 111:250–255

Law MJ, Lower KM, Voon HPJ, Hughes JR, Garrick D, Viprakasit V, Mitson M, De Gobbi M, Marra M, Morris A, Abbott A, Wilder SP, Taylor S, Santos GM, Cross J, Ayyub H, Jones S, Ragoussis J, Rhodes D, Dunham I, Higgs DR, Gibbons RJ (2010) ATR-X syndrome protein targets tandem repeats and influences allele-specific expression in a size-dependent manner. Cell 143:367–378

Lee J, Kitajima TS, Tanno Y, Yoshida K, Morita T, Miyano T, Miyake M, Watanabe Y (2008) Unified mode of centromeric protection by shugoshin in mammalian oocytes and somatic cells. Nat Cell Biol 10:42–52

Leland S, Nagarajan P, Polyzos A, Thomas S, Samaan G, Donnell R, Marchetti F, Venkatachalam S (2009) Heterozygosity for a Bub1 mutation causes female-specific germ cell aneuploidy in mice. Proc Natl Acad Sci USA 106:12776–12781

Levy M, Fernandes A, Tremblay D, Seah C, Berube N (2008) The SWI/SNF protein ATRX co-regulates pseudoautosomal genes that have translocated to autosomes in the mouse genome. BMC Genomics 9:468

Lister LM, Kouznetsova A, Hyslop LA, Kalleas D, Pace SL, Barel JC, Nathan A, Floros V, Adelfalk C, Watanabe Y, Jessberger R, Kirkwood TB, Höög C, Herbert M (2010) Age-related meiotic segregation errors in mammalian oocytes are preceded by depletion of cohesin and Sgo2. Curr Biol 20:1511–1521

Llano E, Gomez R, Gutierrez-Caballero C, Herran Y, Sanchez-Martin M, Vazquez-Quinones L, Hernandez T, de Alava E, Cuadrado A, Barbero JL, Suja JA, Pendas AM (2008) Shugoshin-2 is essential for the completion of meiosis but not for mitotic cell division in mice. Genes Dev 22:2400–2413

Luciani JJ, Depetris D, Usson Y, Metzler-Guillemain C, Mignon-Ravix C, Mitchell MJ, Megarbane A, Sarda P, Sirma H, Moncla A, Feunteun J, Mattei MG (2006) PML nuclear bodies are highly organised DNA-protein structures with a function in heterochromatin remodelling at the G2 phase. J Cell Sci 119:2518–2531

Ma W, Baumann C, Viveiros MM (2010) NEDD1 is crucial for meiotic spindle stability and accurate chromosome segregation in mammalian oocytes. Dev Biol 339:439–450

Maison C, Almouzni G (2004) HP1 and the dynamics of heterochromatin maintenance. Nat Rev Mol Cell Biol 5:296–304

Maison C, Bailly D, Peters A, Quivy J, Roche D, Taddei A, Lachner M, Jenuwein T, Almouzni G (2002) Higher-order structure in pericentric heterochromatin involves a distinct pattern of histone modification and RNA component. Nat Genet 30:329–334

Malik HS, Henikoff S (2009) Major evolutionary transitions in centromere complexity. Cell 138:1067–1082

McDowell TL, Gibbons RJ, Sutherland H, O'Rourke DM, Bickmore WA, Pombo A, Turley H, Gatter K, Picketts DJ, Buckle VJ (1999) Localization of a putative transcriptional regulator (ATRX) at pericentromeric heterochromatin and the short arms of acrocentric chromosomes. Proc Natl Acad Sci USA 96:13983–13988

McGuinness BE, Anger M, Kouznetsova A, Gil-BernabÈ AM, Helmhart W, Kudo NR, Wuensche A, Taylor S, Hoog C, Novak B, Nasmyth K (2009) Regulation of APC/C activity in oocytes by a Bub1-dependent spindle assembly checkpoint. Curr Biol 19:369–380

Mitson M, Kelley LA, Sternberg MJE, Higgs DR, Gibbons RJ (2011) Functional significance of mutations in the Snf2 domain of ATRX. Hum Mol Genet 20:2603–2610

Murphy TD, Karpen GH (1998) Centromeres take flight: alpha satellite and the quest for the human centromere. Cell 93:317–320

Nagaoka S, Hodges CA, Albertini DF, Hunt PA (2011) Oocyte-specific differences in cell-cycle control create an innate susceptibility to meiotic errors. Curr Biol 21:651–657

Nan X, Hou J, Maclean A, Nasir J, Lafuente MJ, Shu X, Kriaucionis S, Bird A (2007) Interaction between chromatin proteins MECP2 and ATRX is disrupted by mutations that cause inherited mental retardation. Proc Natl Acad Sci USA 104(8):2709–2714

Nasmyth K, Haering CH (2009) Cohesin: its roles and mechanisms. Annu Rev Genet 43:525–558

Niault T, Hached K, Sotillo R, Sorger PK, Maro B, Benezra R, Wassmann K (2007) Changing Mad2 levels affects chromosome segregation and spindle assembly checkpoint control in female mouse meiosis-I. PLoS One 2:e1165

Nicklas R, Ward S (1994) Elements of error correction in mitosis: microtubule capture, release, and tension. J Cell Biol 126:1241–1253

Nicklas RB, Waters JC, Salmon ED, Ward SC (2001) Checkpoint signals in grasshopper meiosis are sensitive to microtubule attachment, but tension is still essential. J Cell Sci 114:4173–4183

Nonaka N, Kitajima T, Yokobayashi S, Xiao G, Yamamoto M, Grewal SI, Watanabe Y (2002) Recruitment of cohesin to heterochromatic regions by Swi6/HP1 in fission yeast. Nat Cell Biol 4:89–93

Oegema K, Desai A, Rybina S, Kirkham M, Hyman AA (2001) Functional analysis of kinetochore assembly in *Caenorhabditis elegans*. J Cell Biol 153:1209–1226

Pan H, Ma P, Zhu W, Schultz RM (2008) Age-associated increase in aneuploidy and changes in gene expression in mouse eggs. Dev Biol 316:397–407

Perpelescu M, Fukagawa T (2011) The ABCs of CENPs. Chromosoma 120:425–446

Peters AH, O'Carroll D, Scherthan H, Mechtler K, Sauer S, Schofer C, Weipoltshammer K, Pagani M, Lachner M, Kohlmaier A, Opravil S, Doyle M, Sibilia M, Jenuwein T (2001) Loss of the Suv39h histone methyltransferases impairs mammalian heterochromatin and genome stability. Cell 107:323–337

Petronczki M, Siomos M, Nasmyth K (2003) Un menage a quatre: the molecular biology of chromosome segregation in meiosis. Cell 112:423–440

Picketts DJ, Higgs DR, Bachoo S, Blake DJ, Quarrell OW, Gibbons RJ (1996) ATRX encodes a novel member of the SNF2 family of proteins: mutations point to a common mechanism underlying the ATR-X syndrome. Hum Mol Genet 5:1899–1907

Picketts DJ, Tastan AO, Higgs DR, Gibbons RJ (1998) Comparison of the human and murine ATRX gene identifies highly conserved, functionally important domains. Mamm Genome 9:400–403

Pluta AF, Mackay AM, Ainsztein AM, Goldberg IG, Earnshaw WC (1995) The centromere: hub of chromosomal activities. Science 270:1591–1594

Polo SE, Almouzni G (2006) Chromatin assembly: a basic recipe with various flavours. Curr Opin Genet Dev 16:104–111

Prieto I, Tease C, Pezzi N, Buesa J, Ortega S, Kremer L, Martinez A, Martinez-A C, Hulten M, Barbero JL (2004) Cohesin component dynamics during meiotic prophase I in mammalian oocytes. Chromosome Res 12:197–213

Probst AV, Almouzni G (2011) Heterochromatin establishment in the context of genome-wide epigenetic reprogramming. Trends Genet 27:177–185

Puschendorf M, Terranova R, Boutsma E, Mao X, Isono K, Brykczynska U, Kolb C, Otte AP, Koseki H, Orkin SH, van Lohuizen M, Peters AH (2008) PRC1 and Suv39h specify parental asymmetry at constitutive heterochromatin in early mouse embryos. Nat Genet 40:411–420

Rea S, Eisenhaber F, O'Carroll D, Strahl BD, Sun ZW, Schmid M, Opravil S, Mechtler K, Ponting CP, Allis CD, Jenuwein T (2000) Regulation of chromatin structure by site-specific histone H3 methyltransferases. Nature 406:593–599

Revenkova E, Eijpe M, Heyting C, Gross B, Jessberger R (2001) Novel meiosis-specific isoform of mammalian SMC1. Mol Cell Biol 21:6984–6998

Revenkova E, Eijpe M, Heyting C, Hodges C, Hunt P, Liebe B, Scherthan H, Jessberger R (2004) Cohesin SMC1 beta is required for meiotic chromosome dynamics, sister chromatid cohesion and DNA recombination. Nat Cell Biol 6:555–562

Revenkova E, Herrmann K, Adelfalk C, Jessberger R (2010) Oocyte cohesin expression restricted to predictyate stages provides full fertility and prevents aneuploidy. Curr Biol 20:1529–1533

Ritchie K, Seah C, Moulin J, Issac C, Dick F, Berube N (2008) Loss of ATRX leads to chromosome cohesion and congression defects. J Cell Biol 180:315–324

Roberts R, Iatropoulou A, Ciantar D, Stark J, Becker DL, Franks S, Hardy K (2005) Follicle-stimulating hormone affects metaphase i chromosome alignment and increases aneuploidy in mouse oocytes matured in vitro. Biol Reprod 72:107–118

Santenard A, Torres-Padilla M-E (2009) Epigenetic reprogramming in mammalian reproduction: contribution from histone variants. Epigenetics 4:80–84

Santenard A, Ziegler-Birling C, Koch M, Tora L, Bannister AJ, Torres-Padilla M-E (2010) Heterochromatin formation in the mouse embryo requires critical residues of the histone variant H3.3. Nat Cell Biol 12:853–862

Sarma K, Reinberg D (2005) Histone variants meet their match. Nat Rev Mol Cell Biol 6:139–149

Schneiderman JI, Sakai A, Goldstein S, Ahmad K (2009) The XNP remodeler targets dynamic chromatin in Drosophila. Proc Natl Acad Sci USA 106:14472–14477

Schober CS, Aydiner F, Booth CJ, Seli E, Reinke V (2011) The kinase VRK1 is required for normal meiotic progression in mammalian oogenesis. Mech Dev 128:178–190

Schotta G, Lachner M, Sarma K, Ebert A, Sengupta R, Reuter G, Reinberg D, Jenuwein T (2004) A silencing pathway to induce H3-K9 and H4-K20 trimethylation at constitutive heterochromatin. Genes Dev 18:1251–1262

Serrano E, Lasa A, Perea G, Carnicer M, Brunet S, Aventín A, Sierra J, Nomdedéu J (2006) Acute myeloid leukemia subgroups identified by pathway-restricted gene expression signatures. Acta Hematol 116:77–89

Taddei A, Maison C, Roche D, Almouzni G (2001) Reversible disruption of pericentric heterochromatin and centromere function by inhibiting deacetylases. Nat Cell Biol 3:114–120

Taddei A, Roche D, Bickmore WA, Almouzni G (2005) The effects of histone deacetylase inhibitors on heterochromatin: implications for anticancer therapy? EMBO Rep 6:520–524

Takada Y, Naruse C, Costa Y, Shirakawa T, Tachibana M, Sharif J, Kezuka-Shiotani F, Kakiuchi D, Masumoto H, Shinkai Y-I, Ohbo K, Peters AHFM, Turner JMA, Asano M, Koseki H (2011) HP1 gamma links histone methylation marks to meiotic synapsis in mice. Development 138:4207–4217

Talbert PB, Henikoff S (2010) Centromeres convert but don't cross. PLoS Biol 8:e1000326

Tang J, Wu S, Liu H, Stratt R, Barak OG, Shiekhattar R, Picketts DJ, Yang X (2004) A novel transcription regulatory complex containing death domain-associated protein and the ATR-X syndrome protein. J Biol Chem 279:20369–20377

Torres-Padilla ME, Bannister AJ, Hurd PJ, Kouzarides T, Zernicka-Goetz M (2006) Dynamic distribution of the replacement histone variant H3.3 in the mouse oocyte and preimplantation embryos. Int J Dev Biol 50:455–461

van den Berg IM, Eleveld C, van der Hoeven M, Birnie E, Steegers EAP, Galjaard RJ, Laven JSE, van Doorninck JH (2011) Defective deacetylation of histone 4 K12 in human oocytes is associated with advanced maternal age and chromosome misalignment. Hum Reprod 26 (5):1181–1190

Vialard F, Petit C, Bergere M, Gomes DM, Martel-Petit V, Lombroso R, Ville Y, Gerard H, Selva J (2006) Evidence of a high proportion of premature unbalanced separation of sister chromatids in the first polar bodies of women of advanced age. Hum Reprod 21:1172–1178

Villard L, Lossi AM, Cardoso C, Proud V, Chiaroni P, Colleaux L, Schwartz C, Fonés M (1997) Determination of the genomic structure of the XNP/ATRX gene encoding a potential zinc finger helicase. Genomics 43:149–155
Welburn JPI, Grishchuk EL, Backer CB, Wilson-Kubalek EM, Yates JR, Cheeseman IM (2009) The human kinetochore Ska1 complex facilitates microtubule depolymerization-coupled motility. Dev Cell 16:374–385
Welburn JPI, Vleugel M, Liu D, Yates JR, Lampson MA, Fukagawa T, Cheeseman IM (2010) Aurora B phosphorylates spatially distinct targets to differentially regulate the kinetochore-microtubule interface. Mol Cell 38:383–392
Wiens GR, Sorger PK (1998) Centromeric chromatin and epigenetic effects in kinetochore assembly. Cell 93:313–316
Wood KW, Sakowicz R, Goldstein LSB, Cleveland DW (1997) CENP-E is a plus end-directed kinetochore motor required for metaphase chromosome alignment. Cell 91:357–366
Wood AJ, Severson AF, Meyer BJ (2010) Condensin and cohesin complexity: the expanding repertoire of functions. Nat Rev Genet 11:391–404
Xue Y, Gibbons R, Yan Z, Yang D, McDowell TL, Sechi S, Qin J, Zhou S, Higgs D, Wang W (2003) The ATRX syndrome protein forms a chromatin-remodeling complex with Daxx and localizes in promyelocytic leukemia nuclear bodies. Proc Natl Acad Sci USA 100:10635–10640
Yamagishi Y, Sakuno T, Shimura M, Watanabe Y (2008) Heterochromatin links to centromeric protection by recruiting shugoshin. Nature 455:251–255
Yang F, Baumann C, De La Fuente R (2009) Persistence of histone H2AX phosphorylation after meiotic chromosome synapsis and abnormal centromere cohesion in poly (ADP-ribose) polymerase (Parp-1) null oocytes. Dev Biol 331:326–338
Yang K-T, Li S-K, Chang C-C, Tang C-JC, Lin Y-N, Lee S-C, Tang TK (2010) Aurora-C Kinase deficiency causes cytokinesis failure in meiosis-I and production of large polyploid oocytes in mice. Mol Biol Cell 21:2371–2383

Chapter 4
Cyclin B in Mouse Oocytes and Embryos: Importance for Human Reproduction and Aneuploidy

Zbigniew Polański, Hayden Homer, and Jacek Z. Kubiak

Abstract Oocyte maturation and early embryo development require precise coordination between cell cycle progression and the developmental programme. Cyclin B plays a major role in this process: its accumulation and degradation is critical for driving the cell cycle through activation and inactivation of the major cell cycle kinase, CDK1. CDK1 activation is required for M-phase entry whereas its inactivation leads to exit from M-phase. The tempo of oocyte meiotic and embryonic mitotic divisions is set by the rate of cyclin B accumulation and the timing of its destruction. By controlling when cyclin B destruction is triggered and by co-ordinating this with the completion of chromosome alignment, the spindle assembly checkpoint (SAC) is a critical quality control system important for averting aneuploidy and for building in the flexibility required to better integrate cell cycle progression with development. In this review we focus on cyclin B metabolism in mouse oocytes and embryos and illustrate how the cell cycle-powered clock (in fact cyclin B-powered clock) controls oocyte maturation and early embryo development, thereby providing important insight into human reproduction and potential causes of Down syndrome.

Z. Polański
Department of Genetics and Evolution, Institute of Zoology, Jagiellonian University, 30-387 Kraków, Poland

H. Homer
Mammalian Oocyte and Embryo Research Laboratory, Institute for Women's Health, University College London and University College London Hospitals NHS Foundation Trust, London WC1E 6BT, UK

Department of Obstetrics and Gynaecology, Institute for Women's Health, University College London and University College London Hospitals NHS Foundation Trust, London, WC1E 6BT, UK

J.Z. Kubiak (✉)
Institut de Génétique et Développement de Rennes, CNRS, UMR 6290, Rennes F-35043, France

IFR 140, Faculté de Médecine, Université Rennes 1, UEB, Rennes F-35043, France
e-mail: jacek.kubiak@univ-rennes1.fr

J.Z. Kubiak (ed.), *Mouse Development*, Results and Problems in Cell Differentiation 55, DOI 10.1007/978-3-642-30406-4_4, © Springer-Verlag Berlin Heidelberg 2012

4.1 Introduction

The precise temporal control of key cell cycle events is fundamentally important for the integrity of oocyte maturation and embryonic development. Successive cell divisions, represented by two meiotic divisions in oocytes followed by mitotic divisions in embryos, must be synchronised with developmental transitions in order to ensure harmonious embryonic development. In mammals, the dialogue between the cell cycle regulating machinery and the developmental programme seems to use specific cues since mammalian regulatory development requires particular flexibility absent in other phyla (Tarkowski 1959; Tarkowski 1961; Suwinska 2012; Morris and Zernicka-Goetz 2012). In mammalian oocytes and embryos, contrary to invertebrates and lower vertebrates, this flexibility is achieved among others by the presence of cell cycle-controlling checkpoint systems. The main goal of checkpoints is to ensure the proper order of cell cycle transitions so that the commitment to later events depends on the correct completion of early events (Hartwell and Weinert 1989). In the presence of uncorrected errors, such as misalignment of chromosomes on the spindle apparatus, a checkpoint mechanism becomes activated that prevents the cell cycle from progressing to the subsequent cell cycle stage, thereby furnishing the cell with sufficient time for rectifying the error. Thus, checkpoints in mammalian cells provide opportunities for extending specific cell cycle phases resulting in greater flexibility and unpredictability compared with cell cycles in invertebrates and lower vertebrates (Greenwood et al. 2001). In the latter, the lack of checkpoint control during early development (Gerhart et al. 1984) results in precarious embryonic survival that is compensated for by producing large numbers of female gametes and consequently large numbers of embryos. During evolution, the incorporation of checkpoint control within mammalian oocyte maturation and early embryonic development culminated (synergistically with internal embryonic development and parental care of progeny) in a remarkable improvement in embryonic genetic integrity and consequently in a drastic reduction in the number of embryos produced per mating. Understanding how the embryonic developmental programme became adapted to a highly stochastic pattern of cell cycle progression and the dialogue between the two, remains a critically important focus of investigation.

Strikingly, mammalian oocyte maturation and the first few embryonic divisions (depending on the species) are transcriptionally quiescent. Under these circumstances, cell cycle progression and developmental transitions are orchestrated by maternally derived genetic information in the form of stored maternal mRNAs and proteins that become stock-piled in oocytes during their protracted growth phase (Clarke 2012). Thus, oocytes and early embryos are exclusively dependent upon translational and post-translational modes of regulation. In the mouse, major embryonic genome activation takes place at the two-cell stage introducing a further level of complexity to the coordination of developmental events. Notably, some maternal RNAs and proteins survive at least until the morula stage of preimplantation development (Hamatani et al. 2004). During this transient period therefore, developmental processes are controlled through the concerted actions of both maternal as well as embryonic genetic information (Zheng and Liu 2012).

In this chapter we discuss the metabolism and regulation of cyclin B (largely focussing on cyclin B1) to illustrate how the core machinery of the cell cycle operates in mouse oocytes and early embryos. We discuss how mechanisms for bringing about cyclin B destruction are regulated in order to control the timing of chromosome segregation thereby augmenting the fidelity with which genetic information is transmitted from the oocyte to the embryo at the earliest stages of mammalian development. This review illustrates specialised aspects of cyclin B regulation required for establishing the mammalian oocyte's unique pattern of maturation and fertilisation, thereby providing insight into how cell cycle regulation is integrated with the embryonic programme.

4.1.1 Overview of Cyclins

Cell cycle regulation is critically dependent upon genes encoding a family of proteins called cyclins (Murray 2004). Cyclins are regulatory subunits of major cell cycle kinases known as cyclin-dependent kinases (CDKs). A-type cyclins are involved both in S-phase (along with cyclins D and E) and M-phase regulation, while B-type cyclins play pivotal roles in regulating the G2/M transition and in M-phase progression.

The M-phase-related roles of A-type cyclins (A1 and A2) are still not fully understood. Unlike cyclin A1 whose expression is highly restricted, cyclin A2 is ubiquitously expressed and exerts its major role during S-phase when it associates with CDK2 and mediates phosphorylation of critical S-phase regulating substrates (Wolgemuth 2011). Cyclin A1 is expressed during meiosis in both male and female murine germ cells (Sweeney et al. 1996) but is only essential for male meiosis (Liu et al. 1998). Cyclin A2 is expressed in oocytes, with protein levels exhibiting a decrease between the prophase I and MII arrest stages followed by an increase by the pronuclear stage (Winston et al. 2000). In oocytes, cyclin A2 initially localises to the nuclear membrane, becomes predominantly cytoplasmic following meiotic resumption and then becomes symmetrically enriched on opposite sides of the MII-plate perhaps reflecting localisation to the spindle poles (Winston et al. 2000). Exactly how this pattern of cyclin A2 expression and localisation relates to its role in mouse oocytes and early embryos remains obscure. More is known about cyclin A2 function in *Xenopus laevis* in which it associates both with CDK1 and CDK2 and Cyclin A2/CDK1 seems to potentiate cyclin B/CDK1 activation essential for M-phase entry (Devault et al. 1992).

Cyclins B1 and B2 are pivotal for activating the major M-phase kinase, CDK1, the activity of which can be specifically measured using histone H1 kinase assays (Kubiak et al. 1993). Cyclin B1 is essential for murine embryonic development while cyclin B2 is dispensable (Brandeis et al. 1998). The third mammalian B-type cyclin, B3, is the only other mammalian meiosis-specific cyclin apart from cyclin A1

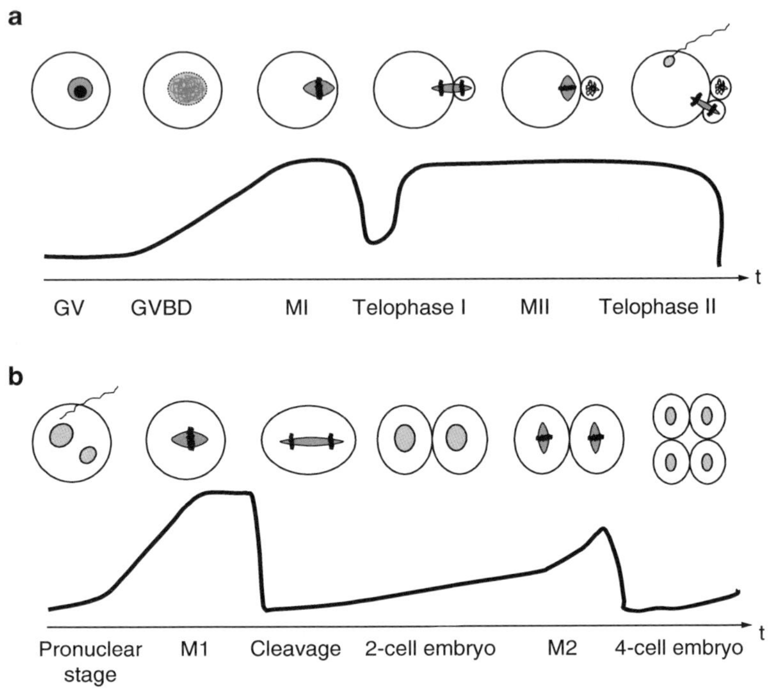

Fig. 4.1 Changes of cyclin B levels during oocyte maturation, fertilisation (**a**) and the first two cleavages of the embryo (**b**)

and associates with CDK2 (Nguyen et al. 2002). Cyclin B3 is expressed in both testes and ovaries becoming maximal at zygotene and decreasing by pachytene, a down-regulation that appears important for normal spermatogenesis (Refik-Rogers et al. 2006).

4.1.2 Time Course of Meiotic Maturation and Fertilisation

Mammalian oocytes are arrested in the ovary at the prophase stage of the first meiotic division characterised by the presence of an intact nucleus (referred to in oocytes as the germinal vesicle or GV). Within each oestrus cycle a cohort of prophase I-arrested oocytes is recruited to undergo further development characterised initially by an extended growth phase during which the oocyte's volume increases over 100-fold. Oocyte growth occurs contemporaneously with development of the oocyte's somatic cell investiture known as the follicle.

In response to hormonal cues, fully grown oocytes contained within dominant antral follicles resume meiotic maturation. The first morphological sign of this process is germinal vesicle breakdown (GVBD, equivalent to nuclear envelope breakdown in somatic cells; Fig. 4.1a) readily visible under the light microscope as the "disappearance" of the GV. Following GVBD, oocytes progress through M-phase of the first meiotic division (MI) and immediately after cytokinesis (marked by extrusion of the first polar body, 1PB), immediately enter the M-phase of the second meiotic division (MII), when they become arrested for a second time at metaphase II, thereby completing meiotic maturation (Masui and Clarke 1979). In mice, the duration of the first meiotic M-phase is strain-specific and usually lasts from 6 to 10 h (Polanski 1997), thus being ~10 times longer than the M-phase of a typical mitotic cell.

Oocyte activation, either by sperm penetration or the application of a parthenogenetic stimulus, relieves the MII arrest, resulting in completion of the second meiotic division (marked by extrusion of the second polar body, 2PB) and entry into interphase of the first embryonic cell cycle manifested by formation of male and female pronuclei (Fig.4.1b). As shown in Fig. 4.1, in the absence of a defined interphase between MI and MII, almost the whole period of meiotic maturation is occupied by the M-phases of the two meiotic divisions. MII arrest is mediated by an activity referred to as cyostatic factor (CSF) and may also last for some hours as there can be a considerable delay before fertilisation occurs (Masui and Markert 1971; Kubiak et al. 1993).

4.2 First Meiotic Division: Modulating CDK1 Activity and Cyclin B1 Levels

4.2.1 Prophase Arrest and Entry Into M-Phase

Associated with the acquisition of meiotic competence during the oocyte's growth phase is a threefold increase in the concentrations of CDK1 and cyclin B1 (Kanatsu-Shinohara et al. 2000). Notably, however, in both meiotically incompetent and competent oocytes, the concentration of cyclin B1 remains about sevenfold in excess of CDK1 (Kanatsu-Shinohara et al. 2000), suggesting that it is the accumulation of CDK1 that is limiting for the acquisition of meiotic competence (Chesnel and Eppig 1995). The repository of CDK1-cyclin B1 is kept inactive in fully grown oocytes in vivo until such time that GVBD and entry into M-phase is triggered by a surge of luteinising hormone (LH).

An important aspect of CDK1-cyclin B1 suppression is to tip the balance in favour of WEE1/MYT kinase family-mediated inhibitory phosphorylation at Thr 14 and Tyr 15 of the CDK1 catalytic subunit over removal of these inhibitory phosporylations by the CDC25 phosphatases (Solc et al. 2010). A key upstream player in setting this inhibitory balance is cyclic adenosine monophosphate (cAMP) whose production depends on the trimeric G-protein (Gs) and the Gs-linked G-protein coupled receptor 3 (GPR3) (Mehlmann et al. 2002; Mehlmann et al. 2004). A direct target of cAMP is protein kinase A (PKA), which serves to inhibit

CDC25B (Zhang et al. 2008; Pirino et al. 2009) on the one hand and to potentiate WEE2 (an oocyte-specific WEE1/MYT kinase family member) (Han et al. 2005) on the other, altogether establishing the inhibitory phosphorylation profile.

Along with inhibitory phosphorylation, another pivotal contributor to suppressing CDK1-cyclin B1 is to restrain the nuclear accumulation of activating cyclin B1. Indeed, forced over-expression of cyclin B1 in GV-stage oocytes compromises the ability to maintain prophase arrest even in the presence of drugs which ordinarily inhibit GVBD by maintaining high levels of cAMP (Ledan et al. 2001; Reis et al. 2006). In recent years it has become evident that a key player involved in preventing unwanted cyclin B1 accumulation is the multi-ubiquitin E3ligase known as the anaphase-promoting complex or cyclosome (APC/C), which earmarks proteins (such as cyclin B1) for destruction by the 26S proteasome (reviewed by Peters 2006). The APC/C in conjunction with its Cdh1 co-activator (APC/C-Cdh1) is indispensable for prophase arrest through nuclear-concentrated cyclin B1 degradation (Reis et al. 2006; Holt et al. 2010). Unexpectedly, this APC/C-Cdh1-dependent mechanism requires BubR1 (Homer et al. 2009), a component of the spindle assembly checkpoint (SAC) whose canonical role centres on regulating the Cdc20-activated APC/C species (APC/C-Cdc20) important for controlling the metaphase-to-anaphase transition (see below). Also required for APC/C-Cdh1 activity and for prophase arrest is the CDC14B phosphatase (Schindler and Schultz 2009) that counters CDK1-mediated inhibitory APC/C-Cdh1 phosphorylation.

M-phase entry triggered by the LH surge is the culmination of a signal transduction cascade conveyed via the surrounding follicular cells that ultimately brings about the activation of nuclear CDK1-cyclin B1 in the oocyte. LH receptor (LHR) expression is not detectable in oocytes (Peng et al. 1991) indicating that the LH effect occurs via the mural granulosa cells where LHR is expressed. LH induces expression of epidermal growth factor (EGF) like factors, notably amphiregulin (AREG) and epiregulin (EREG), which are critical for transducing the LH-mediated signal from granulosa cells to oocytes for inducing GVBD (Park et al. 2004; Hsieh et al. 2007). It appears that a critical downstream target of the LH-EGF pathway is cyclic guanosine monophosphate (cGMP), levels of which decrease in the oocyte after the LH peak resulting in an increase in the activity of the phosphodiesterase, PDE3, and consequently in reduced cAMP levels (Norris et al. 2009; Vaccari et al. 2009).

LH-induced reduction in cAMP levels favours activation of CDK1-cyclin B1 thereby promoting GVBD. Reduced cAMP and hence PKA activity leads to increased CDC25 and reduced WEE1/MYT kinase activities leading to the removal of inhibitory phorphorylations at Thr 14 and Tyr 15 of the CDK1 catalytic subunit (Solc et al. 2010). Added to this, cyclin B1 which is primarily cytoplasmic in GV-intact oocytes translocates to the nucleus (Marangos and Carroll 2004; Reis et al. 2006), an important step in bringing about GVBD (Holt et al. 2010). Furthermore, as CDK1 activity rises, the resulting inhibition of APC/C-Cdh1 would allow for cyclin B1 accumulation and further CDK1 activation.

4.2.2 Regulating M-Phase Progression in Oocytes Through Cyclin B1 Synthesis

As mentioned above, the first meiotic M-phase in mouse oocytes is around ten times longer than a typical mitotic M-phase. This marked discrepancy reflects different levels of cyclin B availability in different cell-types. At the G2/M transition, mitotic cells contain sufficient cyclin B to efficiently and rapidly induce maximal CDK1 activation, which in turn enables fast passage through M-phase. In contrast, in mammalian oocytes (and probably in oocytes of most other phyla), cyclin B availability is restricted so that initial levels are only sufficient to enable entry into M-phase but not to support the execution of all M-phase events. In support of this, initial activation of CDK1 that drives GVBD in mouse oocytes can occur in the absence of protein synthesis (Choi et al. 1991; Hampl and Eppig 1995; Polanski et al. 1998) suggesting that removal of inhibitory phosphorylation from pre-formed inactive CDK1-cyclin B1 complexes is sufficient and that new cyclin B1 synthesis is dispensable for entry into M-phase. Following GVBD, however, protein synthesis inhibitors prevent the attainment of maximal CDK1 activity and severely impair normal spindle assembly and completion of MI (Hampl and Eppig 1995; Polanski et al. 1998). This failure to fully activate CDK1 likely reflects impaired cyclin B1 synthesis since increased cyclin B1 synthesis is strongly correlated with the increase in CDK1 that accompanies maturation (Winston 1997) and over-expressing cyclin B1 augments CDK1 activity (Polanski et al. 1998). Notably, although cyclin B1 destruction is crucial for exit from MI (discussed later), during the CDK1 activation phase of MI, cyclin B1 degradation is extremely low so that net levels are primarily controlled at the level of synthesis (Winston 1997).

The importance of cyclin B1 for determining the nature of meiotic maturation was vividly demonstrated by analyses of oocytes from two inbred strains of mouse, CBA and KE, which differ greatly in the inherent timing of their first meiotic division (Polanski, 1986). In "fast" oocytes of the CBA strain, MI lasts an average of 9 h and is characterised by a constant rate of increase in CDK1 activity over a relatively short interval. In stark contrast, in "slow" oocytes of the KE strain, MI lasts about 12 h during which there is a transient initial increase in CDK1 activity, which then temporarily plateaus before increasing again during a second phase of activation (Polanski et al. 1998). This difference in the kinetics of CDK1 activation during MI in oocytes of these two mouse strains was correlated with significantly greater cyclin B1 accumulation in "fast" CBA oocytes than in "slow" KE oocytes ("Polanski, 1986; Polanski et al. 1998"). Moreover, over-expression of cyclin B1 from exogenous mRNA was sufficient for accelerating passage through MI in "slow" KE oocytes. Thus, the difference in the timing of MI in oocytes from CBA and KE strains is a direct consequence of differences in CDK1 activation patterns that in turn is a direct reflection of de novo cyclin B synthesis. Notably, in keeping with a key role for cyclin B1 synthesis in determining MI progression but less so for entry into M-phase, although CBA and KE oocytes differ markedly in their cyclin B1 and MI progression profiles, they are indistinguishable with regard to the timing of GVBD (Polanski 1997).

On the basis of the foregoing therefore, the mechanism by which oocytes control cyclin B1 synthesis from stored maternal transcripts is critical for regulating maturation. One such mechanism is cytoplasmic polyadenylation-induced translation involving the elongation of short poly(A) tails on dormant maternal mRNAs to up to 100–150 nucleotides. Cytoplasmic polyadenylation is conferred by the cytoplasmic polyadenylation element (CPE) residing in the 3′UTR of responding mRNAs, which is bound by a sequence-specific RNA-binding protein known as CPE binding protein (CPEB). In keeping with polyadenylation-driven cyclin B1 translation, in mouse oocytes, the cyclin B1 3′UTR contains 3 CPE-like sequences, cyclin B1 mRNA undergoes CPE-dependent polyadenylation during MI and sterically impeding access to the CPE-containing region of endogenous cyclin B1 mRNA impairs polyadenylation and progression through MI (Tay et al. 2000). Polyadenylation-dependent translation depends upon Aurora kinase A-mediated CPEB phosphorylation as a dominant negative phosphorylation-resistant CPEB mutant markedly reduces cyclin B1 polyadenylation (Hodgman et al. 2001). Interestingly, in marked contrast to its role in promoting polyadenylation-dependent translation during MI, the CPE appears to exert the opposite role of repressing cyclin B1 translation during the GV stage (Tay et al. 2000) perhaps providing an additional mechanism for preventing CDK1 activation at this stage.

Nuclear-derived factors also appear important for activating CDK1 during MI. Thus, following oocyte bisection to produce nuclear and anuclear fragments, CDK1 undergoes an initial activation which then becomes arrested at an intermediate level in anuclear fragments contrasting with the GV-containing halves in which CDK1 activity rises normally (Hoffmann et al. 2006). The increased CDK1 activation in nuclear fragments was associated with increased levels of cyclin B1 whereas CDK1 levels were similar in nuclear and anuclear halves. Furthermore, the pattern of CDK1 activation in anuclear oocyte fragments is very reminiscent of whole oocytes containing an intact GV which have been exposed to protein synthesis inhibitors. Altogether, this suggests that as-yet-unidentified factor(s) located within the nucleus are crucial for inducing cyclin B1 synthesis in maturing oocytes. This notion is further supported by the finding that the strain-dependent differences in M-phase duration between oocytes from CBA and KE strains requires the presence of the GV (Polanski et al. 1998). Interestingly, further analyses suggest that such nuclear factors increase cyclin B1 translation by targeting the 3′UTR (Hoffmann et al. 2006). Given the contribution of CPEB-dependent polyadenylation to cyclin B1 translation and its dependence upon the 3′UTR, it will be interesting to determine whether components of this machinery reside within the nucleus. It is noteworthy that recent data indicate that CPEB and 3′UTR-mediated translation are critical in mammalian oocytes for regulating the expression not only of cyclin B1 but also of multiple other key regulatory proteins including components of the APC/C and the SAC (Chen et al. 2011). It remains to be seen whether the nuclear dependence illustrated for cyclin B1 expression also applies to these other proteins.

Overall therefore, cyclin B1-dependent progression through the first meiotic M-phase in mouse oocytes relies on two mechanisms to be switched on in a coordinated manner: (1) Activation of a limited store of pre-formed CDK1-cyclin B1 complexes that is sufficient for inducing M-phase entry, and (2) Activation of cyclin B1 synthesis that is important for attaining maximal CDK1 activity necessary for proper execution of all M-phase events and completion of MI.

4.2.3 APC/C-Induced Cyclin B1 Destruction and the Metaphase I-to-Anaphase I Transition and Exit from MI

Whereas M-phase progression depends on cyclin B1 accumulation and consequent CDK1 activation (Winston 1997; Hampl and Eppig 1995; Polanski et al. 1998) as discussed above, exit from M-phase is induced by dramatic cyclin B1 degradation (Kubiak et al. 1993; Herbert et al. 2003; Reis et al. 2007). The mechanism by which cyclin B1 is degraded at M-phase exit involves a highly conserved APC/C-dependent process (Peters 2006; Pesin and Orr-Weaver 2008). The APC/C is itself activated by CDK1-dependent phosphorylation of some of its subunits (e.g. Cdc27) (Kraft et al. 2003). Thus, cyclin B1 accumulation and the resulting CDK1 activation required for M-phase progression also sets in motion the machinery that drives later exit from M-phase through cyclin B1 destruction. This sequence of events underpins the oscillations in CDK1-cyclin B1 responsible for the periodic character of M-phase/interphase (Murray 2004).

Anaphase and M-phase exit are APC/C-mediated processes that must be temporally co-ordinated. Separation of chromosomes at anaphase requires separase-mediated destruction of cohesins, the molecular glue that holds homologous chromosomes (in the case of MI) or sister chromatids (in the case of MII and mitosis) together (Nasmyth 2002; Petronczki et al. 2003). Cohesins resist the forces exerted on chromosomes by spindle microtubules that have become attached to specialised multi-protein complexes assembled on centromeric DNA called kinetochores (Cheeseman and Desai 2008). The APC/C ubiquinates and targets for destruction securin, a molecular chaperone that protects cohesins from cleavage by separase, and cyclin B1, which until this stage had been important for preventing M-phase exit by sustaining CDK1 activity (Peters 2006). Importantly therefore, once the APC/C is kept inactive, both chromosome separation and M-phase exit will be prevented. Following APC/C activation, however, cohesins undergo destruction, thereby resulting in unopposed microtubule-mediated pulling forces on chromosomes and, consequently, in chromosome (or chromatid) movement towards opposite spindle poles. Destruction of the other important APC/C target, cyclin B1, results in CDK1 inactivation (Peters 2006; Pesin and Orr-Weaver 2008) and exit from MI marked by first polar body extrusion.

4.2.4 The SAC in Mammalian Oocytes

In order to achieve correct chromosome segregation, it is imperative that cohesins be destroyed only after kinetochores have acquired properly configured attachments to microtubules (Tanaka 2010). Failure to fulfil this condition could result in improperly attached or unattached kinetochores resulting in the unequal sharing of chromosomes at the time of segregation, culminating in aneuploid daughter cells.

The SAC evolved to prevent aneuploidy by coupling the status of microtubule-kinetochore (MT-K) attachments to the cell cycle engine (Musacchio and Salmon 2007). SAC components are derived primarily from the Mad (for Mitotic arrest-deficient) and Bub (for Budding uninhibited by benzimidazole) protein families and serve to signal from kinetochores to the APC/C. Mad and Bub proteins have the capacity to bind unattached kinetochores and shuttle between such kinetochores and the cytoplasm. Following recruitment to unattached kinetochores, SAC proteins undergo modifications that promote their ability to bind to and sequester the APC/C co-activator, Cdc20 (Musacchio and Salmon 2007). In this way, in the presence of unattached kinetochores, the SAC prevents APC/C activation required for chromosome separation and M-phase exit. Once kinetochores become saturated with microtubules, SAC proteins are displaced from kinetochores thereby making Cdc20 available to bind and activate the APC/C, which triggers anaphase-onset and cell cycle progression.

Results from earlier reports, which questioned the existence of SAC-mediated regulation in mammalian oocytes (LeMaire-Adkins et al. 1997), were subsequently challenged when it was found that cyclin B1 destruction and MI progression were severely delayed by spindle-disrupting agents and that this delay could be overcome by impairing the function of individual SAC components (Brunet et al. 1999; Wassmann et al. 2003; Homer et al. 2005a). Importantly, the functionality of the SAC was indisputably established when it was shown that disruption of any one of a number of different SAC components in unperturbed oocytes accelerated MI progression by 2–4 h as a consequence of premature securin and cyclin B1 destruction, culminating in aneuploidy (Tsurumi et al. 2004; Homer et al. 2005b; Niault et al. 2007; McGuinness et al. 2009; Hached et al. 2011).

Although the foregoing data highlight an absolute requirement of the SAC for accurate chromosome segregation, an important question pertains to whether the oocyte's SAC is physiologically capable of responding to minor kinetochore-microtubule attachment defects. Specifically, a single unattached kinetochore which can efficiently prevent progression to anaphase in small-volume somatic cells (Rieder et al. 1995) might not be capable of generating an inhibitory signal sufficient for suppressing all of the APC/C throughout the much larger oocyte volume (the volume of a mouse oocyte is ~270 pL, whereas that of a PtK1 cell is only ~6 pL). In support of this notion, nocodazole-induced cell cycle arrest in *Xenopus* eggs cannot be invoked until a threshold mass of chromosomes (and hence kinetochores) has been exceeded (Minshull et al. 1994). Furthermore, recent work in recombination-deficient mutant mouse oocytes showed that a minority of misaligned chromosomes did not prevent progression to MII (Nagaoka et al. 2011). One possible explanation for the latter observations is that the SAC signal generated from a small number of incompletely attached kinetochores was not sufficient for halting MI progression, supporting a model of compromised SAC efficiency secondary to SAC signal "dilution" within the oocyte's large volume. However, such an interpretation would be speculative at present as SAC protein recruitment to kinetochores on misaligned chromosomes was not ascertained (Nagaoka et al. 2011). Indeed, a very recent study showed that within oocyte-halves a single bivalent was equally as efficient as 19 bivalents at setting the proper timing of cyclin B1 destruction and at preventing cyclin B1 destruction during MI (Hoffmann et al. 2011). Thus, at least when the oocyte volume is halved, a single unattached bivalent possessing four kinetochores can induce a robust SAC-induced delay—it remains to be seen whether the same holds true in whole oocytes with twice the volume.

Whilst embryonic aneuploidy resulting from suboptimal SAC activity could potentially compromise reproductive performance (Homer 2011), SAC over-activation could also be deleterious. Thus, in the LT/Sv mouse strain, the majority of oocytes become permanently arrested at the first meiotic metaphase (Eppig 1978; O'Neill and Kaufman 1987; Ciemerych and Kubiak 1998). Recent studies have shown that the metaphase arrest underpinning reduced fertility in these oocytes is brought about by the SAC (Hupalowska et al. 2008; Maciejewska et al. 2009; Hoffmann, Krol and Polanski, in preparation). Although one might reasonably assume that such SAC hyper-activation might occur in response to overt kinetochore-microtubule attachments defects, immunofluorescence analysis of LT/Sv oocytes has not revealed any obvious abnormalities in spindle structure (Ciemerych and Kubiak 1998; Hupalowska et al. 2008). It remains possible that much more subtle defects, undetectable by immunofluorescence, are responsible. It will be important to understand the mechanism underpinning the LT/Sv MI arrest phenotype as this could shed light on some of the causes of human infertility that are associated with recurrent oocyte metaphase arrest (e.g. Harrison et al. 2000; Bergère et al. 2001; Schmiady and Neitzel 2002). In another strain of mice, namely 129/Sv, an activated SAC arrests many oocytes at MI during in vitro culture, probably reflecting the specific susceptibility of oocytes from this strain to spindle anomalies under suboptimal culture conditions (Hoffmann, Kubiak and Polanski, unpublished). This phenotype in 129/Sv oocytes is consistent with the protracted SAC-mediated MI arrest that is the general rule in mouse oocytes treated with drugs such as nocodazole which depolymerise the spindle thereby disrupting kinetochore-microtubule interactions (Homer et al. 2005a).

4.2.5 Restraining Cyclin B1 Destruction at the MI-to-MII Transition

Although cyclin B is degraded at the metaphase I-to-anaphase I transition by APC/C-Cdc20, causing a rapid decline in CDK1 activity, cyclin B synthesis does not cease as is the case during all other cell-division types. Notably, marked APC/

C-induced degradation is strictly limited to a short window as a consequence of APC/C inhibition brought about by early mitotic inhibitor 2 (Emi2) (Madgwick et al. 2006).Continuing protein synthesis combined with a restricted period of proteolysis towards the end of MI in the oocyte results in only a transient decrease in cyclin B1. This is crucial for enabling cyclin B levels to re-accumulate after MI thereby restoring high levels of CDK1 activity important for assembling the second meiotic spindle within 1–2 h of exit from MI (Kubiak et al. 1992; Madgwick et al. 2006).

At the MI/MII transition, cyclin B1 is not completely destroyed (Fig.4.1a). Thus, a limited pool of cyclin B1 is detectable at the midbody between the first polar body and the oocyte (Huo et al. 2005). The residual histone H1 kinase activity related to this pool of cyclin B could facilitate reactivation of CDK1 upon MII entry as occurs in Xenopus laevis oocytes at the equivalent meiotic stage (Iwabuchi et al. 2000). Interestingly, however, Huo and colleagues (2005) found that although CDK1 localised to the spindle in mouse oocytes at anaphase/telophase of MI, it was specifically excluded from the midbody. Thus, although detectable levels of cyclin B1 remain upon exit from MI in mouse oocytes, it is unresolved whether this is important for sustaining residual levels of CDK1 activity at the MI/MII transition and for rapid activation of this kinase upon MII entry.

Compared with MI therefore, the time taken for the attainment of an advanced M-phase status and bipolar spindle reformation in MII is considerably shorter. This is perhaps related to the fact that the oocyte does not completely lose its M-phase character during the MI-to-MII transition when the chromatin remains condensed, microtubules remain short and restricted to the spindle area, cyclin B synthesis is sustained and the activity of another important meiotic kinase, MAP kinase, remains high (Verlhac et al. 1994, 1996). A number of additional mechanisms may contribute to the integrity of MII spindle assembly such as the Ran pathway, which could promote efficient bipolar spindle assembly (Dumont et al. 2007) as well as MII-specific factors such as MISS (Lefebvre et al. 2002) and DOC1R (Terret et al. 2003) that stabilise MII spindle structure.

The comparatively long first meiotic M-phase in oocytes might be needed to provide sufficient time for properly bi-orienting bivalent kinetochores prior to anaphase I, a process involving unique kinetochore configurations in which sister kinetochores co-orient towards the same pole and not to opposite poles as in mitosis (Hauf and Watanabe 2004). In support of this, recent high-resolution tracking of kinetochores in living oocytes revealed that bi-orientation at MI is a slow process during acentrosomal spindle assembly often necessitating several unsuccessful attempts (Kitajima et al. 2011).

4.3 Second Meiotic Division

As CDK1 activity rises relatively quickly upon entering MII, one might expect APC/C reactivation and re-establishment of cyclin B1 degradation. However, the degree of APC/C activation is limited as a consequence of CSF, a unique APC/C inhibitory activity present during the second meiotic M-phase in oocytes from higher chordates (including almost all mammals) critical for sustaining cyclin B1 levels and hence CDK1 activity. CSF maintains the oocyte at metaphase II until sperm penetration thereby ensuring that subsequent stages of embryonic development are synchronised with fertilisation. Notably, however, although cyclin B levels are sustained during MII arrest, cyclin B1 is continually turning over, being both synthesised and degraded (Kubiak et al. 1993; Winston 1997; Nixon et al. 2002). This dynamic cyclin B1 equilibrium is rapidly shifted towards net destruction in the presence of puromycin and to net synthesis when the spindle is destroyed by microtubule-depolymerising drugs (Kubiak et al. 1993; Winston 1997; Nixon et al. 2002), the latter showing that cyclin B1 degradation during MII-arrest is dependent on the presence of an intact spindle. These experiments further indicate that during MII arrest, CSF does not completely prevent, but rather slows down cyclin B1 degradation and that the SAC also has the capacity to modulate cyclin B levels.

On the basis of its APC/C inhibitory capacity, the SAC appeared a good candidate for contributing to CSF activity. Indeed, spindle disruption or overexpression of Mad2 in mouse eggs prevent cyclin B1 proteolysis and exit from MII following parthenogenetic activation showing that when activated, the SAC has the capacity to sustain an MII arrest (Tsurumi et al. 2004; Madgwick et al. 2005; Madgwick et al. 2006). Ordinarily, however, when spindle formation is not disrupted, the SAC appears to be dispensable for establishing CSF arrest as expression of dominant negative mutants of Mad2, BubR1 or Bub1 during meiotic maturation do not prevent oocytes from undergoing an MII arrest (Tsurumi et al. 2004). Notably, Mad2 and Bub1 are detectable at kinetochores of MII-arrested oocytes (Kallio et al. 2000; Brunet et al. 2003). Given that kinetochore localisation of SAC proteins is widely considered to be a marker of SAC signalling, these data suggest that the SAC may still have some role in MII arrest. Interestingly, more recent data have shown that kinetochore recruitment of Mad2 is maximal during the initial hours of MII-arrest and then dissipates (Sikora-Polaczek et al. 2006). Taken together therefore, it may be that the SAC helps to improve the efficiency of APC/C inhibition on entry into MII as suggested previously (Madgwick and Jones 2007). Given the importance of cyclin B1 in maintaining MII arrest, it is not surprising that multiple inputs, perhaps including the SAC, along with Emi2 and Mos-MAPK collaborate to prevent its destruction.

Apart from the continuous turn-over of cyclin B1, another aspect that highlights the highly dynamic nature of MII-arrest pertains to the localisation of the MII spindle. Thus, very recently, it was elegantly shown that maintaining an asymmetric spindle position subjacent to the oocyte cortex, whilst appearing to be a quiescent state, in fact requires a dynamic equilibrium of forces (Yi et al. 2011). Using spatiotemporal image correlation spectroscopy (STICS) analysis, the Li lab showed that cortical spindle anchoring involves the Arp2/3-dependent flow of actin filaments that produces streaming of cytoplasmic particles in a direction that pushes the spindle away from the oocyte centre and towards the cortex. This Arp2/3-induced cytoplasmic streaming appears important for counteracting reverse cytoplasmic streaming secondary to myosin II-induced cap contractility that would otherwise disrupt cortical spindle positioning. It has been suggested that such an active mechanism requiring high energy expenditure for sustaining spindle positioning and hence subsequent successful fertilisation might constitute a form of natural selection for the most energetically competent oocytes that are best equipped for sustaining embryonic development (Yi et al. 2011).

4.4 Molecular Correlates of Age-Related Aneuploidy and Reproductive Failure in Humans

Accurate chromosome segregation is crucially important during gametogenesis as aneuploid gametes produce aneuploid embryos, the majority of which are incompatible with life (Hassold and Hunt 2001). The vast majority of aneuploidy is the consequence of errors arising specifically during MI in oocytes with relatively minor contributions from spermatogenic errors (Hassold and Hunt 2001). Notably, the incidence of these female MI segregation errors rises dramatically with advancing female age (Hassold and Hunt 2001; Homer 2011). Given the importance of the SAC in averting aneuploidy during MI in oocytes, one popular hypothesis has been that age-related declines in SAC integrity in oocytes might be an important contributor to the age-related rise in female MI errors (Shonn et al. 2002). In support of declining SAC function with advancing female age, SAC components have been reported to be less abundant in oocytes and ovaries from older mice (Baker et al. 2004; Pan et al. 2008) as well as in oocytes from older women (Steuerwald et al. 2001; Steuerwald et al. 2007). However, the functional significance of this decline in expression with regard to SAC integrity in older oocytes was not borne out by studies which showed that oocytes from older murine females did not exhibit a more rapid transit through MI as would be predicted of SAC compromise (Duncan et al. 2009). Overall, therefore, the contribution of SAC dysfunction to age-related aneuploidy remains unresolved.

More recently, several reports have focused on cohesins and how this molecular glue that is important for the integrity of bivalent chromosomes might be affected during the extremely protracted meiotic prophase I-arrest stage that mammalian oocytes experience (prophase I arrest lasts about 1 year in mice depending on the strain, and for a staggering 40–50 years in humans). Recent data show that the integrity of bivalent-associated cohesins, which are loaded onto chromosomes in foetal life during the predictyate stage of prophase I, is compromised with advancing age in oocytes culminating in aneuploidy (Liu and Keefe 2008; Lister et al. 2010; Chiang et al. 2010). This age-induced loss of chromosomal cohesin is exacerbated by the lack of a mechanism for cohesin regeneration. Thus, bivalents in mice engineered to endogenously express a TEV-containing version of the Rec8 subunit of meiotic cohesin are not protected from TEV protease cleavage by expression of a non-TEV-cleavable Rec8 from a BAC transgene during the oocyte's growth phase (Tachibana-

Konwalski et al. 2010). Furthermore, deleting the gene encoding another meiotic cohesin subunit, SMC1β, shortly after birth, does not affect the structure of bivalents or chromosome segregation during MI (Revenkova et al. 2010).

4.5 Cyclin B1 in Embryonic M-Phase Regulation

The first embryonic division takes place about 24 h post-fertilisation, with the second one following about 18 h later. The M-phases of the first two embryonic divisions bear similarities to M-phases in oocytes insofar as they are longer than M-phases in somatic cells (albeit not as long as that of MI in oocytes) and the mode of spindle formation, at least during the first cleavage division, is similar to the meiotic model (Ciemerych et al. 1999; Kubiak and Ciemerych 2001; Sikora-Polaczek et al. 2006; Maciejewska et al. 2009). Subsequent M-phases beyond the second mitotic division become progressively shorter until, after several cleavage divisions, they have acquired a duration that is comparable with the M-phase of somatic cells. Thus, M-phases during the embryonic divisions exhibit a change in character from the earliest ones, which are the product of mixed meiotic/mitotic regulation to the latter ones, which are exclusively mitotic (see also Courtois and Hiiragi 2012 this volume).

The first embryonic M-phase of the mouse zygote is almost twice as long as the second one (Ciemerych et al. 1999). The striking prolongation of the first mitotic M-phase reflects a specific delay at metaphase (of ~45 min) that may involve meiosis-related activities but appears to be SAC-independent as Mad2 remains undetectable at kinetochores for the duration of the arrest (Kubiak et al. 2008; Sikora-Polaczek et al. 2006). The protracted nature of the first embryonic division compared with the second seems common to embryos from different species including *Xenopus laevis* (Chesnel et al. 2005a), sea urchin and *C. elegans* (Chesnel et al. 2005b) although the difference between the first and second divisions is more subtle than in the mouse and embryos from these species do not arrest transiently in metaphase.

As with MI and MII in oocytes, the timing of M-phases in early embryos is strictly regulated and is dependent upon cyclin B1 metabolism. Thus, the first embryonic divisions in *Xenopus*, sea urchin and *C elegans* are characterised by high levels of cyclin B accumulation resulting in high CDK1 activity and a longer M-phase. Consequently, by changing the levels of cyclin B during the first embryonic mitosis in *X. laevis* one can modify the duration of M-phase; increasing cyclin B levels prolongs and reducing cyclin B levels shortens M-phase (Chesnel et al. 2005a). Consistent with this, during the shorter second embryonic M-phase, cyclin B levels are lower than during the first one (Chesnel et al. 2005a).

M-phase prolongation during early embryonic divisions in the mouse also seems to be related to cyclin B levels. G2-stage zygotes and 2-cell-stage embryos contain comparable levels of cyclin B, indicating that each blastomere of a 2-cell-stage embryo contains approximately half as much cyclin B as that in a single zygote (Ciemerych et al. 1998). Consequently, the proportion of cyclin B per mitotic spindle containing a diploid number of chromosomes is halved by the second mitosis in keeping with which, histone H1 kinase activity is twice as high during the first mitotic division as it is in a single blastomere undergoing the second division (Ciemerych et al. 1999). Given that in mouse oocytes a large proportion of cyclin B (Huo et al. 2004) and histone H1 kinase activity (Kubiak et al. 1993) accumulates in the region of the meiotic spindle, it is conceivable (albeit not directly proven) that the halving of cyclin B and histone H1 kinase activity per blastomere compared with the zygote may result in a lowering of the effective CDK1 levels, thereby accounting for the reduced M-phase duration.

In *Xenopus* and mouse embryos, regulatory pathways that impact cyclin B seem to play a role in modulating M-phase duration. Thus, depletion of Mad2 from cell-free extracts of *X. laevis* embryos shortens the duration of the first embryonic M-phase suggesting that reduced cyclin B levels secondary to increased APC/C-mediated cyclin B degradation culminated in a shortened M-phase altogether supporting the assertion that cyclin B metabolism is a critical factor underpinning M-phase duration (Chesnel et al. 2005b). Notably, since this effect was observed in a cell-free mitotic extract in which chromosomes were absent, this suggests the involvement of cytoplasmic Mad2 and not kinetochore-associated Mad2. Although comparable experiments are lacking in mouse embryos, Mad2 accumulates on kinetochores for a very similar period of approximately 20–30 min during prometaphase of the first two embryonic divisions, suggesting that SAC activity disappears after a similar delay in both mitoses (Sikora-Polaczek et al. 2006). After Mad2 is displaced, the first and second mitoses enter metaphase, but, as mentioned above, only anaphase of the second embryonic division proceeds immediately, while anaphase of the first embryonic division is delayed for further 45 min. Whilst canonical SAC activity seems not to be involved in this 45 min delay insofar as Mad2 is absent from kinetochores, on the basis of findings in *Xenopus* extracts discussed above, one cannot exclude a role for the cytoplasmic pool of Mad2.

Alternatively, the prolongation of the first embryonic M-phase in the mouse could reflect residual meiotic activities derived from the MII-arrest state. One possible candidate is the ERK 1/2 MAP kinase pathway. As judged by mobility shifts on PAGE gels, p90rsk, the downstream target of ERK 1/2 is fully phosphorylated during MII-arrest, but only partially phosphorylated during the first embryonic mitosis ultimately becoming unphosphorylated by the second mitosis (Kalab et al. 1996). This supports the notion of a progressive modification of the mode of regulation in a meiotic-to-mitotic direction during mouse embryonic divisions as has previously been postulated (Kubiak et al. 2008).

Although the SAC itself does not seem to be involved in the specific prolongation of the M-phases of early embryos, it nevertheless exerts its canonical role of delaying anaphase-onset until spindle formation is complete (e.g. in LT/Sv embryos; Maciejewska et al. 2009). This is an important issue since erroneous segregation of chromosomes, both in MI and MII (Fragouli et al. 2011) as well as during the earliest embryonic cell divisions may impact infertility and aneuploidy to similar extents because both affect all or at least a high number of cells of a developing embryo. If the first few cell cycles during development do indeed depend on mixed meiotic/mitotic control mechanisms (Sikora-Polaczek, et al. 2006; Kubiak et al. 2008; Maciejewska et al. 2009), the unresolved questions regarding SAC efficiency in oocytes apply equally well to early embryos.

4.6 Conclusions

Cyclin B1 plays a central role in the cell cycle control of mouse oocytes and embryos via the activation and inactivation of CDK1. Cyclin B1 is absolutely necessary for embryo development, whilst the *CCNB2* gene encoding cyclin B2 is dispensable and seems to be largely compensated for by cyclin B1. The key events of oocyte maturation and early embryonic development are regulated by timely accumulation and degradation of cyclin B1. Thus, GVBD requires sufficient levels of cyclin B1 for establishing a threshold of CDK1 activity capable of promoting M-phase entry. Numerous cytoplasmic and nuclear factors converge to sustain the prophase arrest stage and to facilitate transduction of the LH-induced signal into reduced oocyte cAMP levels that ultimately promote CDK1 activation required for GVBD. Cyclin B1 accumulation during MI is influenced by nuclear factors acting on the 3′UTR region of cyclin B mRNA, whilst APC/C-mediated cyclin B1 degradation is held in check until the MI/MII transition. The dynamics of cyclin B1 accumulation in turn determine the kinetics of oocyte maturation resulting in "fast" and "slow" strains of mice. The timing of the MI/MII transition is set by the SAC, which in oocytes might be somewhat compromised by the inability to detect small numbers of misaligned chromosomes. During MII-arrest cyclin B1 is more dynamic than during MI and is subject to simultaneous synthesis and degradation, the extent of degradation being limited by CSF, which is disabled by fertilisation thereby allowing the transition into the embryonic mitotic divisions. During embryonic mitoses cyclin B accumulates to different levels, which correlate with the duration of different embryonic M-phases. Clearly therefore, cyclin B and its regulators are key determinants of female reproductive capacity by virtue of impacting elements such as the accuracy of chromosome segregation and hence embryonic aneuploidy. As we continue to expand our understanding of how cyclin B regulatory mechanisms operate within the unique context of the large oocyte volume we will undoubtedly get closer to understanding how human aneuploidies, such as those that underpin Down syndrome, come about.

Acknowledgements While writing this article, JZK was supported by a grant from ARC. HH is supported by a Wellcome Trust Fellowship. ZP was supported by the Polish National Science Centre (grant N° DEC-2011/B/NZ3/00190).

References

Baker DJ, Jeganathan KB, Cameron JD, Thompson M, Juneja S, Kopecka A, Kumar R, Jenkins RB, de Groen PC, Roche P, van Deursen JM (2004) BubR1 insufficiency causes early onset of aging-associated phenotypes and infertility in mice. Nat Genet 36:744–749

Bergère M, Lombroso R, Gombault M, Wainer R, Selva J (2001) An idiopathic infertility with oocytes metaphase I maturation block: case report. Hum Reprod 16:2136–2138

Brandeis M, Rosewell I, Carrington M, Crompton T, Jacobs MA, Kirk J, Gannon J, Hunt T (1998) Cyclin B2-null mice develop normally and are fertile whereas cyclin B1-null mice die in utero. Proc Natl Acad Sci USA 95:4344–4349

Brunet S, Santa MA, Guillard P, Dujardin D, Kubiak JZ, Maro B (1999) Kinetochore fibers are not involved in the formation of the first meiotic spindle in mouse oocytes, but control the exit from the first meiotic M-phase. J Cell Biol 146(1):1–11

Brunet S, Pahlavan G, Taylor S, Maro B (2003) Functionality of the spindle checkpoint during the first meiotic division of mammalian oocytes. Reproduction 126:443–450

Cheeseman IM, Desai A (2008) Molecular architecture of the kinetochore-microtubule interface. Nat Rev Mol Cell Biol 9:33–46

Chen J, Melton C, Suh N, Oh J, Horner K, Xie F, Sette C, Blelloch R, Conti M (2011) Genome-wide analysis of translation reveals a critical role for deleted in azoospermia-like (*Dazl*) at the oocyte-to-zygote transition. Genes Dev 25:755–766

Chesnel F, Eppig JJ (1995) Synthesis and accumulation of p34cdc2 and cyclin B in mouse oocytes during acquisition of competence to resume meiosis. Mol Reprod Dev 40:503–508

Chesnel F, Vignaux F, Richard-Parpaillon L, Huguet A, Kubiak JZ (2005a) Differences in regulation of the first two M-phases in Xenopus laevis embryo cell-free extracts. Dev Biol 285:358–375

Chesnel F, Gautier I, Richard-Parpaillon L, Kubiak JZ (2005b) Each mitosis can be different: how the cell cycle machinery modulates early embryonic M-phases. In: Tokumoto T (ed) New impact of protein modifications in the regulation of reproductive system. Research Signpost, Kerala, India

Chiang T, Duncan FE, Schindler K, Schultz RM, Lampson MA (2010) Evidence that weakened centromere cohesion is a leading cause of age-related aneuploidy in oocytes. Curr Biol 20:1522–1528

Choi T, Aoki F, Mori M, Yamashita M, Nagahama Y, Kohmoto K (1991) Activation of p34cdc2 protein kinase activity in meiotic and mitotic cell cycles in mouse oocytes and embryos. Development 113:789–795

Ciemerych MA, Kubiak JZ (1998) Cytostatic activity develops during meiosis I in oocytes of LT/Sv mice. Dev Biol 200:198–211

Ciemerych MA, Tarkowski AK, Kubiak JZ (1998) Autonomous activation of histone H1 kinase, cortical activity and microtubule organization in one- and two-cell cycle mouse embryo. Biol Cell 90:557–564

Ciemerych MA, Maro B, Kubiak JZ (1999) Control of duration of first two mitoses in a mouse embryo. Zygote 7:293–300

Clarke HJ (2012) Post-transcriptional control of gene expression during mouse oogenesis. In: Kubiak JZ (ed) Mouse development, Results Probl Cell Differ. 55., 1–21. Springer, Heidelberg

Courtois A, Hiiragi T (2012) Gradual meiosis-to-mitosis transition in the early mouse embryo. In: Kubiak JZ (ed) Mouse development, Results Probl Cell Differ. 55., 107–114. Springer, Heidelberg

Devault A, Fesquet D, Cavadore JC, Garrigues AM, Labbé JC, Lorca T, Picard A, Philippe M, Dorée M (1992) Cyclin A potentiates maturation-promoting factor activation in the early Xenopus embryo via inhibition of the tyrosine kinase that phosphorylates cdc2. J Cell Biol 118:1109–1120

Dumont J, Petri S, Pellegrin F, Terret ME, Bohnsack MT, Rassinier P, Georget V, Kalab P, Gruss OJ, Verlhac MH (2007) A centriole- and RanGTP-independent spindle assembly pathway in meiosis I of vertebrate oocytes. J Cell Biol 176:295–305

Duncan FE, Chiang T, Schultz RM, Lampson MA (2009) Evidence that a defective spindle assembly checkpoint is not the primary cause of maternal age-associated aneuploidy in mouse eggs. Biol Reprod 81:768–776

Eppig JJ (1978) Developmental potential of LT/Sv parthenotes derived from oocytes matured in vivo and in vitro. Dev Biol 65:244–249

Fragouli E, Alfarawati S, Goodall NN, Sánchez-García JF, Colls P, Wells D (2011) The cytogenetics of polar bodies: insights into female meiosis and the diagnosis of aneuploidy. Mol Hum Reprod 118(3):381

Gerhart J, Wu M, Kirschner M (1984) Cell cycle dynamics of an M-phase-specific cytoplasmic factor in *Xenopus laevis* oocytes and eggs. J Cell Biol 98:1247–1255

Greenwood J, Costanzo V, Robertson K, Hensey C, Gautier J (2001) Responses to DNA damage in Xenopus: cell death or cell cycle arrest. Novartis Found Symp 237:221–330; discussion 230–234

Hached K, Xie SZ, Buffin E, Cladière D, Rachez C, Sacras M, Sorger PK, Wassmann K (2011) Mps1 at kinetochores is essential for female mouse meiosis I. Development 138:2261–2271

Hamatani T, Carter MG, Sharov AA, Ko MS (2004) Dynamics of global gene expression changes during mouse preimplantation development. Dev Cell 6:117–131

Hampl A, Eppig JJ (1995) Translational regulation of the gradual increase in histone H1 kinase activity in maturing mouse oocytes. Mol Reprod Dev 40:9–15

Han SJ, Chen R, Paronetto MP, Conti M (2005) Wee1B is an oocyte-specific kinase involved in the control of meiotic arrest in the mouse. Curr Biol 15:1670–1676

Harrison K, Sherrin D, Keeping J (2000) Repeated oocyte maturation block. J Assist Reprod Genet 17:231–233

Hartwell LH, Weinert TA (1989) Checkpoints: controls that ensure the order of cell cycle events. Science 246:629–634

Hassold T, Hunt P (2001) To err (meiotically) is human: the genesis of human aneuploidy. Nat Rev Genet 2:280–291

Hauf S, Watanabe Y (2004) Kinetochore orientation in mitosis and meiosis. Cell 119:317–327

Herbert M, Levasseur M, Homer H, Yallop K, Murdoch A, McDougall A (2003) Homologue disjunction in mouse oocytes requires proteolysis of securin and cyclin B1. Nat Cell Biol 5:1023–1025

Hodgman R, Tay J, Mendez R, Richter JD (2001) CPEB phosphorylation and cytoplasmic polyadenylation are catalyzed by the kinase IAK1/Eg2 in maturing mouse oocytes. Development 128:2815–2822

Hoffmann S, Tsurumi C, Kubiak JZ, Polanski Z (2006) Germinal vesicle material drives meiotic cell cycle of mouse oocyte through the 3′UTR-dependent control of cyclin B1 synthesis. Dev Biol 292:46–54

Hoffmann S, Maro B, Kubiak JZ, Polanski Z (2011) A single bivalent efficiently inhibits cyclin B1 degradation and polar body extrusion in mouse oocytes indicating robust SAC during female meiosis I. PLoS One 6:e27143

Holt JE, Weaver J, Jones KT (2010) Spatial regulation of APCCdh1-induced cyclin B1 degradation maintains G2 arrest in mouse oocytes. Development 137:1297–1304

Homer H (2011) New insights into the genetic regulation of homologue disjunction in mammalian oocytes. Cytogenet Genome Res 133:209–222

Homer HA, McDougall A, Levasseur M, Murdoch AP, Herbert M (2005a) Mad2 is required for inhibiting securin and cyclin B degradation following spindle depolymerisation in meiosis I mouse oocytes. Reproduction 130:829–843

Homer HA, McDougall A, Levasseur M, Yallop K, Murdoch AP, Herbert M (2005b) Mad2 prevents aneuploidy and premature proteolysis of cyclin B and securin during meiosis I in mouse oocytes. Genes Dev 19:202–207

Homer H, Gui L, Carroll J (2009) A spindle assembly checkpoint protein functions in prophase I arrest and prometaphase progression. Science 326:991–994

Hsieh M, Lee D, Panigone S, Horner K, Chen R, Theologis A, Lee DC, Threadgill DW, Conti M (2007) Luteinizing hormone-dependent activation of the epidermal growth factor network is essential for ovulation. Mol Cell Biol 27:1914–1924

Huo LJ, Fan HY, Zhong ZS, Chen DY, Schatten H, Sun QY (2004) Ubiquitin-proteasome pathway modulates mouse oocyte meiotic maturation and fertilization via regulation of MAPK cascade and cyclin B1 degradation. Mech Dev 121:1275–1287

Huo LJ, Yu LZ, Liang CG, Fan HY, Chen DY, Sun QY (2005) Cell-cycle-dependent subcellular localization of cyclin B1, phosphorylated cyclin B1 and p34cdc2 during oocyte meiotic maturation and fertilization in mouse. Zygote 13:45–53

Hupalowska A, Kalaszczynska I, Hoffmann S, Tsurumi C, Kubiak JZ, Polanski Z, Ciemerych MA (2008) Metaphase I arrest in LT/Sv mouse oocytes involves the spindle assembly checkpoint. Biol Reprod 79(6):1102–1110

Iwabuchi M, Ohsumi K, Yamamoto TM, Sawada W, Kishimoto T (2000) Residual Cdc2 activity remaining at meiosis I exit is essential for meiotic M-M transition in Xenopus oocyte extracts. EMBO J 19:4513–4523

Kalab P, Kubiak JZ, Verlhac MH, Colledge WH, Maro B (1996) Activation of p90rsk during meiotic maturation and first mitosis in mouse oocytes and eggs: MAP kinase-independent and -dependent activation. Development 122:1957–1964

Kallio M, Eriksson JE, Gorbsky GJ (2000) Differences in spindle association of the mitotic checkpoint protein Mad2 in mammalian spermatogenesis and oogenesis. Dev Biol 225:112–123

Kanatsu-Shinohara M, Schultz RM, Kopf GS (2000) Acquisition of meiotic competence in mouse oocytes: absolute amounts of p34(cdc2), cyclin B1, cdc25C, and wee1 in meiotically incompetent and competent oocytes. Biol Reprod 63:1610–1616

Kitajima TS, Ohsugi M, Ellenberg J (2011) Complete kinetochore tracking reveals error-prone homologous chromosome biorientation in mammalian oocytes. Cell 146:568–581

Kraft C, Herzog F, Gieffers C, Mechtler K, Hagting A, Pines J, Peters JM (2003) Mitotic regulation of the human anaphase-promoting complex by phosphorylation. EMBO J 22:6598–6609

Kubiak JZ, Ciemerych MA (2001) Cell cycle regulation in early mouse embryos. Novartis Found Symp 237:79–89; discussion 89–99

Kubiak JZ, Weber M, Géraud G, Maro B (1992) Cell cycle modification during the transitions between meiotic M-phases in mouse oocytes. J Cell Sci 102:457–467

Kubiak JZ, Weber M, de Pennart H, Winston NJ, Maro B (1993) The metaphase II arrest in mouse oocytes is controlled through microtubule-dependent destruction of cyclin B in the presence of CSF. EMBO J 12:3773–3778

Kubiak JZ, Ciemerych MA, Hupalowska A, Sikora-Polaczek M, Polanski Z (2008) On the transition from the meiotic to mitotic cell cycle during early mouse development. Int J Dev Biol 52:201–217

Ledan E, Polanski Z, Terret ME, Maro B (2001) Meiotic maturation of the mouse oocyte requires an equilibrium between cyclin B synthesis and degradation. Dev Biol 232:400–413

Lefebvre C, Terret ME, Djiane A, Rassinier P, Maro B, Verlhac MH (2002) Meiotic spindle stability depends on MAPK-interacting and spindle-stabilizing protein (MISS), a new MAPK substrate. J Cell Biol 157:603–613

LeMaire-Adkins R, Radke K, Hunt PA (1997) Lack of checkpoint control at the metaphase/anaphase transition: a mechanism of meiotic nondisjunction in mammalian females. J Cell Biol 139:1611–1619

Lister LM, Kouznetsova A, Hyslop LA, Kalleas D, Pace SL, Barel JC, Nathan A, Floros V, Adelfalk C, Watanabe Y, Jessberger R, Kirkwood TB, Höög C, Herbert M (2010) Age-related meiotic segregation errors in mammalian oocytes are preceded by depletion of cohesin and Sgo2. Curr Biol 20:1511–1521

Liu L, Keefe DL (2008) Defective cohesin is associated with age-dependent misaligned chromosomes in oocytes. Reprod Biomed Online 16:103–112

Liu D, Matzuk MM, Sung WK, Guo Q, Wang P, Wolgemuth DJ (1998) Cyclin A1 is required for meiosis in the male mouse. Nat Genet 20:377–380

Maciejewska Z, Polanski Z, Kisiel K, Kubiak JZ, Ciemerych MA (2009) Spindle assembly checkpoint-related failure perturbs early embryonic divisions and reduces reproductive performance of LT/Sv mice. Reproduction 137:931–942

Madgwick S, Jones KT (2007) How eggs arrest at metaphase II: MPF stabilisation plus APC/C inhibition equals Cytostatic Factor. Cell Div 2:4

Madgwick S, Levasseur M, Jones KT (2005) Calmodulin-dependent protein kinase II, and not protein kinase C, is sufficient for triggering cell-cycle resumption in mammalian eggs. J Cell Sci 118:3849–3859

Madgwick S, Hansen DV, Levasseur M, Jackson PK, Jones KT (2006) Mouse Emi2 is required to enter meiosis II by reestablishing cyclin B1 during interkinesis. J Cell Biol 174:791–801

Marangos P, Carroll J (2004) The dynamics of cyclin B1 distribution during meiosis I in mouse oocytes. Reproduction 128:153–62

Masui Y, Clarke HJ (1979) Oocyte maturation. Int Rev Cytol 57:185–282

Masui Y, Markert CL (1971) Cytoplasmic control of nuclear behavior during meiotic maturation of frog oocytes. J Exp Zool 177:129–145

McGuinness BE, Anger M, Kouznetsova A, Gil-Bernabé AM, Helmhart W, Kudo NR, Wuensche A, Taylor S, Hoog C, Novak B, Nasmyth K (2009) Regulation of APC/C activity in oocytes by a Bub1-dependent spindle assembly checkpoint. Curr Biol 19:369–380

Mehlmann LM, Jones TL, Jaffe LA (2002) Meiotic arrest in the mouse follicle maintained by a Gs protein in the oocyte. Science 297:1343–1345

Mehlmann LM, Saeki Y, Tanaka S, Brennan TJ, Evsikov AV, Pendola FL, Knowles BB, Eppig JJ, Jaffe LA (2004) The Gs-linked receptor GPR3 maintains meiotic arrest in mammalian oocytes. Science 306:1947–1950

Minshull J, Sun H, Tonks NK, Murray AW (1994) A MAP kinase-dependent spindle assembly checkpoint in Xenopus egg extracts. Cell 79:475–486

Morris SA, Zernicka-Goetz M (2012) Formation of distinct cell types in the mouse blastocyst. In: Kubiak JZ (ed) Mouse development, Results Probl Cell Differ. 55., 203–217. Springer, Heidelberg

Murray AW (2004) Recycling the cell cycle: cyclins revisited. Cell 116:221–234

Musacchio A, Salmon ED (2007) The spindle-assembly checkpoint in space and time. Nat Rev Mol Cell Biol 8:379–393

Nagaoka SI, Hodges CA, Albertini DF, Hunt PA (2011) Oocyte-specific differences in cell-cycle control create an innate susceptibility to meiotic errors. Curr Biol 21(8):651–657

Nasmyth K (2002) Segregating sister genomes: the molecular biology of chromosome separation. Science 297:559–565

Nguyen TB, Manova K, Capodieci P, Lindon C, Bottega S, Wang XY, Refik-Rogers J, Pines J, Wolgemuth DJ, Koff A (2002) Characterization and expression of mammalian cyclin b3, a prepachytene meiotic cyclin. J Biol Chem 277:41960–41969

Niault T, Hached K, Sotillo R, Sorger PK, Maro B, Benezra R, Wassmann K (2007) Changing Mad2 levels affects chromosome segregation and spindle assembly checkpoint control in female mouse meiosis I. PLoS One 2(11):e1165

Nixon VL, Levasseur M, McDougall A, Jones KT (2002) Ca(2+) oscillations promote APC/C-dependent cyclin B1 degradation during metaphase arrest and completion of meiosis in fertilizing mouse eggs. Curr Biol 12:746–750

Norris RP, Ratzan WJ, Freudzon M, Mehlmann LM, Krall J, Movsesian MA, Wang H, Ke H, Nikolaev VO, Jaffe LA (2009) Cyclic GMP from the surrounding somatic cells regulates cyclic AMP and meiosis in the mouse oocyte. Development 136:1869–1878

O'Neill GT, Kaufman MH (1987) Ovulation and fertilization of primary and secondary oocytes in LT/Sv strain mice. Gamete Res 18:27–36

Pan H, Ma P, Zhu W, Schultz RM (2008) Age-associated increase in aneuploidy and changes in gene expression in mouse eggs. Dev Biol 316:397–407
Park JY, Su YQ, Ariga M, Law E, Jin SL, Conti M (2004) EGF-like growth factors as mediators of LH action in the ovulatory follicle. Science 303:682–684
Peng XR, Hsueh AJ, LaPolt PS, Bjersing L, Ny T (1991) Localization of luteinizing hormone receptor messenger ribonucleic acid expression in ovarian cell types during follicle development and ovulation. Endocrinology 129:3200–3207
Pesin JA, Orr-Weaver TL (2008) Regulation of APC/C activators in mitosis and meiosis. Annu Rev Cell Dev Biol 24:475–499
Peters JM (2006) The anaphase promoting complex/cyclosome: a machine designed to destroy. Nat Rev Mol Cell Biol 7:644–656
Petronczki M, Siomos MF, Nasmyth K (2003) Un ménage à quatre: the molecular biology of chromosome segregation in meiosis. Cell 112:423–440
Pirino G, Wescott MP, Donovan PJ (2009) Protein kinase A regulates resumption of meiosis by phosphorylation of Cdc25B in mammalian oocytes. Cell Cycle 8:665–670
Polański Z (1986) In-vivo and in-vitro maturation rate of oocytes from two strains of mice. J Reprod Fertil 78:103–109
Polanski Z (1997) Strain difference in the timing of meiosis resumption in mouse oocytes: involvement of a cytoplasmic factor(s) acting presumably upstream of the dephosphorylation of p34cdc2 kinase. Zygote 5:105–109
Polanski Z, Ledan E, Brunet S, Louvet S, Verlhac MH, Kubiak JZ, Maro B (1998) Cyclin synthesis controls the progression of meiotic maturation in mouse oocytes. Development 125:4989–4997
Refik-Rogers J, Manova K, Koff A (2006) Misexpression of cyclin B3 leads to aberrant spermatogenesis. Cell Cycle 5:1966–1973
Reis A, Chang HY, Levasseur M, Jones KT (2006) APCcdh1 activity in mouse oocytes prevents entry into the first meiotic division. Nat Cell Biol 8:539–540
Reis A, Madgwick S, Chang HY, Nabti I, Levasseur M, Jones KT (2007) Prometaphase APCcdh1 activity prevents non-disjunction in mammalian oocytes. Nat Cell Biol 9:1192–1198
Revenkova E, Herrmann K, Adelfalk C, Jessberger R (2010) Oocyte cohesin expression restricted to predictyate stages provides full fertility and prevents aneuploidy. Curr Biol 20:1529–1533
Rieder CL, Cole RW, Khodjakov A, Sluder G (1995) The checkpoint delaying anaphase in response to chromosome monoorientation is mediated by an inhibitory signal produced by unattached kinetochores. J Cell Biol 130:941–948
Schindler K, Schultz RM (2009) CDC14B acts through FZR1 (CDH1) to prevent meiotic maturation of mouse oocytes. Biol Reprod 80:795–803
Schmiady H, Neitzel H (2002) Arrest of human oocytes during meiosis I in two sisters of consanguineous parents: first evidence for an autosomal recessive trait in human infertility: Case report. Hum Reprod 17:2556–2559
Shonn MA, McCarroll R, Murray AW (2002) Spo13 protects meiotic cohesin at centromeres in meiosis I. Genes Dev 16:1659–1671
Sikora-Polaczek M, Hupalowska A, Polanski Z, Kubiak JZ, Ciemerych MA (2006) The first mitosis of the mouse embryo is prolonged by transitional metaphase arrest. Biol Reprod 74:734–743
Solc P, Schultz RM, Motlik J (2010) Prophase I arrest and progression to metaphase I in mouse oocytes: comparison of resumption of meiosis and recovery from G2-arrest in somatic cells. Mol Hum Reprod 16:654–664
Steuerwald N, Cohen J, Herrera RJ, Sandalinas M, Brenner CA (2001) Association between spindle assembly checkpoint expression and maternal age in human oocytes. Mol Hum Reprod 7:49–55
Steuerwald NM, Bermúdez MG, Wells D, Munné S, Cohen J (2007) Maternal age-related differential global expression profiles observed in human oocytes. Reprod Biomed Online 14:700–708

Suwinska A (2012) Preimplantation mouse embryo: developmental fate and potency of blastomeres. In: Kubiak JZ (ed) Mouse development, Results Probl Cell Differ. 55., 141–164. Springer, Heidelberg

Sweeney C, Murphy M, Kubelka M, Ravnik SE, Hawkins CF, Wolgemuth DJ, Carrington M (1996) A distinct cyclin A is expressed in germ cells in the mouse. Development 122:53–64

Tachibana-Konwalski K, Godwin J, van der Weyden L, Champion L, Kudo NR, Adams DJ, Nasmyth K (2010) Rec8-containing cohesin maintains bivalents without turnover during the growing phase of mouse oocytes. Genes Dev 24:2505–2516

Tanaka TU (2010) Kinetochore-microtubule interactions: steps towards bi-orientation. EMBO J 29:4070–4082

Tarkowski AK (1959) Experiments on the development of isolated blastomers of mouse eggs. Nature 184:1286–1287

Tarkowski AK (1961) Mouse chimaeras developed from fused eggs. Nature 190:857–860

Tay J, Hodgman R, Richter JD (2000) The control of cyclin B1 mRNA translation during mouse oocyte maturation. Dev Biol 221:1–9

Terret ME, Lefebvre C, Djiane A, Rassinier P, Moreau J, Maro B, Verlhac MH (2003) DOC1R: a MAP kinase substrate that control microtubule organization of metaphase II mouse oocytes. Development 130:5169–5177

Tsurumi C, Hoffmann S, Graeser R, Geley S, Polanski Z (2004) The spindle assembly checkpoint is not essential for CSF arrest of mouse oocytes. J Cell Biol 167:1037–1050

Vaccari S, Weeks JL 2nd, Hsieh M, Menniti FS, Conti M (2009) Cyclic GMP signaling is involved in the luteinizing hormone-dependent meiotic maturation of mouse oocytes. Biol Reprod 81:595–604

Verlhac M-H, Kubiak JZ, Clarke HJ, Maro B (1994) Microtubule and chromatin behavior follow MAP kinase activity but not MPF activity during meiosis in mouse oocytes. Development 120:1017–1025

Verlhac M-H, Kubiak JZ, Weber M, Géraud G, Colledge WH, Evans MJ, Maro B (1996) Mos is required for MAP kinase activation and is involved in microtubule organisation during mouse meiosis. Development 122:815–822

Wassmann K, Niault T, Maro B (2003) Metaphase I arrest upon activation of the Mad2-dependent spindle checkpoint in mouse oocytes. Curr Biol 13:1596–1608

Winston NJ (1997) Stability of cyclin B protein during meiotic maturation and the first mitotic cell division in mouse oocytes. Biol Cell 89:211–219

Winston N, Bourgain-Guglielmetti F, Ciemerych MA, Kubiak JZ, Senamaud-Beaufort C, Carrington M, Bréchot C, Sobczak-Thépot J (2000) Early development of mouse embryos null mutant for the cyclin A2 gene occurs in the absence of maternally derived cyclin A2 gene products. Dev Biol 223:139–153

Wolgemuth DJ (2011) Function of the A-type cyclins during gametogenesis and early embryogenesis. Results Probl Cell Differ 53:391–413

Yi K, Unruh JR, Deng M, Slaughter BD, Rubinstein B, Li R (2011) Dynamic maintenance of asymmetric meiotic spindle position through Arp2/3-complex-driven cytoplasmic streaming in mouse oocytes. Nat Cell Biol 13:1252–1258

Zhang Y, Zhang Z, Xu XY, Li XS, Yu M, Yu AM, Zong ZH, Yu BZ (2008) Protein kinase A modulates Cdc25B activity during meiotic resumption of mouse oocytes. Dev Dyn 237:3777–3786

Zheng W, Liu K (2012) Maternal control of mouse preimplantation development. In: Kubiak JZ (ed) Mouse development, Results Probl Cell Differ. 55., 115–139. Springer, Heidelberg

Chapter 5
Src Protein Kinases in Mouse and Rat Oocytes and Embryos

Mattan Levi, Lihi Ninio-Mani, and Ruth Shalgi

Abstract Meiosis of the mammalian oocytes is a specialized cell division, initiated during the female's embryonic life. It arrests at the germinal vesicle (GV) stage and resumes with GV breakdown, followed by segregation of the chromosomes and extrusion of the first polar body in an asymmetric cell division that concludes the first meiotic division, before arresting at metaphase of the second meiotic division (MII). Once fertilized, the oocyte exits from MII, extrudes the second polar body, and the developing zygote will continue dividing to create a blastocyst. Although the two processes of meiosis and mitosis have different developmental functions, it is believed that they share similar mechanisms. Src family kinases (SFKs) are nine non-receptor protein tyrosine kinases that regulate many key cellular functions including meiotic and mitotic cell cycles. In this review we discuss the involvement of SFKs in meiotic and mitotic cell cycle key processes as nuclear envelope breakdown, spindle stabilization, karyokinetic exit from metaphase, regulation of cortical actin, and cytokinetic cleavage furrow ingression.

5.1 The Development from Oocytes to Blastocysts

Meiosis of the mammalian oocytes is a unique process that includes two rounds of chromosome segregation without an S phase, resulting in one haploid daughter cell. Meiosis commences during the female's prenatal life and arrests around birth, at diplotene stage of the first meiotic division prophase, characterized by the presence of a germinal vesicle (GV). During pubertal life, follicle-stimulating hormone (FSH) stimulates the growth of immature ovarian follicles, and the following synergic luteinizing hormone (LH) will induce selected oocytes to resume their first meiotic

---

M. Levi • L. Ninio-Mani • R. Shalgi (✉)
Department of Cell and Developmental Biology, Sackler Faculty of Medicine, Tel Aviv University, Ramat-Aviv, Tel-Aviv 69978, Israel
e-mail: shalgir@post.tau.ac.il

J.Z. Kubiak (ed.), *Mouse Development*, Results and Problems in Cell Differentiation 55,
DOI 10.1007/978-3-642-30406-4_5,

division (RMI) and to arrest at metaphase of the second meiotic division (MII; Wassarman and Albertini 1988). RMI includes GV breakdown (GVBD); condensation of the chromosomes; formation of the first meiotic spindle with aligned chromosomes; segregation of the chromosomes (karyokinesis); extrusion of the first polar body (PBI), a unique case of cytokinesis; and alignment of the chromosomes on the second meiotic spindle during MII. The progress from GV to MII oocyte is referred to as oocyte maturation. LH induces also the process of ovulation that results in transport of mature oocytes to the oviduct where fertilization will take place. The process of fertilization includes penetration of the spermatozoon through the oocyte's glycoprotein extracellular coat (zona pellucida—ZP) and its fusion with the MII oocyte membrane, thus initiating cortical granules exocytosis (CGE), which, in turn, will cause changes in the ZP glycoproteins to establish the block to polyspermy (Miller et al. 1993) and the resumption of the second meiotic division (RMII). This series of events that takes place within the fertilized oocyte is also referred to, as oocyte activation. Extrusion of the second polar body (PBII) indicates the completion of meiosis. The formation of both paternal and maternal haploid pronuclei (PN) within the fertilized oocyte (zygote) marks the beginning of pre-implantation embryonic development (Halet et al. 2003). It continues with S phase in each PN, migration of the two PN to the center of the zygote, nuclear envelope breakdown (NEBD), condensation of the chromosomes, spindle formation, and alignment of the chromosomes at the metaphase plate. The exit from metaphase is followed by segregation of the chromosomes and cytokinesis that will give rise to two embryonic cells, called blastomeres, each containing a new diploid genome. Completion of the first mitotic division is followed by a series of mitotic cleavages that will give rise to embryos of smaller and smaller blastomeres. The segmentation of mammalian embryos is total and asynchronous that produce blastocysts consisting of 32 cells or more (Dard et al. 2008, 2009). Although the two processes of meiosis and mitosis have different functions in development, it is believed that many mitotic mechanisms are similar in meiosis as well, like the establishment and exit from metaphase. Progression through meiotic divisions, like progression through mitotic divisions, is controlled by maturation promoting factor (MPF), a heterodimer consisting of the Cyclin-Dependent Kinase 1 (CDK1 or p34/cdc2) and a regulatory subunit (cyclin-B1; Morgan 1997; Kubiak et al. 2008). MPF activity is low at the GV stage and increases just before GVBD and during the MI. A transient inactivation of MPF accompanies the metaphase–anaphase transition and the extrusion of PBI (Brunet et al. 2003). Under the influence of the reactivated MPF, the MII spindle forms and the oocytes arrest, owing to a cytostatic factor (CSF; Kubiak et al 1993; Verlhac et al. 1996; Ciemerych and Kubiak 1998; Fang et al. 1999; Peters 2002; Tunquist and Maller 2003). Cyclin-B1 degradation occurs in mammalian oocytes during the CSF-induced cell cycle arrest (Nixon et al. 2002), with cyclin-B1 having a half-life of 4 h, thus spontaneous oocyte activation is prevented only by a continued cyclin-B1 synthesis (Markoulaki et al. 2003). Release from the cell cycle arrest is triggered by an increase in intracellular calcium concentration ($[Ca^{2+}]i$) induced by the fertilizing spermatozoon. Inactivation of MPF during exit from both MI and MII is controlled by the anaphase-promoting

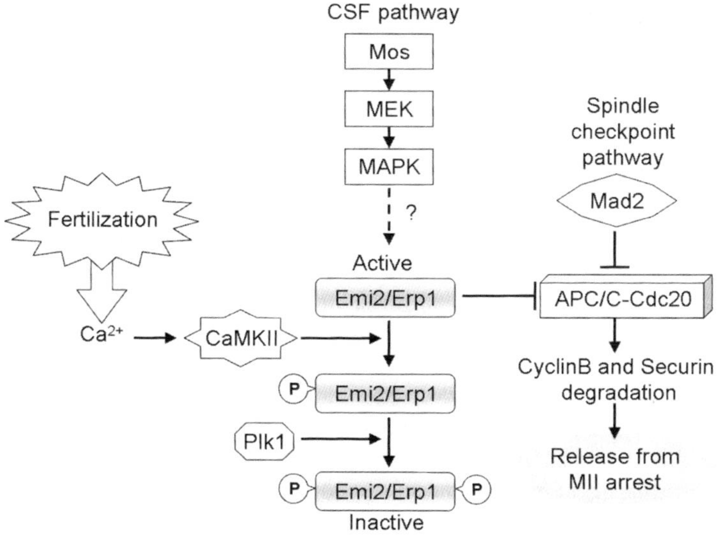

Fig. 5.1 Model of the pathways involved in the release from MII arrest. Oocytes awaiting fertilization are arrested at MII stage. Two pathways of APC/C-Cdc20 inhibition have to be inactivated in order to induce RMII; one of them is the spindle checkpoint. Mitotic arrest-deficient 2 protein (Mad2) is an essential component of the spindle checkpoint pathway. Mad2 inhibits the APC/C when chromosomes are improperly aligned on the metaphase plate. Once the chromosomes alignment is accomplished, Mad2 inhibition is removed (Musacchio and Hardwick 2002). The second pathway of APC/C-Cdc20 inhibition is the CSF that requires the activity of early mitotic inhibitor 2/Emi-related protein 1 (Emi2/Erp1; Liu and Maller 2005; Rauh et al. 2005) as well as the mos-MAP kinase pathway (Tunquist and Maller 2003) that can activate Emi2/Erp1 (Inoue et al. 2007). Upon fertilization, the resulting Ca^{2+} rise activates calmodulin-dependent protein kinase II (CaMKII), which in turn phosphorylates Emi2/Erp1 (*circled P*), thereby creating a docking site for Polo-like kinase 1 (Plk1). The recruited Plk1 re-phosphorylates Emi2/Erp1, thereby targeting it for destruction (Hansen et al. 2006). As a result, APC/C-Cdc20 is no longer inhibited and anaphase inhibitors such as cyclins and securin are targeted for degradation, allowing the release from MII arrest (Peters 2002)

complex (APC/C, Fig. 5.1), a multi-subunit E3 ligase that ubiquitinates cyclin-B1, thereby targeting it for degradation by the 26S proteasome (Kubiak et al. 1993; Fang et al. 1999; Nixon et al. 2002; Peters 2002; Jones 2004).

5.2 Src Family Kinases

Src family kinases (SFKs) are nine non-receptor protein tyrosine kinases that regulate many key cellular functions (Thomas and Brugge 1997). The structural similarities between SFKs suggest that they may act as a family (Kubiak 2010) and compensate for one another (Stein et al. 1994; Roche et al. 1995a). Src, Yes, and Fyn are found in a broad range of mammalian tissues, including oocytes (Schartl and Barnekow 1984; Kinsey 1996; Sato et al. 1996; Talmor-Cohen et al. 2004a;

Mehlmann and Jaffe 2005; Meng et al. 2006; McGinnis et al. 2007; Zheng et al. 2007) while Lck, Fgr, Lyn, Hck, and Blk are restricted to cells of the haematopoietic lineages. SFKs are composed of five distinct functional domains: amino-terminal membrane-binding domain that signals post translational modifications such as myristoylation and palmitoylation, that will in turn lead to membranous localization of the SFKs; Src-homology-3 (SH3) domain capable of binding to proline-rich sequences; Src-homology-2 (SH2) domain capable of binding to phosphotyrosine-containing sequences: Src-homology-1 (SH1) domain that transfers γ-phosphates from ATP molecules to tyrosine residues of target proteins, thereby transmitting extracellular signals to downstream cellular components; carboxy-terminal regulatory domain that contains a phosphorylatable tyrosine residue needed for negative regulation of the kinase activity (Thomas and Brugge 1997). In mammalian cells, Src is normally maintained in an inactive conformation; its SH2 domain engaged with the tyrosine at the carboxy-terminal regulatory domain, and its SH3 domain engaged with the SH2-kinase linker. Dephosphorylation of the tyrosine at the carboxy-terminal regulatory domain, occurring as a result of protein tyrosine phosphatase (PTP) action, disrupts the intramolecular interaction of the SH2 domain and the tyrosine at the carboxy-terminal regulatory domain, thus activating SFKs. Phosphorylation of the carboxy-terminal tyrosine by c-Src tyrosine kinase (Csk) allows reformation of the intramolecular interaction of SH2 and tyrosine, resulting in SFKs inactivation. Binding of SFKs to tyrosine-phosphorylated growth-factor receptors via the receptor-SH2 domain, results in displacement of the carboxy-terminal tyrosine that allows for SFKs activation (Hubbard and Till 2000; Thomas and Brugge 1997).

5.3 SFKs in the Cell Cycle

It was reported that Fyn$^{-/-}$ or Src$^{-/-}$ female mice are sub-fertile (McGinnis et al. 2009; Stein et al. 1994); however, the cause of their sub-fertility is yet to be determined. One hypothesis could be that SFKs are involved in the cell cycle of somatic cells, so that inhibiting SFKs can result in impaired embryonic development. Several lines of evidence from two decades ago indicate the importance of SFKs in the somatic cell cycle control. c-Src, c-Yes, and Fyn are activated in response to various growth factors during the transition from G0 to G1, and at the end of G2 phase of the cell cycle (Kypta et al. 1990; Courtneidge et al. 1993; Roche et al. 1995a), whereas inhibition of SFKs in fibroblasts inhibits mitogenic response to growth factors (Twamley-Stein et al. 1993; Roche et al. 1995b). Fluorescence-activated cell sorter (FACS) analysis of HeLa cells demonstrated that inhibition of SFKs resulted in an arrest at the G2/M stage (Fig. 5.2b). Following a 24 h exposure to 5 μM SU6656, 10 % of the cells were at the G1 stage, 9 % at the S stage, and 32 % at the G2/M stage, whereas 43 % of untreated control cells were at the G1 stage and 11 % at the S stage (Fig. 5.2a). Moreover, it seems that essential cell cycles during early embryogenesis are also affected by SFKs. Inhibition of Fyn in oocytes microinjected with dominant negative form of Fyn (DN-Fyn) cRNA inhibited mitosis in blatomers during the transition from 2-cells stage embryo to blastocyst

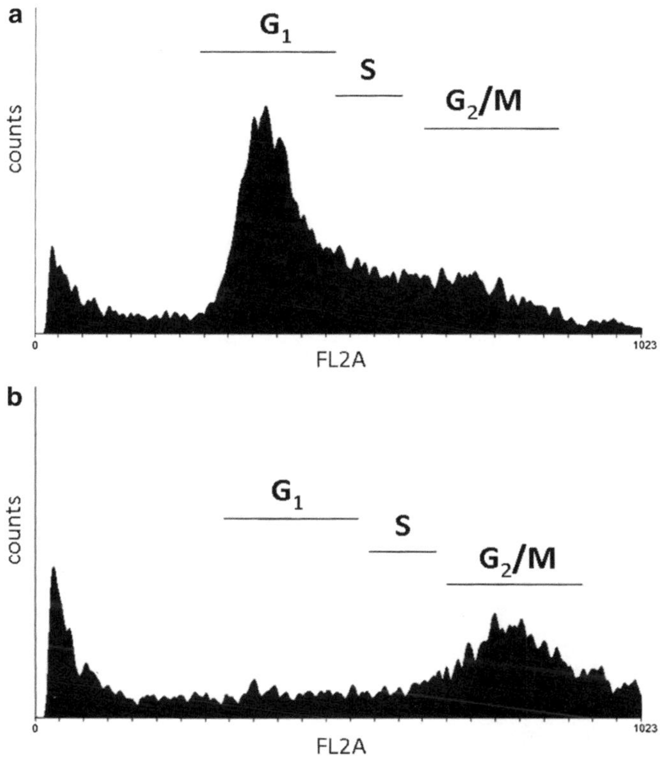

Fig. 5.2 SFKs inhibition causes a cell cycle arrest of HeLa cells. Untreated HeLa cells (**a**; control) and HeLa cells exposed for 24 h to 5 μM SU6656 (**b**) were trypsinized, washed three times in cold phosphate-buffered saline (PBS), re-suspended in 1.0 ml hypotonic buffer (50 μg/ml propidium iodide, 0.1 % sodium citrate and 0.1 % Triton X-100), and incubated for 1 h at 4°C in the dark. Cells were subsequently analyzed by FACSort Flow Cytometer and WinMDI 2.8 software. Single-parameter FL2 (area) histogram illustrate relative DNA content

(Fig. 5.3). However, accumulating data from the last few years suggest that this explanation is not the entire picture and there is another interesting piece to the puzzle. It is possible that SFKs are involved also in cell cycle events that take place at earlier stages of development. This suggestion is based on the evidence regarding SFKs involvement in NEBD, in the exit from metaphase and in cytokinesis during both meiosis and the first mitotic division. In the next part of this review we will elaborate on the involvement of SFKs in these three essential processes.

5.4 SFKs and NEBD

Studies have shown that SFKs take part in GVBD. Mouse oocytes, matured in the presence of the SFK inhibitor, PP2, were unable to undergo the process of GVBD (Zheng et al. 2007); we demonstrated that subjecting oocytes to SU6656, an inhibitor of Fyn, Yes, and Src, or microinjecting them with DN-Fyn, inhibits

a

cRNA injected Into 2-cell embryos	Number of scored embryos	Number of injected cells	Average number of live cells	
			24h after injection	48h after injection
Histone-H2B-RFP	55	1	7.26	48
DN-Fyn-FLAG + Histone-H2B-RFP	28	1	5.62*	23.2*
WT-Fyn-FLAG + Histone-H2B-RFP	35	2	5.51	NA
DN-Fyn-FLAG + Histone-H2B-RFP	41	2	4.05**	NA

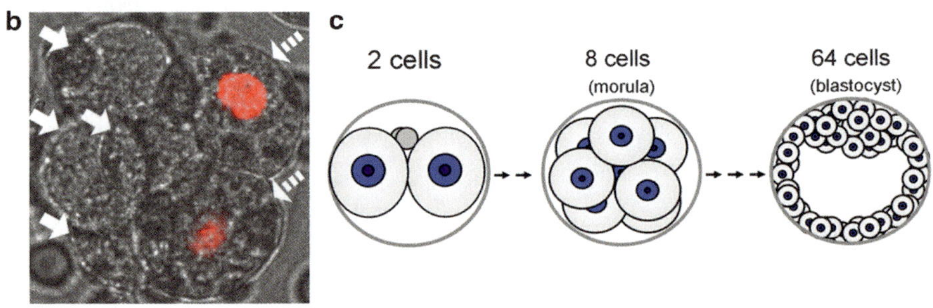

Fig. 5.3 SFKs inhibition impairs early stages of mouse embryogenesis. (**a**) Two-cell stage embryos flushed from the oviducts of mated female mice, 48 h after hCG administration, were microinjected with cRNAs Histone-H2B-RFP cRNA with or without WT-Fyn-FLAG or DN-Fyn-FLAG. The number of live cells (blastomers) was assessed 24 or 48 h later. NA—not assessed. *Number of cells is statistically different from that of Histone-H2B-RFP cRNA injected embryos. **Number of cells is statistically different from that of WT-Fyn-FLAG cRNA injected embryos ($P < 0.05$). (**b**) A representative embryo with four small cells (*solid arrows*) descendants from a non-injected cell and two large cells (*dashed arrows*) descendants from a cell injected with Histone-H2B-RFP and DN-Fyn-FLAG RNAs. (**c**) Illustration of the pre-implantation embryo during the segmentation

GVBD during RMI. Other observations suggest that SFKs involvement is not restricted only to the meiotic GVBD, but they also take part in the mitotic process of NEBD. McGinnis et al. (2007) demonstrated that activated SFKs concentrate around the PN and are in close association with the nuclear envelope. Microinjection of DN-Fyn cRNA into zygotes at the PN stage reduced their ability to develop into two-cell embryos. Meng et al. (2006) showed that MII oocytes injected with protein of the Fyn SH2 remained arrested at the PN stage, while Roche and colleagues (1995a) showed that fibroblasts at the G2 phase, microinjected with antibodies against SFKs do not undergo NEBD. Our recent results reinforced these observations by showing that inhibiting Fyn in oocytes by microinjection of DN-Fyn cRNA caused a dramatic inhibition of NEBD (Fig. 5.4a), while exposing the

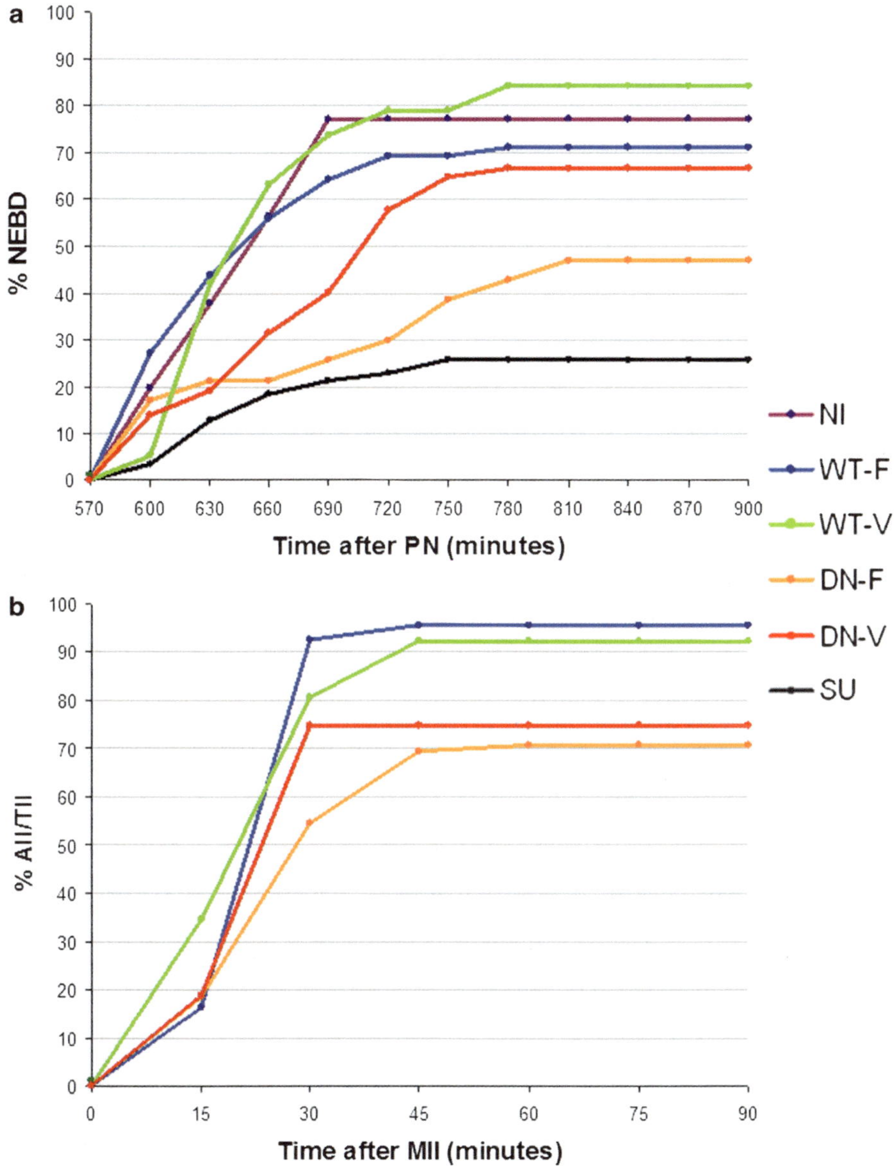

Fig. 5.4 SFKs inhibition prevents exit from MII and NEBD in mouse oocytes. Mouse oocytes at the MII stage were subjected to several treatments: non-injected oocytes incubated with 10 μM SU6656 (SU; 41 oocytes; *black*) or without it (NI—non-injected control; 110 oocytes; *purple*); oocytes microinjected with cRNA of Histone-H2B-RFP and β-tubulin-GFP mixed with either WT-Fyn-FLAG (WT-F; 58 oocytes; *blue*) or DN-Fyn-FLAG (DN-F; 21 oocytes; *orange*); oocytes microinjected with cRNA of Histone-H2B-RFP mixed with either WT-Fyn-Venus (WT-V; 19 oocytes; pale green) or DN-Fyn-Venus (DN-V; 41 oocytes; *red*). Oocytes were cultured in M2 medium for 6 h, allowing translation of the cRNA, triggered parthenogenically with 8 % EtOH for

oocytes to SU6656 at the late PN stage caused an even more significant inhibition of NEBD, implying a functional overlapping of the SFKs members. Altogether, these results suggest that SFKs, in general, and Fyn, in particular, take part in the signaling leading to the meiotic GVBD and the mitotic NEBD. However, the mechanism of SFKs' action in these processes is yet to be determined.

5.5 SFKs and Karyokinesis

The involvement of Fyn in the exit from metaphase during karyokinesis has been reported by several research groups. Inhibition of SFKs by chemical inhibitors or by microinjection of either DN-Fyn or DN-Yes inhibited the exit from MII arrest in rat oocytes (Talmor-Cohen et al. 2004a; Tomashov-Matar et al. 2007, 2008). According to our recent research, the exit from metaphase was inhibited also in mouse oocytes microinjected with DN-Fyn-FLAG or DN-Fyn-Venues cRNA (Fig. 5.4b). Other studies show that the exit from MI was inhibited in Fyn$^{-/-}$ mice oocytes and in oocytes exposed to the SFKs inhibitors, SKI606, PP2 or SU6656 (Zheng et al. 2007; McGinnis et al. 2009; Levi et al. 2010a), implying that SFKs are generally involved in the exit from metaphase, rather than being specifically involved in the special mechanism of exit from MII that includes CSF inactivation. The association of Fyn with the spindle microtubules (MTs) implies a specific site of SFKs action in the exit from metaphase. Fyn co-localizes with the mitotic spindle and spindle poles in immune cells such as T lymphocytes (Ley et al. 1994); it also coimmunoprecipitates, through its SH2 and SH3 domains, with γ-tubulin in activated mast cells (Sulimenko et al. 2006). Pretreatment of mast cells with PP2 inhibits the formation of MTs and reduces the amount of tyrosine-phosphorylated-proteins in γ-tubulin complexes, suggesting that SFKs are involved in the initial stages of MTs formation (Sulimenko et al. 2006).

Fyn localizes at the spindle area in rat and mouse oocytes (Talmor et al. 1998; Talmor-Cohen et al. 2004a, b; Levi et al. 2011b), whereas inhibition of SFKs by SU6656 results in disruption of the spindle structure (Levi and Shalgi 2010). During the exit from metaphase, the total amount of phosphorylated Fyn and the amount of Fyn at the spindle poles were both reduced (Levi et al. 2011b). McGinnis et al. (2007) demonstrated an elevated level of phosphorylated tyrosine at the poles of the MII spindle in mouse oocytes, which eventually disappeared during the exit from MII. It was previously reported that SFKs also take part in MPF pathway; inhibition

Fig. 5.4 (continued) 6 min, washed and imaged with a spinning-disk confocal microscope at 15 min intervals. Non-injected oocytes served as control. Four-dimensional (*x*,*y*,*z*,*t*) image stacks acquired in three channels (488 nm, 561 nm and DIC) were processed and analyzed by the Volocity 5 software. The disappearance of the PN, as observed through the DIC channel, was classified as NEBD stage (**a**). The appearance of segregated chromosomes, as observed through the 561 nm channel, was classified as either anaphase or telophase of the second meiotic division (AII and TII, respectively; **b**)

of SFKs by SU6656 inhibit the degradation of cyclin-B1 and the release from MII arrest in fertilized rat oocytes (Tomashov-Matar et al. 2007). An efficient degradation of Cyclin-B1 is required for the exit of oocytes from metaphase arrest (Jones 2004). In response to mitogenic stimuli, p34cdc2 phosphorylates serine and threonine residues at the amino terminal portion of Src, thus causing dephosphorylation at the carboxylic terminus and activating Src during mitosis (Taylor and Shalloway 1993). Several studies reported an elevated SFKs activity during mitotic metaphase of somatic cells, and its correlation with phosphorylation of SFKs by p34cdc2 on several serine and threonine residues (Shalloway and Shenoy 1991; Fumagalli et al. 1994; Taylor and Shalloway 1996; Thomas et al. 1995). Altogether, these studies suggest a possible negative feedback mechanism during karyokinesis: accumulation of cyclin-B1 during metaphase leads to the high activity of p34cdc2, which phosphorylates SFKs and induces their elevated activity that, in turn, will induce an association with γ-tubulin and the high level of phosphorylated tyrosine at the spindle poles. High activity of SFKs induces degradation of cyclin-B1 and thus, via a negative feedback, the reduction in the level of active SFKs at the spindle poles, allowing the exit from metaphase and karyokinesis.

5.6 SFKs and Cytokinesis

It was suggested that SFKs play also an important role in the intricate pathway of cleavage furrow ingression during cytokinesis. Cytokinesis is blocked in pro-B cells from Fyn-null mice (Yasunaga et al. 1996), and inhibition of SFKs by SU6656 or by microinjection of SH2 domains of Src or Fyn inhibits cytokinesis in sea-urchin embryos (Ng et al. 2005). Moreover, inhibition of Fyn decreased the rate of extrusion of the PBs as well as the first mitotic cleavage in both mice and rats oocytes and embryos (Levi et al. 2010b, a). The ingression of the cleavage furrow in many cell types depends on the constriction of an actomyosin contractile ring (Bompard et al. 2008). Fyn is localized at the cleavage furrow of hematopoietic pro-B cells (Yasunaga et al. 1996) and hybridoma T-cells (Campbell et al. 1998). It also colocalizes with filamentous actin (F-actin) at the meiotic cleavage furrow area of mammalian oocytes (Levi et al. 2010b, a). Recent studies showed that the cortical organization of F-actin is disrupted in oocytes of Fyn-null mice, in oocytes exposed to SFKs inhibitors, SKI606 or SU6656, and in oocytes microinjected with DN-Fyn cRNA (Luo et al. 2009; Levi et al 2011a). Other studies showed that SFKs can regulate actin polymerization by association and activation of cortical actin polymerizing factors such as RhoA (Nozu et al. 1999; Veettil et al. 2006; Knock et al. 2008; Pasquale 2010; Shaifta et al. 2010), Arp2/3, FAK, Wiskott–Aldrich syndrome proteins (WASP), WASP family verprolin homologous (WAVE) family of proteins (Frame and Brunton 2002; Pollard and Borisy 2003; Badour et al. 2004), formins (Uetz et al. 1996; Tominaga et al. 2000), and mDia-interacting protein (DIP; Satoh et al. 2001; Meng et al. 2004). All these studies indicate that SFKs are recruited to the area of cleavage furrow formation and regulate the F-actin stabilization effectors during ingression of the cleavage furrow.

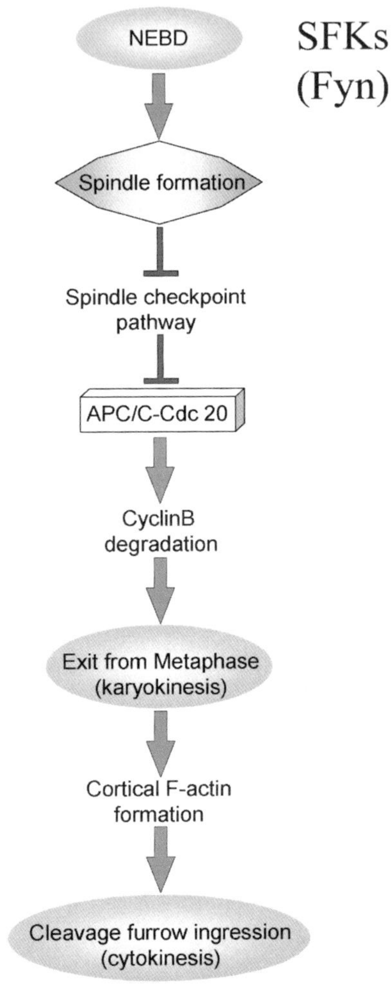

Fig. 5.5 Model for SFKs role in meiosis and mitosis. Based on the data presented in this paper, we suggest a model for the involvement of SFKs, in general, and of Fyn, in particular, in cell cycle events. According to our model SFKs take part in the signaling leading to the meiotic GVBD and the mitotic NEBD. The involvement of SFKs in the stabilization of the division spindle structure causes inactivation of the spindle assembly checkpoint pathway and allows the activation of the anaphase promoting complex (APC/C), a multi-subunit E3 ligase. Active APC/C-Cdc20 induces cyclin-B degradation, exit from metaphase and translocation of SFKs to the cleavage furrow. Cortical SFKs induce the arrangement of F-actin, necessary for cleavage furrow ingression during cytokinesis

5.7 Conclusions

The mammalian oocytes meiotic divisions and the early embryonic mitotic divisions have unique characteristics but they also share many general cell cycle mechanisms. We propose that SFKs takes part in regulating meiotic and mitotic key signaling pathways as NEBD, spindle stabilization during exit from metaphase and F-actin regulation during cytokinetic cleavage furrow ingression (Fig. 5.5). It is also safe to assume that SFKs take part in more than one process within oocytes and embryos.

Acknowledgments We thank Dr. Bernard Maro for his valuable help; Dr. Sophie Louvet-Vallée for microinjection of the mouse embryos and Dr. Marie-Hélène Verlhac for the Histone-H2B-RFP cDNA. We also thank Ruth Kaplan-Kraicer for her help in preparing the manuscript. Authors were supported by a grant from the Israel Science Foundation to R. Shalgi (Grant number 261/09) and by the Eshkol scholarship from the Israel Ministry of Science, Culture and Sport to M. Levi (Grant number 3-5420) while writing this article.

References

Badour K, Zhang J, Shi F, Leng Y, Collins M, Siminovitch KA (2004) Fyn and PTP-PEST-mediated regulation of Wiskott-Aldrich syndrome protein (WASp) tyrosine phosphorylation is required for coupling T cell antigen receptor engagement to WASp effector function and T cell activation. J Exp Med 199:99–112

Bompard G, Rabeharivelo G, Morin N (2008) Inhibition of cytokinesis by wiskostatin does not rely on N-WASP/Arp2/3 complex pathway. BMC Cell Biol 9:42

Brunet S, Pahlavan G, Taylor S, Maro B (2003) Functionality of the spindle checkpoint during the first meiotic division of mammalian oocytes. Reproduction 126:443–450

Campbell KS, Cooper S, Dessing M, Yates S, Buder A (1998) Interaction of p59fyn kinase with the Dynein light chain, Tctex-1, and colocalization during cytokinesis. J Immunol 161:1728–1737

Ciemerych MA, Kubiak JZ (1998) Cytostatic activity develops during meiosis I in oocytes of LT/Sv mice. Dev Biol 200:198–211

Courtneidge SA, Dhand R, Pilat D, Twamley GM, Waterfield MD, Roussel MF (1993) Activation of Src family kinases by colony stimulating factor-1, and their association with its receptor. EMBO J 12:943–950

Dard N, Breuer M, Maro B, Louvet-Vallée S (2008) Morphogenesis of the mammalian blastocyst. Mol Cell Endocrinol 282:70–77

Dard N, Le T, Maro B, Louvet-Vallée S (2009) Inactivation of aPKCλ reveals a context dependent allocation of cell lineages in preimplantation mouse embryos. PLoS One 4:e7117

Fang G, Yu H, Kirschner MW (1999) Control of mitotic transitions by the anaphase-promoting complex. Philos Trans R Soc Lond B Biol Sci 354:1583–1590

Frame MC, Brunton VG (2002) Advances in Rho-dependent actin regulation and oncogenic transformation. Curr Opin Genet Dev 12:36–43

Fumagalli S, Totty NF, Hsuan JJ, Courtneidge SA (1994) A target for Src in mitosis. Nature 368:871–874

Halet G, Marangos P, Fitzharris G, Carroll J (2003) Ca^{2+} oscillations at fertilization in mammals. Biochem Soc Trans 31:907–911

Hansen DV, Tung JJ, Jackson PK (2006) CaMKII and polo-like kinase 1 sequentially phosphorylate the cytostatic factor Emi2/XErp1 to trigger its destruction and meiotic exit. Proc Natl Acad Sci USA 103:608–613

Hubbard SR, Till JH (2000) Protein tyrosine kinases structure and function. Annu Rev Biochem 69:373–398

Inoue D, Ohe M, Kanemori Y, Nobui T, Sagata N (2007) A direct link of the Mos-MAPK pathway to Erp1/Emi2 in meiotic arrest of Xenopus laevis eggs. Nature 446:1100–1104

Jones KT (2004) Turning it on and off: M-phase promoting factor during meiotic maturation and fertilization. Mol Hum Reprod 10:1–5

Kinsey WH (1996) Biphasic activation of Fyn kinase activity upon fertilization of the sea urchin oocyte. Dev Biol 174:281–287

Knock GA, Snetkov VA, Shaifta Y, Drndarski S, Ward JP, Aaronson PI (2008) Role of src-family kinases in hypoxic vasoconstriction of rat pulmonary artery. Cardiovasc Res 80:453–462

Kubiak JZ, Weber M, de Pennart H, Winston NJ, Maro B (1993) The metaphase II arrest in mouse oocytes is controlled through microtubule-dependent destruction of cyclin B in the presence of CSF. EMBO J 12:3773–3778

Kubiak JZ, Bazile F, Pascal A, Richard-Parpaillon L, Polanski Z, Ciemerych MA, Chesnel F (2008) Temporal regulation of embryonic M-phases. Folia Histochem Cytobiol 46:5–9
Kubiak JZ (2010) Yes Fyn can: individual or team player in SFKs? Cell Cycle 9:1460–1461
Kypta RM, Goldberg Y, Ulug ET, Courtneidge SA (1990) Association between the PDGF receptor and members of the src family of tyrosine kinases. Cell 62:481–492
Ley SC, Marsh M, Bebington CR, Proudfoot K, Jordan P (1994) Distinct intracellular localization of Lck and Fyn protein tyrosine kinases in human T lymphocytes. J Cell Biol 125:639–649
Levi M, Shalgi R (2010) The role of Fyn kinase in the release from metaphase in mammalian oocytes. Mol Cell Endocrinol 314:228–233
Levi M, Maro B, Shalgi R (2010a) The involvement of Fyn kinase in resumption of the first meiotic division in mouse oocytes. Cell Cycle 9:1577–1589
Levi M, Maro B, Shalgi R (2010b) Fyn kinase is involved in cleavage furrow ingression during meiosis and mitosis. Reproduction 140:827–834
Levi M, Kaplan-Kraicer R, Shalgi R (2011a) Regulation of division in mammalian oocytes: implications for polar body formation. Mol Hum Reprod 17:328–334
Levi M, Maro B, Shalgi R (2011b) The conformation and activation of Fyn kinase in the oocyte determines its localization to the spindle poles and cleavage furrow. Reprod Fertil Dev 23:846–857
Liu J, Maller JL (2005) Calcium elevation at fertilization coordinates phosphorylation of XErp1/Emi2 by Plx1 and CaMK II to release metaphase arrest by cytostatic factor. Curr Biol 15:1458–1468
Luo J, McGinnis LK, Kinsey WH (2009) Fyn kinase activity is required for normal organization and functional polarity of the mouse oocyte cortex. Mol Reprod Dev 76:819–831
Markoulaki S, Matson S, Abott AL, Ducibella T (2003) Oscillatory CaMKII activity in mouse oocyte activation. Dev Biol 258:464–474
McGinnis LK, Albertini DF, Kinsey WH (2007) Localized activation of Src-family protein kinases in the mouse egg. Dev Biol 306:241–254
McGinnis LK, Kinsey WH, Albertini DF (2009) Functions of Fyn kinase in the completion of meiosis in mouse oocytes. Dev Biol 327:280–287
Mehlmann LM, Jaffe LA (2005) SH2 domain-mediated activation of an SRC family kinase is not required to initiate Ca^{2+} release at fertilization in mouse eggs. Reproduction 129:557–564
Meng L, Luo J, Li C, Kinsey WH (2006) Role of Src homology 2 domain-mediated PTK signaling in mouse zygotic development. Reproduction 132:413–421
Meng W, Numazaki M, Takeuchi K, Uchibori Y, Ando-Akatsuka Y, Tominaga M, Tominaga T (2004) DIP (mDia interacting protein) is a key molecule regulating Rho and Rac in a Src-dependent manner. EMBO J 23:760–771
Miller DJ, Gong X, Decker G, Shur BD (1993) Oocyte cortical granule N-acetylglucosaminidase is required for the mouse zona block to polyspermy. J Cell Biol 123:1431–1440
Morgan DO (1997) Cyclin-dependent kinases: engines, clocks, and microprocessors. Annu Rev Cell Dev Biol 13:261–291
Musacchio A, Hardwick KG (2002) The spindle checkpoint: structural insight into dynamic signaling. Nat Rev Mol Cell Biol 3:731–741
Ng M, Chang F, Burgess D (2005) Movement of membrane domains and requirement of membrane signaling molecules for cytokinesis. Dev Cell 9:9781–9790
Nixon VL, Levasseur M, McDougall A, Jones KT (2002) Calcium oscillations promote APC/C-dependent cyclin B1 degradation during metaphase arrest and completion of meiosis in fertilizing mouse oocytes. Curr Biol 12:746–750
Nozu F, Tsunoda Y, Ibitayo AI, Bitar KN, Owyang C (1999) Involvement of RhoA and its interaction with protein kinase C and Src in CCK-stimulated pancreatic acini. Am J Physiol 276:915–923
Pasquale EB (2010) Eph receptors and ephrins in cancer: bidirectional signalling and beyond. Nat Rev Cancer 10:165–180

Peters JM (2002) The anaphase-promoting complex: proteolysis in mitosis and beyond. Mol Cell 9:931–943

Pollard TD, Borisy GG (2003) Cellular motility driven by assembly and disassembly of actin filaments. Cell 112:453–465

Rauh NR, Schmidt A, Bormann J, Nigg EA, Mayer TU (2005) Calcium triggers exit from meiosis II by targeting the APC/C inhibitor XErp1 for degradation. Nature 437:1048–1052

Roche S, Fumagalli S, Courtneidge SA (1995a) Requirement for Src family protein tyrosine kinases in G2 for fibroblast cell division. Science 269:1567–1569

Roche S, Koegl M, Barone MV, Roussel MF, Courtneidge SA (1995b) DNA synthesis induced by some but not all growth factors requires Src family protein tyrosine kinases. Mol Cell Biol 15:1102–1109

Sato KI, Aoto M, Mori K, Akasofu S, Tokmakov AA, Sahara S, Fukami Y (1996) Purification and characterization of a Src-related p57 protein tyrosine kinase from Xenopus oocytes. J Biol Chem 271:13250–13257

Satoh S, Tominaga T (2001) mDia-interacting protein acts downstream of Rho-mDia and modifies Src activation and stress fiber formation. J Biol Chem 276:39290–39294

Schartl M, Barnekow A (1984) Differential expression of the cellular src gene during vertebrate development. Dev Biol 105:415–422

Shaifta Y, Snetkov VA, Aaronson PI, Ward JPT, Knock GA (2010) Src-family kinases mediate activation of RhoA and constriction in rat pulmonary artery. FASEB J 24:1061.9

Shalloway D, Shenoy S (1991) Oncoprotein kinases in mitosis. Adv Cancer Res 57:185–225

Stein PL, Vogel H, Soriano P (1994) Combined deficiencies of Src, Fyn, and Yes tyrosine kinases in mutant mice. Genes Dev 8:1999–2007

Sulimenko V, Draberova E, Sulimenko T, Macurek L, Richterova V, Draber P (2006) Regulation of microtubule formation in activated mast cells by complexes of gamma-tubulin with Fyn and Syk kinases. J Immunol 176:7243–7253

Talmor A, Kinsey WH, Shalgi R (1998) Expression and immunolocalization of p59c-Fyn tyrosine kinase in rat oocyte. Dev Biol 194:38–46

Talmor-Cohen A, Matar-Tomashov R, Eliyahu E, Shapira R, Shalgi R (2004a) Are Src family kinases involved in cell cycle resumption of rat egg? Reproduction 127:455–463

Talmor-Cohen A, Matar-Tomashov R, Tsai WB, Kinsey WH, Shalgi R (2004b) Fyn kinase-tubulin interaction during meiosis of rat oocytes. Reproduction 128:1–8

Taylor SJ, Shalloway D (1996) Src and the control of cell division. Bioessays 18:9–11

Taylor SJ, Shalloway D (1993) The cell cycle and c-Src. Curr Opin Genet Dev 3:26–34

Thomas SM, Brugge JS (1997) Cellular functions regulated by Src family kinases. Annu Rev Cell Dev Biol 13:513–560

Thomas SM, Soriano P, Imamoto A (1995) Specific and redundant roles of Src and Fyn in organizing the cytoskeleton. Nature 376:267–271

Tomashov-Matar R, Levi M, Tchetchik D, Kraicer-Kaplan R, Shalgi R (2007) The role of Src family kinases in egg activation. Dev Biol 312:77–89

Tomashov-Matar R, Levi M, Shalgi R (2008) The involvement of Src family kinases (SFKs) in the events leading to resumption of meiosis. Mol Cell Endocrinol 282:56–62

Tominaga T, Sahai E, Chardin P, McCormick F, Courtneidge SA, Alberts AS (2000) Diaphanous-related formins bridge Rho GTPase and Src tyrosine kinase signaling. Mol Cell 5:13–25

Tunquist BJ, Maller JL (2003) Under arrest: cytostatic factor-mediated metaphase arrest in vertebrates oocytes. Genes Dev 17:683–710

Twamley-Stein GM, Pepperkok R, Ansorge W, Courtneidge SA (1993) The Src family tyrosine kinases are required for platelet-derived growth factor-mediated signal transduction in NIH 3T3 cells. Proc Natl Acad Sci USA 90:7696–7700

Uetz P, Fumagalli S, James D, Zeller R (1996) Molecular interaction between limb deformity proteins (formins) and Src family kinases. J Biol Chem 271:33525–33530

Veettil MV, Sharma-Walia N, Sadagopan S, Raghu H, Sivakumar R, Naranatt PP, Chandran B (2006) RhoA-GTPase facilitates entry of Kaposi's sarcoma-associated herpesvirus into adherent target cells in a Src-dependent manner. J Virol 80:11432–11446

Verlhac MH, Kubiak JZ, Weber M, Geraud G, Colledge WH, Evans MJ, Maro B (1996) Mos is required for MAP kinase activation and is involved in microtubule organization during meiotic maturation in the mouse. Development 122:815–822

Wassarman PM, Albertini DF (1988) The mammalian ovum. In: Knobil E, Neill JD (eds) The physiology of reproduction. Reven, New york

Yasunaga M, Yagi T, Hanzawa N, Yasuda M, Yamanashi Y, Yamamoto T, Aizawa S, Miyauchi Y, Nishikawa S (1996) Involvement of Fyn tyrosine kinase in progression of cytokinesis of B lymphocyte progenitor. J Cell Biol 132:91–99

Zheng KG, Meng XQ, Yang Y, Yu YS, Liu DC, Li YL (2007) Requirements of Src family kinase during meiotic maturation in mouse oocyte. Mol Reprod Dev 74:125–130

Chapter 6
Gradual Meiosis-To-Mitosis Transition in the Early Mouse Embryo

Aurélien Courtois and Takashi Hiiragi

Abstract The transition from meiosis to mitosis is a fundamental process to guarantee the successful development of the embryo. In the mouse, the transition includes extensive reorganisation of the division machinery, centrosome establishment and changes in spindle proprieties and characteristic. Recent findings indicate that this transition is gradual and lasts until the late blastocyst stage. In-depth knowledge of the mechanisms underlying the transition would provide new insight into de novo centrosome formation and regulation of spindle size and proprieties. Here, we review recent advances in the understanding of acentrosomal spindle formation, centriole establishment and the meiosis-to-mitosis transition in the mouse pre-implantation embryo.

6.1 Introduction

The transition from meiosis to mitosis is an integral part of the oocyte-to-embryo transition programme and is essential for the accurate development during early embryogenesis. The fertilisation by sperm classically defines the transition from meiosis to mitosis. While mitosis-to-meiosis transition during germ line differentiation has been extensively studied (reviewed in Wolgemuth et al. 1995; Kimble 2011), very little is known about the transition from meiosis to mitosis. However,

A. Courtois
Developmental Biology Unit, European Molecular Biology Laboratory, Meyerhofstrasse 1, 69117 Heidelberg, Germany

T. Hiiragi (✉)
Developmental Biology Unit, European Molecular Biology Laboratory, Meyerhofstrasse 1, 69117 Heidelberg, Germany

Institute for Integrated Cell-Material Sciences (WPI-iCeMS), Kyoto University, Sakyo, Kyoto 606-8501, Japan
e-mail: hiiragi@embl.de

J.Z. Kubiak (ed.), *Mouse Development*, Results and Problems in Cell Differentiation 55, DOI 10.1007/978-3-642-30406-4_6,

several studies have shed light on this topic, revealing the presence of a transition phase until establishment of mitosis proper in the early mouse embryo (Calarco-Gillam et al. 1983; Schatten et al. 1985; Schuh and Ellenberg 2007; Kubiak et al. 2008; Fitzharris 2009; Courtois et al. 2012). This 3-day long transition phase is characterised by the absence of centriole, which is the major microtubule organising centre (MTOC) of animal cells and long assumed to be required for accurate mitotic cell divisions. However, it was recently shown that acentriolar spindle can form during animal cell mitosis under experimental conditions (Megraw et al. 2001; Basto et al. 2006; Mahoney et al. 2006).

Another key assumption in the centrosomal cycle has been the requirement of a template for centriole duplication. However, recent studies demonstrated de novo centriole formation in experiments where centrioles were removed or destroyed (Khodjakov et al. 2002; La Terra et al. 2005; Basto et al. 2006). Interestingly, the centriole in the mouse has been shown to be reduced in both gametes (Szollosi et al. 1972; Schatten et al. 1986; Manandhar et al. 1998; discussed in Kubiak and Prigent 2012) and was not found in the early embryo until the 64-cell blastocyst stage (Gueth-Hallonet et al. 1993), the developmental phase that coincides with the final step in meiosis-to-mitosis transition. How can mouse embryonic cells form spindles and carry out mitotic divisions without a centriole? How can the centriole regenerate or form de novo in the mouse embryo? How does the embryonic cell shift from meiotic to mitotic divisions? The mouse pre-implantation embryo provides an attractive system to address these fundamental questions in cell biology, enabling investigation de novo centriole formation under physiological conditions and of the hidden essential component(s) for centriole propagation and spindle formation.

In this review, we summarise recent advances in these topics, with particular focus on understanding the mechanism of acentrosomal spindle formation in mouse oocytes and early embryos and on the potential mechanism(s) underlying de novo centriole formation.

6.2 Acentrosomal Spindle Formation in the Early Mouse Embryo

Given the apparent lack of a centriole during the first 3 days of development, the primary question concerns the mechanism by which mouse embryonic cells assemble the spindle for cell divisions. The shape of the spindle during the first embryonic divisions is distinct from the typical mitotic spindle (Fig. 6.1, metaphase). Immunofluorescence staining analysis in the mouse oocytes (Wassarman and Fujiwara 1978) followed by studies in the early mouse embryo described a similar barrel-shaped form of the spindle (Schatten et al. 1985). Pericentriolar material (PCM) proteins were also visible as a "broad band" at the spindle poles in the early mouse embryonic cells (Calarco-Gillam et al. 1983), consistent with our recent findings that the MTOCs at the bottom of the spindle pole form a circle or two half-circles

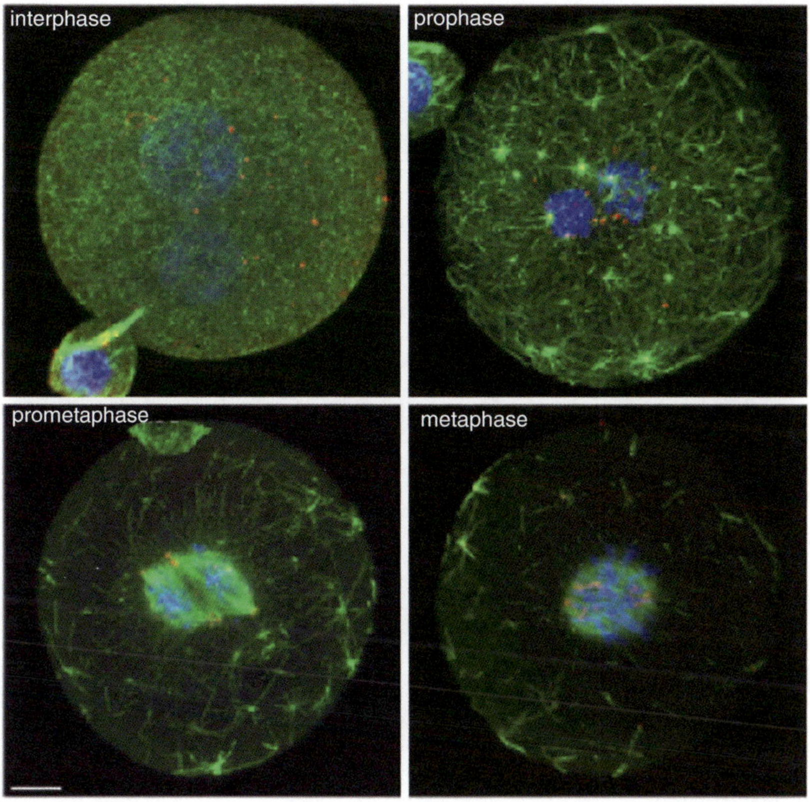

Fig. 6.1 Immunofluorescence staining of the mouse zygote fixed at consecutive stages of development: interphase (*top–left*), prophase (*top–right*), prometaphase (*bottom–left*) and metaphase (*bottom–right*). Z-projected images of confocal sections show microtubules (*green*), pericentrin (*red*) and DNA (*blue*). *Scale bar*: 10 μm

(Courtois et al. 2012), while from the 8-cell stage, the spindle pole becomes more focused (Schatten et al. 1985; Courtois et al. 2012) (Fig. 6.2). The length of the spindle during the first four divisions of the mouse development is around 25 μm, similar to that for the oocyte. The length begins to decrease at the fifth division, becoming around 18 μm at the 16-cell stage and 10 μm at the blastocyst stage. The size of the spindle is well correlated to, and precisely regulated by, cell size, but limited to a maximum size (Courtois et al. 2012). Another interesting property of the mouse embryonic spindle is its dependence on Kinesin-5 to maintain its shape in the first three divisions, which gradually disappears from the 8-cell stage (Fitzharris 2009).

The morphological differences between the acentrosomal spindle in the early mouse embryo and the typical mitotic spindle, with the former having wider spindle poles that progressively focus during subsequent divisions, prompt us to examine how the early embryonic cells manage to establish these spindles in the absence of

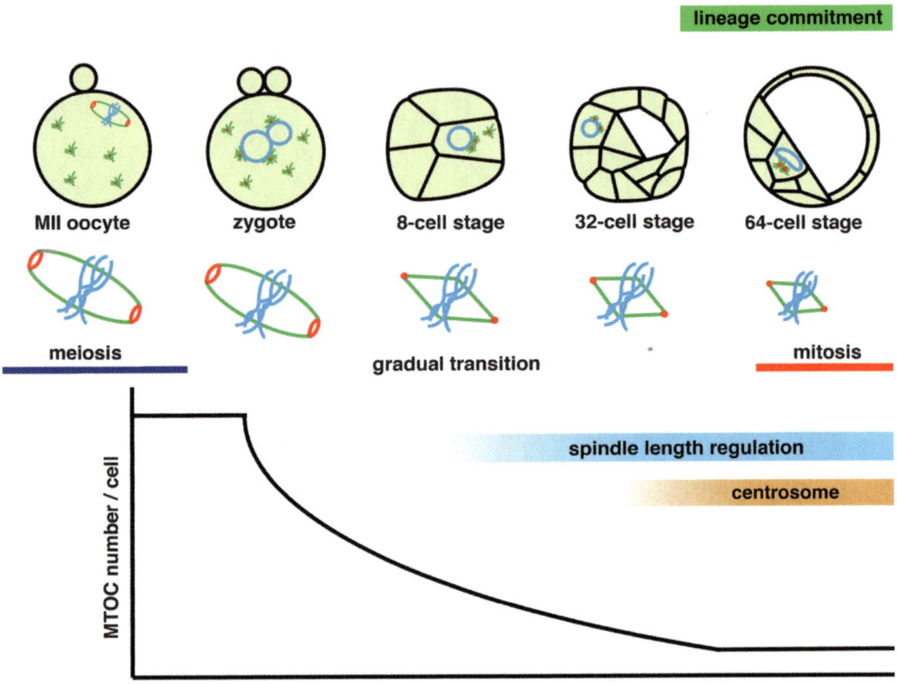

Fig. 6.2 Schematic summary of the meiosis-to-mitosis transition throughout the pre-implantation development. Embryos (*top*) and spindles (*bottom*) are illustrated for different stages. Microtubules (*green*), MTOCs (pericentrin, *red*) and DNA (*blue*) are indicated. See text for details

centriole. At the start of development, the mouse zygote apparently inherits 40–50 cytoplasmic microtubule asters from the oocytes. The cores of these asters are composed of a few PCM components, including pericentrin and gamma-tubulin. Based on this composition and the ability of these structures to polymerise microtubules, we define these structures as MTOCs, as suggested in a recent review (Lüders and Stearns 2007). These non-centriolar MTOCs are randomly distributed throughout the cytoplasm (Courtois et al. 2012). Unlike many other species in which the sperm delivers centrioles to oocytes at fertilisation, no contribution of centrioles or MTOCs by the sperm can be observed in the mouse zygote (Courtois et al. 2012). During interphase, the number of MTOCs visible on the pronuclei membrane increases gradually, to around 20 at the end of interphase in a manner dependent on microtubules and dynein (Courtois et al. 2012) (Fig. 6.1, interphase). Towards the end of the first interphase, the small MTOCs begin to polymerise microtubules and form small asters throughout the cytoplasm and also around the pronuclei (Szollosi et al. 1972; Schatten et al. 1985; Houliston et al. 1987) (Fig. 6.1, prophase). Surprisingly, the cytoplasmic asters are not recruited to the pronuclear membrane, but instead are very limited in their movement. This striking difference from the oocyte (Schuh and Ellenberg 2007) can be explained in part by the presence of kinesin-5 in the mouse zygote and early embryos, which counteracts dynein-based recruitment of MTOCs, and which may be absent in oocytes. At the time of nuclear envelope breakdown (NEBD), the asters on the pronuclear surface massively polymerise microtubules toward the chromosomes due to Ran GTPase, whereas the cytoplasmic asters become destabilised and lose their polymerised microtubule signal, pointing toward an inactivation of their MTOC function. Thus, a multi-polar spindle directly emerges in the middle of the cell (Fig. 6.1, prometaphase) without forming the “microtubule-ball” that is prominent in meiotic division in the oocyte (Schuh and Ellenberg 2007). The minus-end motor protein, dynein, plays a key role in clustering the MTOCs and organising the multi-polar spindle. In its absence, two separate MTOC clouds are formed upon NEBD, which slowly fade away. On the other hand, the plus-end motor protein, kinesin-5, pushes apart the MTOCs, and its lack leads all MTOCs to merge into a monopolar spindle. Thus, the concerted action of dynein and kinesin-5 is required for formation of a bi-polar spindle, reminiscent to the mechanism in oocytes.

Overall, spindle formation during the first three divisions of the mouse embryo appears to be a stochastic self-assembly process involving a number of MTOCs. From the fourth division at the compacted morula stage, MTOCs are often localised at one side of the nucleus, preferentially the apical side (Houliston et al. 1987), and the spindle elongation rates quickens, with the time from NEBD to the beginning of anaphase decreasing from over 40 min during the first three divisions to less than 30 min at subsequent divisions (Kaufman 1973; Ciemerych et al. 1999; Courtois et al. 2012). In some cases the established spindle exhibits rotations (Courtois et al. 2012), raising the possibility of topological cues that direct spindle orientation. In any event, the absence of centriole during the first few “mitotic” divisions implies features and underlying mechanisms that differ substantially from those in typical animal mitosis. Clearly, cells in the mouse embryo establish mitosis proper, i.e., with centrioles, during subsequent development. The question is how the embryo (re-)generates the centriole.

6.3 Changes in MTOCs Proprieties Throughout the Pre-Implantation Stage

The progression from acentrosomal spindle formation in meiosis to eventual establishment of mitosis with bona fide centrioles requires de novo generation of the centriole, a notion that contrasts with the classical view of template-dependence for centriole duplication. Recent studies have begun to shed light on centriole biogenesis under experimental conditions that lead to centriole regeneration after their removal (Khodjakov et al. 2002; La Terra et al. 2005; Basto et al. 2006). In the mouse pre-implantation embryo, the centriole regeneration can be observed and analysed under physiological conditions, making the mouse model uniquely attractive for centriole de novo formation studies.

Non-centriolar MTOCs were initially described as electron-dense fibrillar material (Szollosi et al. 1972) and subsequently confirmed by an anti-PCM antibody (Calarco-Gillam et al. 1983) in mouse early embryonic cells, although their electron-microscopic structure remains to be characterised. Those MTOCs are capable of polymerising microtubules recruited through gamma-tubulin (Gueth-hallonet et al. 1993) and pericentrin (Courtois et al. 2012) and organise the microtubule network. In contrast to the dominant role of the centriole, the small non-centriolar MTOCs act in a concerted fashion and tend to a stochastic organisation of the spindle and microtubule network. In some cases, non-centriolar MTOCs follow the rearranged microtubule network (Houliston et al. 1987), suggesting additional organisation clues after compaction. In the mouse zygote, about 70 non-centriolar MTOCs are identifiable, which gradually increase to around 120 in total in the 16-cell stage embryo. Interestingly, individual MTOCs become progressively smaller during this period, suggesting that these MTOCs may actually be formed by splitting them in the parent cell. Alternatively, MTOC size might be regulated according to cell size or embryo stage.

A previous electron microscopy study demonstrated the presence of centriole in the 64-cell stage mouse embryo at blastocyst (Gueth-Hallonet et al. 1993). Our recent study consistently identified the emergence of centrin-positive MTOCs at E3.5, in the divisions from 32- to 64-cell stage (Courtois et al. 2012). The presence of two intense microtubule asters next to each other on the nuclear surface at prophase also supports the notion that the centrosome regenerates shortly after 32-cell stage, as proposed by Calarco-Gillam and colleagues (1983). This centriole emergence is asynchronous in timing during development, since not all cells exhibit centrin at early E3.5. This asynchronous transition is similar to that in kinesin-5-dependence of the spindle (Fitzharris 2009). In addition, no difference between trophectoderm and inner cell mass cells in the timing of centriole emergence was observed (Courtois et al. 2012). It will be particularly interesting to examine whether the centrosomal components become integrated in each cell in the same or different order; the former would suggest a hierarchical assembly, while in the latter, a stochastic self-assembly to achieve de novo centriole formation.

The cues for centriole reappearance in vivo in the early mouse embryo also remain to be investigated. The progressive decrease in size and number of MTOCs per cell prompts us to consider the possibility that a progressive reduction of the MTOC components at the E3.5 stage might trigger the centrosomal cycle, as might other cellular characteristics that are also changing during this developmental stage, e.g., cleavage duration and interphase length. Alternatively, centriole generation may be induced simply by transcriptional activation of some essential component. However, it remains to be shown whether the centriole is entirely established de novo and not through as yet unknown cues introduced by the sperm or the oocyte.

These progressive changes in MTOCs, spindle assembly and proprieties, accompanied by the establishment of the centriole, indicate a gradual transition from meiosis to mitosis throughout the pre-implantation stage in the early mouse embryo.

6.4 Concluding Remarks

The fertilisation event classically marks the transition from meiosis to mitosis. However, Kubiak et al. (2008) recently proposed the possibility of a meiosis-to-mitosis transition phase, based on their analysis suggesting that while the first mitotic division in the mouse embryo shares cell cycle control such as spindle assembly checkpoint mechanism with oocytes, it differs from meiosis in the activity of cytostatic factors and mechanisms for prolonging the first division. Our recent study (Courtois et al. 2012) provided evidence for this transition phase from meiosis to mitosis during mouse pre-implantation development, based on characteristics and proprieties of the MTOCs and spindle. The first phase until 8-cell stage exhibits similarities with meiosis in many aspects in agreement with Kubiak et al. (2008). The actual transition phase starts at compaction until establishment of the typical mitotic division by late E3.5 blastocyst, a time frame that coincides with the changes in spindle proprieties demonstrated by Fitzharris (2009) and Schatten et al. (1985). This transition phase, during which the centriole reappears de novo under physiological conditions, offers a promising focal point for studies of the mechanisms underlying centrosome biogenesis in vivo.

These recent findings raise several key questions: How is the centriole assembled de novo and what triggers the assembly? Is there a similar transition phase in other mammals? In-depth investigation of the structure and composition of the MTOCs during the transition in the mouse pre-implantation embryo would provide further insight into these fundamental questions in cellular and developmental biology.

References

Basto R, Lau J, Vinogradova T, Gardiol A, Woods CG, Khodjakov A, Raff JW (2006) Flies without centrioles. Cell 125:1375–1386

Calarco-Gillam PD, Siebert MC, Hubble R, Mitchison T, Kirschner M (1983) Centrosome development in early mouse embryos as defined by an autoantibody against pericentriolar material. Cell 35:621–629

Courtois A, Schuh M, Ellenberg J, Hiiragi T. (2012) The transition from meiotic to mitotic spindle assembly proceeds gradually during early mammalian development. J Cell Biol 198(3): 357–370

Ciemerych MA, Maro B, Kubiak JZ (1999) Control of duration of the first two mitoses in a mouse embryo. Zygote 7(4):293–399

Fitzharris G (2009) A shift from kinesin 5-dependent metaphase spindle function during preimplantation development in mouse. Development 136:2111–2119

Gueth-Hallonet C, Antony C, Aghion J, Santa-Maria A, Lajoie-Mazenc I, Wright M, Maro B (1993) gamma-Tubulin is present in acentriolar MTOCs during early mouse development. J Cell Sci 105(Pt 1):157–166

Houliston E, Pickering SJ, Maro B (1987) Redistribution of microtubules and pericentriolar material during the development of polarity in mouse blastomeres. J Cell Biol 104:1299–1308

Kaufman MH (1973) Timing of the first cleavage division of the mouse and the duration of its component stages: a study of living and fixed eggs. J Cell Sci 12:799–808

Khodjakov A, Rieder CL, Sluder G, Cassels G, Sibon O, Wang C-L (2002) De novo formation of centrosomes in vertebrate cells arrested during S phase. J Cell Biol 158:1171–1181
Kimble J (2011) Molecular regulation of the mitosis/meiosis decision in multicellular organisms. Cold Spring Harb Perspect Biol 3:a002683
Kubiak JZ, Ciemerych MA, Hupalowska A, Sikora-Polaczek M, Polanski Z (2008) On the transition from the meiotic to mitotic cell cycle during early mouse development. Int J Dev Biol 52:201–217
Kubiak JZ, Prigent C (2012) The centrosome life story in Xenopus leavis. In: Schatten H (Ed.) The centrosome: cell and molecular mechanisms of functions and dysfunctions in disease. Humana Press (in press)
La Terra S, English CN, Hergert P, McEwen BF, Sluder G, Khodjakov A (2005) The de novo centriole assembly pathway in HeLa cells: cell cycle progression and centriole assembly/maturation. J Cell Biol 168:713–722
Lüders J, Stearns T (2007) Microtubule-organizing centres: a re-evaluation. Nat Rev Mol Cell Biol 8:161–167
Mahoney NM, Goshima G, Douglass AD, Vale RD (2006) Making microtubules and mitotic spindles in cells without functional centrosomes. Curr Biol 16:564–569
Manandhar G, Sutovsky P, Joshi HC, Stearns T, Schatten G (1998) Centrosome reduction during mouse spermiogenesis. Dev Biol 203:424–434
Megraw TL, Kao LR, Kaufman TC (2001) Zygotic development without functional mitotic centrosomes. Curr Biol 11:116–120
Schatten G, Simerly C, Schatten H (1985) Microtubule configurations during fertilization, mitosis, and early development in the mouse and the requirement for egg microtubule-mediated motility during mammalian fertilization. Proc Natl Acad Sci USA 82:4152–4156
Schatten H, Schatten G, Mazia D, Balczon R, Simerly C (1986) Behavior of centrosomes during fertilization and cell division in mouse oocytes and in sea urchin eggs. Proc Natl Acad Sci USA 83:105–109
Schuh M, Ellenberg J (2007) Self-organization of MTOCs replaces centrosome function during acentrosomal spindle assembly in live mouse oocytes. Cell 130:484–498
Szollosi D, Calarco P, Donahue RP (1972) Absence of centrioles in the first and second meiotic spindles of mouse oocytes. J Cell Sci 11:521–541
Wassarman PM, Fujiwara K (1978) Immunofluorescent anti-tubulin staining of spindles during meiotic maturation of mouse oocytes in vitro. J Cell Sci 29:171–188
Wolgemuth DJ, Rhee K, Wu S, Ravnik SE (1995) Genetic control of mitosis, meiosis and cellular differentiation during mammalian spermatogenesis. Reprod Fertil Dev 7(4):669–683

Chapter 7
Maternal Control of Mouse Preimplantation Development

Wenjing Zheng and Kui Liu

Abstract Mammalian preimplantation development is a process of dedifferentiation from the terminally differentiated eggs to the totipotent blastomeres at the cleavage stage, and then to the pluripotent cells at the blastocyst stage. Maternal factors that accumulate during oogenesis dominate early preimplantation development until the embryonic factors gain control after the activation of the embryonic genome. Recently, a handful of maternal factors that are encoded by the maternal-effect genes have been characterized in genetically modified mouse models. These factors are shown to participate in many aspects of preimplantation development, such as the degradation of maternal macromolecues, epigenetic modification, protein translation, cellular signaling transduction, and cell compaction. Even so, little is known about the interactions between different maternal factors. In this chapter, we have summarized the functions of known maternal factors and hopefully this will lead to a better understanding of the regulation of preimplantation embryogenesis by the maternal regulatory network.

7.1 Introduction

In mice, preimplantation development is generally defined as the period from the formation of the one-cell zygote by fertilization to the implantation of the expanded blastocyst into the uterine wall. During this period, the dividing embryo travels down the oviduct to the uterus and goes through three principal phases: (1) from fertilization to the two-cell stage, which is mainly controlled by the maternal factors stored in the oocyte; (2) from the late two-cell stage after the embryonic genome activation (EGA) to the formation of compacted morula, which is controlled by

W. Zheng (✉) • K. Liu (✉)
Department of Cell and Molecular Biology, University of Gothenburg, Medicinaregatan 9C, 40530 Gothenburg, Sweden
e-mail: wenjing.zheng@gu.se; kui.liu@gu.se

J.Z. Kubiak (ed.), *Mouse Development*, Results and Problems in Cell Differentiation 55, DOI 10.1007/978-3-642-30406-4_7, © Springer-Verlag Berlin Heidelberg 2012

declining maternal factors and increasing newly synthesized factors from the embryonic genome; (3) formation of the blastocyst with fluid-filled blastocyst cavity and two distinct cell lineages: the inner cell mass (ICM), which is pluripotent and will give rise to the embryo proper, and the trophectoderm (TE), which will form the extraembryonic tissues.

During follicular development within the ovary, maternal RNAs and proteins that are derived from the maternal genome accumulate within oocytes, with a concomitant ~300-fold increase in volume (Liu et al. 2006). Some of these oocyte-derived macromolecules are believed to be dispensable for oocyte development and fertilization, but are essential for sustaining the early embryogenesis, at least prior to the robust transcription from the embryonic genome. This phenomenon is therefore called the maternal effect (Schultz 2002; Li et al. 2010). In this chapter we focus on the stage-specific control of preimplantation development by the maternal factors.

7.2 Maternal Control of the Egg-to-Embryo Transition

7.2.1 Formation of the Zygotic Nucleus After Fertilization

After the sperm–egg fusion, the haploid DNA from the sperm forms the paternal pronucleus (PN), whereas other subcellular organelles and macromolecules within the cytoplasm of the sperm head do not contribute to the embryogenesis (Latham 1999). Thus, the maternal factors in the egg are believed to dominate the subsequent reunion of maternal and paternal PNs, and finally the constitution of a diploid zygotic genome.

As major architectural components, maternal histone proteins are thought to be crucial for this global genome reorganization. In the mouse, shortly after fertilization, the sperm-specific histone-like proteins (protamines) are rapidly replaced by the oocyte-specific linker histone H1Foo (H1 histone family, member O, oocyte-specific) (Becker et al. 2005). Similarly, during somatic cell nuclear transfer (SCNT) when a nucleus of a somatic cell is introduced into a meiosis II oocyte, the somatic linker histone is replaced by H1Foo within 60 min (Becker et al. 2005). Accordingly, the SCNT-induced genomic reprogramming and cell transdifferentiation are significantly impaired in the absence of H1Foo (Becker et al. 2005; Maki et al. 2010), indicating that H1Foo is required for reprogramming of the somatic-type DNA to a stem cell-type DNA. H1Foo is solely expressed from the germinal vesicle-(GV-) stage oocyte until the late two-cell-stage embryo (Tanaka et al. 2001; Becker et al. 2005). An in vitro chromatin assembly system revealed that maternal H1Foo allows chromatin to be remodeled to an open status by ATP-dependent chromatin remodeling factor, whereas somatic histone H1 prevents this remodeling (Saeki et al. 2005). Thus maternal H1Foo is associated with the establishment of totipotency of the embryonic genome in the one-cell zygote through regulation of the organization of chromatin (Becker et al. 2005). In *Xenopus*, nucleoplasmin, an oocyte-specific nuclear protein, is reported to be essential for sperm chromatin decondensation (SCD) and replacement of the sperm protamines by H1Foo in fertilized eggs (Burglin et al. 1987; Ohsumi and Katagiri 1991). Similarly, null mutation of its ortholog in mouse, NPM2 (nucleoplasmin 2), also delays the SCD (Inoue et al. 2011), revealing the indispensible role of oocyte factors in the processing of paternal PN in one-cell zygotes.

After this reorganization of paternal chromatin, the maternal and paternal PNs migrate to the center of the zygote for DNA replication. A zygotic genome is formed after the breakdown of both PNs, which is the indication of entry into the first cleavage. In mice, a maternal factor called Zar1 (zygote arrest 1) has been found to be essential at this stage. Oocytes from *Zar1*-null females can be fertilized. PN formation and DNA replication also proceed normally in *Zar1*-null one-cell zygotes. However, the maternal and paternal genomes remain separated in discrete PN, arresting the cell cycle progression at the G_2 phase of the first mitotic cycle (Wu et al. 2003). Given that *Zar1* is only detectable in the cytoplasm from oocytes within the primary follicles till one-cell zygotes, it is regarded as the first well-characterized oocyte-specific maternal-effect gene that regulates the association of chromosomes of parental PNs to form the zygotic genome (Table 7.1) (Wu et al. 2003).

In addition, quite apart from its roles in regulating the SCD, NPM2 is required for nuclear and nucleolar organization in oocytes and early embryos. Most *Npm2*-null one-cell zygotes are arrested at the M phase of the first mitotic cell cycle, accompanied by defects in histone H3 deacetylation and heterochromatin formation surrounding the nucleoli (Burns et al. 2003). Thus, NPM2 is probably also involved in the last few steps of the transition from the haploid gamete genomes to the zygotic genome in mice (Table 7.1).

7.2.2 Degradation of Maternal RNAs

In early mouse embryos, maternal macromolecules (mostly proteins and RNAs) are rapidly degraded once their duties have been accomplished, providing a pool of starting materials to generate embryonic macromolecules (Schultz 1993; Li et al. 2010). These cyclic processes are also believed to be governed by the maternal factors and pathways in the oocyte.

In the mouse, the transcription in oocytes has been proposed to be silenced from GV-intact oocytes to late one-cell-stage zygotes (Bachvarova 1985; Bouniol et al. 1995; Aoki et al. 1997). A progressive destruction of maternal RNA is started after the resumption of meiosis. Sixty percent of the maternal RNA is degraded by the late one-cell stage, and more than 90 % is degraded by the mid-two-cell stage (Fig. 7.1) (Bachvarova 1985; Paynton et al. 1988; Alizadeh et al. 2005). Depletion of the maternal RNA has been proposed to take place through the following mechanisms:

Table 7.1 List of maternal-effect genes

Gene	Function	Strategy	References
Phase I: Egg to embryo transition			
H1foo	Replacing sperm protamines	N/A	Becker et al. (2005)
Npm2	1. Sperm chromatin decondensation	KO	Inoue et al. (2011)
	2. Nucleolar organization and heterochromatin formation	KO	Burns et al. (2003)
Zar1	Fuse of parental pronuclei	KO	Wu et al. (2003)
Zfp36l2	Maternal RNA degradation	KO	Ramos et al. (2004)
Dicer	Maternal RNA degradation	Oo-KO	Murchison et al. (2007), Tang et al. (2007)
Ago2	Maternal RNA degradation	Em-RNAi	Lykke-Andersen et al. (2008)
mHR6A	Maternal protein degradation	KO	Roest et al. (2004)
Atg5	Maternal protein degradation	Oo-KO	Tsukamoto et al. (2008)
Phase II: Embryonic genome activation			
Brg1	Chromatin remodeling	Oo-KO	Bultman et al. (2006)
Tif1α	Chromatin remodeling	Em-RNAi	Torres-Padilla and Zernicka-Goetz (2006)
Basonuclin	Transcriptional factor	Oo-RNAi	Ma et al. (2006)
Ctcf	Transcriptional factor	Oo-RNAi	Wan et al. (2008)
Oct4	Transcriptional factor	Em-RNAi	Foygel et al. (2008)
Sox2	Transcriptional factor	KO	Avilion et al. (2003)
		DN	Pan et al. (2011)
Mater	Unknown, components of SCMC	KO	Tong et al. (2000)
Floped	Unknown, components of SCMC	KO	Li et al. (2008a)
Padi6	Formation of cytoplasmic lattices	KO	Yurttas et al. (2008)
Pdk1	Unknown, PI3K signaling pathway	Oo-KO	Zheng et al. (2010)
Tcl1	Unknown, Akt subcellular translocation	KO	Narducci et al. (2002)
Phase III: Beyond embryonic genome activation			
Filia	Maintaining euploidy, components of SCMC	KO	Zheng and Dean (2009)
Cdh1	Junction formation during compaction	KO Oo-KO	Larue et al. (1994), Riethmacher et al. (1995), de Vries et al. (2004)
Dnmt1	Preservation of DMR methylation	Oo-KO	Hirasawa et al. (2008)
Stella	Preservation of DMR methylation	KO	Nakamura et al. (2007)
Zfp57	*De novo* methylation and preservation of DMR methylation	KO	Li et al. (2008b)

N/A no records, *KO* conventional knockout, *Oo-KO* oocyte-specific conditional knockout, *Em-RNAi* introducing siRNA or dsRNA to degrade target mRNAs in embryos, *Oo-RNAi* degradation of target mRNAs through dsRNAs driven by oocyte-specific promoter, *DN* overexpression of the dominant-negative mRNA, *SCMC* subcortical maternal complex, *DMR* differentially methylated region

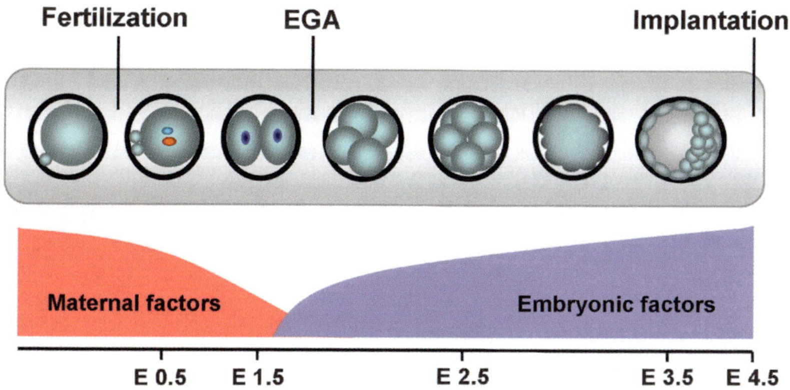

Fig. 7.1 *Preimplantation development in mice*. The time of fertilization is designated embryonic day 0.5 (E0.5). Embryos develop to the two-cell stage at E1.5 when EGA takes place, before they divide to the four- to eight-cell stage at E2.5. The compact morula can be observed at E2.5 or E3.0. A blastocyst with fluid-filled blastocoel cavity is formed at E3.5. The expanded blasocysts become implanted in the uterine wall at E4.5. Note that maternal factors are mostly degraded by EGA in two-cell embryos. Some of the maternal factors persist in the embryos until later stages of development

7.2.2.1 RNA Degradation by Binding of Regulatory Proteins

Tristetraprolin (TTP) is the prototype of the CCCH tandem zinc finger protein family. In somatic cells, TTP has been shown to bind and destabilize mRNAs with an AU-rich region (ARE) at the 3′-UTR (Lai et al. 2000). Disruption of a TTP protein zinc finger protein 36-like 2 (Zfp36l2) in mice results in a cleavage-stage arrest at the two-cell stage. Defectiveness in the degradation of certain types of maternal RNA is associated with developmental arrest, but the mRNAs that are targeted have not been identified (Ramos et al. 2004). The existence of cytoplasmic polyadenylation elements (CPEs) and nuclear polyadenylation signal (NPS) within 60 nucleotides of the 3′-UTR is another property of RNAs that undergo rapid degradation after fertilization, but the maternal factors that target this type of RNA are not yet known (Alizadeh et al. 2005). Efforts are currently being made to identify more maternal factors that act in trans, and their target mRNAs.

7.2.2.2 RNA Degradation by Small Noncoding RNAs

Dicer is a critical factor for the generation of small interfering RNAs (siRNAs) and microRNAs (miRNAs). Fully processed siRNAs and miRNAs are engaged by the RNA-induced silencing complex (RISC) to target certain mRNAs for degradation. Depletion of maternal Dicer from mouse oocytes impairs spindle formation, and the mutant oocytes cease growing in meiosis I (Murchison et al. 2007; Tang et al. 2007). Knockdown of the expression of argonaute 2 (Ago2), a catalytic component of RISC, from the one-cell mouse zygote stabilizes a certain set of maternal RNAs and impairs activation of the embryonic genome, causing developmental arrest at the two-cell stage (Lykke-Andersen et al. 2008). Interestingly, the function of oocytic miRNAs was found to be suppressed during oogenesis and early embryogenesis in a recent study (Ma et al. 2010). Thus, it has been hypothesized that the endogenous siRNA pathway in the oocyte is essential for the degradation of maternal RNAs in oocytes and early embryos, whereas the miRNA pathway may be dispensable at this stage.

7.2.3 Degradation of Maternal Proteins

Clearance of the maternal proteins follows a similar course in mice (Fig. 7.1). The degradation accelerates shortly after fertilization, and only approximately 50 % of the maternal proteins remain in late two-cell embryos (Merz et al. 1981). Evidence is accumulating that the ubiquitin–proteasome system and autophagy are possibly the main pathways for recycling of maternal proteins in early embryos (Table 7.1).

7.2.3.1 Protein Degradation by the Ubiquitin–Proteasome System

In eukaryotic cells, ubiquitination is the prevalent way of selectively targeting proteins for proteolysis. Ubiquitin is activated by E1-activating enzyme and is passed on to E2 ubiquitin-conjugating enzymes. E2 ubiquitin conjugase and E3 ubiquitin ligases selectively target the substrate protein for ubiquitination. The ubiquitinized proteins are finally degraded by proteasomes (Wilkinson 2000; Pickart 2001). Knockout of the mouse homolog of yeast RAD6 (mHR6A), an E2 ubiquitin conjugase, from female mice causes infertility due to the developmental arrest of their embryos after the first cleavage (Roest et al. 2004), thus shedding light on the requirement for a functional maternal ubiquitin system for the early embryonic progression. Ret finger protein-like 4 (Rfpl4) was initially characterized as an E3 ligase exclusively expressed in adult germ cells (Rajkovic et al. 2002). It accumulates in all growing oocytes and disappears in embryos by the eight-cell stage (Suzumori et al. 2003). In vitro assays have indicated that Rfpl4 may bind to mHR6A and target cyclin B1 for degradation (Suzumori et al. 2003). Skp1-Cullin-F-box (SCF) complex is another E3 ligase that targets cell cycle regulators in eukaryotes. The substrate preference of different SCF complexes is determined by F-box proteins. In the mouse, F-box proteins are found to be abundant in both oocytes and early embryos, e.g., F-box/WD repeat-containing protein 15 (Fbxw15) (De La et al. 2008; Wang et al. 2010). Thus, the maternal ubiquitin systems probably contribute to the meiosis-to-mitosis transition by selectively targeting the cell cycle regulation proteins of the oocyte.

7.2.3.2 Autophagy

Autophagy is a major recycling system for massive degradation of damaged or unused cytoplasmic organelles or proteins, and for re-allocation of cellular nutrients under various intracellular and extracellular stimulations (Xie and Klionsky 2007). In metazoans, autophagy takes place in the following steps: signaling and induction; autophagosome nucleation; membrane expansion and vesicle completion; autophagosome targeting, docking, and fusion with the lysosome; and finally, degradation and reexport of breakdown products to the cytoplasm (Melendez and Neufeld 2008). TOR (target of rapamycin) signaling is involved in the induction steps and negatively regulates the autophagy in both yeast and higher eukaryotes (Kamada et al. 2000; Scott et al. 2007). Two ubiquitin-like conjugation pathways—the Atg12 conjugation system (consisting of a complex of Atg12, Atg5, and Atg16) and the Atg8 lipidation system (consisting of Atg8, Atg3, and Atg7)—mediate vesicle expansion and vesicle completion (Ohsumi 2001; Melendez and Neufeld 2008). In the mouse, mTOR signaling is inactivated from the one-cell zygote stage to the four-cell embryo stage, and an experimental fusion protein of LC3 (the mammalian homolog of Atg8) and GFP (green fluorescent protein) has been observed to form dots within the blastomeres, indicating that autophagy is active at this stage (Tsukamoto et al. 2008). Moreover, oocyte-specific *Atg5* knockout females are totally infertile if their eggs are fertilized by *Atg5*-null sperms. All the mutant embryos stop growing at the four- to eight-cell stage even though oogenesis and fertilization were normal in the mother. However, the mutant eggs can give rise to pups if they are fertilized by wild-type (WT) sperms, despite the fact that there is a lower birth rate than in WT females (Tsukamoto et al. 2008). These data indicate that autophagy may be required for preimplantation development, and factors such as Atg5 expressed from the embryonic genome may compensate for the loss of an oocyte-inherited autophagy system in the embryo, although the viability of the embryo is compromised (Tsukamoto et al. 2008).

Collectively, these studies in mice reveal that degradation of RNAs and proteins from the oocyte is supposed to be a prerequisite for the activation of the embryonic genome and preimplantation mitotic progression. A range of oocyte-derived factors have been identified in control of this self-destruction (Table 7.1).

7.3 Maternal Control of Embryonic Genome Activation

In early mouse embryos, clearance of the oocyte-derived factors is coupled to the generation of embryonic products, and the transcripts from the embryonic genome start to dominate the embryonic developmental program after EGA, when the embryo shifts from maternal control to embryonic control (Latham 1999; Latham and Schultz 2001). A minor wave of EGA is detectable as early as in late one-cell mouse embryos and the major wave of EGA takes place in the G_2 phase of two-cell mouse embryos (Schultz 1993; Latham and Schultz 2001). It is generally believed that the epigenetic modifications such as DNA demethylation and chromatin remodeling are the initial steps of EGA; these are followed by activation of RNA polymerase II (pol II) and a range of transcriptional factors (Latham 1999).

7.3.1 The Minor Wave of EGA

Pioneering studies on the expression of plasmid-borne reporter genes within the one-cell mouse zygote first led to the idea that the one-cell zygote is transcriptionally active (Ram and Schultz 1993; Matsumoto et al. 1994; Christians et al. 1995). It has been shown in another study that BrUTP (5-bromouridine 5′-triphosphate) can be incorporated into the newly synthesized RNA and then visualized by immunofluorescence in the one-cell mouse zygote. RNA synthesis is thought to be 30–40 % of that in two-cell embryos, based on the quantification of fluorescence intensity (Aoki et al. 1997). However, addition of α-amanitin, a specific and irreversible inhibitor of RNA pol II in mouse embryos, to a cultured one-cell zygote does not prevent it from dividing into a two-cell embryo (Warner and Hearn 1977; Flach et al. 1982). In addition, it has been found in recent high-throughput studies that basically no α-amanitin-sensitive RNAs (indicating de novo synthesized RNAs) can be detected in one-cell zygotes, leaving the question of whether the minor EGA at the late one-cell stage can generate functional messenger RNAs (Hamatani et al. 2004; Zeng and Schultz 2005).

It has been observed in mouse that a series of epigenetic modifications take place during the one-cell stage, before the first round of DNA replication starts. During gametogenesis, the haploid genomes of the male and female germ cells are highly methylated (Reik 2007). After fertilization, the male PN is actively demethylated by a group of demethylases in the cytoplasm of the oocyte, whereas the female PN is only passively demethylated during the subsequent cleavage events (Oswald et al. 2000; Mayer et al. 2000; Morgan et al. 2005). Meanwhile, histone modifications show a dramatic asymmetry in the pronuclei, with enriched lysine methylation in the female PN and enriched histone H4 acetylation in the male PN, making the latter more accessible for the entry and binding of different transcriptional factors (TFs) (Adenot et al. 1997; Barton et al. 2001; Liu et al. 2004; van der Heijden et al. 2005). Therefore, it has been proposed that the male PN supports a significantly higher level of transcription than the female PN in one-cell zygotes (Aoki et al. 1997). Even so, the asymmetry of transcriptional activity between male and female genomes is found to be lost upon exit from the first mitotic cell cycle (Liu et al. 2005).

7.3.2 The Major Wave of EGA

The robust transcription from the embryonic genome starts at the G2 phase of the second mitotic cell cycle and is crucial for embryonic development beyond the two-cell stage. It has been shown using RNA arrays that genes involved in ribosome biogenesis and assembly, protein synthesis, RNA metabolism, and transcription are transcribed during this major wave of EGA. They are mostly housekeeping genes and are critical for maintaining the cleavage-stage development that follows (Hamatani et al. 2004; Zeng et al. 2004; Zeng and Schultz 2005). Regulation of the major wave of EGA is considered in the following sections.

7.3.2.1 Nuclear Control of EGA

Chromatin Remodeling

Chromatin remodeling refers to the ATP-dependent transient conformational changes of nucleosomes, which alter the accessibility of chromatins to various chromatin proteins that control transcription, DNA replication, recombination, and other biological processes. Thus, chromatin remodeling has long been thought to play an important role in reprogramming of gene expression during EGA (Schultz and Worrad 1995; Thompson et al. 1998). Three classes of chromatin remodeling protein complexes have been characterized in mammalian cells, according to their different catalytic ATPase subunits and associated proteins: SWI/SNF/Brm (switching defective/sucrose non-fermenting/brahma), ISWI (imitation switch), and Mi-2/NuRD (nucleosome remodeling and deacetylase) (Li 2002).

In the mouse, *Brg1* (brahma-related gene 1) encodes a catalytic subunit of the SWI/SNF remodeling complex (Kingston and Narlikar 1999). Oocyte-specific depletion of Brg1 protein has no effect on the oogenesis and fertilization of mutant females, but the resulting embryos all stop growing at the two-cell stage. Expression of TRC (transcription-requiring complex), a 70-kDa protein complex that is commonly accepted as the hallmark of EGA, is significantly downregulated in mutant embryos, accompanied by the defective de novo RNA synthesis at the G2 phase of the two-cell stage. Consequently, the major wave of EGA is impaired in *Brg1* maternal mutant embryos (Bultman et al. 2006). This study, for the first time, directly links the initiation of EGA with the maternal chromatin remodeling effectors.

In addition, Brg1-associated factors have also been found to be essential for the initiation of EGA. TIF1α (transcription intermediary factor 1-alpha) interacts with numerous proteins involved in chromatin structure in *Drosophila* and mammalian somatic cells (Remboutsika et al. 2002; Germain-Desprez et al. 2003). Depletion of TIF1α by microinjection of siRNA leads to aberrant localization of Brg1, RNA pol II, and SNF2H (or Smarca5: SWI/SNF-related, matrix-associated, actin-dependent regulator of chromatin, subfamily a, member 5) in one-cell zygotes. Expression of a subset of genes is downregulated and embryonic development is arrested at the two- to four-cell transition in these injected zygotes (Torres-Padilla and Zernicka-Goetz 2006). These results indicate that TIF1α may be involved in spatial control of the major EGA by influencing the localization of certain remodeling complexes in early embryos.

It is possible that the other two remodeling families may play equally important roles in the initiation of the major wave of EGA, although the mechanisms involved remain elusive. Further studies using the oocyte-specific gene mutation strategy may provide further insights into the temporal and spatial control of chromatin remodeling by maternal factors in early embryos.

Activation of RNA pol II

The notion that RNA pol II is involved as a principal effector of EGA is based on the pioneering studies on embryo culture with α-amanitin. As mentioned in Sect. 6.3.1, α-amanitin selectively and irreversibly inhibits the activity of RNA pol II. Thus, the transcription in early embryos is sensitive to α-amanitin and embryos stop growing at the two-cell stage when cultured with α-amanitin (Flach et al. 1982).

In ovulated mouse eggs, the largest subunit of RNA pol II is predominantly hyperphosphorylated at its carboxy-terminal domain (CTD), which is a sign of permissive transcription. This hyperphosphorylation is lost after fertilization but is regained before the onset of the major wave of EGA, accompanied by gradual translocation of RNA pol II into the nucleus (Bellier et al. 1997). The hyperphosphorylation of CTD is independent of DNA replication but is dependent on protein synthesis, whereas the nuclear importation of RNA pol II is found to be irrelevant to either of these processes (Bellier et al. 1997), revealing the possible existence of an internal zygotic/embryonic clock that regulates the major wave of EGA independent of cell cycle progression. A recent study in mouse embryonic fibroblasts (MEFs) has indicated that Mat1 (menage à trois homolog 1), the kinase subunit of transcription factor IIH (TFIIH), is required for most of the CTD phosphorylation at serine 5 and serine 2 (Helenius et al. 2011). A strong decrease in *RNA pol II* transcription is observed upon deletion of *Mat1*, in combination with defective mRNA capping and splicing (Helenius et al. 2011). Further studies are needed to determine whether oocyte-derived Mat1 is also involved in the CTD phosphorylation and regulation of RNA pol II activity during the major wave of EGA.

Transcriptional Factors

After the epigenetic changes in chromatin structure, the embryonic genome becomes selectively accessible to certain transcription factors to allow the sequential transcription of certain genes during the major wave of EGA.

Four classic transcriptional factors have been documented to have critical roles in early embryogenesis in mice. Sp1 (trans-acting transcription factor 1) belongs to the C_2-H_2 zinc finger family (Briggs et al. 1986). The amount of Sp1 transcript decreases during oocyte maturation and reaches a minimum level at the two-cell stage, after which the amount of this transcript increases progressively to the blastocyst stage, indicating a replacement of oocyte-derived Sp1 by embryo-derived Sp1. The level of Sp1 protein increases gradually during preimplantation development, contributing to an increase in transcription rate at this stage (Worrad and Schultz 1997). TBP (TATA binding protein) is a subunit of the RNA pol II transcription factor TFIID, but it is only essential for transcription by RNA pol I and pol III in mice. Maternal TBP is degraded in embryos at the eight-cell stage, and the embryonic TBP is required for embryos to survive beyond the blastocyst stage (Martianov et al. 2002). ATF1 (activating transcription factor 1) and CREB (cyclic AMP response element binding protein) are basic leucine-zipper transcription factors. Embryos with double knockout of these two genes die before implantation due to developmental arrest, but single knockout of neither gene influences the preimplantation development (Bleckmann et al. 2002). Particularly, the knockout embryos mentioned above are all derived from the fusion of heterozygous oocytes and sperms, due to the embryonic or perinatal lethality. Thus, it is highly possible that the phenotypes of those embryos with null mutations at early developmental stages are actually masked by the proteins inherited from the cytoplasm of the heterozygous oocytes. More studies using the oocyte-specific conditional knockout strategy are needed to elucidate the functions of the above-mentioned transcriptional factors during EGA.

Recently, two more zinc-finger transcriptional factors have been identified using the oocyte-specific siRNA knockdown approach (Table 7.1). Basonuclin is abundant in oocytes and persists in early embryos. Deficiency of basonuclin in mouse oocytes affects both RNA pol I- and II-mediated transcription, and also affects oocyte morphology. Some of these mutant oocytes are fertilizable, but no embryos derived from them grow beyond the two-cell stage (Ma et al. 2006). Similarly, depletion of CTCF (CCCTC-binding factor) in mouse oocytes by siRNA alters the expression of genes related to embryogenesis, and compromises meiotic competence. The resulting embryos experience a restrained EGA and are arrested at various cleavage stages, indicating a strong effect of CTCF on preimplantation development (Wan et al. 2008). These data suggest that the oocyte zinc-finger transcriptional factors are essential for both meiotic maturation in oocytes and EGA in early embryos, functioning either as global regulators of transcription or as regulators of transcription of certain gene families.

The pluripotency-related transcription factors have also been found to be indispensible for the genomic reprogramming during EGA in mice (Table 7.1). Depletion of *Oct4* (octamer-binding transcription factor 4) mRNA in the one-cell zygote of the mouse leads to the arrest of the majority of embryos before the morula stage. The expression of a group of genes involved in translation and RNA processing is altered in the absence of Oct4; these genes are distinct from the Oct4-targeted genes in the embryonic stem cells (ESCs) (Foygel et al. 2008). In addition, Oct4 is also associated with the degradation of two maternal RNAs: *Zar1* and *Nobox* (NOBOX oogenesis homeobox) (Foygel et al. 2008). These data indicate that Oct4 may function in a stage-specific manner during EGA, which is related to the establishment of totipotency in the early mouse blastomeres. *Sox2* (SRY-box containing gene 2) is among the first batch of genes to be expressed during EGA, but its maternally inherited protein persists in embryos throughout the mouse preimplantation stage (Avilion et al. 2003; Pan and Schultz 2011). Maternal SOX2 translocates to the nucleus of two-cell embryos and remains there in the ICM of the blastocyst (Avilion et al. 2003). Intriguingly, overexpression of the WT *Sox2* gene causes more severe defects in EGA and embryonic progression than overexpression of a dominant-negative *Sox2* (*Sox2-D*) (Pan and Schultz 2011). Embryos with WT *Sox2* overexpression are arrested at the two-cell stage, and expression of 15 % of the embryonic genes is altered during the EGA. On the contrary, overexpression of *Sox2-D* arrests the embryos after the two-cell stage but before the eight-cell stage, and has a negative effect on the expression of only 1 % of the embryonic genes (Pan et al. 2011). It is thus highly possible that the activity of these two pluripotency-related transcription factors is fine-tuned to ensure the successful transition from maternal to embryonic control. Probably these factors contribute to the establishment of totipotency during the cleavage stage in a manner that differs from the establishment of pluripotency in ESCs.

7.3.2.2 Cytoplasmic Control of EGA

The cytoplasmic control of EGA relies mainly on the translational and posttranslational regulation of maternally inherited mRNAs and proteins, and of some of the earliest transcripts from the embryonic genome. It apparently plays an equally important role to that played by the nuclear regulatory events at the initiation of EGA, and ensures that it proceeds correctly (Table 7.1) (Latham 1999).

The Subcortical Maternal Complex

Currently, four members have been identified from a protein complex that is prevalent in the subcortical region of oocytes and early embryos in mice: the subcortical maternal complex (SCMC). They are Mater (maternal antigen that embryos require), Floped (factor located in oocytes permitting embryonic development), Tle6 (transducin-like enhancer of split 6) and Filia (RIKEN cDNA 2410004A20) (Li et al. 2008a). Disruption of either of the two core factors, Mater or Floped, may destabilize the SCMC, impair the EGA, and arrest the mouse embryos at the two-cell stage (Tong et al. 2000; Li et al. 2008a). Even so, Filia has been found to be irrelevant to EGA, but is required for normal spindle morphogenesis during preimplantation development in mice (Zheng and Dean 2009). How this oocyte-derived protein complex performs these two seemingly irrelevant functions during early embryogenesis is not known.

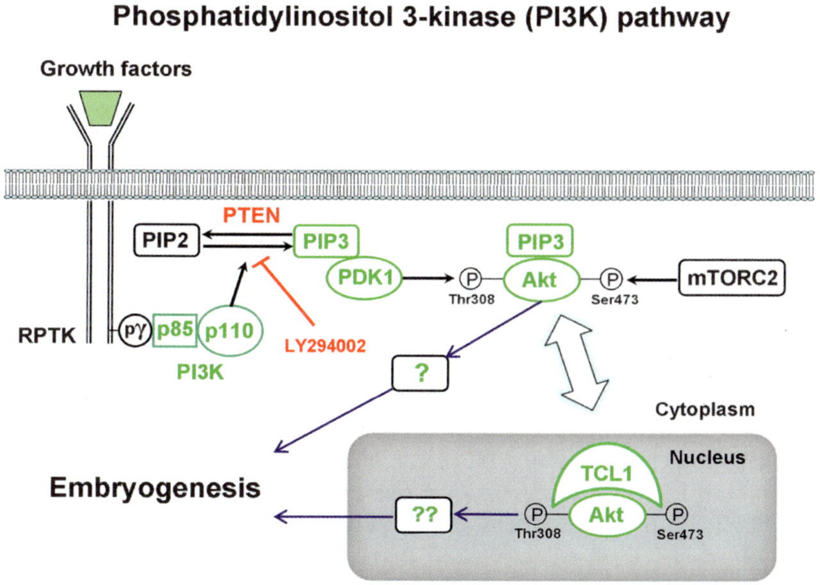

Fig. 7.2 *The phosphatidylinositol 3-kinase (PI3K) pathway*. The class IA PI3Ks are heterodimers consisting of a catalytic subunit (p110) and a regulatory subunit (p85). Upon ligand binding, the RPTK recruits PI3K to the inner membrane area of the cell and activates this lipid kinase to produce PIP3. PIP3 binds to kinases containing PH domains (including PDK1 and Akt) and recruits them to the subcortical area, where PDK1 phosphorylates Akt at threonine 308. Akt is fully activated after being phosphorylated further by mTORC2 at serine 473. The phosphorylated Akt is shuttled to the nucleus during the two-cell stage, with the assistance of TCL1. The maternal PI3K signaling pathway regulates the early embryogenesis through Akt substrates that are yet to be identified

Cytoplasmic lattices (CPLs) store components of the protein synthetic machinery including ribosomes in mouse oocytes, and Padi6 (peptidyl arginine deiminase type VI) has been found to be essential for the formation of CPLs (Hovland et al. 1996; Wright et al. 2003). In *Padi6*-null two-cell embryos, the de novo protein synthesis is altered due to the loss of CPLs, resulting in a defective EGA and developmental arrest at the two-cell stage (Yurttas et al. 2008). Remarkably, Padi6 is located in the subcortical region of early mouse embryos and physically interacts with the SCMC (Li et al. 2008a), giving rise to a notion that early embryos may use the SCMC to organize the temporal and spatial regulation of various maternal factors.

The Maternal Phosphatidylinositol 3-kinase (PI3K) Pathway

In somatic cells, class IA PI3Ks are activated by receptor protein tyrosine kinases (RPTKs) and generate the intracellular second messenger phosphatidylinositol 3,4,5-trisphosphate (PIP3), which promotes cell proliferation and survival.

A phosphatase PTEN (phosphatase and tensin homolog deleted on chromosome 10) converts PIP3 to phosphatidylinositol 4,5-bisphosphate (PIP2), thus negatively regulating the activities of PI3Ks (Fig. 7.2) (Vanhaesebroeck et al. 2001; Cantley 2002). PI3K signaling has been shown to have a dominant role in controlling the oogenesis and folliculogenesis in mice by a series of studies using oocyte-specific conditional knockout mouse models (Reddy et al. 2008, 2009; Zheng et al. 2011). Recently, accumulating evidence strongly suggests that early embryogenesis is also dependent on this oocyte-derived signaling pathway. Embryos from mice and humans can survive and develop in vitro in defined culture media lacking exogenous growth factors or serum (Whitten and Biggers 1968; Edwards et al. 1969), suggesting that early embryos generate intrinsic signals promoting their survival and development (O'Neill 2008). In the mouse, expression of the p85 and p110 subunits of PI3Ks can be detected throughout the preimplantation stage, generating high levels of PIP3 in the subcortical region (Lu et al. 2004; Riley et al. 2005; Halet et al. 2008). Either knockout of PI3K subunits or culture of the embryos with the PI3K inhibitor LY294002 causes embryonic lethality or developmental arrest (Bi et al. 1999, 2002; Brachmann et al. 2005; Halet et al. 2008). Moreover, insulin-like growth factor-I (IGF-I) receptor, IGF-II receptor, and platelet-activating factor (PAF) receptor, which can activate the intracellular PI3K pathway, are expressed in preimplantation embryos (Lighten et al. 1997; Lu et al. 2004), indicating that PI3K signaling is one of the intrinsic signals that the embryos use to maintain their autonomous preimplantation development.

The 3-phosphoinositide-dependent protein kinase-1 (PDK1 or Pdpk1) is a master kinase in PI3K signaling pathway. PIP3 binds to the pleckstrin homology (PH) domain of PDK1 and another master kinase protein kinase B (PKB or Akt), and recruits them to the subcortical region, where PDK1 phosphorylates the activation loop of Akt (Mora et al. 2004). Full activation of Akt is achieved upon further phosphorylation at the hydrophobic motif by mTORC2 (mammalian target of rapamycin complex 2) (Fig. 7.2) (Sarbassov et al. 2005). In the mouse, oocyte-specific deletion of *Pdk1* shows no effects on folliculogenesis, meiosis maturation, or fertilization, but the resultant embryos are arrested at the two-cell stage and EGA is obviously impaired (Zheng et al. 2010). Nonetheless, these developmental defects can be rescued by the restoration of Akt activity through concurrent oocyte-specific deletion of *Pten* (Zheng et al. 2010). In particular, Akt is found to be retained in the cytoplasm in the one-cell mouse zygote but is transiently transported to the nucleus in the mid-two-cell stage with the assistance of TCL1 (T cell leukemia/lymphoma1) (Fiorenza et al. 2008). Disruption of this TCL1-mediated cell cycle dependent Akt shuttling significantly inhibits embryonic progression in mice (Narducci et al. 2002). Thus, it has been postulated that the maternal PI3K signaling pathway plays pivotal roles in the regulation of EGA and early embryonic progression (Fig. 7.2). Further studies are needed to identify the downstream effectors that are directly involved in the onset of EGA and to determine how this signaling pathway interacts with other maternal factors.

7.4 The Influence of Maternal Factors Beyond EGA

7.4.1 Compaction and Formation of the Blastocyst

In mice, after three mitotic cell cycles, the embryos compact and form individually polarized cells that are tightly connected by adhesive junctions (Ducibella et al. 1977). Inhibition of mRNA synthesis at the four-cell stage has no effect on the compaction, while inhibition of protein synthesis even induces premature compaction at this stage (Kidder and McLachlin 1985). Thus, the initiation of compaction has been linked to the posttranslational control of maternal factors, or factors produced before the four-cell stage (Kidder and McLachlin 1985; Cockburn and Rossant 2010). As the major component of adherens junctions (AJs), E-cadherin (or CDH1) becomes localized to the cell boundaries during the process of compaction (Vestweber et al. 1987). Maternal E-cadherin accumulates during oogenesis. Studies using conventional knockout mouse models have shown that the maternal E-cadherin inherited from heterozygous oocytes is sufficient to support the compaction of E-cadherin-null embryos at the eight-cell stage, but the compaction is lost at the blastocyst stage (Larue et al. 1994; Riethmacher et al. 1995). In contrast, oocyte-specific depletion of E-cadherin inhibits the compaction of the resulting eight-cell mouse morula, but the embryonic E-cadherin rescues the compaction before the 16-cell stage (de Vries et al. 2004). Moreover, phosphorylation of both E-cadherin and its binding partner β-catenin is only detectable during compaction, which is known to increase their binding affinity to form AJs (Pauken and Capco 1999). It has been found that forced phosphorylation of these two proteins through artificial activation of protein kinase C (PKC) signaling induces the premature compaction at the four-cell stage, whereas culture of the embryos with PKC inhibitors disrupts the initiation of compaction (Winkel et al. 1990; Pauken and Capco 1999). Based on these data, a hypothesis has been put forward that the initiation of compaction is regulated by PKC signaling through phosphorylation of E-cadherin and β-catenin (Pauken and Capco 1999; Cockburn and Rossant 2010).

After compaction, the blastomeres become polarized and form the apical region (outward-facing) and the basolateral region (inward-facing). Then, at the 16-cell stage, the asymmetrical cell division generates two distinct cell groups: the outer cells, which develop into trophectoderm, and the inner cells, which develop into ICM (Johnson and McConnell 2004; Marikawa and Alarcon 2012; Suwińska 2012; Xenopoulos et al. 2012; Morris and Zernicka-Goetz 2012). Polarization of different cell types is commonly regulated by a set of evolutionarily conserved proteins called the PAR (partition-defective)-aPKC (atypical protein kinase C) system (Ohno 2001). Par genes were initially found to be essential for the asymmetric division of the *Caenorhabditis elegans* zygote (Etemad-Moghadam et al. 1995; Guo and Kemphues 1995; Suzuki and Ohno 2006). It has been reported that in mouse morulas, Par3, Par6, and aPKC all accumulate in the apical region, whereas Par1 and Lethal giant larvae (Lgl) are present in greatest quantity in the basolateral region, indicating that different PAR–aPKC complexes may function mutually to regulate the cell polarization (Suzuki and Ohno 2006). Moreover, injection of dsRNA against *Par3* or mRNA encoding a dominant negative form of aPKC into mouse blastomeres at the four-cell stage has been found to inhibit the injected cells from developing into trophectoderm (Plusa et al. 2005). It is thus highly possible that the embryonic but not the maternal PAR–aPKC is critical for polarization of blastomeres in eight-cell morulas and subsequently influences the specification of cell lineage.

In the mouse, the SCMC becomes localized in the apical region of blastomeres at the morula stage, and trophectoderm cells at the blastocyst stage (Ohsugi et al. 2008; Li et al. 2008a). In contrast, PIP3, the intracellular second messenger, is enriched near the boundaries of the apical region in blastomeres at the morula stage, but is retained in ICM cells at the blastocyst stage. Culture of mouse embryos with LY294002, the inhibitor of PI3K, significantly increases the apoptosis during the morula–blastocyst transition (Halet et al. 2008). This asymmetric but highly organized redistribution of protein probably contributes to the cell polarization and lineage specification, but more experimental evidence is needed to ascertain this hypothesis.

As the majority of maternal factors are degraded after EGA, development of the blastocyst is supposed to be largely dominated by embryonic products (Fig. 7.1). The detailed mechanisms of specification of cell lineage and determination of cell fate are discussed in separate chapters.

7.4.2 DNA Methylation

DNA methylation mainly refers to the addition of a methyl group to the 5′ position of the pyrimidine ring of cytosine. It occurs primarily at CpG dinucleotides in mammals and is linked to the repressive chromatin states (Bird 2002). There are three well-documented universal DNA methyltransferases: Dnmt1, Dnmt3a, and Dnmt3b. Dnmt3a and Dnmt3b are both required for de novo methylation, and Dnmt1 is responsible for the maintenance of methylation during DNA replication (Chen and Li 2004). Dnmt1 has two isoforms, the somatic form (Dnmt1s) and the oocyte-specific form (Dnmt1o). Interestingly, it has been reported that Dnmt1o is inherited from the oocyte and retained in the cytoplasm of early blastomeres all through the preimplantation stage, whereas Dnmt1s is expressed after EGA and is located in the nucleus from the two-cell stage onwards (Hirasawa et al. 2008).

As mentioned in Sect. 6.3.1, the DNA in male and female germ cells is highly methylated and imprinted. Many imprinted genes have differentially methylated regions (DMRs), which are methylated differently between the parental alleles and are critical for stabilization of the genome, inactivation of the X-chromosome in females, and suppression of retroposon transcription (Li et al. 1993; Bestor 2000; Kaneda et al. 2004). Although the genome in early embryos undergoes global epigenetic reprogramming, losing most of its methylation during preimplantation development, the methylations in most DMRs are faithfully maintained (Reik and

Walter 2001; Branco et al. 2008). *Dnmt1o*-null female mice are viable. However, when they are mated with WT males, the resulting heterozygous fetuses die in the uterus during the final third of the gestation period, with a loss of allele-specific expression and methylation at certain imprinted loci (Howell et al. 2001). Indeed, it has been found in mice that imprintings on DMRs are maintained by Dnmt1o before the EGA at the two-cell stage, and by Dnmt1s after that (Hirasawa et al. 2008). In contrast, neither Dnmt3a nor Dnmt3b is essential for preimplantation development, which is consistent with the absence of de novo DNA methylation at this stage (Hirasawa et al. 2008). Thus, *Dnmt1* is regarded as a maternal-effect gene that protects the DNA methylation on DMRs against the global demethylation during preimplantation embryogenesis.

Except for the common DNA methyltransferases mentioned above, two other maternal factors have been found to be critical for the regulation of DNA methylation in early mouse embryos: Stella (or Dppa3, developmental pluripotency-associated 3) and Zfp57 (zinc-finger protein 57) (Table 7.1). The early expression of Stella is detected at late streak stage at approximately E7.0 and remains in the germ line until about E15.5 in male gonads and until about E13.5 in female gonads. Stella protein is again detectable in immature oocytes of newborn female mice and persists in their embryos until the blastocyst stage (Sato et al. 2002; Saitou et al. 2002). Stella contains both a nuclear localization signal (NLS) and a nuclear export signal (NES). It shuttles from the cytoplasm to the pronuclei after fertilization, and is then retained in the embryonic nucleus all through the preimplantation stage (Payer et al. 2003; Nakamura et al. 2007). Loss-of-function studies have indicated that most embryos derived from *Stella*-null oocytes die before reaching the blastocyst stage. Although expression of Stella from the paternal allele starts after EGA, it is insufficient to rescue the abnormalities caused by the absence of maternal Stella during preimplantation development (Payer et al. 2003). It has been shown by later studies that maternal Stella is critical for the maintenance of methylation of most of the maternal genome, and of at least two paternally imprinted genes in early embryos (Nakamura et al. 2007).

Another maternal-effect gene, Zfp57 (zinc-finger protein 57), was initially identified as a key transcriptional factor downstream of STAT3 (signal transducer and activator of transcription 3) and Oct4 in ESCs (Akagi et al. 2005). Its expression is found to be downregulated upon ESC differentiation (Li and Leder 2007). In the mouse oocyte, Zfp57 is essential for de novo methylation at the Snrpn (Small nuclear ribonucleoprotein polypeptide N) region, one of the female DMRs that has been implicated in Prader-Willi and Angelman syndromes in humans (Li et al. 2008b). Oocytic Zfp57 persists in early embryos until the blastocyst stage at E3.5, when the embryonic Zfp57 is both expressed and detectable. Depletion of both maternal and embryonic Zfp57 leads to embryonic lethality, caused by the loss of methylation on both female and male DMRs. Intriguingly, although the methylation of the Snrpn region is lost in oocytes and resulting embryos by depletion of maternal Zfp57, it can be rescued by the expression of paternal Zfp57 in the blastocyst after E3.5, revealing a differential role of maternal and embryonic Zfp57 in the regulation of the DNA methylation of certain DMRs in mice (Li et al. 2008b).

Collectively, the DNA methylation patterns in early embryos have been found to influence the embryonic development both before and after implantation. Both maternal and embryonic DNA methyltransferases are essential for the maintenance of perfect DNA methylation on DMRs, in order to regulate the parent of origin-specific transcription. In particular, methylation of some of the DMRs may be controlled by more than one DNA methyltransferases (Table 7.1), but how these maternal factors interact with each other and whether they directly or indirectly regulate the DNA methylation is still unknown.

7.5 Conclusions

Preimplantation development is regarded as a turnover from terminally differentiated gametes to totipotent or pluripotent blastomeres, which involves the reprogramming of the genome as well as a drastic replacement of cytoplasmic macromolecues. Strikingly, early preimplantation development is dominated by factors that are inherited from oocytes—at least prior to EGA, at the two-cell stage. Some of these factors even persist after EGA, and influence the preimplantation and postimplantation development that follows. Since the first two maternal-effect genes were discovered in 2000, less than 30 maternal-effect genes have been identified in mice participating in PN formation, maternal RNA/protein degradation, histone remodeling, activation of RNA pol II and the transcriptional machinery, morula compaction, DNA methylation, etc. (Table 7.1). Most of the maternal-effect genes have been determined to be essential for normal embryonic progression, and may cause embryonic lethality in their absence. Even so, the exact functions of most maternal-effect genes are still largely unknown, including the genes encoding the components of the SCMC. The emerging roles of maternal signaling transduction indicate that maternal factors may function in coordination, rather than independently, in the regulation of early embryogenesis. More studies employing the conditional knockout strategy and high-throughput approaches such as RNA/siRNA deep sequencing and proteomics may improve our understanding of how maternal factors orchestrate their self-destruction under the zygotic/embryonic clock to accomplish the transition from egg to embryo, and how the embryonic genome is reprogrammed and activated. Hopefully, the knowledge obtained from such studies will also benefit other research fields such as induced pluripotent stem cells (iPSCs), regarding factors that enhance the genomic, and human preimplantation genetic diagnosis (PGD), regarding molecular markers that indicate the viability of the embryo.

References

Adenot PG, Mercier Y, Renard JP, Thompson EM (1997) Differential H4 acetylation of paternal and maternal chromatin precedes DNA replication and differential transcriptional activity in pronuclei of 1-cell mouse embryos. Development 124:4615–4625

Akagi T, Usuda M, Matsuda T, Ko MS, Niwa H, Asano M, Koide H, Yokota T (2005) Identification of Zfp-57 as a downstream molecule of STAT3 and Oct-3/4 in embryonic stem cells. Biochem Biophys Res Commun 331:23–30

Alizadeh Z, Kageyama S, Aoki F (2005) Degradation of maternal mRNA in mouse embryos: selective degradation of specific mRNAs after fertilization. Mol Reprod Dev 72:281–290

Aoki F, Worrad DM, Schultz RM (1997) Regulation of transcriptional activity during the first and second cell cycles in the preimplantation mouse embryo. Dev Biol 181:296–307

Avilion AA, Nicolis SK, Pevny LH, Perez L, Vivian N, Lovell-Badge R (2003) Multipotent cell lineages in early mouse development depend on SOX2 function. Genes Dev 17:126–140

Bachvarova R (1985) Gene expression during oogenesis and oocyte development in mammals. Dev Biol (NY 1985) 1:453–524

Barton SC, Arney KL, Shi W, Niveleau A, Fundele R, Surani MA, Haaf T (2001) Genome-wide methylation patterns in normal and uniparental early mouse embryos. Hum Mol Genet 10:2983–2987

Becker M, Becker A, Miyara F, Han ZM, Kihara M, Brown DT, Hager GL, Latham K, Adashi EY, Misteli T (2005) Differential in vivo binding dynamics of somatic and oocyte-specific linker histones in oocytes and during ES cell nuclear transfer. Mol Biol Cell 16:3887–3895

Bellier S, Chastant S, Adenot P, Vincent M, Renard JP, Bensaude O (1997) Nuclear translocation and carboxyl-terminal domain phosphorylation of RNA polymerase II delineate the two phases of zygotic gene activation in mammalian embryos. EMBO J 16:6250–6262

Bestor TH (2000) The DNA methyltransferases of mammals. Hum Mol Genet 9:2395–2402

Bi L, Okabe I, Bernard DJ, Nussbaum RL (2002) Early embryonic lethality in mice deficient in the p110beta catalytic subunit of PI 3-kinase. Mamm Genome 13:169–172

Bi L, Okabe I, Bernard DJ, Wynshaw-Boris A, Nussbaum RL (1999) Proliferative defect and embryonic lethality in mice homozygous for a deletion in the p110alpha subunit of phosphoinositide 3-kinase. J Biol Chem 274:10963–10968

Bird A (2002) DNA methylation patterns and epigenetic memory. Genes Dev 16:6–21

Bleckmann SC, Blendy JA, Rudolph D, Monaghan AP, Schmid W, Schutz G (2002) Activating transcription factor 1 and CREB are important for cell survival during early mouse development. Mol Cell Biol 22:1919–1925

Bouniol C, Nguyen E, Debey P (1995) Endogenous transcription occurs at the 1-cell stage in the mouse embryo. Exp Cell Res 218:57–62

Brachmann SM, Yballe CM, Innocenti M, Deane JA, Fruman DA, Thomas SM, Cantley LC (2005) Role of phosphoinositide 3-kinase regulatory isoforms in development and actin rearrangement. Mol Cell Biol 25:2593–2606

Branco MR, Oda M, Reik W (2008) Safeguarding parental identity: Dnmt1 maintains imprints during epigenetic reprogramming in early embryogenesis. Genes Dev 22:1567–1571

Briggs MR, Kadonaga JT, Bell SP, Tjian R (1986) Purification and biochemical characterization of the promoter-specific transcription factor, Sp1. Science 234:47–52

Bultman SJ, Gebuhr TC, Pan H, Svoboda P, Schultz RM, Magnuson T (2006) Maternal BRG1 regulates zygotic genome activation in the mouse. Genes Dev 20:1744–1754

Burglin TR, Mattaj IW, Newmeyer DD, Zeller R, Derobertis EM (1987) Cloning of nucleoplasmin from Xenopus-Laevis oocytes and analysis of its developmental expression. Genes Dev 1:97–107

Burns KH, Viveiros MM, Ren Y, Wang P, DeMayo FJ, Frail DE, Eppig JJ, Matzuk MM (2003) Roles of NPM2 in chromatin and nucleolar organization in oocytes and embryos. Science 300:633–636

Cantley LC (2002) The phosphoinositide 3-kinase pathway. Science 296:1655–1657

Chen T, Li E (2004) Structure and function of eukaryotic DNA methyltransferases. Curr Top Dev Biol 60:55–89
Christians E, Campion E, Thompson EM, Renard JP (1995) Expression of the HSP 70.1 gene, a landmark of early zygotic activity in the mouse embryo, is restricted to the first burst of transcription. Development 121:113–122
Cockburn K, Rossant J (2010) Making the blastocyst: lessons from the mouse. J Clin Invest 120:995–1003
De La CE, Kerr B, Paredes A, Merchant-Larios H, Mendez JP, Ojeda SR (2008) Fbxw15/Fbxo12J is an F-box protein-encoding gene selectively expressed in oocytes of the mouse ovary. Biol Reprod 78:714–725
de Vries WN, Evsikov AV, Haac BE, Fancher KS, Holbrook AE, Kemler R, Solter D, Knowles BB (2004) Maternal beta-catenin and E-cadherin in mouse development. Development 131:4435–4445
Ducibella T, Ukena T, Karnovsky M, Anderson E (1977) Changes in cell surface and cortical cytoplasmic organization during early embryogenesis in the preimplantation mouse embryo. J Cell Biol 74:153–167
Edwards RG, Bavister BD, Steptoe PC (1969) Early stages of fertilization in vitro of human oocytes matured in vitro. Nature 221:632–635
Etemad-Moghadam B, Guo S, Kemphues KJ (1995) Asymmetrically distributed PAR-3 protein contributes to cell polarity and spindle alignment in early C. elegans embryos. Cell 83:743–752
Fiorenza MT, Torcia S, Canterini S, Bevilacqua A, Narducci MG, Ragone G, Croce CM, Russo G, Mangia F (2008) TCL1 promotes blastomere proliferation through nuclear transfer, but not direct phosphorylation, of AKT/PKB in early mouse embryos. Cell Death Differ 15:420–422
Flach G, Johnson MH, Braude PR, Taylor RA, Bolton VN (1982) The transition from maternal to embryonic control in the 2-cell mouse embryo. EMBO J 1:681–686
Foygel K, Choi B, Jun S, Leong DE, Lee A, Wong CC, Zuo E, Eckart M, Reijo Pera RA, Wong WH, Yao MW (2008) A novel and critical role for Oct4 as a regulator of the maternal-embryonic transition. PLoS One 3:e4109
Germain-Desprez D, Bazinet M, Bouvier M, Aubry M (2003) Oligomerization of transcriptional intermediary factor 1 regulators and interaction with ZNF74 nuclear matrix protein revealed by bioluminescence resonance energy transfer in living cells. J Biol Chem 278:22367–22373
Guo S, Kemphues KJ (1995) par-1, a gene required for establishing polarity in C. elegans embryos, encodes a putative Ser/Thr kinase that is asymmetrically distributed. Cell 81:611–620
Halet G, Viard P, Carroll J (2008) Constitutive PtdIns(3,4,5)P3 synthesis promotes the development and survival of early mammalian embryos. Development 135:425–429
Hamatani T, Carter MG, Sharov AA, Ko MS (2004) Dynamics of global gene expression changes during mouse preimplantation development. Dev Cell 6:117–131
Helenius K, Yang Y, Tselykh TV, Pessa HK, Frilander MJ, Makela TP (2011) Requirement of TFIIH kinase subunit Mat1 for RNA Pol II C-terminal domain Ser5 phosphorylation, transcription and mRNA turnover. Nucleic Acids Res 39:5025–5035
Hirasawa R, Chiba H, Kaneda M, Tajima S, Li E, Jaenisch R, Sasaki H (2008) Maternal and zygotic Dnmt1 are necessary and sufficient for the maintenance of DNA methylation imprints during preimplantation development. Genes Dev 22:1607–1616
Hovland R, Hesketh JE, Pryme IF (1996) The compartmentalization of protein synthesis: importance of cytoskeleton and role in mRNA targeting. Int J Biochem Cell Biol 28:1089–1105
Howell CY, Bestor TH, Ding F, Latham KE, Mertineit C, Trasler JM, Chaillet JR (2001) Genomic imprinting disrupted by a maternal effect mutation in the Dnmt1 gene. Cell 104:829–838
Inoue A, Ogushi S, Saitou M, Suzuki MG, Aoki F (2011) Involvement of mouse nucleoplasmin 2 in the decondensation of sperm chromatin after fertilization. Biol Reprod 85:70–77
Johnson MH, McConnell JM (2004) Lineage allocation and cell polarity during mouse embryogenesis. Semin Cell Dev Biol 15:583–597
Kamada Y, Funakoshi T, Shintani T, Nagano K, Ohsumi M, Ohsumi Y (2000) Tor-mediated induction of autophagy via an Apg1 protein kinase complex. J Cell Biol 150:1507–1513

Kaneda M, Okano M, Hata K, Sado T, Tsujimoto N, Li E, Sasaki H (2004) Essential role for de novo DNA methyltransferase Dnmt3a in paternal and maternal imprinting. Nature 429:900–903
Kidder GM, McLachlin JR (1985) Timing of transcription and protein synthesis underlying morphogenesis in preimplantation mouse embryos. Dev Biol 112:265–275
Kingston RE, Narlikar GJ (1999) ATP-dependent remodeling and acetylation as regulators of chromatin fluidity. Genes Dev 13:2339–2352
Lai WS, Carballo E, Thorn JM, Kennington EA, Blackshear PJ (2000) Interactions of CCCH zinc finger proteins with mRNA. Binding of tristetraprolin-related zinc finger proteins to Au-rich elements and destabilization of mRNA. J Biol Chem 275:17827–17837
Larue L, Ohsugi M, Hirchenhain J, Kemler R (1994) E-cadherin null mutant embryos fail to form a trophectoderm epithelium. Proc Natl Acad Sci USA 91:8263–8267
Latham KE (1999) Mechanisms and control of embryonic genome activation in mammalian embryos. Int Rev Cytol 193:71–124
Latham KE, Schultz RM (2001) Embryonic genome activation. Front Biosci 6:D748–D759
Li E (2002) Chromatin modification and epigenetic reprogramming in mammalian development. Nat Rev Genet 3:662–673
Li E, Beard C, Jaenisch R (1993) Role for DNA methylation in genomic imprinting. Nature 366:362–365
Li L, Baibakov B, Dean J (2008a) A subcortical maternal complex essential for preimplantation mouse embryogenesis. Dev Cell 15:416–425
Li L, Zheng P, Dean J (2010) Maternal control of early mouse development. Development 137:859–870
Li X, Ito M, Zhou F, Youngson N, Zuo X, Leder P, Ferguson-Smith AC (2008b) A maternal-zygotic effect gene, Zfp57, maintains both maternal and paternal imprints. Dev Cell 15:547–557
Li X, Leder P (2007) Identifying genes preferentially expressed in undifferentiated embryonic stem cells. BMC Cell Biol 8:37
Lighten AD, Hardy K, Winston RM, Moore GE (1997) Expression of mRNA for the insulin-like growth factors and their receptors in human preimplantation embryos. Mol Reprod Dev 47:134–139
Liu H, Kim JM, Aoki F (2004) Regulation of histone H3 lysine 9 methylation in oocytes and early pre-implantation embryos. Development 131:2269–2280
Liu HL, Hara KT, Aoki F (2005) Role of the first mitosis in the remodeling of the parental genomes in mouse embryos. Cell Res 15:127–132
Liu K, Rajareddy S, Liu L, Jagarlamudi K, Boman K, Selstam G, Reddy P (2006) Control of mammalian oocyte growth and early follicular development by the oocyte PI3 kinase pathway: new roles for an old timer. Dev Biol 299:1–11
Lu DP, Chandrakanthan V, Cahana A, Ishii S, O'Neill C (2004) Trophic signals acting via phosphatidylinositol-3 kinase are required for normal pre-implantation mouse embryo development. J Cell Sci 117:1567–1576
Lykke-Andersen K, Gilchrist MJ, Grabarek JB, Das P, Miska E, Zernicka-Goetz M (2008) Maternal Argonaute 2 is essential for early mouse development at the maternal-zygotic transition. Mol Biol Cell 19:4383–4392
Ma J, Flemr M, Stein P, Berninger P, Malik R, Zavolan M, Svoboda P, Schultz RM (2010) MicroRNA activity is suppressed in mouse oocytes. Curr Biol 20:265–270
Ma J, Zeng F, Schultz RM, Tseng H (2006) Basonuclin: a novel mammalian maternal-effect gene. Development 133:2053–2062
Maki N, Suetsugu-Maki R, Sano S, Nakamura K, Nishimura O, Tarui H, Rio-Tsonis K, Ohsumi K, Agata K, Tsonis PA (2010) Oocyte-type linker histone B4 is required for transdifferentiation of somatic cells in vivo. FASEB J 24:3462–3467

Marikawa Y, Alarcon VB (2012) Creation of trophectoderm, the first epithelium, in mouse preimplantation development. In: Kubiak JZ (ed) Results Probl Cell Differ. 55., 165–184. Springer, Heidelberg
Martianov I, Viville S, Davidson I (2002) RNA polymerase II transcription in murine cells lacking the TATA binding protein. Science 298:1036–1039
Matsumoto K, Anzai M, Nakagata N, Takahashi A, Takahashi Y, Miyata K (1994) Onset of paternal gene activation in early mouse embryos fertilized with transgenic mouse sperm. Mol Reprod Dev 39:136–140
Mayer W, Niveleau A, Walter J, Fundele R, Haaf T (2000) Demethylation of the zygotic paternal genome. Nature 403:501–502
Melendez A, Neufeld TP (2008) The cell biology of autophagy in metazoans: a developing story. Development 135:2347–2360
Merz EA, Brinster RL, Brunner S, Chen HY (1981) Protein degradation during preimplantation development of the mouse. J Reprod Fertil 61:415–418
Mora A, Komander D, van Aalten DM, Alessi DR (2004) PDK1, the master regulator of AGC kinase signal transduction. Semin Cell Dev Biol 15:161–170
Morgan HD, Santos F, Green K, Dean W, Reik W (2005) Epigenetic reprogramming in mammals. Hum Mol Genet 14:R47–R58
Morris SA, Zernicka-Goetz M (2012) Formation of distinct cell types in the mouse blastocyst. In: Kubiak JZ (ed) Results Probl Cell Differ. 55., 203–217. Springer, Heidelberg
Murchison EP, Stein P, Xuan Z, Pan H, Zhang MQ, Schultz RM, Hannon GJ (2007) Critical roles for Dicer in the female germline. Genes Dev 21:682–693
Nakamura T, Arai Y, Umehara H, Masuhara M, Kimura T, Taniguchi H, Sekimoto T, Ikawa M, Yoneda Y, Okabe M, Tanaka S, Shiota K, Nakano T (2007) PGC7/Stella protects against DNA demethylation in early embryogenesis. Nat Cell Biol 9:64–71
Narducci MG, Fiorenza MT, Kang SM, Bevilacqua A, Di Giacomo M, Remotti D, Picchio MC, Fidanza V, Cooper MD, Croce CM, Mangia F, Russo G (2002) TCL1 participates in early embryonic development and is overexpressed in human seminomas. Proc Natl Acad Sci USA 99:11712–11717
O'Neill C (2008) The potential roles for embryotrophic ligands in preimplantation embryo development. Hum Reprod Update 14:275–288
Ohno S (2001) Intercellular junctions and cellular polarity: the PAR-aPKC complex, a conserved core cassette playing fundamental roles in cell polarity. Curr Opin Cell Biol 13:641–648
Ohsugi M, Zheng P, Baibakov B, Li L, Dean J (2008) Maternally derived FILIA-MATER complex localizes asymmetrically in cleavage-stage mouse embryos. Development 135:259–269
Ohsumi K, Katagiri C (1991) Characterization of the ooplasmic factor inducing decondensation of and protamine removal from Toad Sperm nuclei - involvement of nucleoplasmin. Dev Biol 148:295–305
Ohsumi Y (2001) Molecular dissection of autophagy: two ubiquitin-like systems. Nat Rev Mol Cell Biol 2:211–216
Oswald J, Engemann S, Lane N, Mayer W, Olek A, Fundele R, Dean W, Reik W, Walter J (2000) Active demethylation of the paternal genome in the mouse zygote. Curr Biol 10:475–478
Pan H, Schultz RM (2011) SOX2 modulates reprogramming of gene expression in two-cell mouse embryos. Biol Reprod 85:409–416
Pauken CM, Capco DG (1999) Regulation of cell adhesion during embryonic compaction of mammalian embryos: roles for PKC and beta-catenin. Mol Reprod Dev 54:135–144
Payer B, Saitou M, Barton SC, Thresher R, Dixon JP, Zahn D, Colledge WH, Carlton MB, Nakano T, Surani MA (2003) Stella is a maternal effect gene required for normal early development in mice. Curr Biol 13:2110–2117
Paynton BV, Rempel R, Bachvarova R (1988) Changes in state of adenylation and time course of degradation of maternal mRNAs during oocyte maturation and early embryonic development in the mouse. Dev Biol 129:304–314
Pickart CM (2001) Mechanisms underlying ubiquitination. Annu Rev Biochem 70:503–533

Plusa B, Frankenberg S, Chalmers A, Hadjantonakis AK, Moore CA, Papalopulu N, Papaioannou VE, Glover DM, Zernicka-Goetz M (2005) Downregulation of Par3 and aPKC function directs cells towards the ICM in the preimplantation mouse embryo. J Cell Sci 118:505–515

Rajkovic A, Lee JH, Yan C, Matzuk MM (2002) The ret finger protein-like 4 gene, Rfpl4, encodes a putative E3 ubiquitin-protein ligase expressed in adult germ cells. Mech Dev 112:173–177

Ram PT, Schultz RM (1993) Reporter gene expression in G2 of the 1-cell mouse embryo. Dev Biol 156:552–556

Ramos SB, Stumpo DJ, Kennington EA, Phillips RS, Bock CB, Ribeiro-Neto F, Blackshear PJ (2004) The CCCH tandem zinc-finger protein Zfp36l2 is crucial for female fertility and early embryonic development. Development 131:4883–4893

Reddy P, Adhikari D, Zheng W, Liang S, Hamalainen T, Tohonen V, Ogawa W, Noda T, Volarevic S, Huhtaniemi I, Liu K (2009) PDK1 signaling in oocytes controls reproductive aging and lifespan by manipulating the survival of primordial follicles. Hum Mol Genet 18:2813–2824

Reddy P, Liu L, Adhikari D, Jagarlamudi K, Rajareddy S, Shen Y, Du C, Tang W, Hamalainen T, Peng SL, Lan ZJ, Cooney AJ, Huhtaniemi I, Liu K (2008) Oocyte-specific deletion of Pten causes premature activation of the primordial follicle pool. Science 319:611–613

Reik W (2007) Stability and flexibility of epigenetic gene regulation in mammalian development. Nature 447:425–432

Reik W, Walter J (2001) Genomic imprinting: parental influence on the genome. Nat Rev Genet 2:21–32

Remboutsika E, Yamamoto K, Harbers M, Schmutz M (2002) The bromodomain mediates transcriptional intermediary factor 1alpha -nucleosome interactions. J Biol Chem 277:50318–50325

Riethmacher D, Brinkmann V, Birchmeier C (1995) A targeted mutation in the mouse E-cadherin gene results in defective preimplantation development. Proc Natl Acad Sci USA 92:855–859

Riley JK, Carayannopoulos MO, Wyman AH, Chi M, Ratajczak CK, Moley KH (2005) The PI3K/ Akt pathway is present and functional in the preimplantation mouse embryo. Dev Biol 284:377–386

Roest HP, Baarends WM, de Wit J, van Klaveren JW, Wassenaar E, Hoogerbrugge JW, van Cappellen WA, Hoeijmakers JH, Grootegoed JA (2004) The ubiquitin-conjugating DNA repair enzyme HR6A is a maternal factor essential for early embryonic development in mice. Mol Cell Biol 24:5485–5495

Saeki H, Ohsumi K, Aihara H, Ito T, Hirose S, Ura K, Kaneda Y (2005) Linker histone variants control chromatin dynamics during early embryogenesis. Proc Natl Acad Sci USA 102:5697–5702

Saitou M, Barton SC, Surani MA (2002) A molecular programme for the specification of germ cell fate in mice. Nature 418:293–300

Sarbassov DD, Guertin DA, Ali SM, Sabatini DM (2005) Phosphorylation and regulation of Akt/ PKB by the rictor-mTOR complex. Science 307:1098–1101

Sato M, Kimura T, Kurokawa K, Fujita Y, Abe K, Masuhara M, Yasunaga T, Ryo A, Yamamoto M, Nakano T (2002) Identification of PGC7, a new gene expressed specifically in preimplantation embryos and germ cells. Mech Dev 113:91–94

Schultz RM (1993) Regulation of zygotic gene activation in the mouse. Bioessays 15:531–538

Schultz RM (2002) The molecular foundations of the maternal to zygotic transition in the preimplantation embryo. Hum Reprod Update 8:323–331

Schultz RM, Worrad DM (1995) Role of chromatin structure in zygotic gene activation in the mammalian embryo. Semin Cell Biol 6:201–208

Scott RC, Juhasz G, Neufeld TP (2007) Direct induction of autophagy by Atg1 inhibits cell growth and induces apoptotic cell death. Curr Biol 17:1–11

Suwińska A (2012) Preimplantation mouse embryo: developmental fate and potency of blastomeres. In: Kubiak JZ (ed) Results Probl Cell Differ. 55., 141–164. Springer, Heidelberg

Suzuki A, Ohno S (2006) The PAR-aPKC system: lessons in polarity. J Cell Sci 119:979–987
Suzumori N, Burns KH, Yan W, Matzuk MM (2003) RFPL4 interacts with oocyte proteins of the ubiquitin-proteasome degradation pathway. Proc Natl Acad Sci USA 100:550–555
Tanaka M, Hennebold JD, Macfarlane J, Adashi EY (2001) A mammalian oocyte-specific linker histone gene H1oo: homology with the genes for the oocyte-specific cleavage stage histone (cs-H1) of sea urchin and the B4/H1M histone of the frog. Development 128:655–664
Tang F, Kaneda M, O'Carroll D, Hajkova P, Barton SC, Sun YA, Lee C, Tarakhovsky A, Lao K, Surani MA (2007) Maternal microRNAs are essential for mouse zygotic development. Genes Dev 21:644–648
Thompson EM, Legouy E, Renard JP (1998) Mouse embryos do not wait for the MBT: chromatin and RNA polymerase remodeling in genome activation at the onset of development. Dev Genet 22:31–42
Tong ZB, Gold L, Pfeifer KE, Dorward H, Lee E, Bondy CA, Dean J, Nelson LM (2000) Mater, a maternal effect gene required for early embryonic development in mice. Nat Genet 26:267–268
Torres-Padilla ME, Zernicka-Goetz M (2006) Role of TIF1alpha as a modulator of embryonic transcription in the mouse zygote. J Cell Biol 174:329–338
Tsukamoto S, Kuma A, Murakami M, Kishi C, Yamamoto A, Mizushima N (2008) Autophagy is essential for preimplantation development of mouse embryos. Science 321:117–120
van der Heijden GW, Dieker JW, Derijck AA, Muller S, Berden JH, Braat DD, van der Vlag J, de Boer P (2005) Asymmetry in histone H3 variants and lysine methylation between paternal and maternal chromatin of the early mouse zygote. Mech Dev 122:1008–1022
Vanhaesebroeck B, Leevers SJ, Ahmadi K, Timms J, Katso R, Driscoll PC, Woscholski R, Parker PJ, Waterfield MD (2001) Synthesis and function of 3-phosphorylated inositol lipids. Annu Rev Biochem 70:535–602
Vestweber D, Gossler A, Boller K, Kemler R (1987) Expression and distribution of cell adhesion molecule uvomorulin in mouse preimplantation embryos. Dev Biol 124:451–456
Wan LB, Pan H, Hannenhalli S, Cheng Y, Ma J, Fedoriw A, Lobanenkov V, Latham KE, Schultz RM, Bartolomei MS (2008) Maternal depletion of CTCF reveals multiple functions during oocyte and preimplantation embryo development. Development 135:2729–2738
Wang S, Kou Z, Jing Z, Zhang Y, Guo X, Dong M, Wilmut I, Gao S (2010) Proteome of mouse oocytes at different developmental stages. Proc Natl Acad Sci USA 107:17639–17644
Warner CM, Hearn TF (1977) The effect of alpha-amanitin on nucleic acid synthesis in preimplantation mouse embryos. Differentiation 7:89–97
Whitten WK, Biggers JD (1968) Complete development in vitro of the pre-implantation stages of the mouse in a simple chemically defined medium. J Reprod Fertil 17:399–401
Wilkinson KD (2000) Ubiquitination and deubiquitination: targeting of proteins for degradation by the proteasome. Semin Cell Dev Biol 11:141–148
Winkel GK, Ferguson JE, Takeichi M, Nuccitelli R (1990) Activation of protein kinase C triggers premature compaction in the four-cell stage mouse embryo. Dev Biol 138:1–15
Worrad DM, Schultz RM (1997) Regulation of gene expression in the preimplantation mouse embryo: temporal and spatial patterns of expression of the transcription factor Sp1. Mol Reprod Dev 46:268–277
Wright PW, Bolling LC, Calvert ME, Sarmento OF, Berkeley EV, Shea MC, Hao Z, Jayes FC, Bush LA, Shetty J, Shore AN, Reddi PP, Tung KS, Samy E, Allietta MM, Sherman NE, Herr JC, Coonrod SA (2003) ePAD, an oocyte and early embryo-abundant peptidylarginine deiminase-like protein that localizes to egg cytoplasmic sheets. Dev Biol 256:73–88
Wu X, Viveiros MM, Eppig JJ, Bai Y, Fitzpatrick SL, Matzuk MM (2003) Zygote arrest 1 (Zar1) is a novel maternal-effect gene critical for the oocyte-to-embryo transition. Nat Genet 33:187–191
Xenopoulos P, Kang M, Hadjantonakis A-K (2012) Cell lineage allocation within the Inner cell mass of the mouse blastocyst. In: Kubiak JZ (ed) Results Probl Cell Differ. 55., 185–202. Springer, Heidelberg

Xie Z, Klionsky DJ (2007) Autophagosome formation: core machinery and adaptations. Nat Cell Biol 9:1102–1109

Yurttas P, Vitale AM, Fitzhenry RJ, Cohen-Gould L, Wu W, Gossen JA, Coonrod SA (2008) Role for PADI6 and the cytoplasmic lattices in ribosomal storage in oocytes and translational control in the early mouse embryo. Development 135:2627–2636

Zeng F, Baldwin DA, Schultz RM (2004) Transcript profiling during preimplantation mouse development. Dev Biol 272:483–496

Zeng F, Schultz RM (2005) RNA transcript profiling during zygotic gene activation in the preimplantation mouse embryo. Dev Biol 283:40–57

Zheng P, Dean J (2009) Role of Filia, a maternal effect gene, in maintaining euploidy during cleavage-stage mouse embryogenesis. Proc Natl Acad Sci USA 106:7473–7478

Zheng W, Gorre N, Shen Y, Noda T, Ogawa W, Lundin E, Liu K (2010) Maternal phosphatidylinositol 3-kinase signalling is crucial for embryonic genome activation and preimplantation embryogenesis. EMBO Rep 11:890–895

Zheng W, Nagaraju G, Liu Z, Liu K (2012) Functional roles of the phosphatidylinositol 3-kinases (PI3Ks) signaling in the mammalian ovary. Mol Cell Endocrinol 356:24–30

Chapter 8
Preimplantation Mouse Embryo: Developmental Fate and Potency of Blastomeres

Aneta Suwińska

Abstract During the past decade we have witnessed great progress in the understanding of cellular, molecular, and epigenetic aspects of preimplantation mouse development. However, some of the issues, especially those regarding the nature and regulation of mouse development, are still unresolved and controversial and raise heated discussion among mammalian embryologists. This chapter presents different standpoints and various research approaches aimed at examining the fate and potency of cells (blastomeres) of mouse preimplantation embryo. In dealing with this subject, it is important to recognize the difference between the fate of blastomere and the prospective potency of blastomere, with the first being its contribution to distinct tissues during normal development, and the second being a full range of its developmental capabilities, which can be unveiled only by experimental perturbation of the embryo. Studies of the developmental potential and the fate of blastomeres are of the utmost importance as they may lead to future clinical application in reproductive and regenerative medicine.

8.1 Introduction

During mammalian embryogenesis early embryonic cells progressively differentiate and form distinct cell lineages. As development proceeds, cells of the embryo gradually lose their full developmental potential, i.e., the ability to give rise to all cell types of an organism. The first fundamental question of developmental biology is: when and how apparently identical cells start to differ from each other within the embryo and acquire a destiny directing them into a specific developmental path? Generally, development of animals is defined as "determined," i.e., regulated by

---

A. Suwińska (✉)
Department of Embryology, Faculty of Biology, University of Warsaw, Miecznikowa 1, 02-096 Warsaw, Poland
e-mail: asuwinska@biol.uw.edu.pl

J.Z. Kubiak (ed.), *Mouse Development*, Results and Problems in Cell Differentiation 55,
DOI 10.1007/978-3-642-30406-4_8,

morphogenetic determinants as, for example, in *Drosophila melanogaster*, *Caenorhabditis elegans*, or *Xenopus laevis*, or "regulative" ("flexible"), which is attributed (among others) to mammals, including the mouse. The next question coming to mind is: to what extent mouse cells and embryos are flexible and what is their potency, i.e., ability to change in response to the environmental changes? The understanding of the mechanisms that underlie this plasticity is extremely important for development of new therapies and assisted reproductive technology and for improving the derivation of embryonic stem cells. Issues of cell lineage allocation and molecular regulation of this process, although closely related, are treated cursorily in this article, because they are thoroughly discussed and reviewed in other chapters of this volume.

8.2 A Brief Overview of Mouse Preimplantation Development

Preimplantation period of mouse development extends from fertilization to implantation of the embryo in uterus that takes place 4.5 days post-fertilization. The fusion of sperm with oocyte, which is arrested in the second meiotic division, results in the formation of a zygote. The zygote then undergoes a series of cleavage divisions, which, because the cell growth is inhibited before preimplantation, give rise to the progressively smaller blastomeres. Epigenetic and morphogenetic events lead to the formation of two different cell types at the 16-cell stage: they have distinct destination and form the first two primary cell lineages. The asymmetric (differentiative) cellular division occurring at the 8- to 16-cell stage transition results in the formation of the inner and an outer population of cells (Johnson and Ziomek 1981). At the 16-cell stage embryo, cell divisions of the outer layer occur either symmetrically, resulting in two progeny of cells that remain in the outside layer of the embryo, or asymmetrically, generating one outer and one inner daughter cell. The outer blastomeres are precursors of trophectoderm (TE), while the inner blastomeres will form the inner cell mass (ICM) of the resulting blastocyst, as proposed in 1967 by Tarkowski in his "inside-outside" hypothesis (Tarkowski and Wroblewska 1967). The trophectoderm covering ICM (polar TE) forms ectoplacental cone and, subsequently, the embryonic part of the placenta, while trophectodermal cells surrounding the blastocyst's cavity (mural TE) stop to divide, replicate DNA, and form trophectodermal giant cells. The mural TE is responsible for implantation of the embryo in the uterus and together with placenta forms the external barrier between the fetus and the mother. ICM is composed of pluripotent cells and occupies one pole of the blastocyst interior.

Specification of both primary cell lineages (TE and ICM) requires interaction of specific transcription factors. CDX2, a caudal-related homeodomain protein, is the regulator of trophectoderm lineage. *Cdx2* gene begins to be expressed at the 8-cell stage and becomes gradually restricted, first to the outside cells of the morula, and eventually to the trophectodermal cells of the blastocyst (Dietrich and Hiiragi 2007; Ralston and Rossant 2008). CDX2 regulates cell polarity and promotes symmetric cell divisions in 8- to 32-cell embryos, thereby contributing to the spatial allocation of the future TE cells within the embryo. On the other hand, blastomere polarity also influences CDX2 expression through asymmetric distribution of CDX2 mRNA to the outer blastomeres (Jedrusik et al. 2008). Among genes required for specification of ICM is *Oct4* (*Octamer-binding transcription factor 4*). It is initially expressed ubiquitously in every cell during period of cleavage, but following blastocyst formation becomes limited to the inside, future ICM, cells. This restriction depends on CDX2 as shown by the fact that in *Cdx2* mutants OCT4 remains expressed in the TE, resulting in the death of the embryo at the time of implantation (Strumpf et al. 2005; Ralston and Rossant 2008). Molecular mechanisms underlying the differentiation of TE and ICM are described in detail in Chaps. 10 and 11.

Just prior to implantation, a third tissue emerges on the surface of the ICM facing the blastocyst cavity—the primitive endoderm (PE) that later contributes to the endoderm layer of extraembryonic tissue, the yolk sac. The remaining, deeper ICM cells form the epiblast (EPI), which is the source of the definite embryo and some of the extraembryonic membranes, such as allantois and amnion.

8.3 In Search of the Best Method to Study the Developmental Potential of Mouse Blastomeres

Developmental biologists argue about the best research approach of studying the course of events in mouse development. Some of them are in favor of examination of cell fate (contribution to tissues) in intact, undisturbed embryos. This method is based on the tracing of cell movements and destinations by time-lapse, live imaging, and thus enables to investigate the progress of development and to assess the fate of marked cells in their relatively natural environment, with minimal interference and harmful influence of any manipulations. Other researchers use different strategy: they observe how the embryonic cells behave when the embryo is subjected to unusual circumstances due to experimental intervention. This "brutal" method makes possible to reveal full range of developmental capabilities of cells and shows their prospective potency. However, this tactics involves the interference in the structure of the embryo. Both methodological approaches have advantages and drawbacks, but only a combination of these two approaches provides complimentary data permitting to understand the mechanism of early development.

8.4 Mouse Embryonic Cells: Determined or Regulative?

Zygote—an egg cell penetrated by a sperm—is a "mother cell" of a complex embryo and finally of the adult organism. It sits atop a lineage hierarchy and can generate multiple types of differentiated cells that contribute to embryo proper and extraembryonic structures supporting its development, i.e., placenta and fetal membranes. In this sense, zygote is a totipotent cell. The progeny cells formed by cleavages just a few divisions down the line become progressively developmentally restricted, i.e., lose totipotency defined as the ability to individually give rise to complete individual, but remain pluripotent, that is, they are able to contribute to all parts of the individual. If embryonic cells are pre-patterned (pre-programmed) and have a specified fate we classify them as "determined", and when their fate is not restricted and they are able to contribute to any organ, they are defined as "regulative" (flexible). In earlier embryological literature (terminology), the first type of embryos was called "mosaic" and the other "regulative".

8.4.1 The Evidence for Existence of Pre-patterning

In many invertebrates and vertebrates, the asymmetries that are established in the oocyte correlate with the segregation of maternally supplied factors, termed cytoplasmic determinants, affecting further cell lineage formation and embryonic axes development. In mammals there is no clear evidence for a polarized distribution of molecules that may function as cytoplasmic determinants. However, some researchers believe that in the undisturbed embryo the spatial information (whatever it might be) recorded in the structure of a fertilized egg determines a strictly defined program, which guides future development. This information would dictate to the arising blastomeres distinct functions and destinies. This hypothesis assumes that differences between cells emerge before cells adopt inside or outside positions within the embryo.

Several factors have been proposed to play a role in the positioning of the plane of first cleavage: the second polar body position, sperm entry site, positions of male and female pronuclei, and shape of the zona pellucida (reviewed in Hiiragi et al. 2006). According to some researchers, the second polar body (2 pb) marks the animal pole, i.e., part of the egg, where the meiotic divisions have occurred. Consequently, the vegetal pole lies on the opposite side. These researchers assume that the 2 pb does not change its position during the cleavage. In the blastocyst, the 2 pb usually localizes at the boundary between embryonic (containing ICM and overlying polar TE) and abembryonic (mural TE) hemispheres (Gardner 1997, 2001; Piotrowska and Zernicka-Goetz 2001; Piotrowska et al. 2001; Plusa et al. 2002a). Such localization would testify that the plane of the first cleavage division is perpendicular to the embryonic–abembryonic axis.

Several studies seem to support the existence of strict correlation between zygote organization and geometry of the blastocyst. By defining the position of sister blastomeres at 2-cell embryo by marking the zona pellucida with mineral oil drops (Gardner 2001) or labeling both blastomeres with different fluorescent dyes (Piotrowska et al. 2001; Fujimori et al. 2003; Plusa et al. 2005) and subsequently tracking their fate in blastocyst, these authors showed that cells derived from the earlier dividing blastomere contribute to the embryonic part of the blastocyst and descendant cells of the second one—to the abembryonic part. In contrast, experimental data from other laboratories indicate that the first two blastomeres are not pre-patterned in their fate, and neither of them is predisposed to form ICM or TE (Alarcón and Marikawa 2003, 2005; Chroscicka et al. 2004; Waksmundzka et al. 2006; Kurotaki et al. 2007). By labeling both blastomeres and following the distribution of their descendants in developing blastocysts or time-lapse photography of intact embryos development from the 2-cell to the blastocyst stage it was shown that they do not form discrete groups, as shown by Piotrowska and Zernicka-Goetz (2001), but undergo mixing and contribute to embryonic as well as abembryonic hemisphere of the blastocyst (Alarcón and Marikawa 2003, 2005; Chroscicka et al. 2004; Waksmundzka et al. 2006; Kurotaki et al. 2007).

Additionally, Zernicka-Goetz and collaborators state that the order of the division of the first two blastomeres is determined by sperm entry site. According to these authors, the sperm preferentially enters the vegetal part of the oocyte and the blastomere that inherits this specific region of the cytoplasmic membrane tends to divide first and produces cells that contribute predominantly to the embryonic part of the blastocyst (Piotrowska and Zernicka-Goetz 2001; Plusa et al. 2002b). However, results of other studies deny that the site of sperm entry has an influence on the establishment of the plane of the zygote division and the sequence of the division of the first two blastomeres (Davies and Gardner 2002; Gardner and Davies 2003; Motosugi et al. 2005).

Aforementioned hypothesis regarding early determination of cell fate was subjected to criticism by many other researchers. The time-lapse observations of preimplantation mouse embryo development from the zygote to the blastocyst stage showed movement of the second polar body towards the furrow during the first cleavage (Hiiragi and Solter 2004; Motosugi et al. 2005). Thus, these authors contest the usefulness of the position of the polar body as a landmark of the first cleavage plane. Instead, Hiiragi and Solter (2004) claim that the first cleavage plane is dependent on the mutual arrangement of two pronuclei and that it passes perpendicularly to the plane of the pronuclei alignment. This discovery was re-confirmed by the visualization of the mitotic spindle at the second cleavage division using tubulin-GFP expressed in zygotes (Louvet-Vallée et al. 2005).

The last of the controversial issues regarding existence of pre-patterning is centered on the question whether the orientation of the first cleavage plane is determined by extrinsic cues, such as the shape of the egg, imposed by the overlying egg membrane—zona pellucida. The zona pellucida is usually ellipsoid in shape and the plane of the first cleavage division is perpendicular to its longer axis (Kurotaki et al. 2007). Experimental change of the shape of eggs resulted in repositioning of the cleavage plane and imposing the division along short axis of the embryo (Gray et al. 2004). Other observations indicate that, due to the mechanical pressure exerted by zona pellucida, the embryonic–abembryonic axis (identified by the site of the nascent blastocyst cavity) is aligned along the long axis of the embryo (Motosugi et al. 2005; Kurotaki et al. 2007). Alarcón and Marikawa (2008) showed that, at the two-cell stage, the shape of the zona pellucida varies among embryos and does not change until the blastocyst stage. In embryos having ellipsoid zona pellucida, the embryonic–abembryonic axis tends to form orthogonally to the

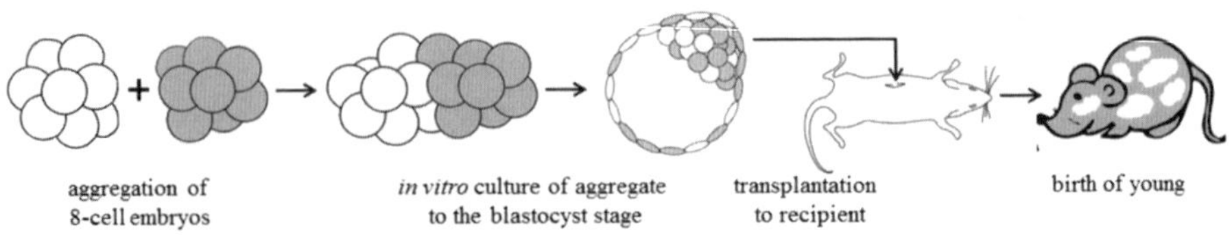

Fig. 8.1 *Regulation of excess of blastomeres—formation of chimeras.* Aggregation of two 8-cell embryos originating from mouse strains that differ in coat color results in the formation of a single chimeric blastocyst that after the transplantation to the foster mother gives rise to the two-colored chimeric individual

first cleavage plane, while in the embryos enclosed in a spherical zona pellucida, there is no such relationship (Alarcón and Marikawa 2008).

8.4.2 The Evidence for Regulative Capacities of Mouse Embryo

Irrespective of whether cell fate is to some extent pre-programmed in intact, developing embryos, as suggested by some studies discussed above, the preimplantation mouse embryo shows a very high degree of flexibility and resistance to the experimentally imposed loss or addition of cells. Mouse embryo is able to accommodate itself by "sensing" missing or added cells and to follow the normal program of embryogenesis culminating in the birth of entirely normal animal.

8.4.2.1 Consequences of the Excess of Blastomeres: Capacity of Combined Embryos to Form a Chimeric Individual

The first reports throwing doubt on the early determination of cell fate in the mouse embryo originate from pioneer experiments performed in early 1960s independently by Tarkowski and by Mintz. Regulative capacities of the mouse embryo have been proven by tracking the development of aggregated whole embryos. Aggregation of two 8-cell mouse embryos resulted in the formation of one giant, but morphologically normal blastocyst, which (following transplantation into the oviduct or uterus of foster mothers) gave rise to a chimera—an individual consisting of both components (Tarkowski 1961, 1963; Mintz 1962; for a review see McLaren 1976; Tarkowski 1998; Fig. 8.1). In order to analyze cell colonization and differentiation, embryos originating from mouse strains differing in coat and eye pigmentation or biochemical markers were used for the creation of chimeras. In addition, it was shown that viable mouse chimeras can be created by aggregation of more than two cleaving embryos: some animals generated from three genetically different 8-cell embryos were composed of cells derived from all three partners (Markert and Petters 1978).

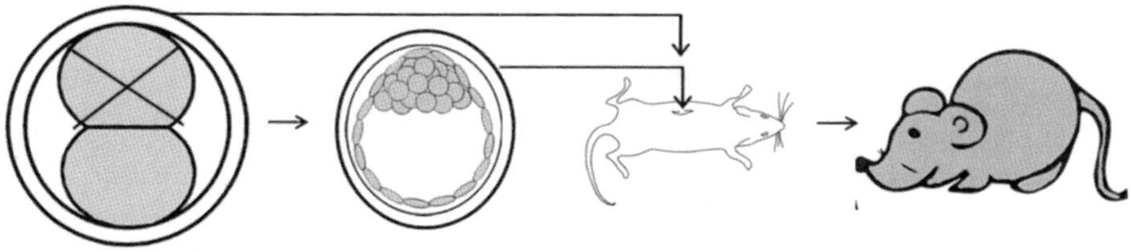

Fig. 8.2 *Compensation for lack of blastomere*. Removing mechanically one blastomere from the 2-cell embryo results in development of the remaining blastomere to the blastocyst that after the transfer to the recipient female gives rise to a viable animal

Mouse chimeras can be produced also from more advanced embryos: the morula composed of 16 cells (Mintz 1964) and even from the blastocysts. In the case of blastocysts, the fusion of trophectoderm of the two adhering blastocysts was obtained with Sendai virus (Tarkowski and Wojewodzka 1982) or electric impulses (Ozdzeński et al. 1997; Tarkowski et al. 2005b). The resulting chimeric blastocysts can have either one common or two separate ICMs developing after implantation into a single chimeric animal or dizygotic twin fetuses, respectively (Tarkowski and Wojewodzka 1982; Ozdzeński et al. 1997; Tarkowski et al. 2005b). All these discoveries strongly indicate that spatial organization of the cleaving embryo is labile and developmental fate of blastomeres is not determined until the late stages of preimplantation development.

8.4.2.2 Consequences of the Reduction of Blastomeres Number: Capacity of Embryo Fragments to Form a Whole Organism

A direct and also the most convincing evidence for the unrestricted developmental potential of each cell in the 2-cell embryo is the birth of live animal derived from a single isolated blastomere. Destroying mechanically one of the blastomeres of a 2-cell embryo (henceforth termed as 1/2 blastomere) inside the zona pellucida and transferring the surviving blastomere individually into reproductive tract of recipient female, Tarkowski (1959a, b) proved that a single 1/2 blastomere is totipotent, i.e., may give rise to an adult, fertile mouse (Fig. 8.2). This exciting discovery proved that the early mouse embryo is characterized by a great developmental flexibility, which allows compensation for not only the excess, but also for the lack of blastomeres. However, Tarkowski did not investigate the compensation capabilities of both "halves" of the same 2-cell embryo. Hence, he did not resolve the dilemma whether the two first blastomeres have an equal developmental potential. This was shown later by Mullen et al. (1970) and Tsunoda and McLaren (1983) who managed to obtain monozygotic mouse twins from both separated sister 1/2 blastomeres of the same embryo.

8.4.2.3 Step-By-Step Attempts to Reveal Totipotency or Pluripotency of Blastomeres at Later Stages of Development

Much work has been done on trying to verify totipotency of single cells originated from mouse embryos at later stages of development (e.g., 4- or 8-cell stage). These attempts turned out to be unsuccessful, as no mouse was born as a result of these experiments (Rossant 1976). Also, the ability to form preimplantation embryos from isolated blastomeres at these stages of development was highly reduced. Most of individual cells isolated at the 4-cell stage (1/4 blastomeres) developed into miniature blastocysts, whereas single cells derived from 8-cell stage (1/8 blastomeres) usually gave rise to trophoblastic vesicles completely devoid of ICM (Tarkowski and Wroblewska 1967). Both of these structures were capable of implanting and forming trophoblast giant cells, but almost never produced egg cylinders (Rossant 1976). However, this failure of development was attributed to the insufficient number of cells required for the formation of functional ICM (future embryo proper), rather than to the identity of blastomeres and the restriction in their developmental potential (Rossant 1976). This conclusion was supported by the data obtained from experiments on other mammalian species. It was previously discovered that single blastomeres of 2- (Seidel 1952, 1960), 4-, and 8-cell rabbit embryo (Moore et al. 1968) are able to develop into viable individuals. In sheep, the single blastomeres of a 4- and sporadically even of an 8-cell embryo also can develop into adult animals (Willadsen 1981). Moreover, four pairs of blastomeres of the 8-cell sheep embryos gave rise to quadruplet lambs (Willadsen 1981), and the separation of blastomeres of 4-cell cow embryo resulted in the birth of four genetically identical quadruplet calves (Johnson et al. 1995). These data support the hypothesis that greater developmental potential of blastomeres of rabbit and cattle cleaving embryo, comparing to that of the mouse, results from the larger total number of cells attained by these embryos at the time of implantation.

One of the ways to overcome the small size of blastomere-derived mouse blastocysts and especially insufficient number of ICM cells at the time of implantation is to restore the normal cell number in the mouse embryo by supporting a single, tested blastomere with helper "carrier" cells. Diploid, polyploid, or parthenogenetic cells originated from other embryos can be used as the supporting cells. When single blastomeres are combined with genetically distinct diploid "carrier" cells, chimeras composed of cells derived from distinct embryos are created (Fig. 8.3a). The application of aggregation method for chimera generation using tetraploid cells (obtained by electrofusion of blastomeres at the 2-cell stage; Kubiak and Tarkowski 1985) results in birth of animals derived entirely from a diploid blastomere. The success of this approach, known as "tetraploid complementation", is due to the elimination of tetraploid cells from the embryonic derivatives of ICM (fetal tissues), which however persist in the extraembryonic structures, where they successfully replace diploid cells (Tarkowski et al. 1977; James et al. 1995; MacKay and West 2005; Fig. 8.3b).

When single blastomeres of 4- and 8-cell mouse embryos were aggregated with sufficient number of "carrier" diploid blastomeres, the progeny of the former was

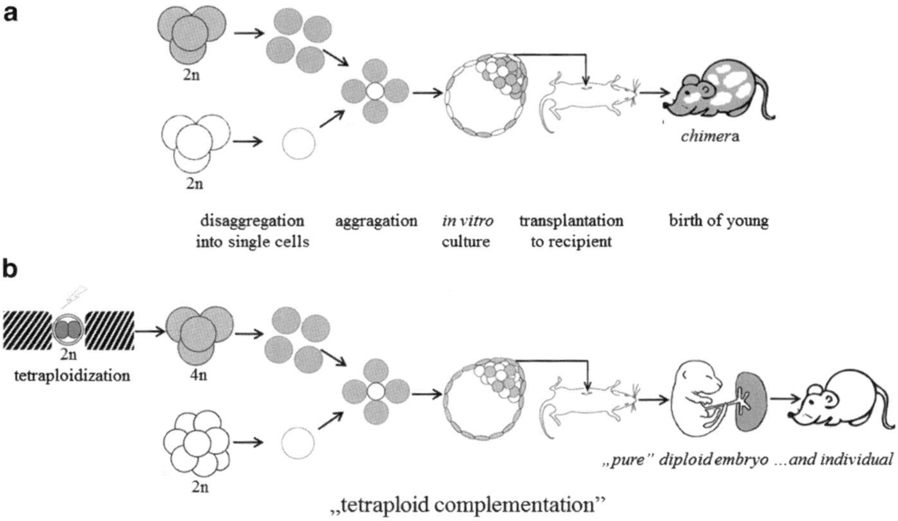

Fig. 8.3 *Methods of testing the potency of individual blastomeres.* (**a**) Aggregation of a single blastomere with genetically distinct diploid "carrier" blastomeres results in formation of chimera composed of cells derived from both types of blastomeres as shown in Fig. 8.1. (**b**) Aggregation of a single diploid blastomere with tetraploid blastomeres, obtained by electrofusion of blastomeres at the 2-cell stage, results in birth of a "pure" diploid animal derived exclusively from the tested blastomere. Tetraploid cells are gradually eliminated from the embryo proper but survive and contribute to the extraembryonic structures (fetal part of placenta and some fetal membranes)

able to contribute to both ICM and TE derivatives in postimplantation E10 chimeric conceptuses, and even successfully born offspring (Kelly 1975, 1977). In another experiment, inner and outer single 1/16 blastomeres were injected under the zona pellucida of diploid 8-cell embryos, transferred to foster mothers and analyzed at midgestation (Rossant and Vijh 1980). Despite preponderance of the embryos with donor-descendant cells solely in the trophectodermal tissues, in several embryos the progeny of donor cells was allocated also in the ICM-derived tissues, thus demonstrating the pluripotency of blastomeres at this stage of development. Similar results were obtained when diploid parthenogenetic cells were used as a source of "carrier" cells (Pinyopummin et al. 1994; Tsunoda et al. 1987). Parthenogenetically activated eggs are not able to develop till term, as both epigenetically different maternal and paternal genomes are required for successful development (Tarkowski et al. 1970; McGrath and Solter 1984; Surani et al. 1984). However, parthenogenetic cells can survive and contribute to both ICM and TE of the chimeric blastocyst if they are aggregated with normal diploid cells. At midgestation they undergo selection, first in extraembryonic, and then in embryonic tissues. Altogether, these data show that diploid parthenogenetic blastomeres can assist single normal blastomeres of 4- (Pinyopummin et al. 1994) and 8-cell stage embryos to develop to term (Tsunoda et al. 1987).

The next step towards verification of developmental capacity of mouse blastomeres was done by the method of "tetraploid complementation". As a result of surrounding a single, diploid blastomere with tetraploid blastomeres of different genotype Tarkowski and colleagues obtained diploid and non-chimeric (i.e., derived exclusively from tested single blastomere) animals from 1/4, 1/8, and even from 1/16 blastomere (Tarkowski et al. 2001a, 2005a; 2010). In addition, they showed that when sister (i.e., originating from one embryo) blastomeres were used to construct separate aggregate embryos, this approach permits to generate genetically identical multiplets (twins, triplets, and quadruplets) from as late blastomeres as those obtained from 16-cell embryos.

Pluripotency of blastomeres from 16-cell embryo was also tested by using disaggregation–reaggregation techniques. Rossant and Vijh (1980) microsurgically removed individual cells from the outside of decompacted 16-cell embryos and then reconstituted morulae by injecting more than 20 of these cells into empty zona pellucida. The majority of thus constructed embryos produced morphologically normal blastocysts that were transplanted to foster mothers. Such reconstituted embryos were able to develop into egg cylinders containing derivatives of both ICM and TE (Rossant and Vijh 1980). Moreover, by separating polar outer cells and apolar inner cells and recombining them as homogeneous aggregates, Ziomek and colleagues (1982) have shown that there were no differences in postimplantation development of such reconstituted embryos and that they were able to develop into 9.5-day-old fetuses. The full developmental potential of at least some cells of the outer and inner population was definitively proved by experiments in which the embryos reconstructed exclusively from inner or outer blastomeres of 16-cell stage embryo were able to develop into fertile mice (Suwinska et al. 2008).

Experimental approaches discussed above proved indirectly that at least some blastomeres of cleaving 4-, 8-, and 16-cell embryo still remain pluripotent. However, the question whether all blastomeres at these stages of development are equivalent and have identical developmental potencies still remains open. Technical limitations of the experimental procedure, such as possible harmful effect of manipulations and low transplantation efficiency, prevent from producing animals from all separated blastomeres of the stages in question. It is therefore practically impossible to ascertain whether this inability is solely related to technical difficulties related to the challenging procedure or to a subtle intrinsic property of the specific blastomeres.

Indeed, some studies indicate the existence of the differences between the blastomeres, especially in the 4-cell stage. The first suggestion comes from the aggregation experiment aimed at evaluation of the number of the founder blastomeres making the animal body. When single mouse blastomeres obtained from 4-cell embryo were aggregated with blastomeres of the same stage, but differing in pigmentation, in the ratio 1:3 and 2:2, the observed proportion of chimeric to non-chimeric animals indicated that usually only two out of four blastomeres are founder cells of ICM, and finally, the animal's body (Tarkowski et al. 2001b). From the above study it did not follow, however, that any particular two of the four blastomeres preferentially contribute to ICM. Another support for the suggestion that each of 1/4 blastomeres has different developmental potential originates from the studies on chimeras comprising of four 1/4 blastomeres of the same origin with respect to the pattern of cleavage (Piotrowska-Nitsche et al. 2005). The most common pattern of division is the one in which one 1/2 blastomere divides meridionally, and the other equatorially. Chimeras constructed entirely from blastomeres resulting from equatorial or oblique division (inheriting either animal or vegetal components of the zygote) showed various developmental abnormalities. The most severe defects were observed in case of cells inheriting vegetal part. In contrast, chimeras made from blastomeres arising from meridional second cleavage division (inheriting both parts of the zygote) developed successfully to term (Piotrowska-Nitsche et al. 2005). Certain cleavage patterns have thus been shown to predict developmental differences, possibly reflecting intrinsic differences between cells that result from epigenetic modifications (Torres-Padilla et al. 2007). According to this hypothesis, the differential methylation of histone H3 at arginines 17 and 26 affects the fate of 4-cell stage blastomeres. The levels of histone H3 methylation at specific arginine residues are highest in those 1/4 blastomeres that will contribute to the ICM and the polar TE. These blastomeres undertake full development when combined with cells of the same type with respect to the pattern of cleavage in chimeras. In contrast, the arginine methylation of histone H3 is at the lower level in cells whose progeny contributes to the mural TE and that show compromised development when combined together in chimeras. Moreover, Jedrusik and collaborators (2008) have shown that both the orientation and the order of divisions at the transition from 2- to 4-cell stage affect the heterogeneity of *Cdx2* expression during the next cell cycle, i.e., in the blastomeres of the 8-cell embryo. Specifically, in embryos in which a meridional division precedes an equatorial one, the blastomeres of 4-cell embryos originating from an equatorial division tend to express elevated levels of *Cdx2* mRNA and protein comparing to other cells, undertake symmetric divisions more frequently, and contribute predominantly to the TE (Jedrusik et al. 2008). Last but not least, the fluorescent decay after photo-activation (FDAP) studies of Plachta and colleagues (2011) investigating the kinetic behavior (nuclear export and import) of the photo-activatable GFP–OCT4 in cells of the live 4- and 8-cell stage embryo showed that the differential dynamics of OCT4 associated with chromatin forecasts cell lineage patterning. Morphologically indistinguishable cells of 4- and 8-cell stage embryos differ in the kinetic behavior of OCT4. Cells with slower OCT4 kinetics preferentially divide asymmetrically giving rise to the pluripotent cell lineage that contributes to the ICM, whereas those with faster OCT4 kinetics preferentially undergo symmetric divisions to contribute mostly to the extraembryonic lineage—TE. The differences in OCT4–GFP movement are due to the differences in the accessibility of the DNA-binding sites for OCT4 in the nuclei of different cells (Plachta et al. 2011).

On the other hand, there are some observations that argue against the heterogeneity of blastomeres of the 4-cell embryo. González et al. (2011) reported that blastomeres isolated from 4-cell embryo are equally competent to derive embryonic stem cells (ES cells). In addition, these authors did not observe any significant differences in the efficiency of ES cell derivation from the first and the second dividing blastomeres of the 2-cell embryos. In contrast, Wakayama et al. (2007) were unable to obtain ES cell lines from all the blastomeres of the same embryo at the 4-cell or 8-cell stage.

8.4.2.4 Pluripotency of Blastocyst Cells: Hanging in the Balance

Mouse embryos reach the blastocyst stage around the 32-cell stage, following five rounds of cell divisions. Although the developmental potency of cells at the blastocyst stage was evaluated by various approaches, the results obtained as well as the interpretation and conclusions drawn from these experiments are often very conflicting.

Inside and outside cells derived from 32-cell embryos differ in their behavior and potency when separated and arranged in monotypic groups. Suwinska et al. (2008) showed that homogeneous aggregates composed of 32 outer blastomeres from 32-cell embryos are able to form normally structured blastocysts built of TE and ICM only if the source of blastomeres is the embryo that had just completed the fifth cleavage division in all or nearly all blastomeres, but had not yet initiated cavitation. The aggregated outer blastomeres originating from nascent blastocysts that had already commenced cavitation gave rise only to clumps of trophectoderm vesicles devoid of ICM and, hence, were unable to form an embryo (Suwinska et al 2008). In contrast, clusters of exclusively inner cells of the same stage developed into morphologically normal blastocysts. Externally located cells in such aggregate embryos were able to switch on *Cdx2* expression and, albeit with a certain delay, to form TE. In in vitro cultured outgrowths TE cells differentiated into giant cells expressing cytokeratin Endo-A (a marker of differentiated trophectoderm). However, the transfer of such embryos to recipients did not result in birth of young, probably due to their inability to implant correctly (Suwinska et al. 2008). These results point to the restricted potential of blastomeres at this stage of development. On the other hand, when single inner and outer 1/32 blastomeres were supported with tetraploid cells, they located in all three primary cell lineages: PE, EPI, and TE and resulting embryos, after the transfer to recipients, sporadically generated fetuses (Tarkowski et al. 2010). These results indicate that some of the blastomeres of 32-cell stage are pluripotent.

Another method of testing developmental potency of the inner cells of blastocyst is an isolation of ICMs and checking its ability to undergo trophectodermal differentiation. ICM isolation can be performed either by immunosurgery (cytotoxic immune lysis of outer cells by sequential treatment with antiserum and active complement) (Solter and Knowles 1975), or microsurgically by the mechanical separation of ICM from the trophectoderm (Gardner and Johnson 1972). There are reports indicating that ICMs isolated immunosurgically from early blastocysts (not fully expanded, ca 32–50 cells) still retain the ability to form the outer layer of trophectoderm. However, they lose this capacity after reaching the stage of approximately 60 cells (Handyside 1978; Hogan and Tilly 1978a; Spindle 1978; Nichols and Gardner 1984; Pierce et al. 1988). In these studies regeneration of TE was judged mainly by the formation of the blastocyst-like structures and outgrowths of trophectodermal giant cells in in vitro cultures as well as in ectopic grafts. Further evidence arguing in favor of pluripotency of blastocyst cells was provided by Rossant and Lis (1979) who showed that ICM cells of early blastocysts (32–50 cell) when aggregated with morulae occasionally contributed not only to the ICM but also to TE tissues of chimeric fetuses examined at mid-pregnancy. Moreover, when four ICMs derived from the early blastocysts were aggregated together to restore normal embryonic cell number, they were able to induce a decidual reaction and developed into normal E5.5 egg cylinders. In contrast, the aggregates composed of four late ICMs (derived from 50–70-cell blastocysts) produced trophectoderm in vivo very rarely (Rossant and Lis 1979).

The main problem with the interpretation of the data from the above studies is that the method of immunosurgery does not guarantee complete removal of all TE cells. Various approaches were undertaken to confirm the reliability of this method, including labeling of TE cells with horseradish peroxidase, granules of melanin, or antibodies against TE-specific surface proteins (Handyside and Barton 1977; Fleming et al. 1984; Louvet-Vallee et al. 2001). However, it was not possible to detect for certain single trophectodermal cells that could have escaped complement-mediated lysis and could become a source of regenerated TE. Indeed, recent report of Szczepanska et al. (2011) contradicted results presented in many previous studies. These authors evaluated the effectiveness of the lysis of TE cells by labeling cells of the original TE with the fluorescent latex microspheres before immunosurgery, and detecting surviving TE cells, after isolation of ICM, on the basis of the presence of TE-specific marker—CDX2. This study showed that the reconstructed outer layer of TE develops from the labeled cells of the original TE that survived immunosurgery and remained on the surface of isolated ICM. Thus, these results indicate that after segregation of TE and ICM (in advanced blastocysts composed of 50–70 cells) the ICM cells lose their potential to differentiate into TE (Szczepanska et al. 2011). This conclusion finds confirmation in similar experiments in which microsurgery rather than immunosurgery was used for the isolation of ICMs, and in which ICM cells isolated from early blastocysts were not able to differentiate into TE (Gardner and Johnson 1972; Rossant 1975a, b). Most likely, the microsurgery was more efficient in removing TE cells. These data are also in agreement with the experiments in which single, GFP-positive cells from ICM of 3.5-day-old blastocysts (containing ca 64 cells) were aggregated with 8-cell stage embryos (Chazaud et al. 2006). These cells colonized TE only in 8 % of the resulting blastocysts and after microinjection into the embryos at the blastocyst stage they did not contribute to the TE layer at all (Chazaud et al. 2006).

These studies demonstrate that at the early blastocyst stage (i.e., between the 32- and 64-cell stage) cells of two lineages, ICM and TE, become committed to their prospective fates and their pluripotency is lost.

8.4.2.5 Time of Ultimate Determination of Cell Fate

All the data reported here suggest that the restriction of developmental potencies occurs earlier in cells of the future TE, than of ICM. Whereas pluripotency of outer cells appears to be lost around 32-cell stage, the plasticity of inner cells of the blastocyst, especially in regard to the ultimate choice of one of the two alternative paths of differentiation—EPI or PE, is maintained for relatively long period during the preimplantation development. Microinjection of cells derived from ICM of 3.5-day-old blastocysts into the embryos at the blastocyst stage revealed that while the minority of ICM cells was bipotential and capable of generating both EPI and PE cells of ICM, the majority of them had restricted fate. This suggests that ICM of mid-blastocyst is a mosaic of morphologically indistinguishable EPI and PE progenitors having different developmental potential (Chazaud et al. 2006). The heterogeneity of ICM cells relies on diversified expression of markers specific for EPI and PE, such as NANOG (Nanog homeobox) and GATA6 (Gata-binding protein), respectively. Before PE and EPI layers become clearly morphologically distinct, the cells expressing these genes are intermingled and localized in both deeper and surface compartments of the early ICM (Chazaud et al. 2006; Plusa et al. 2008). The presence of two populations of ICM cells: progenitors of PE and EPI having diversified gene profiling was confirmed by a single-cell gene expression analysis (Kurimoto et al. 2006; Guo et al. 2010). In late (80–120 cells) blastocyst PE and EPI progenitor cells sort out into the appropriate cell layers. It was shown that multiple factors, such as actin-dependent cell movement, adhesion, selective apoptosis and positional signals participate in PE and EPI lineage segregation (Plusa et al. 2008; Meilhac et al. 2009; Morris et al. 2010). However, the precise timing of irreversible determination of ICM cells, i.e., definitive choice of the routes of differentiation—into EPI or PE—has not yet been established. It seems that it takes place prior to implantation, because in 4.5-day-old fully expanded blastocyst the EPI and PE layers are already separated and developmentally committed. This conclusion is supported by the observation that after microinjection of dissociated EPI and PE cells of 4.5-day-old blastocyst into recipient blastocyst, their contribution in resulting chimeras was restricted to the lineage of their origin (Gardner and Rossant 1979). On the other hand, it was also shown that the inner cells isolated from expanded, late blastocysts either by immunosurgery or by microdissection were able to regenerate a layer of PE cells on the blastocyst surface (Rossant 1975a, b; Pedersen et al. 1977; Hogan and Tilly 1978a, b; Dziadek 1979; Gardner 1985). However, it still is not clear whether regenerated PE originated from the EPI cells that, in response to outside environment, reversed their fate towards PE, or from the original PE cells that escaped lysis during ICM isolation procedure. Until one of these alternatives is confirmed, these experiments do not help to define the precise stage at which the final and irrevocable determination of ICM cells occurs.

8.4.3 Possible Mechanisms of Embryo Regulation

Experimental studies involving disaggregation of embryos into single cells and their re-aggregation in various configurations revealed great regulative properties and plasticity of the preimplantation mouse embryo. Depending on circumstances, embryonic cells may use diverse strategies to adapt, switch their fate, and regulate their development. Homogeneous aggregates of inner or outer blastomeres derived

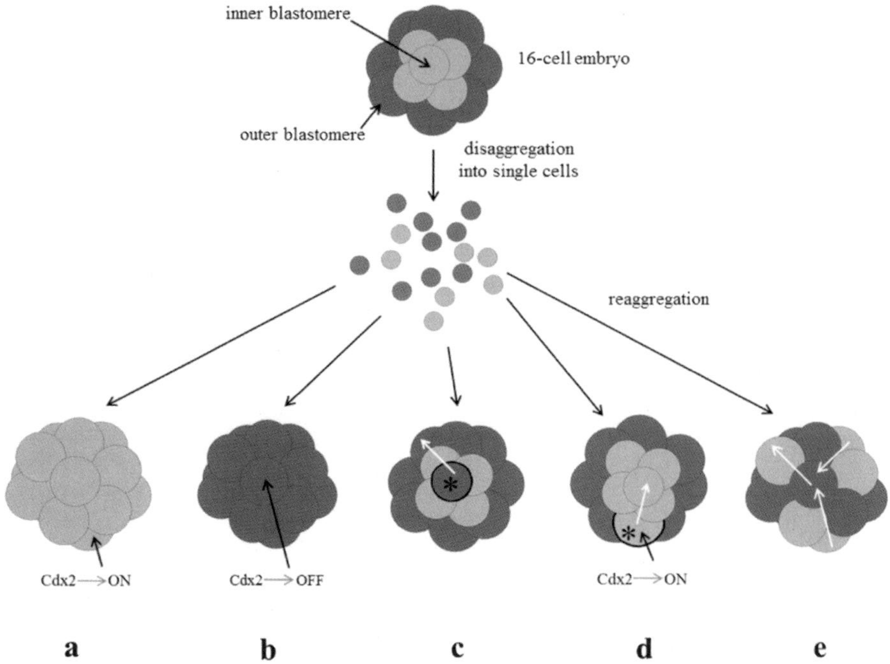

Fig. 8.4 *Mechanisms involved in embryo regulation after experimental manipulation.* Sixteen-cell embryos use different strategies to regulate their development when they are subjected to disaggregation into individual cells and subsequently re-aggregated in various configurations. (**a**) In the aggregates composed exclusively of inner or outer cells the blastomeres undergo molecular reprogramming of lineage-specific genes, e.g., *Cdx2*. Outer cells of "inner" aggregates start to synthesize CDX2. (**b**) The inner cells of "outer" aggregates downregulate the expression of CDX2. (**c**) Single originally outer cell (marked with asterisk) placed experimentally inside the embryo translocates to the outside. (**d**) Originally inner blastomere (marked with *asterisk*) forced to locate at the surface of the embryo either returns to its original inside position, or remains outside and undergoes reprogramming according to a new position. (**e**) In aggregates consisting of spatially mixed sister inner and outer blastomeres cells tend to migrate to their original positions. *White arrows* point to the direction of cell movement; *black arrows* indicate cells undergoing molecular reprogramming

from 16-cell embryos continue their embryogenesis till term, despite disruption of the original position, polarity, and connections between cells within the embryo (Suwinska et al. 2008). At the time of re-aggregation, in the aggregates composed of the outer cells, most of the blastomeres started to synthesize CDX2, while in the aggregates built of the inner cells the blastomeres were CDX2-negative. Thus, the regulation of development of newly formed embryos required a switch in the direction of differentiation in accordance with a new position occupied by the cells within the aggregate. Former inner blastomeres, which became allocated externally, and former outer blastomeres, that happened to be located in the inside position, had to undergo developmental reprogramming, i.e., regulation of expression of genes involved in cell lineage differentiation, such as upregulation and downregulation of

Cdx2 gene, respectively (Fig. 8.4a, b). This observation supports the belief that, at least until certain stage of development, the position of cells determines their future fate.

The consequences of experimental alteration of the arrangement of blastomeres were also examined by determining the contribution of labeled single cells in the re-aggregates, which were cultured until the blastocyst stage. Most of these experiments indicate that originally outer cells placed internally have a tendency to relocate to the surface and do not change their phenotype (Ziomek and Johnson 1982; Randle 1982; Surani and Handyside 1983; Fig. 8.4c), while initially inner cells placed on the outside of embryo either differentiate according to the new position, thus acquiring a new phenotype (Ziomek and Johnson 1982), or migrate inside and participate in the formation of the ICM (Randle 1982; Surani and Handyside 1983; Fig. 8.4d). Similar cell behavior was also observed in our laboratory (Suwinska et al. 2008). In the aggregates composed of spatially intermingled inner and outer sister cells derived from 16- and 32-cell embryos, the blastomeres used sorting mechanism rather than molecular reprogramming concordant with a new location within the embryo, to regulate their development (Suwinska et al. 2008). Irrespective of implemented strategy of regulation the final outcome turned out to be successful, as evidenced by the birth of a young (Suwinska et al. 2008). By labeling the outside cells of embryos before disaggregation with a lipophilic dye we followed repositioning of cells in newly formed aggregates, and found that the majority of cells occupied their original, precedent position. This indicates that the cells proceeded along their original route of differentiation (Fig. 8.4e). Solitary cells that were misplaced and occupied “wrong” position probably underwent a re-programming (or underwent apoptosis). These observations advocate the thesis that both phenotype and the position of cells affect the cell fate. It seems that differentiative division generates blastomeres differing in properties and tendency to differentiate into one specific cell lineage. However, it seems that the localization within the embryo and the positional signals coming from other cells give additional support for the choice between different differentiation pathways. In summary, all these data strongly suggest that the multiple regulative mechanisms act in concert to assure proper development of embryo.

8.5 Pre-patterning Versus Flexibility Model: Do They Rule Out or Complement Each Other?

At the first glance the pre-patterning of the mouse embryo might appear at odds with the experimental data, which demonstrate the apparent pluripotency of the early embryonic cells. However, even if an intact, undisturbed embryo possesses an internal pattern of development, as it is suggested by some researchers (Gardner 1997, 2001; Piotrowska et al. 2001; Piotrowska and Zernicka-Goetz 2001), this does not preclude the regulation of development. Although various manipulations performed on the embryos, such as the alteration of the position, number, or composition of the blastomeres, should severely perturb the potential pattern of the embryo, they do not have deleterious effect on further development; the manipulated embryos recover and complete embryogenesis successfully. The results of a variety of experimental approaches, such as reduction of blastomere number or an aggregation of cleaving embryos to create chimeras, testify to the extraordinary regulative abilities of the mouse embryo. This, however, does not necessarily mean that during a normal development the embryo will always activate and use its inherent regulative mechanisms, as under normal circumstances there is no need to use them.

The followers of pre-patterning model claim that although the existence of pre-patterning (developmental bias) is evident, meaning that some regions of the zygote or some blastomeres are predisposed to contribute to specific parts of the embryo, it guides rather than determines cell fate. The fact that during normal development the blastomeres follow their delineated paths, and in the case of emergency they can be re-directed and re-programmed indicates their compatibility with the regulative nature of development. However, the advocates of the regulative model believe that as long as the fate of the blastomeres can be altered in response to experimental manipulation, these cells are not determined, i.e., irreversibly committed to specific fates.

In summary, although a new technology of time-lapse observation of cells in intact embryos allows monitoring of such subtle changes as dynamic differences in the mobility of specific transcription factors, it does not tell when the final and irreversible cell fate decisions are made. Only the integration of emerging technologies with the traditional methods of experimental embryology allows the whole outlook of developmental processes and indicates that there is a relationship between these two seemingly incompatible models of development.

8.6 Do the Same Developmental Mechanisms Operate in Human and Mouse Preimplantation Embryo?

For ethical and legal reasons most of the experiments performed on animal embryos cannot be repeated on human embryos. Very limited experimental work performed on human embryos indicates that blastomeres isolated from human 4-cell embryo can individually develop in vitro into the blastocysts with TE and ICM (Van de Velde et al. 2008). For obvious reasons, such blastocysts could not be transferred into a uterus in order to determine whether they were totipotent, i.e., whether they could develop into babies. At any rate the above study shows that all four blastomeres of the 4-cell human embryo are not yet irreversibly allocated to either cell lineage, i.e., ICM or TE. Additional evidence in favor of the pluripotency of blastomeres of preimplantation human embryo is provided by derivation of human embryonic stem (hES) cells from single blastomeres of 4-cell (Geens et al. 2009) and 8-cell embryos (Klimanskaya et al. 2007). The establishment of embryonic and extraembryonic stem cell lines from 8-cell embryos was first demonstrated in the mouse (Chung et al. 2005). As both mouse and human stem cells generated in these ways have the capacity to differentiate into the cell types representing all three embryonic germ layers both in vitro (i.e., in embryoid bodies) and in vivo (i.e., in teratomas formed in immunodeficient mice), it indicates that at least some human blastomeres at these stages are indeed pluripotent.

On the other hand, it was found by reverse transcription polymerase chain reaction (RT-PCR) studies that individual blastomeres of 2- to 5-cell embryos express differently lineage-specific genes: marker of ICM—*Oct4*, and TE markers—*β-HCG* and *β-LH* (Hansis et al. 2002, 2004). These data suggest, in turn, that molecular divergence of blastomeres into inner and outer cell lineages (ICM and TE, respectively) begins in human embryos earlier than in the mouse. In conclusion, the scarcity of information on human preimplantation embryos does not permit to make any sound generalization.

8.7 Summary: Significance of Experimental Approach in Studying Early Mouse Development

The knowledge emerging from experiments described in this chapter gives us better understanding of mechanisms controlling preimplantation mammalian development, and is also of key importance for practical medical applications. The information provided by experimental embryology of the mouse undoubtedly helps in developing techniques presently used in assisted human reproduction, preimplantation genetic diagnosis, and in establishing ES cell lines.

Experiments initiated as early as in the 1950s of the last century became foundations for the biopsy procedure employed in the preimplantation genetic diagnosis (PGD), which is carried out in the clinics worldwide to identify the genetic defects in embryos created through in vitro fertilization before they are implanted to the uterus (Handyside et al. 1990). This procedure consists of biopsy of one blastomere from 8-cell stage embryo for genetic analysis and the transfer of the remaining 7-cell embryo that has no abnormalities to the mother's uterus for implantation and pregnancy. More than a decade of experience during which many healthy babies were born following PGD procedure indicates that the removal of a single blastomere does not affect embryo viability, reduction in pregnancy rate, or the potential increase in the rate of congenital malformations comparing to their prevalence in the general population (Hardy et al 1990; Staessen et al. 2004). Moreover, as human ES cells can be derived from single blastomeres of 4-cell (Geens et al. 2009) and 8-cell stage embryos (Klimanskaya et al. 2007), this procedure may potentially provide additional benefits for overcoming ethical barrier in isolation of hES cells. If therapies based on patient-derived hES cells come to fruition, the blastomere-derived human ES cell lines, obtained without of the harm to the embryo, would be of great potential benefit for regenerative medicine.

Acknowledgments I would like to dedicate this work to Professor A.K. Tarkowski, my mentor, as an expression of respect for his work and scientific achievements. I am grateful to Professor A.K. Tarkowski, Professor M. Kloc, Dr. J. Kubiak, and Professor M. Maleszewski for critical reading and valuable suggestions. During the preparation of this work Aneta Suwinska was supported by the grant in the framework of PARENT-BRIDGE Programme from the Foundation for Polish Science POMOST/2010-1/9 and a grant from Polish Ministry of Science and Higher Education N N301 311637.

References

Alarcón VB, Marikawa Y (2003) Deviation of the blastocyst axis from the first cleavage plane does not affect the quality of mouse postimplantation development. Biol Reprod 69:1208–1212

Alarcón VB, Marikawa Y (2005) Unbiased contribution of the first two blastomeres to mouse blastocyst development. Mol Reprod Dev 72:354–361

Alarcón VB, Marikawa Y (2008) Spatial alignment of the mouse blastocyst axis across the first cleavage plane is caused by mechanical constraint rather than developmental bias among blastomeres. Mol Reprod Dev 75:1143–1153

Chazaud C, Yamanaka Y, Pawson T, Rossant J (2006) Early lineage segregation between epiblast and primitive endoderm in mouse blastocysts through the Grb2-MAPK pathway. Dev Cell 10:615–624

Chroscicka A, Komorowski S, Maleszewski M (2004) Both blastomeres of the mouse 2-cell embryo contribute to the embryonic portion of the blastocyst. Mol Reprod Dev 68:308–312

Chung Y, Klimanskaya I, Becker S, Marh J, Lu SJ, Johnson J, Meisner L, Lanza R (2005) Embryonic and extraembryonic stem cell lines derived from single mouse blastomeres. Nature 439:216–219

Davies TJ, Gardner RL (2002) The plane of first cleavage is not related to the distribution of sperm components in the mouse. Hum Reprod 17:2368–2379

Dietrich JE, Hiiragi T (2007) Stochastic patterning in the mouse pre-implantation embryo. Development 134:4219–4231

Dziadek M (1979) Cell differentiation in isolated inner cell masses of mouse blastocysts in vitro: onset of specific gene expression. J Embryol Exp Morphol 53:367–379

Fleming TP, Warren PD, Chisholm JC, Johnson MH (1984) Trophectodermal processes regulate the expression of totipotency within the inner cell mass of the mouse expanding blastocyst. J Embryol Exp Morphol 84:63–90

Fujimori T, Kurotaki Y, Miyazaki J, Nabeshima Y (2003) Analysis of cell lineage in two- and four-cell mouse embryos. Development 130:5113–5122

Gardner RL (1985) Regeneration of endoderm from primitive ectoderm in the mouse embryo: fact or artifact. J Embryol Exp Morphol 88:303–326

Gardner RL (1997) The early blastocyst is bilaterally symmetrical and its axis of symmetry is aligned with the animal-vegetal axis of the zygote in the mouse. Development 124:289–301

Gardner RL (2001) Specification of embryonic axes begins before cleavage in normal mouse development. Development 128:839–847

Gardner RL, Davies TJ (2003) Is the plane of first cleavage related to the point of sperm entry in the mouse? Reprod Biomed Online 6:157–160

Gardner RL, Johnson MH (1972) An investigation of inner cell mass and trophoblast tissues following their isolation from the mouse blastocyst. J Embryol Exp Morphol 28:279–312

Gardner R, Rossant J (1979) Investigation of the fate of 4,5 day post coitum mouse inner cell mass by blastocyst injection. J Embryol Exp Morphol 52:141–152
Geens M, Mateizel I, Sermon K, De Rycke M, Spits C, Cauffman G, Devroey P, Tournaye H, Liebaers I, Van de Velde H (2009) Human embryonic stem cell lines derived from single blastomeres of two 4-cell stage embryos. Hum Reprod 24:2709–2717
González S, Ibáñez E, Santaló J (2011) Influence of early fate decisions at the two-cell stage on the derivation of mouse embryonic stem cell lines. Stem Cell Res 7:54–65
Gray D, Plusa B, Piotrowska K, Na J, Tom B, Glover DM, Zernicka-Goetz M (2004) First cleavage of the mouse embryo responds to change in egg shape at fertilization. Curr Biol 14:397–405
Guo G, Huss M, Tong GQ, Wang C, Sun LL, Clarke ND, Robson P (2010) Resolution of cell fate decisions revealed by single-cell gene expression analysis from zygote to blastocyst. Dev Cell 18:675–685
Handyside AH (1978) Time of commitment of inside cells isolated from preimplantation mouse embryos. J Embryol Exp Morphol 45:37–53
Handyside AH, Barton SC (1977) Evaluation of the technique of immunosurgery for the isolation of inner cell masses from mouse blastocysts. J Embryol Exp Morphol 37:217–226
Hansis C, Grifo JA, Tang Y, Krey LC (2002) Assessment of beta-HCG, beta-LH mRNA and ploidy in individual human blastomeres. Reprod Biomed Online 5:156–161
Hansis C, Grifo JA, Krey LC (2004) Candidate lineage marker genes in human preimplantation embryos. Reprod Biomed Online 8:577–583
Hardy K, Martin KL, Leese HJ, Winston RM, Handyside AH (1990) Human preimplantation development in vitro is not adversely affected by biopsy at the 8-cell stage. Hum Reprod 5:708–714
Hiiragi T, Solter D (2004) First cleavage plane of the mouse egg is not predetermined but defined by the topology of the two apposing pronuclei. Nature 430:360–364
Hiiragi T, Alarcon VB, Fujimori T, Louvet-Vallee S, Maleszewski M, Marikawa Y, Maro B, Solter D (2006) Where do we stand now? Mouse early embryo patterning meeting in Freiburg, Germany. Int J Dev Biol 50:581–586
Hogan B, Tilly R (1978a) In vitro development of inner cell masses isolated immunosurgically from mouse blastocysts. II. Inner cell masses from 3,5- to 4,0-day p.c. blastocysts. J Embryol Exp Morphol 45:107–21
Hogan B, Tilly R (1978b) In vitro development of inner cell masses isolated immunosurgically from mouse blastocysts. I. Inner cell masses from 3,5-day p.c. blastocysts incubated for 24h before immunosurgery. J Embryol Exp Morphol 45:93–105
Handyside AH, Kontogianni EH, Hardy K, Winston RM (1990) Pregnancies from biopsied human preimplantation embryos sexed by Y-specific DNA amplification. Nature 344:768–770
James RM, Klerkx AHEM, Keighren M, Flockhart JH, West JD (1995) Restricted distribution of tetraploid cells in mouse tetraploid <==> diploid chimaeras. Dev Biol 167:213–226
Jedrusik A, Parfitt DE, Guo G, Skamagki M, Grabarek JB, Johnson MH, Robson P, Zernicka-Goetz M (2008) Role of Cdx2 and cell polarity in cell allocation and specification of trophectoderm and inner cell mass in the mouse embryo. Genes Dev 22:2692–2706
Johnson MH, Ziomek CA (1981) The foundation of two distinct cell lineages within the mouse morula. Cell 24:71–80
Johnson WH, Loskutoff NM, Plante Y, Betteridge KJ (1995) Production of four identical calves by the separation of blastomeres from an in vitro derived four-cell embryo. Vet Rec 137:15–16
Kelly SJ (1975) Studies of the potency of the early cleavage blastomeres of the mouse. In: Balls M, Wild AE (eds) The early development of mammals. Cambridge University Press, Cambridge, pp 97–105
Kelly SJ (1977) Studies of the developmental potential of 4- and 8-cell stage mouse blastomeres. J Exp Zool 200:365–376
Klimanskaya I, Chung Y, Becker S, Lu SJ, Lanza R (2007) Derivation of human embryonic stem cells from single blastomeres. Nat Protoc 2:1963–1972
Kubiak JZ, Tarkowski AK (1985) Electrofusion of mouse blastomeres. Exp Cell Res 157:561–566

Kurimoto K, Kabuta Y, Ohinata Y, Ono Y, Uno KD, Hamada RG, Ueda HR, Saitou M (2006) An improved single-cell cDNA amplification method for efficient high-density oligonucleotide microarray analysis. Nucleic Acids Res 34:e42

Kurotaki Y, Hatta K, Nakao K, Nabeshima Y, Fujimori T (2007) Blastocyst axis is specified independently of early cell lineage but aligns with the ZP shape. Science 316:719–723

Louvet-Vallee S, Dard N, Santa-Maria A, Aghion J, Maro B (2001) A major posttranslational modification of ezrin takes place during epithelial differentiation in the early mouse embryo. Dev Biol 231:190–200

Louvet-Vallée S, Vinot S, Maro B (2005) Mitotic spindles and cleavage planes are oriented randomly in the two-cell mouse embryo. Curr Biol 15:464–469

Markert CL, Petters RM (1978) Manufactured hexaparental mice show that adults are derived from three embryonic cells. Science 202:56–58

MacKay GE, West JD (2005) Fate of tetraploid cells in 4n↔2n chimeric mouse blastocysts. Mech Dev 122:1266–1281

McGrath J, Solter D (1984) Completion of mouse embryogenesis requires both the maternal and paternal genomes. Cell 37:179–183

McLaren A (1976) Mammalian chimaeras. Cambridge University Press, Cambridge, London, New York, Melbourne

Mintz B (1962) Formation of genotypically mosaic mouse embryos. Am Zool 2:432

Mintz B (1964) Formation of genetically mosaic mouse embryos and early development of "lethal (t^{12}/t^{12})-normal" mosaics. J Exp Zool 157:273–292

Meilhac SM, Adams RJ, Morris SA, Danckaert A, Le-Garrec JF, Zernicka-Goetz M (2009) Active cell movements coupled to positional induction are involved in lineage segregation in the mouse. Dev Biol 331:210–221

Moore NW, Adams CE, Rowson LEA (1968) Developmental potential of single blastomeres of the rabbit egg. J Reprod Fertil 17:527–531

Morris SA, Teo RT, Li H, Robson P, Glover DM, Zernicka-Goetz M (2010) Origin and formation of the first two distinct cell types of the inner cell mass in the mouse embryo. Proc Natl Acad Sci USA 107:6364–6369

Motosugi N, Bauer T, Polanski Z, Solter D, Hiiragi T (2005) Polarity of the mouse embryo is established at blastocyst and is not prepatterned. Genes Dev 19:1081–1092

Mullen RJ, Whitten WK, Carter SC (1970) Studies on chimeric mice and half-embryos. In: Annual report of the Jackson Laboratory. Bar harbor, Maine, pp 67–68

Nichols J, Gardner RL (1984) Heterogeneous differentiation of external cells in individual isolated early mouse inner cell masses in culture. J Embryol Exp Morphol 80:225–240

Ozdzeński W, Szczesny E, Tarkowski AK (1997) Postimplantation development of mouse blastocysts with two separate inner cell masses. Anat Embryol (Berl) 195:467–471

Pedersen RA, Spindle AI, Wiley LM (1977) Regeneration of endoderm by ectoderm isolated from mouse blastocysts. Nature 270:453–457

Pierce GB, Arechaga J, Muro C, Wells RS (1988) Differentiation of ICM cells into trophectoderm. Am J Pathol 132:356–364

Pinyopummin A, Takahashi Y, Hishinuma M, Kanagawa H (1994) Development of single blastomeres from 4-cell stage embryos after aggregation with parthenogenones in mice. Jpn J Vet Res 42:119–126

Piotrowska K, Zernicka-Goetz M (2001) Role for sperm in spatial patterning of the early mouse embryo. Nature 409:517–521

Piotrowska K, Wianny F, Pedersen RA, Zernicka-Goetz M (2001) Blastomeres arising from the first cleavage division have distinguishable fates in normal mouse development. Development 128:3739–3748

Piotrowska-Nitsche K, Perea-Gomez A, Haraguchi S, Zernicka-Goetz M (2005) Four-cell stage mouse blastomeres have different developmental properties. Development 132:479–490

Plachta N, Bollenbach T, Pease S, Fraser SE, Pantazis P (2011) Oct4 kinetics predict cell lineage patterning in the early mammalian embryo. Nat Cell Biol 13:117–123

Plusa B, Grabarek JB, Piotrowska K, Glover DM, Zernicka-Goetz M (2002a) Site of the previous meiotic division defines cleavage orientation in the mouse embryo. Nat Cell Biol 4:811–815
Plusa B, Piotrowska K, Zernicka-Goetz M (2002b) Sperm entry position provides a surface marker for the first cleavage plane of the mouse zygote. Genesis 32:193–198
Plusa B, Hadjantonakis AK, Gray D, Piotrowska-Nitsche K, Jedrusik A, Papaioannou VE, Glover DM, Zernicka-Goetz M (2005) The first cleavage of the mouse zygote predicts the blastocyst axis. Nature 434:391–395
Plusa B, Piliszek A, Frankenberg S, Artus J, Hadjantonakis AK (2008) Distinct sequential cell behaviours direct primitive endoderm formation in the mouse blastocyst. Development 135:3081–3091
Ralston A, Rossant J (2008) Cdx2 acts downstream of cell polarization to cell-autonomously promote trophectoderm fate in the early mouse embryo. Dev Biol 313:614–629
Randle BJ (1982) Cosegregation of monoclonal antibody reactivity and cell behaviour in the mouse preimplantation embryo. J Embryol Exp Morphol 70:261–278
Rossant J (1975a) Investigation of the determinative state of the mouse inner cell mass. II. The fate of isolated inner cell masses transferred to the oviduct. J Embryol Exp Morphol 33:991–1001
Rossant J (1975b) Investigation of the determinative state of the mouse inner cell mass. I. Aggregation of isolated inner cell masses with morulae. J Embryol Exp Morphol 33:979–990
Rossant J (1976) Postimplantation development of blastomeres isolated from 4- and 8-cell mouse eggs. J Embryol Exp Morphol 36:283–290
Rossant J, Lis WT (1979) Potential of isolated mouse inner cell masses to form trophectoderm derivatives in vivo. Dev Biol 70:255–261
Rossant J, Vijh KM (1980) Ability of outside cells from preimplantation mouse embryos to form inner cell mass derivatives. Dev Biol 76:475–482
Seidel F (1952) Die entwicklugspotenzen einen isolierten blastomere des zweizellen-stadiums im saugetierei. Naturwissenschaften 39:355–356
Seidel F (1960) Die entwicklungsfahigkeiten isolierter furchungszellen aus dem ei des kaninchens Oryctolagus cuniculus. Roux Arch Entw Mech 152:43–130
Solter D, Knowles BB (1975) Immunosurgery of mouse blastocyst. Proc Natl Acad Sci USA 72:5099–5102
Spindle AI (1978) Trophoblast regeneration by inner cell masses isolated from cultured mouse embryos. J Exp Zool 203:483–489
Staessen C, Platteau P, Van Assche E, Michiels A, Tournaye H, Camus M, Devroey P, Liebaers I, Van Steirteghem A (2004) Comparison of blastocyst transfer with or without preimplantation genetic diagnosis for aneuploidy screening in couples with advanced maternal age: a prospective randomized controlled trial. Hum Reprod 19:2849–2858
Strumpf D, Mao CA, Yamanaka Y, Ralston A, Chawengsaksophak K, Beck F, Rossant J (2005) Cdx2 is required for correct cell fate specification and differentiation of trophectoderm in the mouse blastocyst. Development 132:2093–2102
Surani MAH, Handyside AH (1983) Reassortment of cells according to position in mouse morulae. J Exp Zool 225:505–511
Surani MA, Barton SC, Norris ML (1984) Development of reconstituted mouse eggs suggests imprinting of the genome during gametogenesis. Nature 308:548–550
Suwinska A, Czolowska R, Ozdzenski W, Tarkowski AK (2008) Blastomeres of the mouse embryo lose totipotency after the fifth cleavage division: expression of Cdx2 and Oct4 and developmental potential of inner and outer blastomeres of 16- and 32-cell embryos. Dev Biol 322:133–44
Szczepanska K, Stanczuk L, Maleszewski M (2011) Isolated mouse inner cell mass is unable to reconstruct trophectoderm. Differentiation 82:1–8
Tarkowski AK (1959a) Experiments on the development of isolated blastomeres of mouse eggs. Nature 184:1286–1287
Tarkowski AK (1959b) Experimental studies on regulation in the development of isolated blastomeres of mouse eggs. Acta Theriol 3:191–267

Tarkowski AK (1961) Mouse chimaeras developed from fused eggs. Nature 190:857–860
Tarkowski AK (1998) Mouse chimaeras revisited: recollections and reflections. Int J Dev Biol 42:903–908
Tarkowski AK (1963) Studies on mouse chimeras developed from eggs fused in vitro. Natl Cancer Inst Monogr 11:51–71
Tarkowski AK, Witkowska A, Nowicka J (1970) Experimental partheonogenesis in the mouse. Nature 226:162–165
Tarkowski AK, Wojewodzka M (1982) A method for obtaining chimaeric mouse blastocysts with two separate inner cell masses: a preliminary report. J Embryol Exp Morphol 71:215–221
Tarkowski AK, Wroblewska J (1967) Development of blastomeres of mouse eggs isolated at the 4- and 8-cell stage. J Embryol Exp Morphol 18:155–180
Tarkowski AK, Witkowska A, Opas J (1977) Development of cytochalasin B-induced tetraploid and diploid/tetraploid mosaic mouse embryos. J Embryol Exp Morphol 41:47–64
Tarkowski AK, Ozdzenski W, Czołowska R (2001a) Mouse singletons and twins developed from isolated diploid blastomeres supported with tetraploid blastomeres. Int J Dev Biol 45:591–596
Tarkowski AK, Ozdzenski W, Czołowska R (2001b) How many blastomeres of the 4-cell embryo contribute cells to the mouse body? Int J Dev Biol 45:811–816
Tarkowski AK, Ozdzenski W, Czolowska R (2005a) Identical triplets and twins developed from isolated blastomeres of 8- and 16-cell mouse embryos supported with tetraploid blastomeres. Int J Dev Biol 49:825–832
Tarkowski AK, Jagiello K, Czolowska R, Ozdzenski W (2005b) Mouse chimaeras developed from electrofused blastocysts: new evidence for developmental plasticity of the inner cell mass. Int J Dev Biol 49:909–914
Tarkowski AK, Suwinska A, Czolowska R, Ozdzeński W (2010) Individual blastomeres of 16- and 32-cell mouse embryos are able to develop into foetuses and mice. Dev Biol 348:190–198
Torres-Padilla ME, Parfitt DE, Kouzarides T, Zernicka-Goetz M (2007) Histone arginine methylation regulates pluripotency in the early mouse embryo. Nature 445:214–218
Tsunoda Y, McLaren A (1983) Effect of various procedures on the viability of mouse embryos containing half the normal number of blastomeres. J Reprod Fertil 69:315–322
Tsunoda Y, Yasui T, Okubo Y, Nakamura K, Sugie T (1987) Development of one or two blastomeres from eight-cell mouse embryos to term in the presence of parthenogenetic eggs. Theriogenology 28:615–623
Wakayama S, Hikichi T, Suetsugu R, Sakaide Y, Bui H, Mizutani E, Wakayama T (2007) Efficient establishment of mouse embryonic stem cell lines from single blastomeres and polar bodies. Stem Cells 25:986–993
Willadsen SM (1981) The developmental capacity of blastomeres from 4- and 8- cell sheep embryos. J Embryol Exp Morphol 65:165–172
Waksmundzka M, Wisniewska A, Maleszewski M (2006) Allocation of cells in mouse blastocyst is not determined by the order of cleavage of the first two blastomeres. Biol Reprod 75:582–587
Van de Velde H, Cauffman G, Tournaye H, Devroey P, Liebaers I (2008) The four blastomeres of a 4-cell stage human embryo are able to develop individually into blastocysts with inner cell mass and trophectoderm. Hum Reprod 23:1742–1747
Ziomek CA, Johnson MH (1982) The roles of phenotype and position in guiding the fate of 16-cell mouse blastomeres. Dev Biol 91:440–447
Ziomek CA, Johnson MH, Handyside AH (1982) The developmental potential of mouse 16-cell blastomeres. J Exp Zool 221:345–355

Chapter 9
Creation of Trophectoderm, the First Epithelium, in Mouse Preimplantation Development

Yusuke Marikawa and Vernadeth B. Alarcon

Abstract Trophectoderm (TE) is the first cell type that emerges during development and plays pivotal roles in the viviparous mode of reproduction in placental mammals. TE adopts typical epithelium morphology to surround a fluid-filled cavity, whose expansion is critical for hatching and efficient interaction with the uterine endometrium for implantation. TE also differentiates into trophoblast cells to construct the placenta. This chapter is an overview of the cellular and molecular mechanisms that control the critical aspects of TE formation, namely, the formation of the blastocyst cavity, the expression of key transcription factors, and the roles of cell polarity in the specification of the TE lineage. Current gaps in our knowledge and challenging issues are also discussed that should be addressed in future investigations in order to further advance our understanding of the mechanisms of TE formation.

9.1 Introduction

The very first morphogenetic event in animal development is epithelialization of the outer cells. After several rounds of cleavages, cells located on the surface of the embryo become firmly adhered to each other and establish junctional connections to create a seamless epithelial sheet. This epithelium serves as structural as well as chemical barrier to separate the embryo from the environment. In many species, the epithelium surrounds a fluid-filled cavity called the blastocoel. The blastocoel provides space to allow cell movement during gastrulation, which brings presumptive mesoderm and endoderm interiorly.

Y. Marikawa (✉) • V.B. Alarcon
Institute for Biogenesis Research, University of Hawaii John A. Burns School of Medicine, Biosciences Building, 651 Ilalo Street, Honolulu, HI, USA
e-mail: marikawa@hawaii.edu; vernadet@hawaii.edu

J.Z. Kubiak (ed.), *Mouse Development*, Results and Problems in Cell Differentiation 55,
DOI 10.1007/978-3-642-30406-4_9, © Springer-Verlag Berlin Heidelberg 2012

Like other animals, the first morphogenetic event in development of eutherians (placental mammals, such as human and mouse) is epithelialization of the outer cells. However, unlike other animals, the first epithelium of eutherian embryos is specifically called the trophectoderm (TE), and it plays unique roles that are inherent to the viviparous nature of eutherian reproduction. TE physically interacts with the endometrium of the mother's uterus to initiate implantation and differentiates into extraembryonic tissues, namely, trophoblast, to construct the placenta. In addition, the fluid-filled cavity surrounded by TE (which is specifically referred to as the blastocyst cavity, rather than the blastocoel, in this chapter) is neither functionally nor developmentally equivalent to the blastocoel in other animals; the blastocyst cavity is not a space for gastrulation to bring in presumptive mesoderm or endoderm. Instead, the blastocyst cavity expands robustly during preimplantation development, which facilitates hatching of the embryo from the egg shell, or the zona pellucida. In addition, expansion of the blastocyst cavity enlarges the surface area of the embryo, which increases the efficiency of interaction with the uterine endometrium during implantation. Thus, TE formation is one of the evolutionary inventions that are customized for the specialized reproductive mode of placental mammals.

Fertilization and preimplantation development of eutherian embryos naturally occur in the oviduct of the mother. However, for many species, including the human and mouse, fertilization and preimplantation development can be recreated in vitro in a chemically defined culture medium under a controlled environment without significantly impairing the developmental potential of embryos (Summers and Biggers 2003; Gardner 2008; Vajta et al. 2010). Thus, a large portion of our current knowledge on the morphological as well as molecular aspects of preimplantation development is derived from the studies using in vitro cultured embryos. In the case of the mouse, the first cleavage occurs at about 16–20 h after fertilization, and the following cleavages take place at an interval of roughly 12 h. At late 8-cell stage, all blastomeres become attached firmly to each other, and the embryo adopts a spherical appearance (Fig. 9.1). This process is known as compaction. The following cleavages result in emergence of the two types of cells with respect to their positions in the embryo: outer and inner cells. Generally, outer cells give rise to TE, whereas the inner cells become the inner cell mass (ICM), the latter of which is a population of pluripotent stem cells that eventually contribute to all cell types in the body. Between 16- and 32-cell stages, the blastocyst cavity starts to form, which gradually expands. At this point, the embryo is called the blastocyst (Fig. 9.1).

The purpose of this chapter is to review the current knowledge on the mechanisms of TE formation in the mouse embryo, with a specific focus on the molecular processes controlling the generation of the blastocyst cavity, the transcriptional regulators that are involved in TE development, and the initial signals that separate the TE lineage from the ICM lineage. The chapter is particularly devoted to highlight elusive aspects and gaps in our knowledge as well as challenging issues that may be hindering progress in our understanding of the mechanisms of TE formation.

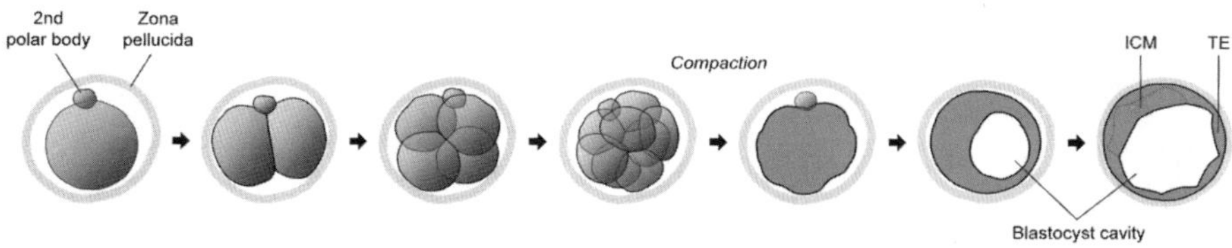

Fig. 9.1 Morphological transformation of the mouse embryo from the 1-cell stage to the blastocyst stage, which takes place during the first 3 days after fertilization

9.2 Formation, Expansion, and Maintenance of the Blastocyst Cavity

The creation of the blastocyst cavity is the morphological landmark of TE epithelialization. The following four steps are fundamental to the formation and maintenance of an expanded blastocyst cavity: (1) formation of microlumen as the initial small cavity, (2) generation of osmotic gradient between the cavity and the external environment, (3) water influx into the cavity, and (4) establishment of paracellular sealing in TE (Fig. 9.2a). Failure in any of these steps would compromise the blastocyst cavity formation.

9.2.1 Formation of Microlumen

The blastocyst cavity starts as a small cavity, or microlumen, which emerges in the intercellular space after the fifth cleavages, i.e., between 16- and 32-cell stages. In some embryos, multiple microlumens appear which eventually coalesce to form a single cavity during expansion. Microlumens originate from intracellular vesicles, or vacuoles, that form in the outer cells (presumptive TE) and are released by exocytosis at the basal membrane (Fig. 9.2b; Aziz and Alexandre 1991; Motosugi et al. 2005). Various questions regarding the molecular mechanisms of microlumen formation during blastocyst development are yet to be answered. What signals control the timing of vacuole production and exocytosis? What machinery directs vacuoles specifically to the basal membrane?

The molecular mechanisms of microlumen formation have been intensively studied using in vitro cell culture systems, such as the MDCK (Madin-Darby canine kidney) cell line (Wang et al. 1990; Apodaca 2010; Engelberg et al. 2011). By culturing three-dimensionally in a collagen gel, individual MDCK cells proliferate and assemble into a cyst, which is a spherical epithelial layer surrounding a central cavity. Using the MDCK cyst as a model for luminogenesis, several molecules have

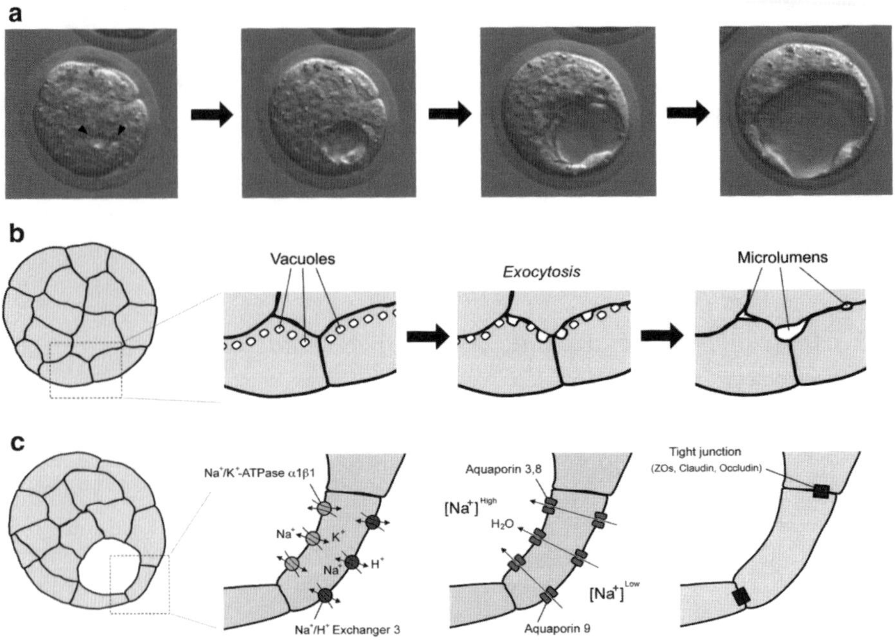

Fig. 9.2 (**a**) Snapshot images taken by time-lapse cinematography of a developing mouse embryo, showing the initial formation of microlumens and their expansion to generate the blastocyst cavity. (**b**) Schematic diagrams, depicting the microlumen formation by exocytosis of vacuoles at the basal membrane in the outer cells. (**c**) Schematic diagrams, portraying three critical steps to expand and maintain the blastocyst cavity, namely, directional sodium ion transport, water influx, and paracellular sealing. See text for details

been identified that are involved in the generation of the initial intracellular vesicles and their targeted delivery and exocytosis (Bryant et al. 2010). These molecules may also regulate microlumen formation in mouse preimplantation development. Nonetheless, it is of particular importance to note that the orientation of epithelial cell polarity is totally reversed between the MDCK cyst and the blastocyst: the apical side is facing the lumen in the MDCK cyst, whereas the basal side of TE is facing the blastocyst cavity. In the MDCK cyst formation, intracellular vesicles are targeted to the apical membrane for exocytosis, in which conserved apical components, such as Cdc42, Par3, and atypical PKC, play essential roles (Martin-Belmonte et al. 2008; Delous et al. 2009; Schluter et al. 2009; Yoshihama et al. 2011). In contrast, the apical components in TE are located exteriorly, that is, facing away from the cavity. Thus, these apical components are unlikely to be involved in targeted vesicle transport and exocytosis in the blastocyst cavity formation. Despite that these results could be instructive, it is critical to investigate the mechanisms of initial lumen formation in the context of blastocyst formation since they seem to involve other mechanisms than the ones described in the MDCK cells.

9.2.2 Generation of Osmotic Gradient

After the initial microlumen formation by vacuole exocytosis, further expansion is mediated by water influx from the external environment into the microlumen. This water influx is driven by an osmotic gradient, which is established by differential sodium ion (Na^+) concentration, such that it is higher in the cavity than in the external environment (Fig. 9.2c; Borland et al. 1977; Biggers et al. 1988; Manejwala et al. 1989; Watson and Barcroft 2001). Elevated Na^+ concentration in the cavity is created by active transport of Na^+ through transmembrane pumps (Fig. 9.2c). Na^+/H^+ exchangers and Na^+/K^+-ATPases play key roles in Na^+ influx at the apical membrane and Na^+ efflux at the basal membrane, respectively (Wiley 1984; Manejwala et al. 1989; Watson 1992; Barr et al. 1998; Barcroft et al. 2004; Kawagishi et al. 2004). One of the Na^+/H^+ exchangers, NHE3 (Slc9a3; Mouse Genome Informatics), is enriched in the apical membrane of TE. The inhibition of NHE3 using a specific inhibitor S3226 interferes with blastocyst cavity formation, whereas cariporide, an inhibitor for NHE1, does not (Kawagishi et al. 2004). Importantly, S3226 also interferes with the re-expansion of the blastocyst cavity after its collapse by a transient treatment with cytochalasin D (Kawagishi et al. 2004). This indicates that S3226 inhibits the blastocyst cavity formation by interfering with its expansion rather than the initial microlumen formation. While the inhibitor experiment strongly suggests the importance of NHE3, surprisingly the homozygous NHE3 mutant embryo does not appear to have abnormalities in preimplantation development (Schultheis et al. 1998). Further investigations are required to test whether S3226 affects other Na^+/H^+ exchangers that may function redundantly with NHE3. It also needs to be determined whether maternally supplied NHE3 is sufficient to support preimplantation development, which would explain the lack of a phenotype in the homozygous mutant embryo.

Multiple isoforms of alpha- and beta-subunits of Na^+/K^+-ATPase are expressed in mouse preimplantation embryos (MacPhee et al. 2000; Kidder 2002; Kidder and Watson 2005). In particular, α1-subunit (Atp1a1) and β1-subunit (Atp1b1) are enriched in the basal membrane of TE. Atp1a1 null-mutant embryos fail to develop to the blastocyst stage when cultured in vitro, although they develop normally to the blastocyst stage in vivo (Barcroft et al. 2004). Blastocyst cavity formation is also impaired when Atp1b1 is knocked down by specific siRNA (Madan et al. 2007). While these studies suggest that the basally localized Na^+/K^+-ATPases are essential for blastocyst cavity formation, further studies are essential to determine whether the cavity defect is caused primarily by the lack of osmotic gradient. It has been shown that knockdown of Atp1b1 causes abnormal distributions of tight junction components, namely, ZO-1 and occludin, raising the possibility that Atp1b1 is also essential for establishing paracellular sealing of TE (see Sect. 9.2.4). Indeed, in addition to the ion transport activity, Atp1b1 has been shown to function as a cell adhesion molecule (Rajasekaran and Rajasekaran 2009; Tokhtaeva et al. 2011), which may be linked to the tight junction defects in Atp1b1 knockdown embryos.

The phenotypes of the Atp1a1 null-mutant embryo, as described above, pose an interesting question about critical differences between in vitro and in vivo conditions for preimplantation development. Evidently, the absence of Atp1a1 is not detrimental enough to cause an abnormality in the blastocyst formation in an in vivo environment within the oviduct. In contrast, conventional medium for in vitro embryo culture, such as KSOM (Lawitts and Biggers 1993), cannot support the preimplantation development of Atp1a1 null-mutant embryos, although KSOM can support blastocyst formation in wild-type embryos. The chemical composition of KSOM is much simpler than that of the endogenous oviduct fluid. Thus, it is possible that KSOM may not contain certain types or levels of specific ions that are essential to compensate the lack of the Na^+/K^+-ATPase. A recent study shows that a complex medium that is originally formulated to culture mouse embryonic stem cells can promote the expansion of the blastocyst cavity more robustly than KSOM (Gelber et al. 2011). Thus, culture medium for mouse preimplantation embryo may be further modified to support blastocyst formation in slightly impaired embryos, such as Atp1a1 null-mutant.

9.2.3 Water Influx

Although osmotic gradient is the driving force for water influx into the cavity, the actual osmotic pressure is not high enough to move water molecules directly across the lipid bilayers of the plasma membrane (Borland et al. 1977; Biggers et al. 1988). Thus, water channels, namely, aquaporins (AQPs), are required to facilitate water influx through the outer cells (Fig. 9.2c). In mouse preimplantation development, various AQP isoforms are expressed, and each isoform has a distinct temporal and spatial expression patterns (Edashige et al. 2000; Offenberg et al. 2000; Barcroft et al. 2003). For example, AQP9 is enriched in the apical membrane, whereas AQP3 and AQP8 are mainly present in the basal membrane of TE (Barcroft et al. 2003; Bell et al. 2009), raising the possibility that each isoform plays unique roles. However, currently, functional significance of specific isoforms for preimplantation development has not been clearly delineated, as the knockout of individual AQP gene has no apparent defect in blastocyst formation (Ma et al. 1997b, 1998, 1999, 2000; Krane et al. 2001; Hara-Chikuma et al. 2005; Morishita et al. 2005; Sohara et al. 2005; Yang et al. 2005; Rojek et al. 2006, 2007; Kitaura et al. 2009; Ohta et al. 2009; Thrane et al. 2011). It is possible that the lack of one AQP gene may be compensated by other AQPs or maternally supplied mRNA and/or protein, although these speculations need to be tested by further investigations.

9.2.4 Paracellular Sealing

The blastocyst cavity expansion requires retaining of Na^+ and water molecule that had moved into the cavity. This task is achieved by paracellular sealing of TE (Fig. 9.2c). For the first time in the embryo life, the junctional complexes, namely, adherens junction, tight junction (TJ), desmosome junction, and gap junction, are established between TE cells. This is an important step in embryo development since it enables to form the first tissue-like structures within the embryo. Among these junctions, TJ is mainly responsible for paracellular sealing, which acts as permeability barrier for ions and water molecules. TJ is composed of various components, such as claudins, occludin, and zonula occludens proteins (ZO-1, ZO-2, and ZO-3), and these components are assembled sequentially after compaction to build the mature permeability barrier of TE (Eckert and Fleming 2008). However, which TJ components play essential roles in the paracellular sealing of TE are not fully elucidated. None of the single knockout of the TJ components has resulted in the inhibition of the blastocyst cavity formation (Saitou et al. 2000; Adachi et al. 2006; Katsuno et al. 2008; Xu et al. 2008; Escudero-Esparza et al. 2011). We can speculate that the lack of clear phenotypes in single knockouts during preimplantation development is most likely due to functional redundancy with the other TJ components, which still need to be confirmed through further investigations. Nonetheless, experiments using RNAi to knock down specific TJ components have yielded phenotypes in blastocyst development. For example, knockdown of ZO-1 and/or ZO-2 causes delayed expansion of the blastocyst cavity, suggesting their roles in paracellular sealing (Sheth et al. 2008). In another study, RNAi-mediated knockdown of ZO-1 results in a much severe phenotype, such as delayed cell division and abnormal gene expressions that are involved in both TE and ICM development, which implicates that ZO-1 plays other functions in addition to TJ formation (Wang et al. 2008). The cause of the discrepancy between these studies is currently unclear, although the specificity of the RNAi-mediated knockdown needs to be carefully evaluated (e.g., by the use of different siRNA sequences and/or rescue experiment) to confirm that the observed phenotypes are caused solely by the lack of ZO-1.

Another approach to test the role of TJ components, specifically claudins, has been taken by using *Clostridium perfringens* enterotoxin (CPE), which specifically binds claudins in the intestinal epithelium to cause food poisoning (Katahira et al. 1997). While the C-terminal half of CPE (C-CPE) is noncytotoxic, it can still bind and sequester claudins to increase paracellular permeability of the epithelium (Sonoda et al. 1999). C-CPE-treated mouse embryos exhibit impaired blastocyst cavity expansion (Moriwaki et al. 2007), suggesting that claudins play an essential role in TE permeability sealing. C-CPE can bind to various claudin isoforms, namely, claudins 3, 4, 6, 7, 8, and 14 (Sonoda et al. 1999; Fujita et al. 2000). Among these, however, only claudins 4, 6, and 7 are detectable in the mouse blastocyst (Moriwaki et al. 2007), suggesting that these isoforms are specifically involved in TE paracellular sealing.

9.3 Transcriptional Regulators Involved in TE Development

Transcription factors can coordinate the expressions of many other genes, and thereby function as master regulators of complex events, such as cell differentiation. Several transcription factors have been identified that are involved in TE development, as described below. Each factor appears to function in a distinct manner based on their knockout or knockdown phenotypes in mouse preimplantation as well as early postimplantation development. However, the identities of their transcriptional targets are mostly unclear, particularly with respect to how and which aspects of TE development are controlled by these transcription factors. Further studies involving global gene expression analyses, in conjunction with loss-of-function as well as gain-of-function of each transcription factor, should reveal the transcriptional regulatory networks that control TE development.

9.3.1 Cdx2

Cdx2 encodes a caudal-type homeodomain protein, and is specifically expressed in TE of the mouse blastocyst (Strumpf et al. 2005). Homozygous Cdx2 knockout mutant embryos develop defective TE, as they fail to implant in vivo and are unable to give rise to trophoblast giant cells as well as trophoblast stem cells in vitro. However, Cdx2 knockout embryos develop normal ICM, from which embryonic stem cells can be derived. In addition, ICM marker genes Nanog and Oct4 (Pou5f1; Mouse Genome Informatics) are ectopically expressed in the outer cells of Cdx2 knockout embryos, suggesting that Cdx2 promotes TE development and suppresses the ICM lineage formation (Strumpf et al. 2005; Ralston and Rossant 2008). Nonetheless, Cdx2 knockout embryos are capable of forming the blastocyst cavity, indicating that epithelialization of TE is not dependent on Cdx2 (Strumpf et al. 2005; Ralston and Rossant 2008). At later stages, however, TE epithelium starts to exhibit defects, such as a disorganized ZO-1 distribution pattern, resulting in the collapse of the blastocyst cavity. Thus, Cdx2 may be involved in gene regulations that are essential for the maintenance of epithelial integrity at later stages of the blastocyst. Alternatively, significant increase in apoptosis is observed in the outer cells of Cdx2 knockout embryos (Strumpf et al. 2005), which may be contributing to the degeneration of epithelial integrity. It is unclear how the loss of Cdx2 is mechanistically linked to the increased incidence of apoptosis.

Transcriptional targets of Cdx2 that are involved in trophoblast differentiation are still elusive. As mentioned above, loss of Cdx2 during preimplantation development results in ectopic expression of Pou5f1 and Nanog (Strumpf et al. 2005; Wang et al. 2010). Also, Cdx2 directly binds to the promoters of Pou5f1 and Nanog to repress their expressions in embryonic stem cells (Niwa et al. 2005; Chen et al. 2009). Thus, Pou5f1 and Nanog may be direct transcriptional targets of Cdx2 to repress the key regulators of ICM formation. However, the expression pattern of Cdx2 does not necessarily correlate with reduced expression of Pou5f1 or Nanog during the early stages of blastocyst development (Dietrich and Hiiragi 2007), implicating that Cdx2 expression may not be sufficient to repress Pou5f1 and Nanog. A recent study shows that Brg1, which encodes a regulator of chromatin remodeling, cooperates with Cdx2 to repress the expression of Pou5f1 (Wang et al. 2010). Thus, Cdx2 is necessary, but may not be sufficient by itself to repress Pou5f1 and Nanog. In contrast, Cdx2 may act as a positive regulator of the transcription of Eomes, which encodes another TE-specific transcription factor (see Sect. 9.3.2). The Eomes transcript level is markedly reduced in Cdx2 knockout embryos (Strumpf et al. 2005). However, whether the expression of Eomes is directly regulated by Cdx2 is yet to be determined.

Recent conflicting reports argue for, and against, a distinct functional role for maternally supplied Cdx2 mRNA in preimplantation development. In one group of studies, siRNA- or antisense oligo-mediated knockdown of both maternal and zygotic Cdx2 mRNA causes much severe defects than the phenotype of homozygous Cdx2 mutant embryos: delayed cell division as well as developmental arrest even at 4-cell stage (Jedrusik et al. 2008, 2010). In these studies, the expression of cell polarity regulators has been suggested to be under the control of maternal Cdx2, which may be linked to the early developmental defects. However, the other study shows that effective removal of maternal and zygotic Cdx2 mRNA by RNAi results in defects only at later stages that are comparable to the homozygous Cdx2 mutant embryos (Wu et al. 2010). The cause of this contradiction has not been clarified at this point (Bruce 2011), although it would be essential to investigate the role of maternal Cdx2 through other approaches, such as conditional gene knockout in developing oocytes (Sun et al. 2008).

9.3.2 Eomes

Eomes encodes a T-box domain transcription factor and is expressed specifically in TE of the mouse blastocyst (Russ et al. 2000; McConnell et al. 2005). In contrast to the Cdx2 mutant embryos, Eomes homozygous knockout embryos can form expanded blastocysts, hatch from the zona pellucida, and maintain the TE epithelial integrity for an extended period in culture (Russ et al. 2000). Although the Eomes mutant embryos can implant, they are unable to differentiate into trophoblast cells (Russ et al. 2000; Strumpf et al. 2005). Thus, the function of Eomes is required in TE, particularly for trophoblast differentiation but not for epithelialization or blastocyst cavity formation.

9.3.3 *Klf5*

Klf5 is a Kruppel-like zinc-finger transcription factor and is expressed ubiquitously during preimplantation development, although slightly enriched in the outer cells at the blastocyst stage (Ema et al. 2008; Lin et al. 2010). Klf5 homozygous knockout embryos are unable to implant. The expressions of Cdx2 and Eomes are markedly reduced in Klf5 mutant embryos, suggesting that Klf5 acts upstream of Cdx2 and Eomes in TE development (Ema et al. 2008; Lin et al. 2010). Although a blastocyst cavity initially forms in the Klf5 mutant embryos, its expansion is significantly impaired (Lin et al. 2010). Thus, the morphological phenotype of Klf5 mutant embryos is reminiscent of the Cdx2 mutant embryos. However, the gene expression profiles are strikingly different between these two types of mutant embryos: the expressions of Nanog and Pou5f1 are markedly reduced in the Klf5 mutant, while they are upregulated in the Cdx2 mutant embryos (Ema et al. 2008; Lin et al. 2010). Embryonic stem cells can be derived consistently from the Cdx2 mutant but not from the Klf5 mutant embryos (Strumpf et al. 2005; Ema et al. 2008). Thus, while Klf5 is essential for TE development, it is also required for normal ICM development.

9.3.4 *Gata3*

Gata3 is a member of the GATA family of transcription factors and is specifically expressed in TE of the mouse blastocyst (Home et al. 2009; Ralston et al. 2010). Gata3 homozygous knockout embryos appear to develop into normal blastocysts, which can implant and form trophoblast, although their placenta exhibits abnormal gene expression profiles (Pandolfi et al. 1995; Ting et al. 1996; Ma et al. 1997a). In contrast, shRNA-mediated knockdown of Gata3 results in much earlier and severer phenotypes; only about half of the knockdown embryos are able to form the blastocyst cavity (Home et al. 2009). One of the potential Gata3 targets contributing to normal TE development is Cdx2. Cdx2 expression is significantly reduced in Gata3-knockdown embryos. Also, the first intron of Cdx2 gene contains a conserved Gata3-binding site, which is sufficient to drive the reporter expression in trophoblast stem cells (Home et al. 2009). In contrast, Gata3 expression in TE is independent from Cdx2 expression (Ralston et al. 2010). These studies suggest that Gata3 acts upstream of Cdx2, although the apparent lack of preimplantation phenotypes in Gata3-knockout embryos needs to be further investigated with respect to the Cdx2 expression.

9.3.5 *Sox2*

Sox2 is an HMG-domain transcription factor and is expressed both in TE and ICM of the mouse blastocyst (Avilion et al. 2003; Keramari et al. 2010). Sox2 homozygous knockout embryos form the blastocyst, which can implant but cannot develop the epiblast. This suggests that Sox2 is essential for ICM development but is dispensable for TE development in the blastocyst (Avilion et al. 2003). However, a recent study employing siRNA to knock down both maternal and zygotic Sox2 mRNA shows that Sox2 may also play an essential role at earlier stages of preimplantation development; many of the knockdown embryos are unable to reach the blastocyst stage (Keramari et al. 2010). Interestingly, the expression of Cdx2 and the nuclear localization of Yap protein (see Sect. 9.3.6) are markedly reduced, whereas Pou5f1 is still robustly expressed in the knockdown embryos (Keramari et al. 2010). This implicates that maternal Sox2 may be required specifically for TE formation, which would be in striking contrast with its role in the maintenance of pluripotency as well as in suppression of trophoblast lineages in embryonic stem cells (Li et al. 2007; Masui et al. 2007). Further studies, particularly with the use of oocyte-specific conditional knockout, would illuminate the functional significance of maternal Sox2.

9.3.6 Tead4

Tead4 is a TEA-domain transcription factor. Although Tead4 is ubiquitously expressed during mouse preimplantation development, the homozygous knockout embryos exhibit TE-specific defects (Yagi et al. 2007; Nishioka et al. 2008). Tead4-knockout embryos are unable to form the blastocyst cavity at all. The Cdx2 expression is markedly reduced in the Tead4 mutant embryos, whereas Nanog and Pou5f1 are ectopically expressed in the outer cells. Furthermore, the Tead4 mutant embryos can give rise to embryonic stem cells but not to trophoblast stem cells (Yagi et al. 2007; Nishioka et al. 2008). Thus, Tead4 is specifically required for the critical aspects of TE formation: blastocyst cavity formation as well as trophoblast differentiation. TE-specific action of Tead4 is mediated by its binding protein, Yap, which acts as a transcriptional co-activator. Yap protein accumulates in the nucleus only in the outer cells, starting at 16- to 32-cell stage, before the onset of the blastocyst cavity formation. In contrast, Yap is phosphorylated by Lats kinases in the inner cells, which results in exclusion from the nucleus (Nishioka et al. 2009). Currently, however, it is not clear whether the activity of Lats kinases is differentially regulated between the inner and outer cells.

The action of Yap/Tead4 complex as a transcriptional activator is both necessary and sufficient to induce the expression of Cdx2 (Nishioka et al. 2009), although whether Tead4 directly binds to the promoter or enhancer of Cdx2 gene is yet to be determined. On the other hand, the activation of Yap/Tead4 appears to be insufficient to induce epithelialization, because morphologically normal blastocysts, including a distinct ICM, still form when Yap or Tead4 is ectopically activated in the inner cells (Nishioka et al. 2009). Regardless, Tead4 is essential for the blastocyst cavity formation, based on knockout as well as knockdown phenotypes (Yagi et al. 2007; Nishioka et al. 2008; Alarcon 2010). How Tead4 is involved in the regulation of the blastocyst cavity formation, particularly with respect to the four critical steps discussed above, is currently under investigation.

9.4 Initial Signals for TE Formation

What is the first cue that segregates the TE lineage from the ICM lineage during preimplantation development? Cell lineage tracing as well as cell transplantation experiments have demonstrated that cell positions at around 32-cell stage dictate the cell fate: cells on the outside of the embryo give rise to TE, whereas cells in the inside become ICM (Rossant and Vijh 1980; Ziomek et al. 1982; Ziomek and Johnson 1982; Johnson and Ziomek 1983; Suwinska et al. 2008). How does a cell interpret its relative position and translate it into distinct cellular morphogenesis (i.e., epithelialization) and activation of TE-specific transcription factors? The establishment of apical–basal cell polarity plays a key role in interpreting cell position. Basically, only the outside cells can form the apical domain on the cell surface that is not in contact with other cells (Fig. 9.3a). Various molecules have been identified that are specifically localized at, or excluded from, the apical domain in the outside cells (Fig. 9.3b; Yamanaka et al. 2006; Eckert and Fleming 2008; Johnson 2009). Do these molecules play any essential roles in TE formation? If so, which aspects of TE development are controlled by these molecules?

A recent study shows that one of the apically localized molecules, Pard6b, is crucial for TE formation (Alarcon 2010). Pard6b is one of the three mouse homologs of the nematode par-6, which was originally identified as the gene essential for specifying the oocyte polarity linked to the anterior–posterior body axis (Watts et al. 1996; Hung and Kemphues 1999). In the mouse preimplantation embryo, Pard6b starts to localize to the apical domain around the time of compaction at the 8-cell stage and remains enriched in the apical domain in the outer cells up to the blastocyst stage (Vinot et al. 2005). Knockdown of Pard6b by microinjection of specific shRNA plasmid causes cavitation failure in the embryo, without impairing cell division cycles. Lack of blastocyst cavity formation in the Pard6b-knockdown embryo is due to defective TJs, which is responsible for paracellular sealing (Alarcon 2010). Consistent with this finding, impairment of TJs is also induced by the knockdown of aPKCλ, which is a functional partner of Par6 homologs (Dard et al. 2009). Furthermore, Pard6b knockdown causes significant reduction of Cdx2 expression as well as upregulation of Nanog expression in the outer cells. Thus, the polarity regulator Pard6b is essential for both epithelial morphogenesis as well as activation of Cdx2 (Alarcon 2010).

The phenotype of Pard6b-knockdown embryos is strikingly similar to that of Tead4-knockout or knockdown embryos, both of which fail to form the blastocyst cavity and to upregulate Cdx2 (Yagi et al. 2007; Nishioka et al. 2008; Alarcon 2010). It is currently unclear whether there is a mechanistic link between Pard6b and Lats/Yap/Tead4. It is unlikely that the transcriptional regulation of the Pard6b gene is under the control of Tead4, because the level of Pard6b mRNA is unchanged in response to Tead4 knockdown (Alarcon 2010). Also, Tead4-knockout embryos still exhibit a distinct apical localization of aPKCζ (Nishioka et al. 2008), which is dependent on Pard6b (Alarcon 2010). These observations raise the possibility that Pard6b may act upstream of, or parallel to, Tead4.

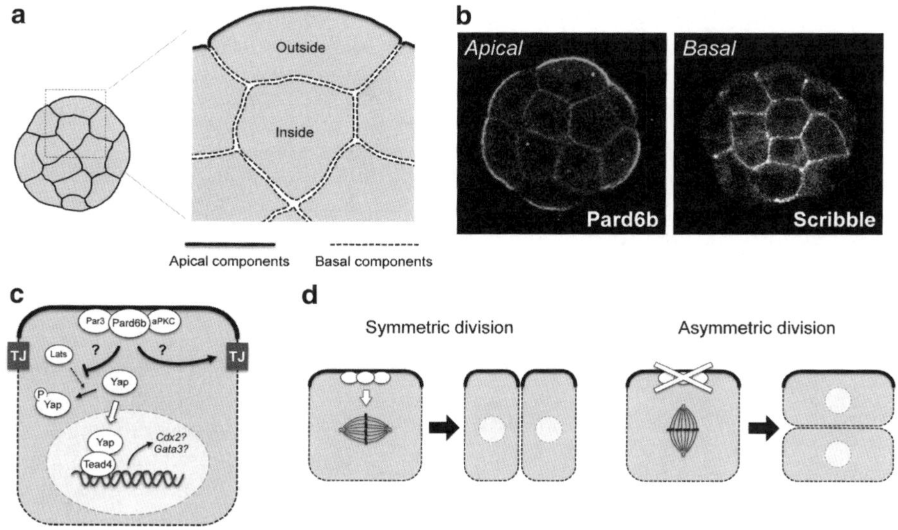

Fig. 9.3 (**a**) Schematic drawing, showing the formation of the apical–basal cell polarity in the outer cells. (**b**) Confocal optical sections of morula-stage embryos that are immunofluorescently stained for polarity proteins, Pard6b (apical component) or Scribble (basal component). (**c**) A schematic diagram of a model for the regulation of tight junction formation and TE-specific gene activation by the apical polarity regulators. See text for details. (**d**) Schematic diagrams, showing the control of cell division patterns by the polarity regulators to influence the positions of daughter cells

Further investigations are required to elucidate the mechanistic links between these molecules that are fundamental for TE formation.

The apical–basal cell polarity may control morphogenesis and gene expression in a cell-autonomous manner, as shown in Fig. 9.3c. However, several studies have also proposed a more indirect mechanism for cell polarity to control TE specification (Fig. 9.3d; Plusa et al. 2005; Jedrusik et al. 2008). In the latter case, cell polarity affects the cell division plane, namely, by regulating the orientation of the mitotic spindle, which determines whether daughter cells contribute to the outside or inside position. Specifically, the presence of the apical–basal polarity induces symmetric cell division, which creates two outside cells. In contrast, the absence of polarity results in asymmetric cell division, which generates one outside and one inside daughter cells. Thus, cells with the apical–basal polarity preferentially give rise to outside cells, that is, presumptive TE cells.

The above two mechanisms by which cell polarity regulates TE formation (i.e., direct cell-autonomous regulation and more indirect regulation through the control of cell division orientation) are not exclusive of each other. However, the dichotomy of symmetric versus asymmetric cell divisions, though it may appear straightforward theoretically, is more complex and obscure during actual embryo development. Recent advancements in live imaging techniques have revealed that cells change their positions after cleavages more dynamically than previously expected (Kurotaki et al. 2007; Dard et al. 2009; Morris et al. 2010; Yamanaka et al. 2010; McDole et al. 2011). For example, even when both daughter cells are initially positioned outside soon after cleavage (i.e., appearing "symmetric"), one of the daughter cells is later internalized by surrounding cells and ends up inside (i.e., appearing "asymmetric") (Yamanaka et al. 2010; McDole et al. 2011). Thus, ultimate positions of daughter cells may not be determined by the orientation of the mitotic spindle. Furthermore, in Pard6b-knockdown embryos, cells that are located outside fail to form TE and instead exhibit a characteristic of ICM (Alarcon 2010). This suggests that the cell polarity regulator plays more direct roles in specifying the TE lineage, even though it could also influence cell division pattern to bias the contribution to outside positions.

While epithelialization and expression of TE-specific transcription factors take place simultaneously, these two events may be uncoupled from each other. A recent study shows that the lack of cell adhesion molecule E-cadherin severely impairs epithelialization even though Cdx2 is still robustly expressed, suggesting that epithelialization is not obligatory for TE-specific gene expression (Stephenson et al. 2010). Even in the absence of E-cadherin, membrane enrichment of apical domain components, such as Par6 and aPKCζ, occurs, which correlates with the expression of Cdx2 and nuclear localization of Yap protein (Stephenson et al. 2010). This suggests that the polarity regulators may control the expression of TE-specific transcription factors independently from epithelialization.

9.5 Concluding Remarks

TE is the first cell type to differentiate in the development of placental mammals. In most nonmammalian animals, the first cell differentiation events are the generation of the three germ layers (i.e., ectoderm, mesoderm, and endoderm) and the specification of the body axes (i.e., anterior–posterior and dorsal–ventral axes). These body patterning events are preceded by TE formation in placental mammals, reflecting their evolutionary transition to a unique viviparous mode of reproduction. Embryo development takes place in utero, relying on nutrients supplied by the mother through the placenta, instead of depending on the yolk stored in the egg. This often serves as a disadvantage in studying mammalian development, because embryos in utero are generally inaccessible for experimental manipulations and observations. However, preimplantation development is an exception, because the fertilized egg gives rise to the blastocyst without depending on the mother. The availability of methods to recapitulate the whole preimplantation development in vitro has significantly advanced our knowledge on the mechanisms of TE development.

But, there are still several challenges that have been hindering further progress in our understanding of the molecular mechanisms of preimplantation development, as discussed in this chapter. First, there are subtle but significant differences between in vivo and in vitro conditions. This is most strikingly exemplified by the phenotype of Atp1a1 mutant embryos, which develop normally in vivo but fail to form the blastocyst in vitro. It is critical to elucidate the causes of the differences in order to interpret the experimental results appropriately. It may also be necessary to further modify the in vitro culture media to better mimic the in vivo condition. Secondly, maternally supplied molecules that may play essential roles in preimplantation development need to be analyzed more systematically and carefully. In one example, RNAi-mediated knockdown of maternal and zygotic Sox2 mRNA results in the unexpected abnormal development of TE, which opens up the possibility of a new function for this well-known pluripotency regulator. However, in another example, knockdown of maternal and zygotic Cdx2 mRNA has resulted in two conflicting results. Further investigations are pivotal to elucidate maternal contribution of this transcription factor to preimplantation development. Thirdly, functional redundancy between multiple genes often makes elucidation of gene function very difficult, because knockout or knockdown of a single gene may not exhibit a clear phenotype. For example, there are more than 20 claudin genes in the mouse genome (Mouse Genome Informatics), and the loss of a single claudin does not appear to cause TJ defects during TE development. Thus, simultaneous knockout or knockdown of multiple genes, or the use of unique inhibitors or dominant-negative constructs, is essential to determine the functional significance of a group of genes.

Nonetheless, significant progress has been made in the past decade to elucidate the molecular and cellular mechanisms of TE formation, particularly with the assistance of new methodologies, such as gene-specific knockdown and live imaging. Further studies will surely reveal the mechanisms that create TE, in which epithelial morphogenesis and cell differentiation events are intricately regulated.

Acknowledgments Figures used in this chapter were obtained through the authors' research, which is supported by NIH grant P20RR024206 and 8P20GM103457.

References

Adachi M, Inoko A, Hata M, Furuse K, Umeda K, Itoh M, Tsukita S (2006) Normal establishment of epithelial tight junctions in mice and cultured cells lacking expression of ZO-3, a tight-junction MAGUK protein. Mol Cell Biol 26:9003–15

Alarcon VB (2010) Cell polarity regulator PARD6B is essential for trophectoderm formation in the preimplantation mouse embryo. Biol Reprod 83:347–58

Apodaca G (2010) Opening ahead: early steps in lumen formation revealed. Nat Cell Biol 12:1026–8

Avilion AA, Nicolis SK, Pevny LH, Perez L, Vivian N, Lovell-Badge R (2003) Multipotent cell lineages in early mouse development depend on SOX2 function. Genes Dev 17:126–40

Aziz H, Alexandre H (1991) The origin of the nascent blastocoele in preimplantation mouse embryos: Ultrastructural cytochemistry and effect of chloroquine. Roux's Arch Dev Biol 200:77–85

Barcroft LC, Moseley AE, Lingrel JB, Watson AJ (2004) Deletion of the Na/K-ATPase alpha1-subunit gene (Atp1a1) does not prevent cavitation of the preimplantation mouse embryo. Mech Dev 121:417–26

Barcroft LC, Offenberg H, Thomsen P, Watson AJ (2003) Aquaporin proteins in murine trophectoderm mediate transepithelial water movements during cavitation. Dev Biol 256:342–54
Barr KJ, Garrill A, Jones DH, Orlowski J, Kidder GM (1998) Contributions of Na+/H+ exchanger isoforms to preimplantation development of the mouse. Mol Reprod Dev 50:146–53
Bell CE, Lariviere NM, Watson PH, Watson AJ (2009) Mitogen-activated protein kinase (MAPK) pathways mediate embryonic responses to culture medium osmolarity by regulating Aquaporin 3 and 9 expression and localization, as well as embryonic apoptosis. Hum Reprod 24:1373–86
Biggers JD, Bell JE, Benos DJ (1988) Mammalian blastocyst: transport functions in a developing epithelium. Am J Physiol 255:C419–32
Borland RM, Biggers JD, Lechene CP (1977) Studies on the composition and formation of mouse blastocoele fluid using electron probe microanalysis. Dev Biol 55:1–8
Bruce AW (2011) What is the role of maternally provided Cdx2 mRNA in early mouse embryogenesis? Reprod Biomed Online 22:512–5
Bryant DM, Datta A, Rodríguez-Fraticelli AE, Peränen J, Martín-Belmonte F, Mostov KE (2010) A molecular network for de novo generation of the apical surface and lumen. Nat Cell Biol 12:1035–45
Chen L, Yabuuchi A, Eminli S, Takeuchi A, Lu CW, Hochedlinger K, Daley GQ (2009) Cross-regulation of the Nanog and Cdx2 promoters. Cell Res 19:1052–61
Dard N, Le T, Maro B, Louvet-Vallée S (2009) Inactivation of aPKClambda reveals a context dependent allocation of cell lineages in preimplantation mouse embryos. PLoS One 4:e7117
Delous M, Hellman NE, Gaudé HM, Silbermann F, Le Bivic A, Salomon R, Antignac C, Saunier S (2009) Nephrocystin-1 and nephrocystin-4 are required for epithelial morphogenesis and associate with PALS1/PATJ and Par6. Hum Mol Genet 18:4711–23
Dietrich JE, Hiiragi T (2007) Stochastic patterning in the mouse pre-implantation embryo. Development 134:4219–31
Eckert JJ, Fleming TP (2008) Tight junction biogenesis during early development. Biochim Biophys Acta 1778:717–28
Edashige K, Sakamoto M, Kasai M (2000) Expression of mRNAs of the aquaporin family in mouse oocytes and embryos. Cryobiology 40:171–5
Ema M, Mori D, Niwa H, Hasegawa Y, Yamanaka Y, Hitoshi S, Mimura J, Kawabe Y, Hosoya T, Morita M, Shimosato D, Uchida K, Suzuki N, Yanagisawa J, Sogawa K, Rossant J, Yamamoto M, Takahashi S, Fujii-Kuriyama Y (2008) Krüppel-like factor 5 is essential for blastocyst development and the normal self-renewal of mouse ESCs. Cell Stem Cell 3:555–67
Engelberg JA, Datta A, Mostov KE, Hunt CA (2011) MDCK cystogenesis driven by cell stabilization within computational analogues. PLoS Comput Biol 7:e1002030
Escudero-Esparza A, Jiang WG, Martin TA (2011) The Claudin family and its role in cancer and metastasis. Front Biosci 16:1069–83
Fujita K, Katahira J, Horiguchi Y, Sonoda N, Furuse M, Tsukita S (2000) Clostridium perfringens enterotoxin binds to the second extracellular loop of claudin-3, a tight junction integral membrane protein. FEBS Lett 476:258–61
Gardner DK (2008) Dissection of culture media for embryos: the most important and less important components and characteristics. Reprod Fertil Dev 20:9–18
Gelber K, Tamura AN, Alarcon VB, Marikawa Y (2011) A potential use of embryonic stem cell medium for the in vitro culture of preimplantation embryos. J Assist Reprod Genet 28:659–68
Hara-Chikuma M, Sohara E, Rai T, Ikawa M, Okabe M, Sasaki S, Uchida S, Verkman AS (2005) Progressive adipocyte hypertrophy in aquaporin-7-deficient mice: adipocyte glycerol permeability as a novel regulator of fat accumulation. J Biol Chem 280:15493–6
Home P, Ray S, Dutta D, Bronshteyn I, Larson M, Paul S (2009) GATA3 is selectively expressed in the trophectoderm of peri-implantation embryo and directly regulates Cdx2 gene expression. J Biol Chem 284:28729–37
Hung TJ, Kemphues KJ (1999) PAR-6 is a conserved PDZ domain-containing protein that colocalizes with PAR-3 in Caenorhabditis elegans embryos. Development 126:127–35

Jedrusik A, Bruce AW, Tan MH, Leong DE, Skamagki M, Yao M, Zernicka-Goetz M (2010) Maternally and zygotically provided Cdx2 have novel and critical roles for early development of the mouse embryo. Dev Biol 344:66–78

Jedrusik A, Parfitt DE, Guo G, Skamagki M, Grabarek JB, Johnson MH, Robson P, Zernicka-Goetz M (2008) Role of Cdx2 and cell polarity in cell allocation and specification of trophectoderm and inner cell mass in the mouse embryo. Genes Dev 22:2692–706

Johnson MH, Ziomek CA (1983) Cell interactions influence the fate of mouse blastomeres undergoing the transition from the 16- to the 32-cell stage. Dev Biol 95:211–18

Johnson MH (2009) From mouse egg to mouse embryo: polarities, axes, and tissues. Annu Rev Cell Dev Biol 25:483–512

Katahira J, Inoue N, Horiguchi Y, Matsuda M, Sugimoto N (1997) Molecular cloning and functional characterization of the receptor for Clostridium perfringens enterotoxin. J Cell Biol 136:1239–47

Katsuno T, Umeda K, Matsui T, Hata M, Tamura A, Itoh M, Takeuchi K, Fujimori T, Nabeshima Y, Noda T, Tsukita S, Tsukita S (2008) Deficiency of zonula occludens-1 causes embryonic lethal phenotype associated with defected yolk sac angiogenesis and apoptosis of embryonic cells. Mol Biol Cell 19:2465–75

Kawagishi R, Tahara M, Sawada K, Morishige K, Sakata M, Tasaka K, Murata Y (2004) Na^+/H^+ exchanger-3 is involved in mouse blastocyst formation. J Exp Zool A Comp Exp Biol 301:767–75

Keramari M, Razavi J, Ingman KA, Patsch C, Edenhofer F, Ward CM, Kimber SJ (2010) Sox2 is essential for formation of trophectoderm in the preimplantation embryo. PLoS One 5:e13952

Kidder GM (2002) Trophectoderm development and function: the roles of Na^+/K^+-ATPase subunit isoforms. Can J Physiol Pharmacol 80:110–5

Kidder GM, Watson AJ (2005) Roles of Na+, K + -ATPase in early development and trophectoderm differentiation. Semin Nephrol 25:352–5

Kitaura H, Tsujita M, Huber VJ, Kakita A, Shibuki K, Sakimura K, Kwee IL, Nakada T (2009) Activity-dependent glial swelling is impaired in aquaporin-4 knockout mice. Neurosci Res 64:208–12

Krane CM, Melvin JE, Nguyen HV, Richardson L, Towne JE, Doetschman T, Menon AG (2001) Salivary acinar cells from aquaporin 5-deficient mice have decreased membrane water permeability and altered cell volume regulation. J Biol Chem 276:23413–20

Kurotaki Y, Hatta K, Nakao K, Nabeshima Y, Fujimori T (2007) Blastocyst axis is specified independently of early cell lineage but aligns with the ZP shape. Science 316:719–23

Lawitts JA, Biggers JD (1993) Culture of preimplantation embryos. Methods Enzymol 225:153–64

Li J, Pan G, Cui K, Liu Y, Xu S, Pei D (2007) A dominant-negative form of mouse SOX2 induces trophectoderm differentiation and progressive polyploidy in mouse embryonic stem cells. J Biol Chem 282:19481–92

Lin SC, Wani MA, Whitsett JA, Wells JM (2010) Klf5 regulates lineage formation in the preimplantation mouse embryo. Development 137:3953–63

Ma GT, Roth ME, Groskopf JC, Tsai FY, Orkin SH, Grosveld F, Engel JD, Linzer DI (1997a) GATA-2 and GATA-3 regulate trophoblast-specific gene expression in vivo. Development 124:907–14

Ma T, Yang B, Gillespie A, Carlson EJ, Epstein CJ, Verkman AS (1997b) Generation and phenotype of a transgenic knockout mouse lacking the mercurial-insensitive water channel aquaporin-4. J Clin Invest 100:957–62

Ma T, Song Y, Gillespie A, Carlson EJ, Epstein CJ, Verkman AS (1999) Defective secretion of saliva in transgenic mice lacking aquaporin-5 water channels. J Biol Chem 274:20071–4

Ma T, Yang B, Gillespie A, Carlson EJ, Epstein CJ, Verkman AS (1998) Severely impaired urinary concentrating ability in transgenic mice lacking aquaporin-1 water channels. J Biol Chem 273:4296–9

Ma T, Song Y, Yang B, Gillespie A, Carlson EJ, Epstein CJ, Verkman AS (2000) Nephrogenic diabetes insipidus in mice lacking aquaporin-3 water channels. Proc Natl Acad Sci USA 97:4386–91

MacPhee DJ, Jones DH, Barr KJ, Betts DH, Watson AJ, Kidder GM (2000) Differential involvement of Na+, K + -ATPase isozymes in preimplantation development of the mouse. Dev Biol 222:486–98

Madan P, Rose K, Watson AJ (2007) Na^+/K^+-ATPase beta1 subunit expression is required for blastocyst formation and normal assembly of trophectoderm tight junction-associated proteins. J Biol Chem 282:12127–34

Manejwala FM, Cragoe EJ Jr, Schultz RM (1989) Blastocoel expansion in the preimplantation mouse embryo: role of extracellular sodium and chloride and possible apical routes of their entry. Dev Biol 133:210–20

Martin-Belmonte F, Yu W, Rodríguez-Fraticelli AE, Ewald AJ, Werb Z, Alonso MA, Mostov K (2008) Cell-polarity dynamics controls the mechanism of lumen formation in epithelial morphogenesis. Curr Biol 18:507–13

Masui S, Nakatake Y, Toyooka Y, Shimosato D, Yagi R, Takahashi K, Okochi H, Okuda A, Matoba R, Sharov AA, Ko MS, Niwa H (2007) Pluripotency governed by Sox2 via regulation of Oct3/4 expression in mouse embryonic stem cells. Nat Cell Biol 9:625–35

McConnell J, Petrie L, Stennard F, Ryan K, Nichols J (2005) Eomesodermin is expressed in mouse oocytes and pre-implantation embryos. Mol Reprod Dev 71:399–404

McDole K, Xiong Y, Iglesias PA, Zheng Y (2011) Lineage mapping the pre-implantation mouse embryo by two-photon microscopy, new insights into the segregation of cell fates. Dev Biol 355:239–49

Morishita Y, Matsuzaki T, Hara-chikuma M, Andoo A, Shimono M, Matsuki A, Kobayashi K, Ikeda M, Yamamoto T, Verkman A, Kusano E, Ookawara S, Takata K, Sasaki S, Ishibashi K (2005) Disruption of aquaporin-11 produces polycystic kidneys following vacuolization of the proximal tubule. Mol Cell Biol 25:7770–9

Moriwaki K, Tsukita S, Furuse M (2007) Tight junctions containing claudin 4 and 6 are essential for blastocyst formation in preimplantation mouse embryos. Dev Biol 312:509–22

Morris SA, Teo RT, Li H, Robson P, Glover DM, Zernicka-Goetz M (2010) Origin and formation of the first two distinct cell types of the inner cell mass in the mouse embryo. Proc Natl Acad Sci USA 107:6364–9

Motosugi N, Bauer T, Polanski Z, Solter D, Hiiragi T (2005) Polarity of the mouse embryo is established at blastocyst and is not prepatterned. Genes Dev 19:1081–92

Nishioka N, Inoue K, Adachi K, Kiyonari H, Ota M, Ralston A, Yabuta N, Hirahara S, Stephenson RO, Ogonuki N, Makita R, Kurihara H, Morin-Kensicki EM, Nojima H, Rossant J, Nakao K, Niwa H, Sasaki H (2009) The Hippo signaling pathway components Lats and Yap pattern Tead4 activity to distinguish mouse trophectoderm from inner cell mass. Dev Cell 16:398–410

Nishioka N, Yamamoto S, Kiyonari H, Sato H, Sawada A, Ota M, Nakao K, Sasaki H (2008) Tead4 is required for specification of trophectoderm in pre-implantation mouse embryos. Mech Dev 125:270–83

Niwa H, Toyooka Y, Shimosato D, Strumpf D, Takahashi K, Yagi R, Rossant J (2005) Interaction between Oct3/4 and Cdx2 determines trophectoderm differentiation. Cell 123:917–29

Offenberg H, Barcroft LC, Caveney A, Viuff D, Thomsen PD, Watson AJ (2000) mRNAs encoding aquaporins are present during murine preimplantation development. Mol Reprod Dev 57:323–30

Ohta E, Itoh T, Nemoto T, Kumagai J, Ko SB, Ishibashi K, Ohno M, Uchida K, Ohta A, Sohara E, Uchida S, Sasaki S, Rai T (2009) Pancreas-specific aquaporin 12 null mice showed increased susceptibility to caerulein-induced acute pancreatitis. Am J Physiol Cell Physiol 297: C1368–78

Pandolfi PP, Roth ME, Karis A, Leonard MW, Dzierzak E, Grosveld FG, Engel JD, Lindenbaum MH (1995) Targeted disruption of the GATA3 gene causes severe abnormalities in the nervous system and in fetal liver haematopoiesis. Nat Genet 11:40–4

Plusa B, Frankenberg S, Chalmers A, Hadjantonakis AK, Moore CA, Papalopulu N, Papaioannou VE, Glover DM, Zernicka-Goetz M (2005) Downregulation of Par3 and aPKC function directs cells towards the ICM in the preimplantation mouse embryo. J Cell Sci 118:505–15

Rajasekaran SA, Rajasekaran AK (2009) Na, K-ATPase and epithelial tight junctions. Front Biosci 14:2130–48

Ralston A, Cox BJ, Nishioka N, Sasaki H, Chea E, Rugg-Gunn P, Guo G, Robson P, Draper JS, Rossant J (2010) Gata3 regulates trophoblast development downstream of Tead4 and in parallel to Cdx2. Development 137:395–403

Ralston A, Rossant J (2008) Cdx2 acts downstream of cell polarization to cell-autonomously promote trophectoderm fate in the early mouse embryo. Dev Biol 313:614–29

Rojek A, Füchtbauer EM, Kwon TH, Frøkiaer J, Nielsen S (2006) Severe urinary concentrating defect in renal collecting duct-selective AQP2 conditional-knockout mice. Proc Natl Acad Sci USA 103:6037–42

Rojek AM, Skowronski MT, Füchtbauer EM, Füchtbauer AC, Fenton RA, Agre P, Frøkiaer J, Nielsen S (2007) Defective glycerol metabolism in aquaporin 9 (AQP9) knockout mice. Proc Natl Acad Sci USA 104:3609–14

Rossant J, Vijh KM (1980) Ability of outside cells from preimplantation mouse embryos to form inner cell mass derivatives. Dev Biol 76:475–82

Russ AP, Wattler S, Colledge WH, Aparicio SA, Carlton MB, Pearce JJ, Barton SC, Surani MA, Ryan K, Nehls MC, Wilson V, Evans MJ (2000) Eomesodermin is required for mouse trophoblast development and mesoderm formation. Nature 404:95–9

Saitou M, Furuse M, Sasaki H, Schulzke JD, Fromm M, Takano H, Noda T, Tsukita S (2000) Complex phenotype of mice lacking occludin, a component of tight junction strands. Mol Biol Cell 11:4131–42

Schluter MA, Pfarr CS, Pieczynski J, Whiteman EL, Hurd TW, Fan S, Liu CJ, Margolis B (2009) Trafficking of Crumbs3 during cytokinesis is crucial for lumen formation. Mol Biol Cell 20:4652–63

Schultheis PJ, Clarke LL, Meneton P, Harline M, Boivin GP, Stemmermann G, Duffy JJ, Doetschman T, Miller ML, Shull GE (1998) Targeted disruption of the murine Na^+/H^+ exchanger isoform 2 gene causes reduced viability of gastric parietal cells and loss of net acid secretion. J Clin Invest 101:1243–53

Sheth B, Nowak RL, Anderson R, Kwong WY, Papenbrock T, Fleming TP (2008) Tight junction protein ZO-2 expression and relative function of ZO-1 and ZO-2 during mouse blastocyst formation. Exp Cell Res 314:3356–68

Sohara E, Rai T, Miyazaki J, Verkman AS, Sasaki S, Uchida S (2005) Defective water and glycerol transport in the proximal tubules of AQP7 knockout mice. Am J Physiol Renal Physiol 289:F1195–200

Sonoda N, Furuse M, Sasaki H, Yonemura S, Katahira J, Horiguchi Y, Tsukita S (1999) Clostridium perfringens enterotoxin fragment removes specific claudins from tight junction strands: Evidence for direct involvement of claudins in tight junction barrier. J Cell Biol 147:195–204

Stephenson RO, Yamanaka Y, Rossant J (2010) Disorganized epithelial polarity and excess trophectoderm cell fate in preimplantation embryos lacking E-cadherin. Development 137:3383–91

Strumpf D, Mao CA, Yamanaka Y, Ralston A, Chawengsaksophak K, Beck F, Rossant J (2005) Cdx2 is required for correct cell fate specification and differentiation of trophectoderm in the mouse blastocyst. Development 132:2093–102

Summers MC, Biggers JD (2003) Chemically defined media and the culture of mammalian preimplantation embryos: historical perspective and current issues. Hum Reprod Update 9:557–82

Sun QY, Liu K, Kikuchi K (2008) Oocyte-specific knockout: a novel in vivo approach for studying gene functions during folliculogenesis, oocyte maturation, fertilization, and embryogenesis. Biol Reprod 79:1014–20

Suwinska A, Czolowska R, Ozdzenski W, Tarkowski AK (2008) Blastomeres of the mouse embryo lose totipotency after the fifth cleavage division: expression of Cdx2 and Oct4 and developmental potential of inner and outer blastomeres of 16- and 32-cell embryos. Dev Biol 322:133–44

Thrane AS, Rappold PM, Fujita T, Torres A, Bekar LK, Takano T, Peng W, Wang F, Thrane VR, Enger R, Haj-Yasein NN, Skare Ø, Holen T, Klungland A, Ottersen OP, Nedergaard M, Nagelhus EA (2011) Critical role of aquaporin-4 (AQP4) in astrocytic $Ca2^{+}$ signaling events elicited by cerebral edema. Proc Natl Acad Sci USA 108:846–51

Ting CN, Olson MC, Barton KP, Leiden JM (1996) Transcription factor GATA-3 is required for development of the T-cell lineage. Nature 384:474–8

Tokhtaeva E, Sachs G, Souda P, Bassilian S, Whitelegge JP, Shoshani L, Vagin O (2011) Epithelial junctions depend on intercellular trans-interactions between the Na, K-ATPase β subunits. J Biol Chem 286:25801–12

Vajta G, Rienzi L, Cobo A, Yovich J (2010) Embryo culture: can we perform better than nature? Reprod Biomed Online 20:453–69

Vinot S, Le T, Ohno S, Pawson T, Maro B, Louvet-Vallée S (2005) Asymmetric distribution of PAR proteins in the mouse embryo begins at the 8-cell stage during compaction. Dev Biol 282:307–19

Wang AZ, Ojakian GK, Nelson WJ (1990) Steps in the morphogenesis of a polarized epithelium. I. uncoupling the roles of cell-cell and cell-substratum contact in establishing plasma membrane polarity in multicellular epithelial (MDCK) cysts. J Cell Sci 95:137–51

Wang H, Ding T, Brown N, Yamamoto Y, Prince LS, Reese J, Paria BC (2008) Zonula occludens-1 (ZO-1) is involved in morula to blastocyst transformation in the mouse. Dev Biol 318:112–25

Wang K, Sengupta S, Magnani L, Wilson CA, Henry RW, Knott JG (2010) Brg1 is required for Cdx2-mediated repression of Oct4 expression in mouse blastocysts. PLoS One 5:e10622

Watson AJ, Barcroft LC (2001) Regulation of blastocyst formation. Front Biosci 6:D708–30

Watson AJ (1992) The cell biology of blastocyst development. Mol Reprod Dev 33:492–504

Watts JL, Etemad-Moghadam B, Guo S, Boyd L, Draper BW, Mello CC, Priess JR, Kemphues KJ (1996) par-6, a gene involved in the establishment of asymmetry in early C. elegans embryos, mediates the asymmetric localization of PAR-3. Development 122:3133–40

Wiley LM (1984) Cavitation in the mouse preimplantation embryo: Na/K-ATPase and the origin of nascent blastocoele fluid. Dev Biol 105:330–42

Wu G, Gentile L, Fuchikami T, Sutter J, Psathaki K, Esteves TC, Araúzo-Bravo MJ, Ortmeier C, Verberk G, Abe K, Scholer HR (2010) Initiation of trophectoderm lineage specification in mouse embryos is independent of Cdx2. Development 137:4159–69

Xu J, Kausalya PJ, Phua DC, Ali SM, Hossain Z, Hunziker W (2008) Early embryonic lethality of mice lacking ZO-2, but Not ZO-3, reveals critical and nonredundant roles for individual zonula occludens proteins in mammalian development. Mol Cell Biol 28:1669–78

Yagi R, Kohn MJ, Karavanova I, Kaneko KJ, Vullhorst D, DePamphilis ML, Buonanno A (2007) Transcription factor TEAD4 specifies the trophectoderm lineage at the beginning of mammalian development. Development 134:3827–36

Yamanaka Y, Ralston A, Stephenson RO, Rossant J (2006) Cell and molecular regulation of the mouse blastocyst. Dev Dyn 235:2301–14

Yamanaka Y, Lanner F, Rossant J (2010) FGF signal-dependent segregation of primitive endoderm and epiblast in the mouse blastocyst. Development 137:715–24

Yang B, Song Y, Zhao D, Verkman AS (2005) Phenotype analysis of aquaporin-8 null mice. Am J Physiol Cell Physiol 288:C1161–70

Yoshihama Y, Sasaki K, Horikoshi Y, Suzuki A, Ohtsuka T, Hakuno F, Takahashi S, Ohno S, Chida K (2011) KIBRA suppresses apical exocytosis through inhibition of aPKC kinase activity in epithelial cells. Curr Biol 21:705–11

Ziomek CA, Johnson MH, Handyside AH (1982) The developmental potential of mouse 16-cell blastomeres. J Exp Zool 221:345–55

Ziomek CA, Johnson MH (1982) The roles of phenotype and position in guiding the fate of 16-cell mouse blastomeres. Dev Biol 91:440–7

Chapter 10
Cell Lineage Allocation Within the Inner Cell Mass of the Mouse Blastocyst

Panagiotis Xenopoulos, Minjung Kang, and Anna-Katerina Hadjantonakis

Abstract At the time of implantation, the early mouse embryo consists of three distinct cell lineages: the epiblast (EPI), primitive endoderm (PrE), and trophectoderm (TE). Here we will focus on the EPI and PrE cell lineages, which arise within the inner cell mass (ICM) of the blastocyst. Though still poorly understood, our current understanding of the mechanisms underlying this lineage allocation will be discussed. It was originally thought that lineage choice was strictly controlled by the position of a cell within the ICM. However, it is now believed that the EPI and PrE lineages are defined both by their position and by the expression of lineage-specific transcription factors. Interestingly, these lineage-specific transcription factors are initially co-expressed in early ICM cells, suggesting an initial multi-lineage priming state. Thereafter, lineage-specific transcription factors display a mutually exclusive salt-and-pepper distribution that reflects cell specification of the EPI or PrE fates. Later on, lineage segregation and likely commitment are completed with the sequestration of PrE cells to the surface of the ICM, which lies at the blastocyst cavity roof. We discuss recent advances that have focused on elucidating how the salt-and-pepper pattern is established and then resolved within the ICM, leading to the correct apposition of cell lineages in preparation for implantation.

10.1 Preimplantation Development Involves Two Cell Fate Decisions

Before implanting into the maternal uterus, the mouse embryo consists of three molecularly distinct spatially segregated cell lineages: the epiblast (EPI) that lies within the interior of the inner cell mass (ICM) of the blastocyst. The EPI is

P. Xenopoulos • M. Kang • A.-K. Hadjantonakis (✉)
Developmental Biology Program, Sloan-Kettering Institute, New York, NY 10065, USA
e-mail: hadj@mskcc.org

J.Z. Kubiak (ed.), *Mouse Development*, Results and Problems in Cell Differentiation 55,
DOI 10.1007/978-3-642-30406-4_10, © Springer-Verlag Berlin Heidelberg 2012

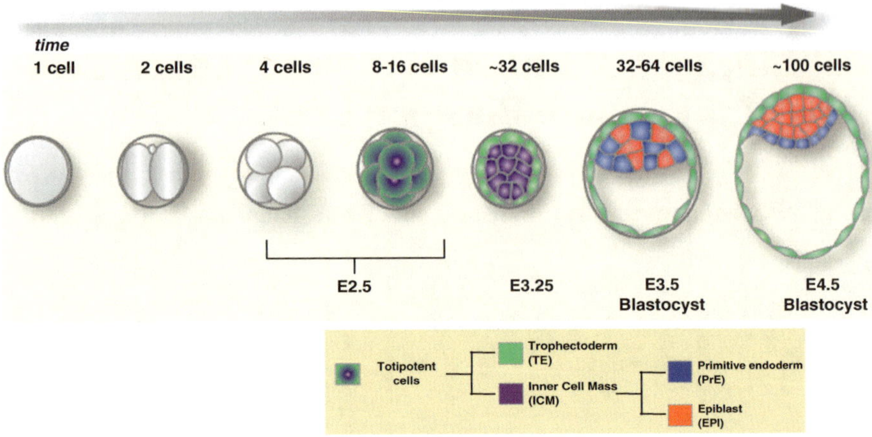

Fig. 10.1 *Mouse preimplantation development leading to blastocyst formation.* During preimplantation development, the mouse embryo undergoes cleavages that culminate in the proper segregation of three lineages at the blastocyst stage. This process involves two cell fate decisions: the first decision occurs when the inner cell mass (ICM) is segregated from the extraembryonic trophectoderm (TE); the second decision occurs within the ICM and involves the segregation of the pluripotent epiblast (EPI) from the extraembryonic primitive endoderm (PrE)

encapsulated by two tissues: the primitive endoderm (PrE) an epithelium located on the surface of the ICM, which lies in contact with the blastocyst cavity; and the trophectoderm (TE) comprising the epithelial surface of the blastocyst, which lies in contact with the external environment (Fig. 10.1) (reviewed by Arnold and Robertson 2009; Nowotschin and Hadjantonakis 2010; Rossant and Tam 2009; Zernicka-Goetz et al. 2009). The EPI is the pluripotent lineage within mammalian embryos and so will give rise to most of the fetus, whereas the TE and PrE predominantly give rise to extraembryonic tissues, namely, the fetal portion of the placenta and the endodermal component of the visceral and parietal yolk sacs, respectively. Thus two cell fate decisions take place before blastocyst formation to ensure the proper specification and spatial segregation of the extraembryonic lineages from the pluripotent epiblast.

After fertilization, the embryo undergoes three rounds of cell division, leading to the eight-cell stage. At this time, the blastomeres generally appear morphologically indistinguishable and have the ability to contribute to any of the three lineages of the blastocyst, as has been assessed in chimera experiments (Kelly 1977; Suwinska et al. 2008; Tarkowski and Wroblewska 1967). However, several studies have argued for an existing heterogeneity between the blastomeres at the four- and eight-cell stages, resulting from previous asymmetric cell divisions at the animal part of the oocyte (Gardner 1996). This heterogeneity is evident at the level of differential epigenetic modifications at the four-cell stage, as well as at the level of expression and kinetics of certain transcription factors (e.g., Oct4; discussed later) (Plachta et al. 2011; Torres-Padilla et al. 2007). Moreover, each of these eight blastomeres has acquired an apical-basal polarity as a result of compaction, a process in which cell–cell contacts increase (Johnson and Ziomek 1981). The first fate decision, involving the segregation of the TE lineage from the ICM, takes place after the third embryonic cell division. It relies on a cascade of cell divisions taking place at the 8- to 16- and 16- to 32-cell stage transitions. These divisions can be either symmetric or asymmetric, depending on the orientation of the mitotic spindle with respect to the apical–basal (inside–outside) polarity of the blastomeres. Symmetric divisions generate two daughter cells that remain on the outer surface of the embryo and contribute to TE, whereas asymmetric divisions produce one cell that stays on the outer surface, giving rise to TE and one cell that becomes internalized and contributes to the ICM.

This "inside–outside" model is based on the two rounds of asymmetric divisions; it was first introduced more than 40 years ago and could explain observations from experiments where spatial rearrangements have an effect on cell fate (Tarkowski and Wroblewska 1967). However, more recent studies have challenged the positional model suggesting that it may only provide a mechanism underlying the first fate choice and have thus argued for more determinants to be taken into account. First, acquisition of cell polarity affects cell fate, perhaps earlier than the emergence of inside and outside cells (Jedrusik et al. 2008; Johnson and Ziomek 1981; Plusa et al. 2005). Second, TE and ICM identities are linked to key molecular determinants such as differential Hippo signaling, which depends on the position of the cell and regulates the expression of certain lineage-specific transcription factors. Hippo signaling remains inactive in outer cells, leading to the translocation of the transcriptional cofactor Yap/Taz in the nucleus, which then binds the transcription factor TEAD4 and results in the expression of the TE-specific transcription factor Cdx2; importantly, an additional target gene directed by TEAD4 is *Gata3*, which encodes for another TE-specific transcription factor (Ralston et al. 2010). Conversely, the Hippo signaling pathway becomes activated in inside cells, promoting phosphorylation and exclusion of Yap/Taz from the nucleus, which results in repression of *Cdx2* (Nishioka et al. 2009, 2008; Ralston and Rossant 2008; Strumpf et al. 2005; Yagi et al. 2007). Third, expression of the pluripotency-linked transcription factor Oct4, which has been shown to repress Cdx2 activity, ensures the TE versus ICM segregation (Chambers and Smith 2004; Niwa et al. 2005). To this end, a recent study suggested that seemingly "equivalent" blastomeres at the eight-cell stage display differential Oct4 transcription factor kinetics; those with slow kinetics mostly undergo asymmetric divisions, contributing mostly to ICM, whereas those with fast kinetics divide symmetrically, giving rise to outer TE cells (Plachta et al. 2011). It should however be noted that this study involved the widespread misexpression of a photoactivatable Oct4–GFP fusion protein, which might not fully recapitulate the endogenous behavior of the Oct4 promoter or protein. Therefore, the identity of the two lineages emerging from the first fate decision is linked to the levels of expression as well as the kinetics of lineage-specific transcription factors, which play an equally important role for the second fate decision, as will be discussed extensively in this chapter.

The second fate decision occurs within the ICM and results in the segregation of the extraembryonic PrE lineage from the pluripotent EPI. The segregated PrE lineage acquires an epithelial morphology and lies on the surface of the ICM, contacting the blastocyst cavity. On the other hand, the EPI lineage is located in the interior of the ICM likely having no contact with the blastocyst cavity or the outer environment. In this chapter, we focus on recent findings aimed at elucidating the mechanisms underlying this lineage allocation event within the ICM.

10.2 Challenging the Positional Model for Determining Cell Fate Choice in the ICM: The Role of Lineage-Specific Transcription Factors

Drawing parallels with the "inside–outside" model underlying the first cell fate decision, it was originally thought that cells within the ICM are segregated in corresponding lineages based on their positions, such that inner cells give rise to EPI, whereas outer ICM cells, which face the blastocyst cavity, give rise to PrE. Moreover, PrE specification was thought to be linked with positional cues acting on the ICM cells at the blastocyst cavity roof (Fig. 10.2) (Enders et al. 1978). This positional model was supported by studies which have demonstrated that during the formation of embryonic bodies from cultured embryonic stem (ES) cells, a PrE layer is generated on the surface (Becker et al. 1992; Martin and Evans 1975; Murray and Edgar 2001). Recently however, this strictly positional-based model has been challenged from several studies over the last years (Chazaud et al. 2006; Plusa et al. 2008). These studies have shown that, cells within the ICM express certain EPI-specific or PrE-specific transcription factors. Interestingly, these transcription factors start to be expressed early, at around the 32-cell stage, irrespectively of the position of cells within the ICM, perhaps suggesting a multi-lineage priming state (Fig. 10.2). As the two nascent lineages emerge, PrE cells begin to exclusively express transcription factors such as Gata4, Gata6, Sox7, and Sox17, whereas EPI cells express pluripotency-associated factors such as Nanog, Sox2, and Oct4.

Gata4 and Gata6 are two transcription factors of the Gata family, which are expressed in the PrE. Gata6 is one of the first lineage-specific transcription factors to be observed, being expressed at around the 16- to 32-cell stage, in an overlapping manner with the expression of the EPI-specific factor Nanog. Gata4 is expressed later, at around the 64-cell stage, when the "salt-and-pepper" distribution of transcription factors is evident and thus, when cells are more likely fated to form PrE. Mouse mutant embryos for either *Gata4* or *Gata6* usually form PrE and die at postimplantation stages and, in the case of *Gata6* mutants, they exhibit defects in the visceral endoderm, a tissue derivative of PrE (Koutsourakis et al. 1999; Kuo et al. 1997; Molkentin et al. 1997; Morrisey et al. 1998). Therefore, PrE formation is not compromised in either single mutant embryos, perhaps hinting at a functional redundancy between these two transcription factors. One might predict that a double *Gata4; Gata6* mutant might exhibit a phenotype in PrE formation. Gata6-expressing cells also co-express the transcription factor Sox17, which is required

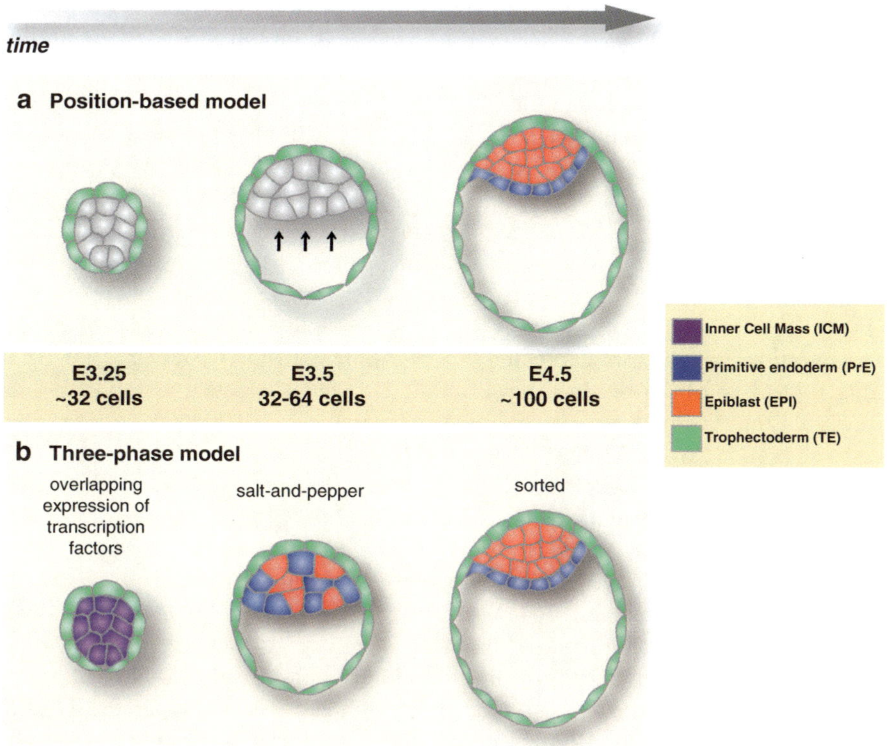

Fig. 10.2 *Models for lineage allocation within the ICM.* (**a**) In the positional-based model, all ICM cells are morphologically and molecularly equal, being able to contribute to either EPI or PrE. Lineage allocation is strictly controlled by cell position: cells positioned in the interior of the ICM give rise to EPI, whereas cells that underlie the blastocyst cavity form PrE, possibly as a result of positional induction. (**b**) In the three-phase-based model, lineage allocation is controlled by the sequential activation of key transcription factors as well as dynamic cell movements within the ICM. Initially, EPI- and PrE-specific transcription factors are co-expressed in all cells of the ICM. That is followed by a mutually exclusively expression pattern, where cells expressing EPI- or PrE-specific transcription factors are distributed in a salt-and-pepper fashion irrespectively of their position within the ICM. Finally, randomly positioned PrE precursor cells will sort to the surface of the ICM ensuring the spatial segregation of the EPI and PrE lineages before implantation

for the maintenance of the PrE lineage (Artus et al. 2011; Niakan et al. 2010). Another transcription factor, Sox7, is expressed at the final stages of preimplantation development, when PrE cells are spatially segregated on the blastocyst cavity roof (Artus et al. 2011). This observation reveals that PrE cells positioned on the cavity are molecularly distinct with those that have yet to sort to their final position. In addition to these PrE-specific transcription factors, other PrE markers have been identified. For example, Pdgfrα is a marker of PrE cells, and a *Pdgfr*α*::H2B-GFP* reporter has been used as a tool for single-cell resolution live imaging experiments (Plusa et al. 2008).

The EPI-specific factors Oct4 and Nanog start being expressed in all cells at the eight-cell stage, and at around the 32-cell stage, they are co-expressed in all cells along with the PrE-specific Gata6. Nanog persists within the ICM, and thereafter becomes exclusive in EPI cells until the late blastocyst stage when its expression declines and later becomes re-established in the germline (Chambers et al. 2003; Mitsui et al. 2003). Nanog has been shown to be required for the maintenance of pluripotency in mouse ES cells, as well as in the ICM (Chambers et al. 2007). Interestingly, Nanog expression is also required for the maintenance of the PrE lineage (Messerschmidt and Kemler 2010). Along with Nanog, Oct4 is another key transcriptional regulator, expressed by pluripotent cells (Chambers and Smith 2004). Oct4 is a POU-domain transcription factor that is co-expressed with Cdx2 in all blastomeres of the early embryo. Oct4 regulates the activity of Cdx2, thus promoting ICM over TE fate during the first fate decision (Niwa et al. 2005). Moreover, Oct4 mutant embryos do not form an ICM and inner blastomeres acquire a TE identity (Nichols et al. 1998). As mentioned previously, the kinetics of Oct4 were suggested to play a significant role during the first cell fate choice involving cells dividing symmetrically or asymmetrically and thus contributing to TE or ICM, respectively (Plachta et al. 2011). As with Nanog, Oct4 expression becomes restricted within the ICM and is enriched in EPI cells; however, unlike Nanog, low levels of expression Oct4 are observed in PrE cells, even at the late blastocyst stage. Oct4 does eventually become restricted and is expressed specifically in the germline (Palmieri et al. 1994). Another early ICM-specific, and thereafter EPI-specific, transcription factor is Sox2. A recent study showed that, along with Oct4, Sox2 is one of the first markers of the emerging ICM. However unlike Oct4 and Nanog, Sox2 is not expressed by all blastomeres early on, instead it was shown to be specifically upregulated in cells that internalize first during the divisions occurring between the 8- to 16-cell stages (Guo et al. 2010). Later on, Sox2 is downregulated in ICM cells that contribute to PrE but remains expressed in TE cells (as opposed to Oct4). Disruption of *Sox2* results in preimplantation lethality, emphasizing its importance on the formation and maintenance of multipotent cell lineages (Avilion et al. 2003).

It is now widely accepted that the EPI versus PrE lineage decision is likely to be linked to the expression of lineage-specific transcription factors rather than strictly determined by cell position alone. Several studies, supported by experimental data obtained by live embryo imaging, have indicated that lineage allocation within the ICM involves three distinct, but successive phases involving: (1) initial co-expression of lineage-specific transcription factors (at around the 32-cell stage), (2) subsequent mutually-exclusive expression and salt-and-pepper distribution of EPI- and PrE-precursor cells (at around the 64-cell stage), and (3) finally dynamic cell movements leading to the final sorting and spatial segregation of the EPI and PrE cell lineages (at around the 100-cell stage) (Fig. 10.2) (Chazaud et al. 2006; Meilhac et al. 2009; Plusa et al. 2008). Therefore, cells within the ICM initiate expression of lineage-specific transcription factors prior to the formation of the blastocyst cavity, suggesting that initially they may acquire a state of multi-lineage priming, and thereafter develop a propensity to form the EPI or PrE lineage before the induction of any positional cues. An important step during this process occurs when transcription factors transition their expression from a homogeneous to a mutually-exclusive pattern, where cells expressing PrE-specific factors cease to express EPI-specific ones and vice versa, resulting in the emergence of a salt-and-pepper distribution of nascent PrE and EPI progenitors.

During the next sections, we will focus on how this mutually exclusive pattern, which marks the first point on EPI versus PrE cell fate decision, takes place and is acquired within the ICM cell population. Specifically, recent studies have aimed at answering the following questions: Is the establishment of a salt-and-pepper distribution influenced by signaling cues? Do ICM cells acquire this pattern of expression randomly, or is there a lineage bias that influences the subsequent cell fate decision?

10.3 Salt-and-Pepper Expression of Lineage-Specific Transcription Factors

At around the morula stage (16–32 cells), the EPI-specific transcription factor Nanog starts being expressed homogeneously within inner blastomeres and is excluded from outer cells (Chazaud et al. 2006; Plusa et al. 2008). A similar pattern of expression has also been observed for the earliest observed PrE-specific transcription factor, Gata6. This overlapping expression is evident until the 64-cell stage. By this stage, the EPI and PrE markers are being expressed in a nonoverlapping mutually exclusive manner, which is also referred to as a salt-and-pepper distribution (Chazaud et al. 2006; Plusa et al. 2008). By using the PrE-specific marker Pdgfrα, live imaging experiments in embryos carrying a *Pdgfrα::H2B-GFP* knock-in reporter, comprising a human histone H2B fusion to GFP targeted to the Pdgfrα locus (Hadjantonakis and Papaioannou 2004), have confirmed that Gata6-positive PrE precursors do not express Nanog and are positioned randomly within the ICM in a salt-and-pepper distribution (Plusa et al. 2008). At the 64- to 100-cell stage transition, the majority of GFP-positive cells not already residing on the cavity roof move toward the surface of the ICM, subsequently becoming committed to the PrE. A minor population of internal GFP-positive cells will likely either downregulate the *Pdgfrα::H2B-GFP* reporter or apoptose, and so be eliminated from the embryo. In this way, the salt-and-pepper distribution precedes the formation of the PrE epithelium, suggesting that the specification of EPI and PrE precursor cells takes place before the subsequent cell sorting and epithelialization of the PrE lineage, which in turn depends on positional cues relative to the blastocyst cavity.

The salt-and-pepper distribution marks the first step of PrE versus EPI lineage allocation within the ICM. Cells expressing PrE but not EPI markers are fated to contribute to the PrE lineage rather than the EPI during normal development

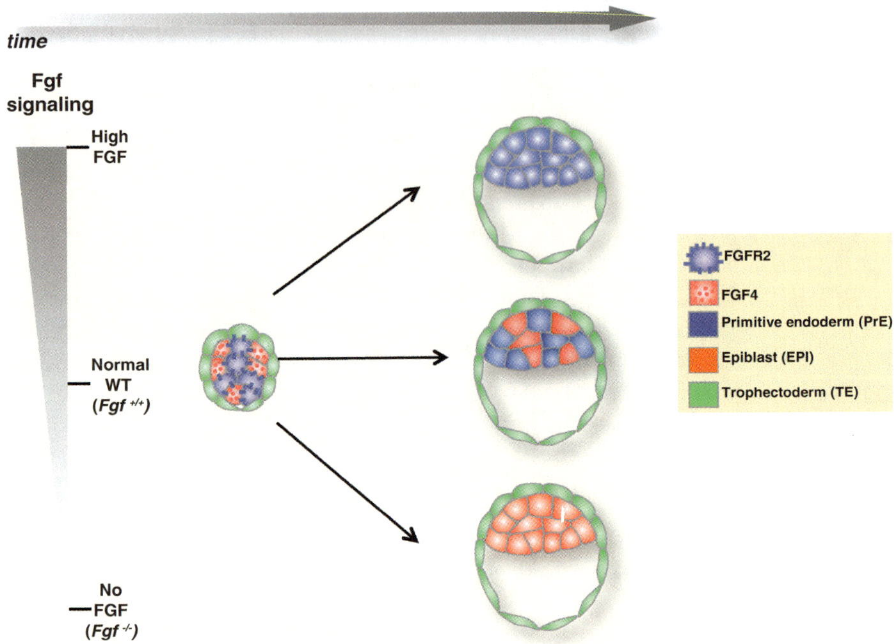

Fig. 10.3 *Lineage allocation in the ICM depends on FGF/MAPK signaling*. At around the 32-cell stage, when EPI- and PrE-specific transcription factors are expressed homogeneously in all ICM cells, a differential expression of the ligand Fgf4 and the receptor Fgfr2 starts to become evident. EPI precursor cells express and secrete Fgf4, whereas PrE precursor cells express Fgfr2. Binding of Fgf4 to Fgfr2 activates the FGF/MAPK signaling pathway, which is required for the induction of PrE-specific transcription factors, thus promoting PrE formation. When the signaling is not active, such as in *Fgf4* mutants or by inhibiting the receptor Fgfr2, PrE is not formed and all ICM cells contribute to EPI. On the contrary, in the presence of excess Fgf4, all ICM cells contribute to PrE

(Chazaud et al. 2006). It is however debatable whether these cells though committed to a specific lineage may exhibit plasticity dependent on context. This raises the question of whether cells possess a less restricted developmental potential, not necessarily exclusively reflecting the lineage for which they exhibit marker-specific expression. Recent experiments support such a hypothesis, indicating that, at the salt-and-pepper stage, individual ICM cells, and in particular those of the PrE, exhibit greater plasticity than generally appreciated when isolated from embryos and reintroduced into chimeras, and that this plasticity is lost once the cells have sorted to their respective tissue layers and the PrE begins to epithelialize (Grabarek et al. 2012).

Lineage allocation and the developmental potential within the ICM have been shown to be influenced by FGF/MAPK signaling (Fig. 10.3) (reviewed in Lanner and Rossant 2010). Activation of FGF/MAPK signaling induces the expression of PrE-specific transcription factors, such as Gata6 (Li et al. 2004; Morrisey et al. 1998).

The analysis of mouse mutants has revealed that the ligand Fgf4, the receptor Fgfr2 and the adaptor protein Grb2, which mediate FGF/MAPK signaling, are all required for PrE formation (Arman et al. 1998; Cheng et al. 1998; Feldman et al. 1995; Wilder et al. 1997). Perturbations in FGF/MAPK signaling greatly influence the balance between the EPI and PrE lineages in the ICM: excess of Fgf4 converts all ICM cells to adopt a PrE identity (Yamanaka et al. 2010), whereas in *Fgf4* mutants, all ICM cells become Nanog positive and PrE is not formed (Piliszek A, MK and AKH unpublished observations). A comparable defect is observed in embryos lacking the adaptor protein Grb2 (Chazaud et al. 2006). In support of these observations on mutants with perturbations in Fgf signaling, chemical inhibition of FGF signaling results in all ICM cells adopting an EPI fate (Nichols et al. 2009). Therefore, it has been proposed that cells expressing Fgfr2 and thus receiving signal through binding of the Fgf4 ligand are fated to form PrE, whereas cells that do not express Fgfr2 but instead secrete Fgf4 will be EPI precursors.

One interesting question pertains the stage at which this differential FGF signaling takes place with respect to the expression of EPI- and PrE-specific transcription factors. A recent single-cell expression analysis study showed that, at around the 32-cell stage, there is a reciprocal *Fgf4*/*Fgfr2* ratio of expression in the ICM cells with approximately half cells predominantly expressing the ligand and the other half expressing the receptor (Fig. 10.3) (Guo et al. 2010). Interestingly, this reciprocal ligand/receptor expression is evident prior to the emergence of the salt-and-pepper distribution of transcription factors, indicating that FGF/MAPK signaling might act upstream of the differential expression of lineage-specific transcription factors. Consistent with this hypothesis, in *Fgf4* mutant embryos, even though the initial phase of transcription factor co-expression is observed, the salt-and-pepper distribution is not established and all ICM cells shift to the EPI lineage (Piliszek A, MK and AKH unpublished observations). This suggests that FGF signaling is not required for the initial expression of lineage-specific transcription factors, although it does regulate the salt-and-pepper patterning and thus cell fate choice.

It is tempting to speculate how the heterogeneity in FGF signaling might be established prior to the emergence of a salt-and-pepper distribution of lineage-fated precursors. First, might it be possible that the initial homogeneous expression of factors such as Nanog, Oct4 and Gata6 play a role? It is worth mentioning that Fgf4 is produced in the ICM under the control of the early ICM factors Oct4 and Sox2 (Yuan et al. 1995), suggesting that an initial expression of these factors is needed for the ligand to be expressed and secreted. Second, could this heterogeneity in Fgf4/Fgfr2 expression be stochastic? This hypothesis fits well with the heterogeneity observed in ES populations, which exhibit stochastic fluctuations in the levels of pluripotency-linked transcription factors, such as Nanog (Chambers et al. 2007; Kalmar et al. 2009; Singh et al. 2007). Third, might it be possible that FGF signaling is influenced by cues from other signaling pathways? As mentioned earlier, the Hippo signaling pathway plays a role in the first (TE versus ICM) cell fate decision, by converting cell-density signals into cell growth control and gene activity (Nishioka et al. 2009; Yagi et al. 2007). In low-density/outside cells, the transcriptional cofactor Yap/Taz mediates the Hippo signaling pathway by its translocation to the nucleus, which results in the establishment of Tead4 activity. Tead4 will then direct the expression of Cdx2, which is the first lineage-specific transcription factor marking the TE. By contrast, in high density/inside cells, Yap/Taz remains in the cytoplasm, and thus Cdx2 is not expressed. Therefore, Oct4 is not downregulated and these cells adopt an ICM identity. Interestingly, recent studies have connected the Hippo signaling pathway with the TGF-b/Smad activity (Varelas et al. 2010). Considering the significant degree of cross-talk between signaling pathways during late development (Guo and Wang 2009), it is tempting to speculate that these pathways might influence the initial ratio of expression of Fgf4/Fgfr2 in early ICM cells. Furthermore, if existing, such a link would connect the first and second cell fate decisions, a hypothesis that has been put forward over the last years (Bruce and Zernicka-Goetz 2010). Finally, does the heterogeneity in FGF signaling reflect a lineage bias? An interesting observation was that cells that are internalized early (between the 8- and 16-cell stages) were shown to express Sox2 (Guo et al. 2010). Since Sox2 directs the expression of the ligand Fgf4 in these cells (Chen et al. 2008; Yuan et al. 1995), this might then signal to Fgfr2-expressing cells that are internalized later setting up a lineage bias earlier than believed (i.e., in the transition between the 16- and 32-cell stages). Therefore, at around the 32-cell stage, the reciprocal Fgf4/Fgfr2 pattern of expression starts to become evident. Consistent with these observations, cells that are internalized first are more likely to contribute to EPI, whereas cells that are internalized later are fated to become PrE. This attractive "time inside–time outside" model was recently put forward by Zernicka-Goetz and colleagues and will be discussed in the next section in greater detail (Morris et al. 2010). It is worth mentioning that, although this model may explain the upregulation of Fgf4 in the early-internalized cells, the mechanisms underlying its downregulation as well as the upregulation of Fgfr2 in the latter internalized cells remain unknown.

10.4 Two Models for the Establishment of the "Salt-and-Pepper" Pattern of Expression in the ICM

EPI and PrE lineage-restricted precursors are first identified at the 64-cell stage, when EPI- and PrE-specific transcription factors, such as Nanog and Gata6, respectively, are observed in a mutually exclusive pattern. As mentioned previously, the establishment of this salt-and-pepper distribution of lineage progenitors greatly depends on signaling cues, in particular the FGF/MAPK pathway. Nevertheless, an interesting observation is that nascent EPI and PrE cells appear to be positioned in a salt-and-pepper distribution within the ICM at this stage. Recent studies have attempted to elucidate how this distribution emerges: does it occur in a random way, or is it biased by the developmental history of a cell? Two models have been put forward with studies providing support for both the random-based as well as the lineage bias-based models (Fig. 10.4).

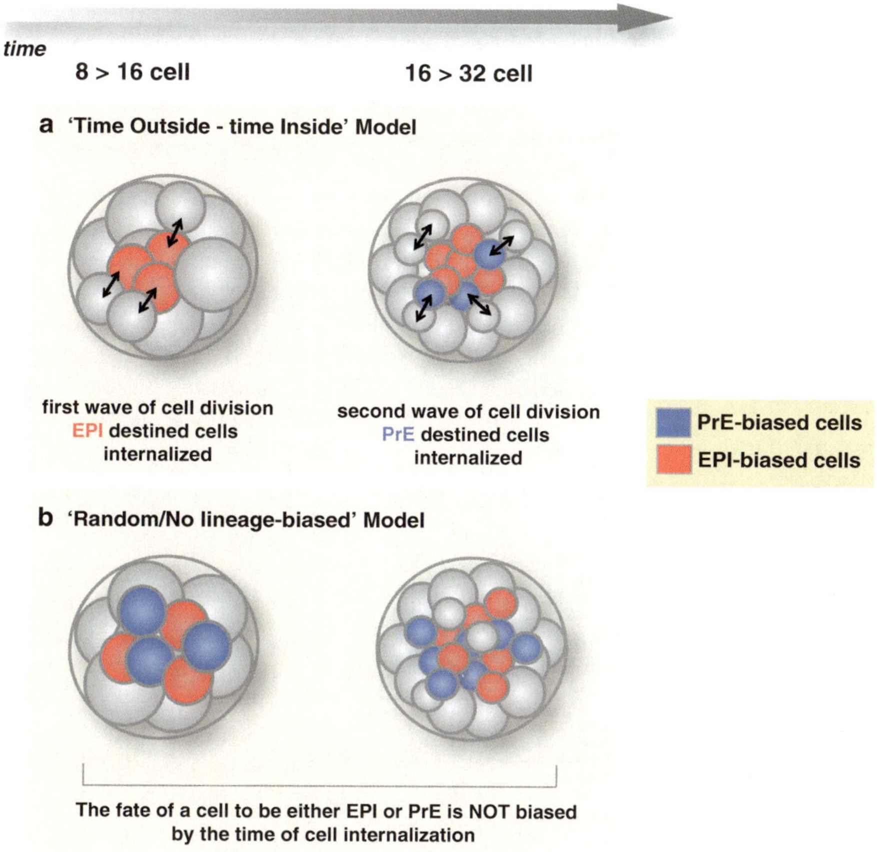

Fig. 10.4 *Two models for the establishment of the salt-and-pepper distribution of cells in the ICM*. The two "waves" of asymmetric cell divisions, which take place at the 8- to 16-cell and 16- to 32-cell stages, generate the inner cells of the ICM. (**a**) In the "time-outside/time-inside" model, cells that are internalized with the first wave are biased to form EPI, whereas cells that are internalized with the second wave are biased to form PrE. (**b**) In the "random/no lineage-biased" model, cells could contribute to EPI or PrE irrespectively of whether they have internalized during the first or the second wave

Two "waves" of asymmetric cell divisions (8- to 16-cell and 16- to 32-cell stages) generate the inner cells that contribute to the ICM (Chazaud et al. 2006; Chisholm and Houliston 1987). It has been proposed that the history of a cell, namely whether it was internalized during the first or the second wave of cell divisions, could influence its propensity to contribute to EPI or PrE. Experimental support for this hypothesis (Morris et al. 2010) was reported through experiments which used noninvasive lineage tracing of cells from the eight-cell to late blastocyst stage. This study demonstrated that cells internalized during the first wave exhibited a greater bias toward the EPI, whereas cells internalized with the second were biased to form PrE. Moreover, the few cells that are internalized during the third wave of asymmetric divisions (32- to 64-cell stage) invariably contribute to PrE. Contributions to EPI or PrE were scored based on the morphology of cells by the late blastocyst stage. Based on these observations, Morris and colleagues put forward the "time-outside/time-inside model," suggesting that the greater the time an ICM cell spends on the outside of an embryo the greater its propensity to differentiate, whereas with more time spent inside, their pluripotency is preserved and thus, they contribute to EPI.

The "time-outside/time-inside" model invoking a cellular memory was challenged by the findings of another study that showed that the waves of asymmetric divisions generate EPI- and PrE-precursors in an apparently random fashion, irrespectively of whether the cells were internalized early or late (Yamanaka et al. 2010). Yamanaka et al. injected blastomeres at the eight-cell stage with mRNA for nuclear and membrane markers, which allowed them to monitor whether inner cells were generated during the first or the second wave of asymmetric divisions. The fate of these cells in postimplantation stages (i.e., contribution to either established EPI or PrE-derived tissues) was then assayed after uterine transfer to pseudo-pregnant females. In contrast to Morris and colleagues, Yamanaka et al. reported that cells internalized from either the first or the second wave contributed to either EPI or PrE without being biased to form a certain lineage.

Recently, an informed debate has taken place over the conclusions drawn from different studies (Bruce and Zernicka-Goetz 2010). It has been suggested that the discrepancies between studies, may be due to technical differences, for example different methods have been used for lineage tracing. However, an absolute correlation between lineage bias and timing of internalization is not evident. Morris et al. reported that 25 % of cells internalized with the first wave (i.e., cells biased to form EPI) in fact contribute to PrE, whereas a 15 % of cells internalized with the second wave (i.e., cells biased to form PrE) can instead contribute to EPI (Morris et al. 2010). Therefore, even though a lineage bias could certainly be a factor, the cells of the early ICM seem to exhibit some developmental plasticity, and additional mechanisms are likely required for their final commitment to a specific lineage. Most notably, as mentioned previously, FGF/MAPK signaling influences cell fate choice within the ICM. The study of Yamanaka and colleagues lent further support for the critical role of FGF signaling by showing that inhibition of FGF/MAPK signaling results in all ICM cells shifting to a Nanog-positive EPI fate. Conversely, removal of the FGF/MAPK inhibitor resulted in the restoration of the PrE lineage, indicating that the ICM cells retain a highly dynamic and plastic state. Interestingly, the effects of FGF signaling inhibition could be observed after the salt-and-pepper pattern of expression, suggesting that the developmental plasticity in the ICM is evident even after that stage. Nevertheless, the plasticity was progressively lost after E4.0, indicating that by the late blastocyst stage, cells have committed to a certain lineage (Yamanaka et al. 2010). Therefore, commitment likely becomes fixed as, or soon after, cells have undergone their final sorting process (discussed in the next section). Consistent with these observations, recent experimental data have further shown that lineage plasticity is retained at the salt-and-pepper stage and is lost much later, just before implantation (Grabarek et al. 2012). Interestingly, this stage coincides with when Oct4 has begun to be excluded from cells of the PrE lineage after they have sorted (Grabarek et al. 2012).

10.5 The Final Cell Sorting and Spatial Segregation

After the salt-and-pepper distribution at around the 64-cell stage, dynamic cell movements and rearrangements occur within the ICM and lead to the spatial segregation (aka "sorting") of the PrE precursor cells toward the surface of the ICM, facing the blastocyst cavity (Fig. 10.5) (Gerbe et al. 2008; Meilhac et al. 2009). By using a PrE-specific GFP reporter for expression of Pdgfrα, live imaging experiments have unraveled the cell behaviors that take place during the sorting process (Plusa et al. 2008): (1) GFP-positive cells (i.e., PrE precursors) that are positioned at an inner location in the ICM, tend to move toward the surface of the cavity. These cells later contribute to form the established PrE epithelial layer. However, some other GFP-positive cells, located at similar inner positions, do not move close to the cavity and instead undergo apoptosis. Interestingly, a substantial number of these cells have also been shown to evade the apoptotic pathway downregulating GFP expression and convert to an EPI fate. (2) GFP-positive cells that are located close to the surface of the cavity already from an early stage rarely move. These observations suggest that a positional signal might be important to retain these cells to their location. (3) Some GFP-negative cells (i.e., EPI precursors) that lie next to the cavity have been shown to upregulate GFP expression and eventually contribute to PrE. Therefore, at this case, cells have shifted their fate from EPI to PrE, possibly due to positional cues originating at the surface of the cavity.

Based on the studies described above, it is becoming widely accepted that dynamic cell rearrangements, involving actin-dependent cell movements and/or apoptosis, as well as possible PrE-to-EPI or EPI-to-PrE cell lineage conversions occur during the final sorting. After sorting, cells are committed to their prospective lineages, confirmed by their inability to contribute to chimeras. Indeed, the sorted PrE cells are morphologically and molecularly distinct from the unsorted PrE precursor cells. The sorting cells gradually polarize as they move next to the surface of the blastocyst cavity, showing a characteristic epithelial morphology (Gerbe et al. 2008); moreover, they start to express the transcription factor Sox7, which is an exclusive marker for these PrE-committed cells (Artus et al. 2011). Activation of Sox7 in these cells (as well as exclusion of Oct4) marks the point where developmental plasticity has been lost and commitment to PrE has occurred (Grabarek et al. 2012).

The mechanisms underlying the final cell sorting have not yet been fully elucidated. The fact that sorted cells are polarized is critical. Indeed, important markers were shown to localize in the apical surface (positioned adjacent to the blastocyst cavity) of the sorted cells (Gerbe et al. 2008). One of them, the low-density lipoprotein receptor-related protein Lrp2, is expressed by PrE-precursor cells at around the salt-and-pepper stage and then localizes specifically at the apical surface of the sorted PrE-committed cells (Gerbe et al. 2008). The cargo protein adaptor Dab2 also shows a distinct localization on the apical surface of the sorted cells (Yang et al. 2002). Dab2 binds to Lrp2 and it has been suggested that Dab2 recruits Lrp2 to the apical surface after sorting. Interestingly, disruption of Dab2

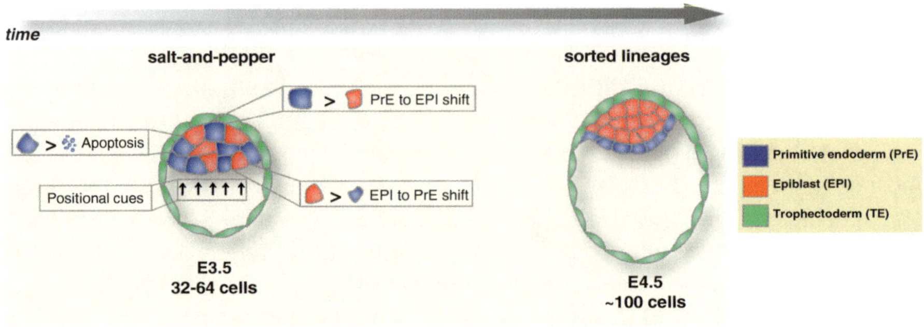

Fig. 10.5 *Cell sorting ensures the spatial segregation of PrE and EPI.* After the salt-and-pepper distribution of cells within the ICM, dynamic cell rearrangements take place, involving the movement of EPI and PrE precursor cells to inner locations of the ICM and surface of the blastocyst cavity, respectively. Some PrE precursor cells that are located deep into the ICM may undergo apoptosis or convert to an EPI fate. On the contrary, some EPI precursor cells that lie next to the blastocyst cavity might convert to a PrE identity. Moreover, positional signals from the blastocyst cavity (such as Wnt9α) might induce the migration of PrE precursor cells toward the cavity and the maintenance of PrE identity in these cells

results in preimplantation embryos in which PrE precursor cells specified within the ICM are unable to sort to their final position adjacent to the cavity (Yang et al. 2002). However, the mechanisms through which Dab2 facilitates sorting remain unknown. It is possible that it could play a critical role in transporting protein(s) important for the movement of these cells to the surface of the ICM. To this end, mutation of the Dab2 interaction partner integrin β1 leads to PrE formation failure, suggesting that cell adhesion changes, mediated by Dab2 and its integrin partners, play a significant role in the sorting process (Fassler and Meyer 1995; Stephens et al. 1995). Overall, the apical localization of several markers in the sorted cells and the gradual polarization of these cells seem to be essential for the sorting process.

Signaling cues might also be important for the final sorting step. A possible candidate could be the PDGF signaling. As mentioned previously, Pdgfrα is one of the first markers to be expressed in the PrE precursor cells (along with Gata6) and PDGF signaling plays a crucial role for the expansion of PrE-committed cells in late blastocysts (Artus et al. 2010). Nevertheless, the sorting process in *Pdgfrα* mutant embryos seems to occur normally, even though the PrE epithelium consists of fewer cells compared to wild-type blastocysts (Artus et al. 2010). However, positional-induced signaling cues seem to have a more direct effect on the sorting process. To this end, Wnt9α, which is expressed on the surface of the blastocyst, was shown to facilitate repositioning of the Gata6-expressing cells (Meilhac et al. 2009). In conclusion, it is becoming widely accepted that the sorting process does not simply involve cell movement governed by the expression of lineage-specific transcription factors. Instead, several other parameters, such as gradual polarization, changes in cell adhesion properties, and signaling cues induced by position, all likely have a cumulative effect on this final sorting step, which culminates in lineage commitment and tissue segregation.

10.6 Making a Compromise: Is Lineage Allocation in the ICM a Stochastic Process with a Lineage Bias?

Many aspects regarding lineage allocation in the ICM remain puzzling. However, seemingly conflicting reports might actually be revealing complementary mechanisms governing the EPI versus PrE fate choice. The initial EPI- and PrE-specification at the onset of the salt-and-pepper distribution noticeably occurs in a spatially disorganized manner. This observation could point to a stochastic process underlying the second fate decision. Indeed, the stem cell population of the ICM might exhibit a similar dynamically heterogeneous state as their in vitro counterparts (Chambers et al. 2007; Kalmar et al. 2009; Singh et al. 2007). Moreover, heterogeneities in the levels of Cdx2 and Nanog expression have been observed at even earlier stages in the embryo, indicating that stochastic processes play a significant role during the first ICM versus TE fate decision as well (Dietrich and Hiiragi 2007). However, amongst this apparent randomness, the fate of an individual ICM cell could be biased; to this end, stochastic does not necessarily mean unbiased (Zernicka-Goetz and Huang 2010). Indeed, if lineage allocation were solely based on random and unbiased mechanisms, its phenomenally certain outcome would have been unpredictable. Moreover, the final cell sorting is clearly influenced by deterministic cues, such as positional signaling originating from the cavity surface, which stabilizes lineage commitment. Therefore, the early embryo provides an experimental model in order to investigate in vivo the crosstalk between stochastic and deterministic processes and how intercellular signaling pathways, such as the FGF pathway, influence these processes. Future experiments will likely elucidate the order of these events, and the mechanisms driving them during embryogenesis, which ultimately give rise to the remarkably structured blastocyst stage embryo.

Acknowledgements We thank Jerome Artus, Silvia Munoz-Descalzo and Marilena Papaioannou for discussions and comments on this review. Work in our laboratory is supported by the HFSP, NIH, and NYSTEM.

References

Arman E, Haffner-Krausz R, Chen Y, Heath JK, Lonai P (1998) Targeted disruption of fibroblast growth factor (FGF) receptor 2 suggests a role for FGF signaling in pregastrulation mammalian development. Proc Natl Acad Sci USA 95:5082–5087

Arnold SJ, Robertson EJ (2009) Making a commitment: cell lineage allocation and axis patterning in the early mouse embryo. Nat Rev Mol Cell Biol 10:91–103

Artus J, Panthier JJ, Hadjantonakis AK (2010) A role for PDGF signaling in expansion of the extra-embryonic endoderm lineage of the mouse blastocyst. Development 137:3361–3372

Artus J, Piliszek A, Hadjantonakis AK (2011) The primitive endoderm lineage of the mouse blastocyst: sequential transcription factor activation and regulation of differentiation by Sox17. Dev Biol 350:393–404

Avilion AA, Nicolis SK, Pevny LH, Perez L, Vivian N, Lovell-Badge R (2003) Multipotent cell lineages in early mouse development depend on SOX2 function. Genes Dev 17:126–140

Becker S, Casanova J, Grabel L (1992) Localization of endoderm-specific mRNAs in differentiating F9 embryoid bodies. Mech Dev 37:3–12
Bruce AW, Zernicka-Goetz M (2010) Developmental control of the early mammalian embryo: competition among heterogeneous cells that biases cell fate. Curr Opin Genet Dev 20:485–491
Chambers I, Colby D, Robertson M, Nichols J, Lee S, Tweedie S, Smith A (2003) Functional expression cloning of Nanog, a pluripotency sustaining factor in embryonic stem cells. Cell 113:643–655
Chambers I, Silva J, Colby D, Nichols J, Nijmeijer B, Robertson M, Vrana J, Jones K, Grotewold L, Smith A (2007) Nanog safeguards pluripotency and mediates germline development. Nature 450:1230–1234
Chambers I, Smith A (2004) Self-renewal of teratocarcinoma and embryonic stem cells. Oncogene 23:7150–7160
Chazaud C, Yamanaka Y, Pawson T, Rossant J (2006) Early lineage segregation between epiblast and primitive endoderm in mouse blastocysts through the Grb2-MAPK pathway. Dev Cell 10:615–624
Chen X, Xu H, Yuan P, Fang F, Huss M, Vega VB, Wong E, Orlov YL, Zhang W, Jiang J, Loh YH, Yeo HC, Yeo ZX, Narang V, Govindarajan KR, Leong B, Shahab A, Ruan Y, Bourque G, Sung WK, Clarke ND, Wei CL, Ng HH (2008) Integration of external signaling pathways with the core transcriptional network in embryonic stem cells. Cell 133:1106–1117
Cheng AM, Saxton TM, Sakai R, Kulkarni S, Mbamalu G, Vogel W, Tortorice CG, Cardiff RD, Cross JC, Muller WJ, Pawson T (1998) Mammalian Grb2 regulates multiple steps in embryonic development and malignant transformation. Cell 95:793–803
Chisholm JC, Houliston E (1987) Cytokeratin filament assembly in the preimplantation mouse embryo. Development 101:565–582
Dietrich JE, Hiiragi T (2007) Stochastic patterning in the mouse pre-implantation embryo. Development 134:4219–4231
Enders AC, Given RL, Schlafke S (1978) Differentiation and migration of endoderm in the rat and mouse at implantation. Anat Rec 190:65–77
Fassler R, Meyer M (1995) Consequences of lack of beta 1 integrin gene expression in mice. Genes Dev 9:1896–1908
Feldman B, Poueymirou W, Papaioannou VE, DeChiara TM, Goldfarb M (1995) Requirement of FGF-4 for postimplantation mouse development. Science 267:246–249
Gardner RL (1996) Can developmentally significant spatial patterning of the egg be discounted in mammals? Hum Reprod Update 2:3–27
Gerbe F, Cox B, Rossant J, Chazaud C (2008) Dynamic expression of Lrp2 pathway members reveals progressive epithelial differentiation of primitive endoderm in mouse blastocyst. Dev Biol 313:594–602
Grabarek JB, Żyżyńska K, Saiz N, Piliszek A, Frankenberg S, Nichols J, Hadjantonakis AK, Plusa B (2012) Differential plasticity of epiblast and primitive endoderm precursors within the ICM of the early mouse embryo. Development 139:129–139
Guo G, Huss M, Tong GQ, Wang C, Li Sun L, Clarke ND, Robson P (2010) Resolution of cell fate decisions revealed by single-cell gene expression analysis from zygote to blastocyst. Dev Cell 18:675–685
Guo X, Wang XF (2009) Signaling cross-talk between TGF-beta/BMP and other pathways. Cell Res 19:71–88
Hadjantonakis AK, Papaioannou VE (2004) Dynamic in vivo imaging and cell tracking using a histone fluorescent protein fusion in mice. BMC Biotechnol 4:33
Jedrusik A, Parfitt DE, Guo G, Skamagki M, Grabarek JB, Johnson MH, Robson P, Zernicka-Goetz M (2008) Role of Cdx2 and cell polarity in cell allocation and specification of trophectoderm and inner cell mass in the mouse embryo. Genes Dev 22:2692–2706
Johnson MH, Ziomek CA (1981) The foundation of two distinct cell lineages within the mouse morula. Cell 24:71–80
Kalmar T, Lim C, Hayward P, Munoz-Descalzo S, Nichols J, Garcia-Ojalvo J, Martinez Arias A (2009) Regulated fluctuations in nanog expression mediate cell fate decisions in embryonic stem cells. PLoS Biol 7:e1000149

Kelly SJ (1977) Studies of the developmental potential of 4- and 8-cell stage mouse blastomeres. J Exp Zool 200:365–376

Koutsourakis M, Langeveld A, Patient R, Beddington R, Grosveld F (1999) The transcription factor GATA6 is essential for early extraembryonic development. Development 126:723–732

Kuo CT, Morrisey EE, Anandappa R, Sigrist K, Lu MM, Parmacek MS, Soudais C, Leiden JM (1997) GATA4 transcription factor is required for ventral morphogenesis and heart tube formation. Genes Dev 11:1048–1060

Lanner F, Rossant J (2010) The role of FGF/Erk signaling in pluripotent cells. Development 137:3351–3360

Li L, Arman E, Ekblom P, Edgar D, Murray P, Lonai P (2004) Distinct GATA6- and laminin-dependent mechanisms regulate endodermal and ectodermal embryonic stem cell fates. Development 131:5277–5286

Martin GR, Evans MJ (1975) Differentiation of clonal lines of teratocarcinoma cells: formation of embryoid bodies in vitro. Proc Natl Acad Sci USA 72:1441–1445

Meilhac SM, Adams RJ, Morris SA, Danckaert A, Le Garrec JF, Zernicka-Goetz M (2009) Active cell movements coupled to positional induction are involved in lineage segregation in the mouse blastocyst. Dev Biol 331:210–221

Messerschmidt DM, Kemler R (2010) Nanog is required for primitive endoderm formation through a non-cell autonomous mechanism. Dev Biol 344:129–137

Mitsui K, Tokuzawa Y, Itoh H, Segawa K, Murakami M, Takahashi K, Maruyama M, Maeda M, Yamanaka S (2003) The homeoprotein Nanog is required for maintenance of pluripotency in mouse epiblast and ES cells. Cell 113:631–642

Molkentin JD, Lin Q, Duncan SA, Olson EN (1997) Requirement of the transcription factor GATA4 for heart tube formation and ventral morphogenesis. Genes Dev 11:1061–1072

Morris SA, Teo RT, Li H, Robson P, Glover DM, Zernicka-Goetz M (2010) Origin and formation of the first two distinct cell types of the inner cell mass in the mouse embryo. Proc Natl Acad Sci USA 107:6364–6369

Morrisey EE, Tang Z, Sigrist K, Lu MM, Jiang F, Ip HS, Parmacek MS (1998) GATA6 regulates HNF4 and is required for differentiation of visceral endoderm in the mouse embryo. Genes Dev 12:3579–3590

Murray P, Edgar D (2001) Regulation of the differentiation and behaviour of extra-embryonic endodermal cells by basement membranes. J Cell Sci 114:931–939

Niakan KK, Ji H, Maehr R, Vokes SA, Rodolfa KT, Sherwood RI, Yamaki M, Dimos JT, Chen AE, Melton DA, McMahon AP, Eggan K (2010) Sox17 promotes differentiation in mouse embryonic stem cells by directly regulating extraembryonic gene expression and indirectly antagonizing self-renewal. Genes Dev 24:312–326

Nichols J, Silva J, Roode M, Smith A (2009) Suppression of Erk signalling promotes ground state pluripotency in the mouse embryo. Development 136:3215–3222

Nichols J, Zevnik B, Anastassiadis K, Niwa H, Klewe-Nebenius D, Chambers I, Scholer H, Smith A (1998) Formation of pluripotent stem cells in the mammalian embryo depends on the POU transcription factor Oct4. Cell 95:379–391

Nishioka N, Inoue K, Adachi K, Kiyonari H, Ota M, Ralston A, Yabuta N, Hirahara S, Stephenson RO, Ogonuki N, Makita R, Kurihara H, Morin-Kensicki EM, Nojima H, Rossant J, Nakao K, Niwa H, Sasaki H (2009) The Hippo signaling pathway components Lats and Yap pattern Tead4 activity to distinguish mouse trophectoderm from inner cell mass. Dev Cell 16:398–410

Nishioka N, Yamamoto S, Kiyonari H, Sato H, Sawada A, Ota M, Nakao K, Sasaki H (2008) Tead4 is required for specification of trophectoderm in pre-implantation mouse embryos. Mech Dev 125:270–283

Niwa H, Toyooka Y, Shimosato D, Strumpf D, Takahashi K, Yagi R, Rossant J (2005) Interaction between Oct3/4 and Cdx2 determines trophectoderm differentiation. Cell 123:917–929

Nowotschin S, Hadjantonakis AK (2010) Cellular dynamics in the early mouse embryo: from axis formation to gastrulation. Curr Opin Genet Dev 20:420–427

Palmieri SL, Peter W, Hess H, Scholer HR (1994) Oct-4 transcription factor is differentially expressed in the mouse embryo during establishment of the first two extraembryonic cell lineages involved in implantation. Dev Biol 166:259–267

Plachta N, Bollenbach T, Pease S, Fraser SE, Pantazis P (2011) Oct4 kinetics predict cell lineage patterning in the early mammalian embryo. Nat Cell Biol 13:117–123

Plusa B, Hadjantonakis AK, Gray D, Piotrowska-Nitsche K, Jedrusik A, Papaioannou VE, Glover DM, Zernicka-Goetz M (2005) The first cleavage of the mouse zygote predicts the blastocyst axis. Nature 434:391–395

Plusa B, Piliszek A, Frankenberg S, Artus J, Hadjantonakis AK (2008) Distinct sequential cell behaviours direct primitive endoderm formation in the mouse blastocyst. Development 135:3081–3091

Ralston A, Cox BJ, Nishioka N, Sasaki H, Chea E, Rugg-Gunn P, Guo G, Robson P, Draper JS, Rossant J (2010) Gata3 regulates trophoblast development downstream of Tead4 and in parallel to Cdx2. Development 137:395–403

Ralston A, Rossant J (2008) Cdx2 acts downstream of cell polarization to cell-autonomously promote trophectoderm fate in the early mouse embryo. Dev Biol 313:614–629

Rossant J, Tam PP (2009) Blastocyst lineage formation, early embryonic asymmetries and axis patterning in the mouse. Development 136:701–713

Singh AM, Hamazaki T, Hankowski KE, Terada N (2007) A heterogeneous expression pattern for Nanog in embryonic stem cells. Stem Cells 25:2534–2542

Stephens LE, Sutherland AE, Klimanskaya IV, Andrieux A, Meneses J, Pedersen RA, Damsky CH (1995) Deletion of beta 1 integrins in mice results in inner cell mass failure and peri-implantation lethality. Genes Dev 9:1883–1895

Strumpf D, Mao CA, Yamanaka Y, Ralston A, Chawengsaksophak K, Beck F, Rossant J (2005) Cdx2 is required for correct cell fate specification and differentiation of trophectoderm in the mouse blastocyst. Development 132:2093–2102

Suwinska A, Czolowska R, Ozdzenski W, Tarkowski AK (2008) Blastomeres of the mouse embryo lose totipotency after the fifth cleavage division: expression of Cdx2 and Oct4 and developmental potential of inner and outer blastomeres of 16- and 32-cell embryos. Dev Biol 322:133–144

Tarkowski AK, Wroblewska J (1967) Development of blastomeres of mouse eggs isolated at the 4- and 8-cell stage. J Embryol Exp Morphol 18:155–180

Torres-Padilla ME, Parfitt DE, Kouzarides T, Zernicka-Goetz M (2007) Histone arginine methylation regulates pluripotency in the early mouse embryo. Nature 445:214–218

Varelas X, Samavarchi-Tehrani P, Narimatsu M, Weiss A, Cockburn K, Larsen BG, Rossant J, Wrana JL (2010) The Crumbs complex couples cell density sensing to Hippo-dependent control of the TGF-beta-SMAD pathway. Dev Cell 19:831–844

Wilder PJ, Kelly D, Brigman K, Peterson CL, Nowling T, Gao QS, McComb RD, Capecchi MR, Rizzino A (1997) Inactivation of the FGF-4 gene in embryonic stem cells alters the growth and/or the survival of their early differentiated progeny. Dev Biol 192:614–629

Yagi R, Kohn MJ, Karavanova I, Kaneko KJ, Vullhorst D, DePamphilis ML, Buonanno A (2007) Transcription factor TEAD4 specifies the trophectoderm lineage at the beginning of mammalian development. Development 134:3827–3836

Yamanaka Y, Lanner F, Rossant J (2010) FGF signal-dependent segregation of primitive endoderm and epiblast in the mouse blastocyst. Development 137:715–724

Yang DH, Smith ER, Roland IH, Sheng Z, He J, Martin WD, Hamilton TC, Lambeth JD, Xu XX (2002) Disabled-2 is essential for endodermal cell positioning and structure formation during mouse embryogenesis. Dev Biol 251:27–44

Yuan H, Corbi N, Basilico C, Dailey L (1995) Developmental-specific activity of the FGF-4 enhancer requires the synergistic action of Sox2 and Oct-3. Genes Dev 9:2635–2645

Zernicka-Goetz M, Huang S (2010) Stochasticity versus determinism in development: a false dichotomy? Nat Rev Genet 11:743–744

Zernicka-Goetz M, Morris SA, Bruce AW (2009) Making a firm decision: multifaceted regulation of cell fate in the early mouse embryo. Nat Rev Genet 10:467–477

Chapter 11
Formation of Distinct Cell Types in the Mouse Blastocyst

Samantha A. Morris and Magdalena Zernicka-Goetz

Abstract Early development of the mouse comprises a sequence of cell fate decisions in which cells are guided along a pathway of restricted potential and increasing specialisation. The first choice faced by cells of the embryo is whether to become trophectoderm (TE) or inner cell mass (ICM); TE is an extra-embryonic tissue which will form the embryonic portion of the placenta, whilst ICM gives rise to cells responsible for generating the foetus. In the second cell fate decision, the ICM is further refined into pluripotent cells forming the future body of the embryo, epiblast (EPI) and extra-embryonic primitive endoderm (PE), a tissue essential for patterning the embryo and establishing the developmental circulation. Understanding this early lineage segregation is critical for informing attempts to capture pluripotency and direct cell fate in vitro. Unlike the predictability of nonmammalian cell fate, development of the mouse embryo retains the flexibility to adapt to changing circumstances during development. Here we describe these first cell fate decisions, how they can be biased whilst maintaining flexibility and, finally, some of the molecular circuitry underlying early fate choice.

S.A. Morris
Department of Biological Chemistry and Molecular Pharmacology, Harvard Medical School, Boston Children's Hospital, Boston, MA, USA

M. Zernicka-Goetz (✉)
Wellcome Trust/Cancer Research UK Gurdon Institute, Cambridge CB2 1QN, UK

Department of Physiology, Development and Neuroscience, University of Cambridge, Downing Street, Cambridge CB2 3DY, UK
e-mail: m.zernicka-goetz@gurdon.cam.ac.uk

J.Z. Kubiak (ed.), *Mouse Development*, Results and Problems in Cell Differentiation 55,
DOI 10.1007/978-3-642-30406-4_11,

11.1 Introduction

Embryonic development comprises a series of fate decisions whereupon cells decrease in potential and increase in specialisation. In contrast to nonmammalian species where first fate decisions are predictable, development of the mouse embryo is regulative; it has the flexibility to adapt to changing circumstances. For example, one or two cells (blastomeres) from an eight-cell stage human embryo can be removed for pre-implantation genetic diagnosis without negatively impacting on developmental outcome (Handyside et al. 1992). The same is true in more drastic circumstances: following destruction of one blastomere of the two-cell stage mouse embryo, the surviving blastomere can develop to term (Tarkowski 1959). To further demonstrate this developmental flexibility, the position of cells can be altered experimentally and they adapt to their new position, taking on the appropriate fate (Suwinska et al. 2008). This is a unique feature of mammalian development relying upon the fact that cells at this stage are not yet fully committed to a specific lineage (Rossant and Lis 1979; Rossant and Vijh 1980). This flexibility in early development is thought to facilitate formation of the essential first three lineages, which will support implantation and continued development of the embryo to term. Although there remain many unanswered questions as to how the embryo is endowed with this remarkable regulative capacity, we shall see in this chapter that we have a grasp of the molecular circuitry rooted in this early event and later mechanisms which help maintain flexibility in development, thus creating a robust system to establish the first lineages.

Three distinct cell types arise in the 4.5 days of development between the time of fertilisation and implantation: the epiblast (EPI), primitive endoderm (PE), and trophectoderm (TE). EPI will give rise to all the cells of the body; PE forms the yolk sac, a structure responsible for patterning the embryo and initiating developmental circulation prior to establishment of an internal circulatory system; the TE will form the embryonic portion of the placenta (Fig. 11.1). The priority of the embryo at this early stage centres on implantation where the TE takes on the role of providing chemical and physical integration with the uterus. The TE is the first distinct lineage, whereas the EPI and PE are initially mixed within the inner cell mass (ICM) of the embryo. ICM and TE lineages are physically separated by 3.5 days after fertilisation in a structure referred to as the blastocyst (from Greek *blastos*, meaning *bud*). Visual inspection of the blastocyst reveals a fluid filled ball of cells, inside which lies a mass of cells asymmetrically positioned to one side. These inside cells constitute the pluripotent ICM, whereas the surrounding outside cell population forms the differentiated extra-embryonic TE. The formation of these inside and outside cell populations are the foundation of the first cell fate decision, but how do cells become different? To understand the establishment and properties of the first three lineages, we must understand the developmental events leading up to and contributing to their formation.

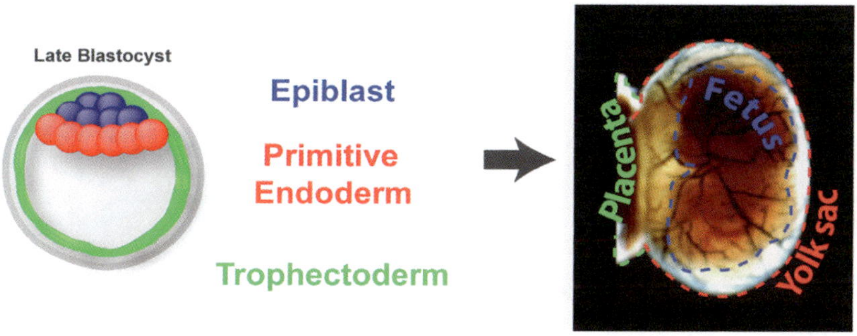

Fig. 11.1 Three distinct lineages of the pre-implantation mouse embryo: epiblast, primitive endoderm and trophectoderm, giving rise to the foetus, yolk sac and placenta, respectively

11.2 Events Leading to Formation of the Blastocyst

Prior to implantation, the first seven cell cycles are cleavage divisions in which cells halve in size, due to the absence of growth (Lehtonen 1980). After fertilisation in the ampulla of the oviduct, a region in close proximity to the ovary, the first cleavage begins around a day later as the embryo travels towards the uterus.

Mammalian cleavage is unique; in addition to the slow, asynchronous divisions (mammalian embryos frequently contain odd numbers of blastomeres) and early activation of the genome (Schultz 2002), the second cleavage in particular has been a subject of much interest due to its unique orientation. The first cleavage is a regular meridional division, following the direction of the animal–vegetal axis. However, in the second cleavage, it is common for one blastomere to divide meridionally, and the other to divide equatorially, this is termed *rotational cleavage* (Fig. 11.2) (Gulyas 1975). The order and orientation of the second cleavage divisions has been found to influence later patterning. This was a provocative finding as the traditional view of early mammalian development considered it to be random, since pre-implantation development can withstand perturbation and there is no clear morphological axis determination until after implantation.

Following a further round of cleavage to the eight-cell stage, where all blastomeres are in contact with the extracellular space between the embryo and zona pellucida (ZP: a glycoprotein membrane surrounding the embryo, produced during oogenesis, functioning to prevent polyspermy), the embryo begins a transformation marked by the formation of an inside cell population as it transitions to the 16-cell stage. After a fifth round of cleavage to produce a 32-cell stage embryo, there is an osmotic accumulation of water in-between cells which establishes the blastocyst cavity, driven by a trans-trophectoderm sodium ion gradient (Watson 1992). The ICM is asymmetrically positioned to one end within this cavity (Fig. 11.3). Concerning the formation of the blastocyst, three models exist to account for its generation, which we will discuss in detail. Each model centres on the establishment of the inside and outside cell populations—is this totally random, or do initial differences between cells exist which influence their final fate?

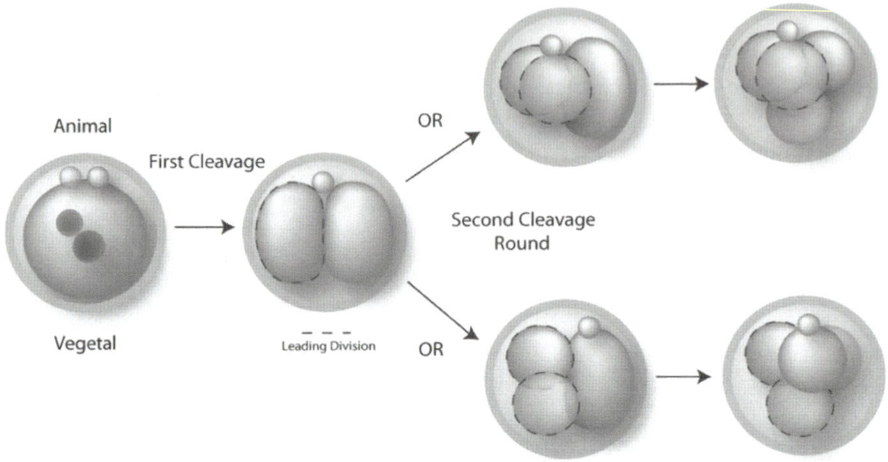

Fig. 11.2 Rotational cleavage. The first cleavage is a regular meridional division, following the direction of the animal–vegetal axis. In the second cleavage, it is common for one blastomere to divide meridionally, and the other to divide equatorially

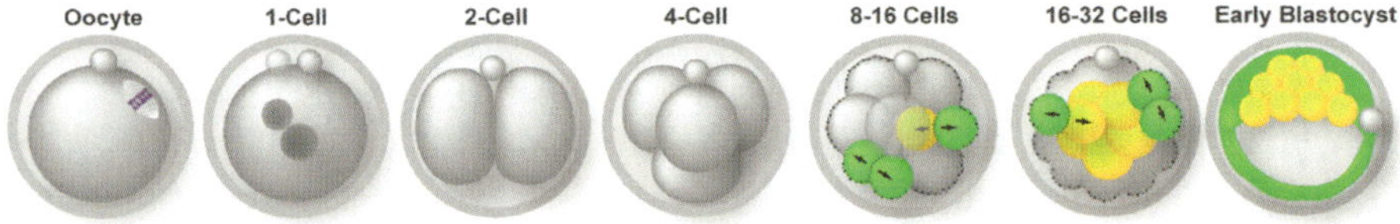

Fig. 11.3 Formation of the blastocyst. Generation of inside cells commences from the 8- to 16-cell stage transition. At the 32-cell stage, fluid accumulates in a cavity of the embryos, completing formation of the blastocyst

11.2.1 The "Early-Asymmetry" Hypothesis

The "early-asymmetry" hypothesis in its most extreme form was rooted in studies of nonmammalian embryos, where partitioning of determinants in combination with standardised cleavage patterns was held responsible for determination of cell fate. This view proposed that asymmetry of the egg would generate differences between cells that would dictate their final fate (Dalcq 1957). Support for this hypothesis came from studies where asymmetric distribution of the leptin protein hormone and STAT3 transcription factor was discovered along the animal–vegetal (AV) axis in oocytes and embryos (Antczak and Van Blerkom 1997). In addition to this, earlier studies where silicon oil droplets were injected to mark central or peripheral cytoplasm in two- and four-cell stage embryos (Graham and Deussen 1978; Wilson et al. 1972) suggested a relationship between early position and later cell fate. However, this was only correlative, and proof for the early-asymmetry hypothesis in its most drastic interpretation would rely on maintenance of cell fate following manipulation of cells into alternate positions. Rather than support the early-asymmetry hypothesis, when such manipulative experiments were eventually performed, the hypothesis was ruled out to account for lineage establishment in the early mouse embryo.

11.2.2 The "Inside–Outside" Hypothesis

This is perhaps the most conceptually simple of the models to account for formation of inside and outside cell populations. The "inside–outside" model hypothesised that blastomeres are equivalent and totipotent until around the 32-cell stage, at which point some blastomeres are surrounded by other blastomeres, resulting in microenvironmental positional differences which would then dictate cell fate (Tarkowski and Wroblewska 1967). This model was supported by experiments where labelled blastomeres were placed into inside or outside positions within an unlabelled embryo. When positioned outside, the labelled blastomeres contributed to TE the reciprocal was true with inside-placed labelled blastomeres contributing to ICM (Hillman et al. 1972), thus demonstrating that blastomeres respond to positional cues. Still though, it remained unknown what, if anything, governed cell position, and with the discovery that eight-cell stage blastomeres become polarised along their apical–basal axes (Johnson and Ziomek 1981), prior to establishment of inside and outside populations, it seemed the model for blastocyst formation would have to be revised.

11.2.3 The Polarisation Hypothesis

Two and a half days after fertilisation, the embryo consists of eight loosely arranged blastomeres. Through the duration of the eight-cell stage, the physical properties of cell–cell contacts undergo a dramatic phenotypic change, with blastomeres transitioning from a spherical appearance to flatten against each other, giving the embryo a smooth appearance. At the cellular level, blastomere polarity is established, resulting in an apical face covered in microvilli and smooth basolateral cell–cell contacts (Handyside 1980; Reeve and Ziomek 1981; Ziomek and Johnson 1980). This polarity is maintained throughout the subsequent two cleavage divisions to the 32-cell stage. It was proposed that the polarisation of blastomeres was a critical event in creating differences between cells which would lead to lineage segregation (Johnson and Ziomek 1981); the apical pole of microvilli is structurally stable, and it is the inheritance of this pole which functions as an outer determinant, in that any cell inheriting it becomes polarised (Wiley and Obasaju 1988). Inside cells are generated when some of these eight-cell blastomeres divide to generate an inside and an outside cell in a so-called differentiative division, which can be considered as an asymmetric division since the resulting cells take different position and fate. This contrasts to conservative (or symmetric) divisions responsible for the production of two outside cells (Johnson and Ziomek 1981). The question is, which cells divide asymmetrically? What this model does not address is if it is random or pre-determined to any extent.

11.2.4 Integration of the Models

There has traditionally been an inclination to favour one of the above models over the others, but are all three models of blastocyst formation mutually exclusive? For example, even though inside and outside cells differ from the moment they are established, with polarity clearly involved, they are still able to respond to an experimental change in position. The links between polarity and position are clear, but are there any findings that early asymmetry is integrated to enhance our understanding of blastocyst development. Indeed, there is evidence that the order and pattern of the second cleavage divisions can influence cell fate.

The majority of embryos (over 80 %) divide in the two- to four-cell transition to form a tetrahedral structure, where one cell has divided meridionally and the other equatorially (Gardner 2002). Within this major group, several key discoveries of developmental bias have been uncovered where the first cleavage is meridional (M), and the second equatorial (E, a so called "ME" embryo). Firstly, it has been demonstrated that blastomeres in four-cell stage ME embryos are not equivalent; when chimeras of a like cells were constructed, the blastomeres inheriting material from both the animal and vegetal pole in the meridional division (AV blastomeres) were fully competent to develop into a live mouse. In contrast to this, the animal (A) and vegetal (V) blastomeres produced upon the subsequent equatorial cleavage (which divides the animal and vegetal cytoplasm) lack full developmental potential. This is demonstrated by animal chimeras producing live mice in only 25 % of cases, whereas vegetal chimeras are not viable (Piotrowska-Nitsche et al. 2005). Further support for the relation between developmental bias and early cleavages comes from live-imaging of unmanipulated embryos; AV blastomeres of ME embryos were found to contribute to the ICM whereas V blastomeres preferentially contributed to the extra-embryonic lineages of PE and TE (Bischoff et al. 2008).

As we will see in the following discussion, understanding molecular regulation helps to resolve the importance of each model in understanding blastocyst formation, and now with the advent of molecular biology and live-imaging technologies we have gained a better understanding of how all three models, early asymmetry, position, and polarity, can be integrated.

11.3 Interplay Between Cell Polarity, Position and Fate

11.3.1 Polarity: PAR Proteins

Partitioning defective (PAR) proteins, originally discovered in the worm, *Caenorhabditis elegans*, have been implicated in the regulation of cell polarisation and asymmetric division. PAR homologues, and their interactors, atypical protein kinase C (aPKCs), are expressed asymmetrically in mouse oocytes and embryos. For example, members of the Par complex, JAM1 (Thomas et al. 2004), aPKC, and

PAR3 (Plusa et al. 2005) become apically localised at the eight-cell stage, whilst Par1 is localised basolaterally (Vinot et al. 2005). Tight junctions will progressively form between the blastomeres to eventually establish the epithelium of the TE (Fleming et al. 2001). Downregulation of polarity complex members, aPKC or PAR3, in individual blastomeres drives their progeny into an inside position where they develop as ICM, either by promoting asymmetric division or engulfment by more polarised neighbours (Plusa et al. 2005). The reciprocal is also true; transplantation of inside cells to an outside position results in their polarisation and contribution to TE (Handyside 1978; Rossant and Lis 1979; Spindle 1978), thus demonstrating the interplay between polarity and position. Moreover, when expression of pluripotency-related genes is enhanced in a blastomere, its progeny are directed to the pluripotent ICM (Torres-Padilla et al. 2007), facilitated by downregulation of cell polarity (Parfitt and Zernicka-Goetz 2010). Is the same also true in that genes regulating TE formation are implicated in polarity?

11.3.2 Cdx2: Interplay of Cell Fate, Polarity and Early Asymmetry

Can polarity and position affect expression of genes driving lineage segregation in the early mouse embryo? Cdx2 is a transcription factor (TF) central to TE formation; Cdx2 protein is restricted to the TE of the blastocyst, where in the absence of Cdx2, TE identity cannot be maintained (Strumpf et al. 2005). The initiation of Cdx2 protein expression was found to be heterogeneous in the eight-cell stage embryo (Dietrich and Hiiragi 2007; Ralston and Rossant 2008), and was therefore suggested to arise at random (Dietrich and Hiiragi 2007). In a series of experiments addressing the role of Cdx2 in segregation of the TE lineage, Jedrusik et al. (2008) found that Cdx2 expression levels related to early cleavage divisions; in ME embryos, the progeny of the vegetal blastomere were found to express higher levels of *Cdx2* mRNA and Cdx2 protein. This was in agreement with earlier findings of such vegetal cells contributing more frequently to TE (Bischoff et al. 2008) and having little potential to develop to term (Piotrowska-Nitsche et al. 2005). Probing the mechanism in further detail, blastomeres in which *Cdx2* expression was experimentally enhanced, prior to inside cell generation, exhibited a higher frequency of symmetric divisions, and hence contribution to the outside TE population. Moreover, these blastomeres in which *Cdx2* was elevated appeared to be more highly polarised, on the basis of apical aPKC localisation (Jedrusik et al. 2008). In addition to this, polarity was found to influence localisation of *Cdx2* mRNA, thus creating a mutually reinforcing loop between polarity and Cdx2 expression to establish TE fate. Together, this demonstrates the interplay between early asymmetry, gene expression and polarity, which in turn directs cell allocation to either the inside or outside lineages. A question remained though: how do cells sense their position in the embryo to respond to it; by up- or downregulating polarity to elicit a position change, or by adjusting lineage-specific gene expression?

11.3.3 Hippo Signalling: Unlocking Position, Polarity and Fate?

How are differences translated into differential gene expression in inside and outside cells? An ideal candidate could function by sensing cell position through cell–cell contacts. In mammals, Hippo signalling controls growth through cell contact-mediated control of proliferation (Pan 2007). Cell–cell contact regulates the nuclear accumulation of Yes-associated protein 1 (Yap1) through Hippo signalling and controls cell proliferation by regulating transcriptional activity of Tead proteins (Ota and Sasaki 2008). Interestingly, Tead4 null mice die shortly after implantation due to reduced cell proliferation and increased apoptosis (Sawada et al. 2008) and Cdx2 expression is controlled by the transcriptional regulator, TEAD4 (Nishioka et al. 2008; Yagi et al. 2007), thus establishing a link between cell contact/position and lineage-specific gene expression. Prior to the blastocyst stage, Yap1 is localised to the nucleus of outside cells (cytoplasm of inside cells), regulated by phosphorylation via the Hippo signalling pathway member kinases Lats1/2. Here, in the nucleus, Yap can directly interact with its transcriptional co-activator Tead4 to stimulate transcription of *Cdx2* (Nishioka et al. 2009). As yet, the identity of the Yap-regulating signals that can sense cell position remain unknown, but likely involve the Hippo signalling pathway and possibly proteins involved in cell contact such as cadherins. Nonetheless, this recent advance represents a promising direct link between cell position and gene expression.

Taken together, the above examples demonstrate the interplay between early asymmetry, position, polarisation and cell fate. First of all, blastomeres are not identical at their inception; gene expression relates to developmental history, with Cdx2 expressed in a distinct blastomere population. This gene expression is in turn able to effect a change in polarisation, which itself is able to translate into specific positioning within the embryo, either executed through orientation of division or physical movement of entire cells. Furthermore, polarity feeds into gene expression, reinforcing cell identity. We must remember though that developmental flexibility is a hallmark of the mouse embryo, thus this sequence of events as described is not absolute. This is demonstrated by the capacity for polarity and gene expression to adapt to a change in blastomere position within the embryo, and vice versa. We have recently been able to garner a deeper understanding of this elegant interplay through improved technologies to uncover the molecular circuitry underlying developmental flexibility. To understand the segregation of embryonic (ICM) and extra-embryonic fates (TE) in greater detail, we must address the second cell fate decision in pre-implantation development where the ICM is further refined into the EPI and PE lineages.

11.4 The Second Cell Fate Decision: Lineage Segregation of the Inner Cell Mass

A second cell fate decision distinguishes two ICM cell types: pluripotent EPI, stem cells for the future foetus; and PE, the second extra-embryonic tissue that becomes visceral endoderm and parietal yolk sac after implantation (Gardner 1982). Of the

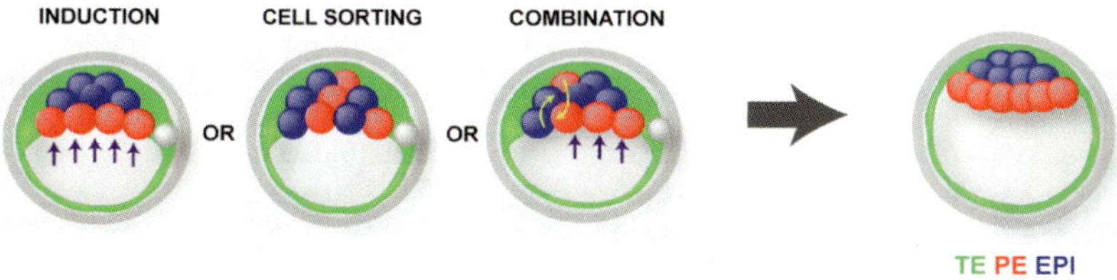

Fig. 11.4 Models of PE and EPI segregation in the second cell fate decision. Induction: PE is induced in the cells contacting the blastocyst cavity, where they will finally reside. Cell sorting: EPI and PE are specified randomly, or in relation to developmental history and subsequently sort into their correct positions. Alternatively, the induction and cell sorting models may be combined

numerous possibilities to account for this fate decision, one of the earliest was the Induction (or Positional) hypothesis. In this, the fate of ICM cells was thought to reflect their position: surface ICM cells next to the blastocyst cavity would become PE, and deeper cells, epiblast. Consistent with this, the outside cells of isolated ICM or embryoid bodies formed from embryonic stem (ES) cells differentiate into endoderm when cultured in vitro (Martin and Evans 1975; Rossant 1975; Solter and Knowles 1975; Dziadek 1979; Murray and Edgar 2001). Later studies of the differential distribution of cytokeratin filaments in blastomeres suggested that inner cells generated in two successive rounds of asymmetric divisions might physically differ (Chisholm and Houliston 1987) leading to speculation, without definitive evidence, that this might be the route towards PE and epiblast formation (Rossant et al. 2003). The subsequent finding that the PE marker Gata6 and the epiblast marker Nanog "were expressed in a random 'salt-and-pepper' pattern" in early blastocysts in advance of PE formation (Chazaud et al. 2006; Kurimoto et al. 2006) gave rise to the idea of a mixed population of epiblast and PE progenitors that would then segregate into their composite layers: the Sorting model. The origins of this pattern had never been addressed, but it was widely taken to have stochastic origins. The apparent lack of any relationship of this heterogeneity to cell position seemed to signal the demise of the Positional/Induction model. However, neither the positional nor the sorting model alone appears sufficient to account for PE origins because when single surface ICM cells were injected with lineage markers in the early blastocyst, they gave rise predominantly to PE but a minor proportion gave epiblast or even both epiblast and PE lineages (Fig. 11.4) (Perea-Gomez et al. 2007; Weber et al. 1999).

Despite this multitude of theoretical proposals, a definitive explanation of the heterogeneity of the early ICM had been lacking because all interpretations were of fixed rather than living, dynamic preparations. The first, recently published time-lapse study, starting at the early blastocyst stage, demonstrated that at least some deep ICM cells expressing Pdgfrα, a marker of the late PE, move to the ICM surface (Plusa et al. 2008). Unfortunately, this work could not follow the behaviour of all ICM cells and thus the question of genesis and behaviour of the epiblast was left open. The implication was, however, that the intermingling of epiblast and PE progenitors was the consequence of a stochastic process. There are reasons to be

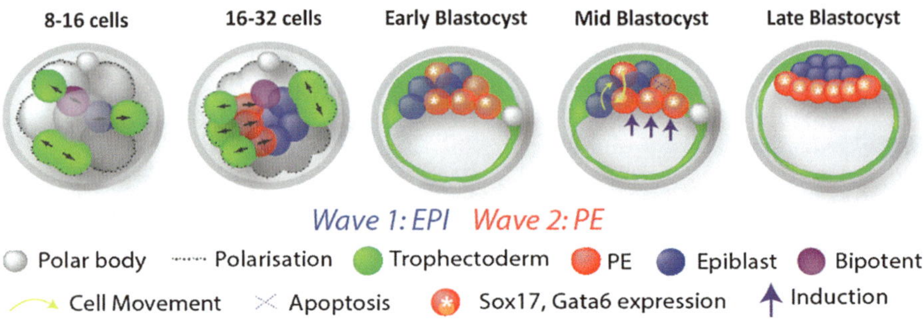

Fig. 11.5 The influence of developmental history on lineage segregation in the ICM. EPI is derived from wave 1 internalised cells, whereas PE is derived from the later, wave 2 internalisations

cautious about this interpretation. First, these particular time-lapse studies commenced when it was only possible to observe the resolution of the "salt-and-pepper" pattern and not how the pattern originated. Second, the "PE-reporter" gene used is initially expressed in a broad domain before becoming restricted to PE, making it impossible to unambiguously identify PE progenitors by this criterion. Thus, it had remained unknown how intermingling of epiblast and PE progenitors first arises and exactly how it becomes resolved into both epiblast and PE lineages.

To resolve these different views and to uncover the genesis of ICM lineages, the origin, behaviour and fate of each and every individual cell was followed as it developed into either PE or epiblast. This revealed that the developmental timing with which cells become set aside in the inside part of the embryo by successive waves of asymmetric division dictates their fate. Thus, inside cells generated by the first wave during the 8- to 16-cell transition give rise to the majority of epiblast cells. Conversely, the majority of PE cells are progeny of inside cells generated by the second wave (Morris et al. 2010). This relationship between cell fate and developmental history is most apparent when the ICM is generated from equal proportions of early and late asymmetric divisions, as is most often the case. When the ICM is derived mostly from the early first wave, these cells have the flexibility to generate both the EPI and PE (Morris et al. 2010) (Yamanaka et al. 2010). Thus, the asymmetric divisions, which play such an important part in the first cell fate decision, also have a major impact upon the second cell fate decision (Fig. 11.5).

11.4.1 Molecular Circuitry of the Second Cell Fate Decision

The transcription factors Gata4 and Gata6 contribute to development of extra-embryonic endoderm after implantation (Morrisey et al. 1998; Koutsourakis et al. 1999). Both genes are expressed at the early blastocyst stage (Wang et al. 2004) and drive PE formation when overexpressed in ES cells (Fujikura et al. 2002; Shimosato et al. 2007). However, these might not be the sole transcription factors required to specify the PE and indeed the Sox17 protein appears important for endoderm differentiation in different model systems (Kanai-Azuma et al. 2002) (Niakan et al. 2010). Sox17 levels rise dramatically when PE becomes specified at the blastocyst stage (Wang et al. 2004) beginning in PE progenitor cells of the second wave of asymmetric divisions (Morris et al. 2010). Consistently, downregulating levels of Sox17 in individual blastomeres prevents their forming PE and promotes their development to EPI. Conversely, overexpression of Sox17, enhanced by Gata6, directs cells to the PE lineage.

11.4.2 The Role of FGF Signalling in the Second Cell Fate Decision

Intact FGF signalling is essential for PE formation, where loss of the FGF receptor, or ligand, FGF-4 results in the absence of PE differentiation and peri-implantation lethality (Arman et al. 1998; Feldman et al. 1995). The exact mechanism by which FGF signals participate in ICM patterning is currently unknown, although attempts to integrate it into stochastic and origin-dependent models have been made.

In the stochastic model of ICM pattering—it is proposed that individual ICM cells randomly respond to different levels of FGF signalling, leading to the "salt-and-pepper" distribution of EPI and PE progenitors as described above. This model was developed from the finding that chemical inhibition of FGF signalling with small molecules blocks differentiation into PE and promotes the formation of an ICM consisting purely of EPI cells (Nichols et al. 2009; Yamanaka et al. 2010). Conversely, when embryos are treated with high-level exogenous FGF-4 ligand, the ICM differentiates entirely into PE (Yamanaka et al. 2010). This lead to the conclusion that cells of the early ICM have the potential to become either EPI or PE—and stochastic fluctuations in FGF signal intensity drive the switch between pluripotency and differentiation.

The finding that ICM cells possess early flexibility does not preclude the influence of developmental history on the capacity of a cell to transduce FGF signals. For example, the FGF4 receptor, FGFR2, has been found to be expressed at higher levels in outside cells of the 16-cell stage embryo (Guo et al. 2010). As these are the cells which generate the second wave divisions giving rise to PE progenitors, raising the possibility that such cells inherit FGFR2 conferring competence to form PE. In contrast to this, those early wave 1 internalised cells would have little or no FGFR2 receptor, rendering them "blind" to FGF4 ligand and fated to EPI. In support of this, in the early blastocyst, PE markers Sox17 and Gata6 expression cluster with FGFR2 expression whereas Nanog expression clusters with that of FGF4 (Kurimoto et al. 2006).

In summary, the first and second cell fate decisions both demonstrate the remarkable capacity for the early mouse embryo to maintain the flexibility to adjust to changing circumstances during its development. Even though these adaptive mechanisms exist, it does not rule out the existence of biases, in which developmental history influences cell fate. Through examining these facets and flexibilities of patterning, we may build towards a more complete understanding of patterning in the early embryo, and learn how to direct it in vitro in stem cells for clinical use in regenerative medicine.

References

Antczak M, Van Blerkom J (1997) Oocyte influences on early development: the regulatory proteins leptin and STAT3 are polarized in mouse and human oocytes and differentially distributed within the cells of the preimplantation stage embryo. Mol Hum Reprod 3:1067–1086

Arman E, Haffner-Krausz R, Chen Y, Heath JK, Lonai P (1998) Targeted disruption of fibroblast growth factor (FGF) receptor 2 suggests a role for FGF signaling in pregastrulation mammalian development. Proc Natl Acad Sci USA 95:5082–5087

Bischoff M, Parfitt DE, Zernicka-Goetz M (2008) Formation of the embryonic-abembryonic axis of the mouse blastocyst: relationships between orientation of early cleavage divisions and pattern of symmetric/asymmetric divisions. Development 135:953–962

Chazaud C, Yamanaka Y, Pawson T, Rossant J (2006) Early lineage segregation between epiblast and primitive endoderm in mouse blastocysts through the Grb2-MAPK pathway. Dev Cell 10:615–624

Chisholm JC, Houliston E (1987) Cytokeratin filament assembly in the preimplantation mouse embryo. Development 101:565–582

Dalcq AM (1957) Terminology of induction. Acta Anat (Basel) 30:242–253

Dietrich JE, Hiiragi T (2007) Stochastic patterning in the mouse pre-implantation embryo. Development 134:4219–4231

Dziadek M (1979) Cell differentiation in isolated inner cell masses of mouse blastocysts in vitro: onset of specific gene expression. J Embryol Exp Morphol 53:367–379

Feldman B, Poueymirou W, Papaioannou VE, DeChiara TM, Goldfarb M (1995) Requirement of FGF-4 for postimplantation mouse development. Science 267:246–249

Fleming TP, Sheth B, Fesenko I (2001) Cell adhesion in the preimplantation mammalian embryo and its role in trophectoderm differentiation and blastocyst morphogenesis. Front Biosci 6: D1000–1007

Fujikura J, Yamato E, Yonemura S, Hosoda K, Masui S, Nakao K, Miyazaki Ji J, Niwa H (2002) Differentiation of embryonic stem cells is induced by GATA factors. Genes Dev 16:784–789

Gardner RL (1982) Investigation of cell lineage and differentiation in the extraembryonic endoderm of the mouse embryo. J Embryol Exp Morphol 68:175–198

Gardner RL (2002) Experimental analysis of second cleavage in the mouse. Hum Reprod 17:3178–3189

Graham CF, Deussen ZA (1978) Features of cell lineage in preimplantation mouse development. J Embryol Exp Morphol 48:53–72

Guo G, Huss M, Tong GQ, Wang C, Li Sun L, Clarke ND, Robson P (2010). Resolution of cell fate decisions revealed by single-cell gene expression analysis from zygote to blastocyst. Dev Cell 18(4):675–85

Gulyas BJ (1975) A reexamination of cleavage patterns in eutherian mammalian eggs: rotation of blastomere pairs during second cleavage in the rabbit. J Exp Zool 193:235–248

Handyside AH (1978) Time of commitment of inside cells isolated from preimplantation mouse embryos. J Embryol Exp Morphol 45:37–53

Handyside AH (1980) Distribution of antibody- and lectin-binding sites on dissociated blastomeres from mouse morulae: evidence for polarization at compaction. J Embryol Exp Morphol 60:99–116

Handyside AH, Lesko JG, Tarin JJ, Winston RM, Hughes MR (1992) Birth of a normal girl after in vitro fertilization and preimplantation diagnostic testing for cystic fibrosis. N Engl J Med 327:905–909

Hillman N, Sherman MI, Graham C (1972) The effect of spatial arrangement on cell determination during mouse development. J Embryol Exp Morphol 28:263–278

Jedrusik A, Parfitt DE, Guo G, Skamagki M, Grabarek JB, Johnson MH, Robson P, Zernicka-Goetz M (2008) Role of Cdx2 and cell polarity in cell allocation and specification of trophectoderm and inner cell mass in the mouse embryo. Genes Dev 22:2692–2706

Johnson MH, Ziomek CA (1981) The foundation of two distinct cell lineages within the mouse morula. Cell 24:71–80

Kanai-Azuma M, Kanai Y, Gad JM, Tajima Y, Taya C, Kurohmaru M, Sanai Y, Yonekawa H, Yazaki K, Tam PP et al (2002) Depletion of definitive gut endoderm in Sox17-null mutant mice. Development 129:2367–2379

Koutsourakis M, Langeveld A, Patient R, Beddington R, Grosveld F (1999) The transcription factor GATA6 is essential for early extraembryonic development. Development 126:723–732

Kurimoto K, Yabuta Y, Ohinata Y, Ono Y, Uno KD, Yamada RG, Ueda HR, Saitou M (2006) An improved single-cell cDNA amplification method for efficient high-density oligonucleotide microarray analysis. Nucleic Acids Res 34:e42

Lehtonen E (1980) Changes in cell dimensions and intercellular contacts during cleavage-stage cell cycles in mouse embryonic cells. J Embryol Exp Morphol 58:231–249

Martin GR, Evans MJ (1975) Differentiation of clonal lines of teratocarcinoma cells: formation of embryoid bodies in vitro. Proc Natl Acad Sci USA 72:1441–1445

Morris SA, Teo RT, Li H, Robson P, Glover DM, Zernicka-Goetz M (2010) Origin and formation of the first two distinct cell types of the inner cell mass in the mouse embryo. Proc Natl Acad Sci USA 107:6364–6369

Morrisey EE, Tang Z, Sigrist K, Lu MM, Jiang F, Ip HS, Parmacek MS (1998) GATA6 regulates HNF4 and is required for differentiation of visceral endoderm in the mouse embryo. Genes Dev 12:3579–3590

Murray P, Edgar D (2001) The regulation of embryonic stem cell differentiation by leukaemia inhibitory factor (LIF). Differentiation 68:227–234

Niakan KK, Ji H, Maehr R, Vokes SA, Rodolfa KT, Sherwood RI, Yamaki M, Dimos JT, Chen AE, Melton DA et al (2010) Sox17 promotes differentiation in mouse embryonic stem cells by directly regulating extraembryonic gene expression and indirectly antagonizing self-renewal. Genes Dev 24:312–326

Nichols J, Silva J, Roode M, Smith A (2009) Suppression of Erk signalling promotes ground state pluripotency in the mouse embryo. Development 136:3215–3222

Nishioka N, Inoue K, Adachi K, Kiyonari H, Ota M, Ralston A, Yabuta N, Hirahara S, Stephenson RO, Ogonuki N et al (2009) The Hippo signaling pathway components Lats and Yap pattern Tead4 activity to distinguish mouse trophectoderm from inner cell mass. Dev Cell 16:398–410

Nishioka N, Yamamoto S, Kiyonari H, Sato H, Sawada A, Ota M, Nakao K, Sasaki H (2008) Tead4 is required for specification of trophectoderm in pre-implantation mouse embryos. Mech Dev 125:270–283

Ota M, Sasaki H (2008) Mammalian Tead proteins regulate cell proliferation and contact inhibition as transcriptional mediators of Hippo signaling. Development 135:4059–4069

Pan D (2007) Hippo signaling in organ size control. Genes Dev 21:886–897

Parfitt DE, Zernicka-Goetz M (2010) Epigenetic modification affecting expression of cell polarity and cell fate genes to regulate lineage specification in the early mouse embryo. Mol Biol Cell 21:2649–2660

Perea-Gomez A, Meilhac SM, Piotrowska-Nitsche K, Gray D, Collignon J, Zernicka-Goetz M (2007) Regionalization of the mouse visceral endoderm as the blastocyst transforms into the egg cylinder. BMC Dev Biol 7:96

Piotrowska-Nitsche K, Perea-Gomez A, Haraguchi S, Zernicka-Goetz M (2005) Four-cell stage mouse blastomeres have different developmental properties. Development 132:479–490

Plusa B, Frankenberg S, Chalmers A, Hadjantonakis AK, Moore CA, Papalopulu N, Papaioannou VE, Glover DM, Zernicka-Goetz M (2005) Downregulation of Par3 and aPKC function directs cells towards the ICM in the preimplantation mouse embryo. J Cell Sci 118:505–515

Plusa B, Piliszek A, Frankenberg S, Artus J, Hadjantonakis AK (2008) Distinct sequential cell behaviours direct primitive endoderm formation in the mouse blastocyst. Development 135:3081–3091

Ralston A, Rossant J (2008) Cdx2 acts downstream of cell polarization to cell-autonomously promote trophectoderm fate in the early mouse embryo. Dev Biol 313:614–629

Reeve WJ, Ziomek CA (1981) Distribution of microvilli on dissociated blastomeres from mouse embryos: evidence for surface polarization at compaction. J Embryol Exp Morphol 62:339–350

Rossant J (1975) Investigation of the determinative state of the mouse inner cell mass. II. The fate of isolated inner cell masses transferred to the oviduct. J Embryol Exp Morphol 33:991–1001

Rossant J, Chazaud C, Yamanaka Y (2003) Lineage allocation and asymmetries in the early mouse embryo. Philos Trans R Soc Lond B Biol Sci 358:1341–1348, discussion 1349

Rossant J, Lis WT (1979) Potential of isolated mouse inner cell masses to form trophectoderm derivatives in vivo. Dev Biol 70:255–261

Rossant J, Vijh KM (1980) Ability of outside cells from preimplantation mouse embryos to form inner cell mass derivatives. Dev Biol 76:475–482

Sawada A, Kiyonari H, Ukita K, Nishioka N, Imuta Y, Sasaki H (2008) Redundant roles of Tead1 and Tead2 in notochord development and the regulation of cell proliferation and survival. Mol Cell Biol 28:3177–3189

Schultz RM (2002) The molecular foundations of the maternal to zygotic transition in the preimplantation embryo. Hum Reprod Update 8:323–331

Shimosato D, Shiki M, Niwa H (2007) Extra-embryonic endoderm cells derived from ES cells induced by GATA factors acquire the character of XEN cells. BMC Dev Biol 7:80

Solter D, Knowles BB (1975) Immunosurgery of mouse blastocyst. Proc Natl Acad Sci USA 72:5099–5102

Spindle AI (1978) Trophoblast regeneration by inner cell masses isolated from cultured mouse embryos. J Exp Zool 203:483–489

Strumpf D, Mao CA, Yamanaka Y, Ralston A, Chawengsaksophak K, Beck F, Rossant J (2005) Cdx2 is required for correct cell fate specification and differentiation of trophectoderm in the mouse blastocyst. Development 132:2093–2102

Suwinska A, Czolowska R, Ozdzenski W, Tarkowski AK (2008) Blastomeres of the mouse embryo lose totipotency after the fifth cleavage division: expression of Cdx2 and Oct4 and developmental potential of inner and outer blastomeres of 16- and 32-cell embryos. Dev Biol 322:133–144

Tarkowski AK (1959) Experiments on the development of isolated blastomers of mouse eggs. Nature 184:1286–1287

Tarkowski AK, Wroblewska J (1967) Development of blastomeres of mouse eggs isolated at the 4- and 8-cell stage. J Embryol Exp Morphol 18:155–180

Thomas FC, Sheth B, Eckert JJ, Bazzoni G, Dejana E, Fleming TP (2004) Contribution of JAM-1 to epithelial differentiation and tight-junction biogenesis in the mouse preimplantation embryo. J Cell Sci 117:5599–5608

Torres-Padilla ME, Parfitt DE, Kouzarides T, Zernicka-Goetz M (2007) Histone arginine methylation regulates pluripotency in the early mouse embryo. Nature 445:214–218

Vinot S, Le T, Ohno S, Pawson T, Maro B, Louvet-Vallee S (2005) Asymmetric distribution of PAR proteins in the mouse embryo begins at the 8-cell stage during compaction. Dev Biol 282:307–319

Wang QT, Piotrowska K, Ciemerych MA, Milenkovic L, Scott MP, Davis RW, Zernicka-Goetz M (2004) A genome-wide study of gene activity reveals developmental signaling pathways in the preimplantation mouse embryo. Dev Cell 6:133–144

Watson AJ (1992) The cell biology of blastocyst development. Mol Reprod Dev 33:492–504

Weber RJ, Pedersen RA, Wianny F, Evans MJ, Zernicka-Goetz M (1999) Polarity of the mouse embryo is anticipated before implantation. Development 126:5591–5598

Wiley LM, Obasaju MF (1988) Induction of cytoplasmic polarity in heterokaryons of mouse 4-cell-stage blastomeres fused with 8-cell- and 16-cell-stage blastomeres. Dev Biol 130:276–284

Wilson IB, Bolton E, Cuttler RH (1972) Preimplantation differentiation in the mouse egg as revealed by microinjection of vital markers. J Embryol Exp Morphol 27:467–469

Yagi R, Kohn MJ, Karavanova I, Kaneko KJ, Vullhorst D, DePamphilis ML, Buonanno A (2007) Transcription factor TEAD4 specifies the trophectoderm lineage at the beginning of mammalian development. Development 134:3827–3836
Yamanaka Y, Lanner F, Rossant J (2010) FGF signal-dependent segregation of primitive endoderm and epiblast in the mouse blastocyst. Development 137:715–724
Ziomek CA, Johnson MH (1980) Cell surface interaction induces polarization of mouse 8-cell blastomeres at compaction. Cell 21:935–942

Chapter 12
Cell Movements in the Egg Cylinder Stage Mouse Embryo

Bradley Joyce and Shankar Srinivas

12.1 Introduction

Embryonic development does not simply consist of the growth of a preformed miniature foetus or "homunculus". It is a tremendously dynamic process, characterised by a great deal of cell movement and tissue rearrangement. For example, during embryogenesis the forming heart starts "ahead" of the forming brain and only comes to lie in its more familiar position with respect to the brain following extensive tissue movements. This chapter looks at the movement of the anterior visceral endoderm, a signalling centre in the early embryo that is responsible for the correct orientation of the anterior–posterior axis.

12.2 Formation of the Mouse Egg Cylinder

By about 3.5 days post-coitum (dpc), the zygote has undergone several rounds of cleavage divisions to give rise to the blastocyst, a roughly spherical structure with an asymmetrically positioned fluid-filled space, the blastocyst cavity (reviewed in Rossant and Tam 2009). The blastocyst is initially composed of two cell types: the inner cell mass (ICM) surrounded by a layer of trophectoderm (TE). The ICM subsequently differentiates to give rise to the primitive endoderm (PE) abutting the blastocyst cavity, and the epiblast, positioned between the PE and overlying TE (Plusa et al. 2008). Over the course of the next day, some cells of the PE differentiate into the parietal endoderm, which migrate away to line the blastocyst cavity. The remaining cells of the PE form the visceral endoderm (VE).

B. Joyce • S. Srinivas (✉)
Department of Physiology Anatomy and Genetics, University of Oxford, South Parks Road, Oxford OX1 3QX, UK
e-mail: shankar.srinivas@dpag.ox.ac.uk

J.Z. Kubiak (ed.), *Mouse Development*, Results and Problems in Cell Differentiation 55, DOI 10.1007/978-3-642-30406-4_12, © Springer-Verlag Berlin Heidelberg 2012

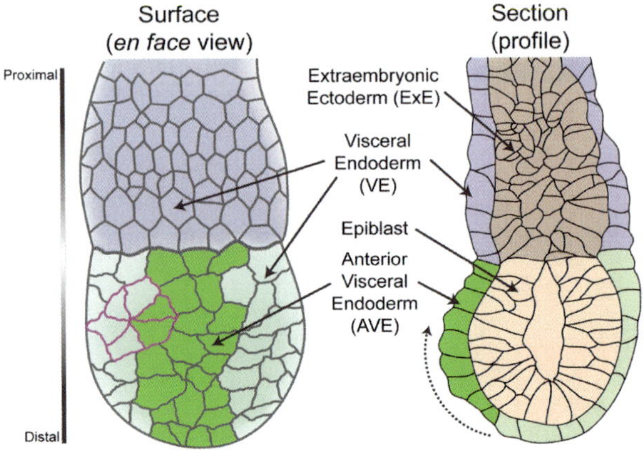

Fig. 12.1 Diagram of an egg cylinder stage mouse embryo, showing the major tissues. The AVE is marked in green. A multicellular rosette is outlined in magenta in the surface view at *left*

From about 4.0 dpc onwards, the embryo undergoes not only implantation in the uterus but also profound morphological changes. It is transformed from a roughly spherical blastocyst to a more elongated "egg cylinder" (Fig. 12.1). The formation of the egg cylinder from the late blastocyst occurs as a result of rapid proliferation of cells of the epiblast and overlying TE, causing them to grow into the blastocyst cavity. The TE derived portion of the egg cylinder is called the extra-embryonic ectoderm (ExE). The VE comes to envelope both the epiblast and ExE. The anterior visceral endoderm (AVE) is a specialised subset of cells in the VE that is responsible for the correct positioning of the primitive streak in the epiblast, thereby specifying the orientation of the anterior–posterior (A–P) axis (Arnold and Robertson 2009; Rossant and Tam 2009; Srinivas 2006).

At around 5.5 dpc, cells at the distal tip of the VE differentiate to form the distinct subpopulation of the AVE, characterised by the expression of genetic markers such as *Hex*, *Lefty1* and *Cer1* (Thomas et al. 1998; Belo et al. 1997; Yamamoto et al. 2004). The TGF-β protein Nodal is a key factor required for the specification of the AVE (Brennan et al. 2001). Though *Nodal* is expressed throughout the epiblast at this stage, induction of the AVE is restricted to the distal tip of the egg cylinder by repressive signals from the ExE (Rodriguez et al. 2005). Proliferation of epiblast cells results in the elongation of the egg cylinder, presumably moving the distal VE cells beyond the repressive influence of the ExE and allowing them to differentiate into AVE cells (Mesnard et al. 2006).

12.3 The AVE Shows Directional Movement

The importance of the AVE as a signalling centre was identified by microsurgical removal, which results in the loss of forebrain markers (Thomas and Beddington 1996). Early experiments quickly identified the AVE as a characteristic feature of the prospective anterior, although the role of the AVE in the generation of the A–P axis was not identified until later. Further analysis identified the expression of the homeobox containing transcription factor *Hex* as a marker of the AVE (Thomas et al. 1998). The expression of *Hex* indicated that the AVE first formed at the distal tip of the egg cylinder at early 5.5 dpc before becoming localised to the prospective anterior at late 5.5 dpc. At the time it was identified, the asymmetric expression of *Hex* was the earliest indicator of the orientation of the A–P axis which until this point had been defined by the formation of the primitive streak or the expression of streak markers like *Brachyury (T)* in the prospective posterior as much as a day later, at about 6.5 dpc. Lineage tracing experiments using DiI labelling of the distal VE showed that the shift in *Hex* expression from the distal tip of the egg cylinder to the prospective anterior was the result of cells moving, as opposed to changes in the pattern of expression of *Hex* (Thomas et al. 1998).

12.4 Cell Context for AVE Migration

Using a GFP transgenic reporter for *Hex* (Rodriguez et al. 2001) to mark the cells of the AVE, it was shown by time-lapse imaging of cultured embryos that AVE cells actively migrated over a 4–5 h period, often projecting cellular processes in the direction of migration. As AVE cells reach the border of the ExE, they move laterally, remaining in the VE overlying the epiblast rather than move into the VE overlying the ExE. The border with the ExE therefore represents an endpoint to the proximal migratory movement of the AVE (Srinivas et al. 2004).

From this position, the AVE induces anterior pattern in the underlying epiblast by restricting expression of posterior markers to the opposite side of the epiblast cup (Perea-Gomez et al. 2001; Arnold and Robertson 2009). In mutants such as $Nodal^{\Delta 600/LacZ}$ and $Cripto^{-/-}$, the AVE is correctly induced at the distal tip of the egg cylinder but fails to migrate, leading consequently to posterior markers in the epiblast being incorrectly localised. Such embryos show severe gastrulation defects and fail to develop further (Ding et al. 1998; Norris et al. 2002).

The VE remains a monolayer during AVE migration, suggesting that AVE cells migrate through the surrounding VE cells rather than on top of or beneath them. Recently, Trichas and collaborators (2011) have characterised in further detail the migratory behaviour of the AVE and surrounding cells of the VE. This was done using differential interference contrast (DIC) microscopy to visualise all VE cells, in combination with confocal microscopy to identify fluorescently labelled AVE cells. Time-lapse sequences were used to track and characterise cellular behaviour across the VE, of both GFP labelled AVE cells as well as surrounding non-fluorescently labelled VE cells. Parameters such as cell shape characteristics and neighbour exchanges between cells were quantified.

The resulting analysis indicated that the AVE migrates by neighbour exchange and intercalation with surrounding (non-AVE) VE cells. Furthermore this approach identified differences in migratory behaviour between cells in the regions of the VE overlying the epiblast (Epi–VE) and the ExE (ExE–VE). Epi–VE cells exhibited a great deal of movement as well as neighbour exchange events and distorted cell shapes, indicative of motile behaviour. In comparison ExE–VE cells remained relatively static and showed very little cell mixing. The barrier to AVE migration therefore appears to be a region of VE (the ExE–VE) that is non-permissive of the cell rearrangements required for AVE migration (Trichas et al. 2011).

12.5 Migration in an Epithelial Context

During the process of AVE migration the VE remains a coherent, single-layer epithelium (Trichas et al. 2011). Since AVE cells migrate to the prospective anterior by neighbour exchange within the context of an intact epithelium, they presumably undergo remodelling of apical junctional complexes.

The remodelling of cell shape required for neighbour exchange in epithelia is achieved through the actions of F-actin and non-muscle myosin (Bertet et al. 2004; Blankenship et al. 2006). The two behaviourally distinct regions of VE show dramatic differences in the localisation of F-actin and myosin IIA. In the Epi–VE, F-actin is localised to cortical rings both before and after AVE migration. In contrast in the ExE–VE, F-actin is localised in cortical rings prior to migration, but during and after migration, it is greatly enriched in the apical cell cortex which in opacity rendering appears as a shroud across this region (Fig. 12.2 and Trichas et al. 2011). Myosin IIA shows a similar dynamic localisation pattern. It is notable that mutants for Myosin IIA have severe VE defects by 6.5 dpc and die by 7.5 dpc (Conti et al. 2004).

Wnt-planar cell polarity (PCP) signalling is responsible for coordinating morphogenetic events across fields of cells, influencing cell movement by regulating actin dynamics (Schlessinger et al. 2009). Dishevelled (Dvl) is a key mediator of Wnt signalling through both canonical and PCP pathways. Dvl translocation to the cell membrane is a hallmark of PCP signalling (Axelrod et al. 1998). The two behaviourally distinct regions of the VE show dramatic differences in Dishevelled-2 (Dvl-2) localisation. Dvl-2 is membrane localised specifically in the Epi–VE, suggestive of active PCP signalling in this region that shows robust cell movement activity. In contrast Dvl-2 is excluded from the plasma membrane in the behaviourally static ExE–VE. Interestingly, in embryos, mutant for the key patterning molecule Nodal, membrane localisation of Dvl-2 is reduced and in mutants for the *Nodal* inhibitor *Lefty1*, Dvl-2 is ectopically membrane localised, establishing a role for Nodal in modulating PCP signalling in the VE and providing a mechanism for integrating cell differentiation and morphogenetic cues (Trichas et al. 2011).

12.6 AVE Migration and Multicellular Rosettes

AVE migration is accompanied by changes in the packing of cells in the VE (Trichas et al. 2012). A hexagonal arrangement of cells (where the mean number of sides or “polygon number” of cells is close to six) is considered to be the

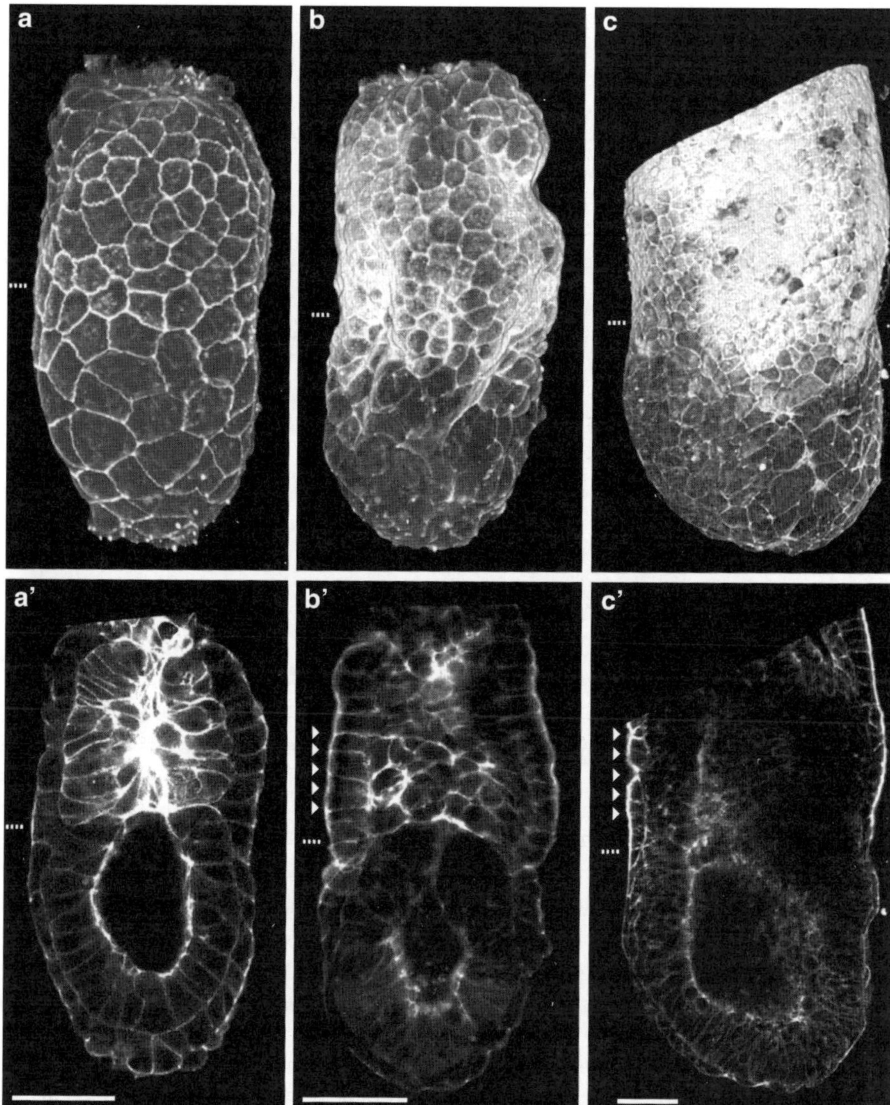

Fig. 12.2 Opacity rendering (**a**–**c**) and confocal optical sections (**a**′–**c**′) of F-actin localisation in fixed embryos at different stages of AVE migration. Prior to migration F-actin is localised to the cortical regions (**a**, **a**′). During migration and post-migration at E6.5 (**b**, **b**′ and **c**, **c**′, respectively), F-actin becomes enriched at the apical surface of the ExE-VE (*arrow heads* in **b**′ and **c**′). *Dotted lines* indicate the border of the Epi-VE and ExE-VE. *Scale bars* represent 50 μm

preferred or equilibrium packing of cells in an epithelium and deviations from this are indicative of increased disequilibrium (Gibson et al. 2006; Zallen and Zallen 2004). Before the AVE starts to migrate, the distribution of cell polygon number is comparable in the Epi–VE and ExE–VE, with a peak between five and six sides.

During AVE migration, however, mean polygon number drops and polygon distribution shifts towards three- and four-sided cells, but only in the Epi–VE. The ExE–VE in contrast does not show so marked a reduction in mean polygon number. This type of shift in polygon numbers is not seen in mutant embryos in which the AVE fails to migrate, suggesting that it is specifically linked to AVE migration. Time-lapse studies show that the change in mean polygon number in the course of AVE migration is due to dynamic changes in the packing of existing VE cells rather than, for example, new cells with fewer cell edges arising through division. These findings suggest that during AVE migration the Epi–VE is in a state of increased disequilibrium with respect to cell packing (Trichas et al. 2012).

Multicellular rosettes, a striking conformation of cells where five or more cells meet at a common central point, are also observed in the Epi–VE. In the Drosophila germ-band, rosettes have been shown to be transient intermediaries of the long-range coordinated cell movements of convergent-extension (Blankenship et al. 2006). There are no convergent-extension movements evident in the mouse VE, and rosettes appear to play a different role in this context. There is a significant increase in rosettes during AVE migration, found almost exclusively in the Epi–VE, the region of the VE to which AVE migration is restricted. Mutants with a failure of AVE migration show greatly reduced rosette numbers, indicating a specific link between rosettes and AVE migration (Trichas et al. 2012).

Mathematical models have been helpful in probing the role of rosettes during AVE migration. A model has been made where the cells of the VE are represented as a vertex model implemented on an ellipsoidal surface. Simulations of AVE migration using parameters where rosettes are allowed to form result in AVE migration that is very similar to that observed experimentally. However, simulations in which rosettes are not allowed to form show abnormally disordered AVE migration (Trichas et al. 2012). This is backed up by experiments using a mutant line in which PCP signalling is disrupted by expressing a membrane tethered C-terminal fragment of the core PCP molecule Celsr1 (*ROSA26*$^{Lyn\text{-}Celsr1}$) (Trichas et al. 2011). Genetically perturbing PCP signalling leads to a dramatic reduction in rosettes and also disrupts AVE migration. Interestingly, the mean polygon number in such mutant embryos is significantly lower than that in wild-type embryos at the equivalent stage, indicating that there is increased epithelial disequilibrium in the absence of rosettes. This suggests that while rosettes are not essential for AVE migration, they are essential for the *orderliness* of this migration and that they might somehow buffer the epithelial disequilibrium that accompanies AVE migration, allowing AVE cells to migrate as a coherent group (Trichas et al. 2012).

12.7 Formation of Cellular Projections

Migrating AVE cells show cellular projections, typically orientated in the direction of AVE migration (Srinivas et al. 2004). These projections often extend greater than one cell radius in length (Srinivas et al. 2004). Time-lapse data collected by Migeotte and colleagues (2010) show that these long projections form along the basal surface of the VE and are often "tipped with filopodia-like structures that appear to sample the environment" (Migeotte et al. 2010). These projections have so far only been observed in living embryos in which the outline of AVE cells can be discerned by the expression of GFP. They are not observed in the AVE of fixed embryos, possibly because they are delicate structures that do not fix well. Even in live embryos, such cellular projections have so far not been identified in non-AVE cells of the VE imaged by DIC optics. It is not clear if this is because these projections are a feature unique to AVE cells or because of limitations of imaging cellular characteristics using non-fluorescence microscopy.

The formation of cellular projections is dependent upon the small GTPase Rac1, which modulates cytoskeletal dynamics in response to PCP signals. *Rac1* mutants do not exhibit the ability to remodel the actin cytoskeleton in order to form lamellipodia or change shape and exchange neighbours. As a result the AVE of Rac1 null embryos fails to migrate (Migeotte et al. 2010). Similarly, approximately 50% of mutants for Nap1, a component of the WAVE complex that regulates Actin branching, show a failure of AVE migration (Rakeman and Anderson 2006).

12.8 Molecular Basis for AVE Migration

Mutants with reduced levels of Nodal activity show a failure of the AVE to migrate, indicating that Nodal is required for AVE migration (Norris et al. 2002). However, it is not completely understood how Nodal exerts a control on the migratory behaviour of the AVE. Nodal antagonists *Lefty1* and *Cer1* are both expressed in the AVE and inhibit Nodal signalling by either competitively binding the Nodal receptor as in the case of Lefty1 (Sakuma et al. 2002), or by binding Nodal itself as in the case of Cer1 (Piccolo et al. 1999). Both *Lefty1* and *Cer1* are expressed at the distal tip of the egg cylinder at the time of AVE formation. The expression domain of both is slightly tilted towards that prospective anterior even prior to AVE migration (Yamamoto et al. 2004). This asymmetric localisation of *Lefty1* and *Cer1* expression would be expected to create a bias in the activity of Nodal signalling, with higher levels in the posterior VE. One possibility is that this increased Nodal activity in the posterior leads to increased levels of proliferation in the VE in the prospective posterior relative to the anterior, causing directional movement of AVE cells (Yamamoto et al. 2004). However, more recent work has shown that there is no difference in levels of cell proliferation in the anterior and posterior VE, but that Nodal-dependent proliferation of epiblast cells is essential for AVE migration (Stuckey et al. 2011).

The asymmetric expression of *Lefty1* in the VE has been traced back to asymmetric expression of *Lefty1* in the primitive endoderm of the pre-implantation blastocyst, providing an intriguing mechanism by which the direction of AVE migration might already be specified at least as early as the pre-implantation blastocyst (Takaoka et al. 2011, 2006).

Experiments by Yamamoto and collaborators (2004) showed that the AVE migrates towards ectopically expressed Nodal antagonists, indicating that they play an important role in the regulation of AVE migration. However, genetic knock-outs of *Lefty1* (Meno et al. 1998) and *Cer1* (Belo et al. 2000; Simpson et al. 1999; Stanley et al. 2000) exhibit seemingly normal AVE migration, possibly indicating that these genes exhibit some functional redundancy. When both *Lefty1* and *Cer1* are removed, embryos show three broad classes of phenotypes (Perea-Gomez et al. 2002). In roughly 30% of double mutants ("Class I") the streak appears to be correctly positioned (though it shows other defects), suggesting that in these embryos the AVE migrated to the prospective anterior and that Nodal antagonists are not required absolutely for AVE migration. On the other hand, the majority of double mutants show very severe defects in the positioning of the streak and importantly, in the AVE itself, which is seen to form an unusual accumulation at the prospective anterior, which argues for a major role for these two antagonists in AVE migration, even if they are not absolutely required. It is not clear what the cause of this variability in phenotype is, but it would appear that there are multiple levels of control over AVE migration. In addition, recent observations have shown that the majority of *Lefty1* null embryos exhibit a subtle abnormality of AVE migration with cells moving onto the ExE–VE, pointing to a role for *lefty1* and Nodal signalling in maintaining the differences in behaviour between the Epi–VE and ExE–VE that normally results in a "barrier" to migration (Trichas et al. 2011).

The secreted Wnt antagonist Dickkopf1 (Dkk1) causes reduced Wnt activity by binding the Wnt receptors LRP5/6. This consequently allows phosphorylation and subsequent degradation of β-Catenin, thus preventing canonical Wnt signalling (Bafico et al. 2001). *Dkk1* has been shown to be sufficient to rescue the AVE arrest phenotype of *Otx2* mutants and act as a guidance cue for AVE migration in bead experiments (Kimura-Yoshida et al. 2005). However, despite acting as a guidance cue when provided exogenously, the in vivo role of Dkk1 remains unclear given that *Dkk1* null embryos have correctly positioned primitive streaks (though they show a later anterior neuroectoderm defect), indicating that the AVE is able to migrate in such embryos (Mukhopadhyay et al. 2001).

Given that Dkk1, in addition to its well-established role as an inhibitor of canonical signalling, can also act as an agonist of Wnt-PCP signalling (Caneparo et al. 2007; van Amerongen and Nusse 2009), and PCP signalling is important in regulating the orderliness of AVE migration by facilitating rosette formation (Trichas et al. 2011; Trichas et al. 2012), it is tempting to speculate that in the context of the VE, Dkk1 might be acting not directly as a guidance cue for AVE migration, but more subtly, to modulate AVE migration so that it occurs in an ordered manner. If this were the case, one might find that in *Dkk1* null embryos, the AVE migrates sufficiently for it to fulfil its function of positioning the primitive streak, but does so abnormally, as in *Lefty1* or $ROSA26^{Lyn\text{-}Celsr1}$ mutants.

12.9 Conclusion

The VE is a uniquely accessible epithelium, located on the outside of the developing embryo, which makes it relatively easy to image at high resolution. In addition the peri-implantation embryo lends itself to live imaging both due to its size and because the embryos can be cultured to allow the imaging of normal developmental processes. Study of the VE and in particular the migratory behaviour of the AVE provides useful insights not only into the process of anterior patterning, but also regulation of cell migration in other epithelial tissues. The AVE therefore makes a good model to study the process of epithelial cell movements in mammals.

The past decade has seen many advances in our understanding of AVE migration, but much of it understandably has focused on the AVE itself. It is becoming increasingly clear that to fully understand the movement of AVE cells, we need to also understand the behaviour of the surrounding VE cells. Future research will hopefully provide insights into how AVE migration is coordinated with the movement of surrounding cells so that they can reproducibly negotiate their way through the epithelial VE to the future anterior.

References

Arnold SJ, Robertson EJ (2009) Making a commitment: cell lineage allocation and axis patterning in the early mouse embryo. Nat Rev Mol Cell Biol 10(2):91–103

Axelrod JD, Miller JR, Shulman JM, Moon RT, Perrimon N (1998) Differential recruitment of dishevelled provides signaling specificity in the planar cell polarity and Wingless signaling pathways. Genes Dev 12(16):2610–2622

Bafico A, Liu G, Yaniv A, Gazit A, Aaronson SA (2001) Novel mechanism of Wnt signalling inhibition mediated by Dickkopf-1 interaction with LRP6/Arrow. Nat Cell Biol 3(7):683–686

Belo JA, Bachiller D, Agius E, Kemp C, Borges AC, Marques S, Piccolo S, De Robertis EM (2000) Cerberus-like is a secreted BMP and nodal antagonist not essential for mouse development. Genesis 26(4):265–270

Belo JA, Bouwmeester T, Leyns L, Kertesz N, Gallo M, Follettie M, De Robertis EM (1997) Cerberus-like is a secreted factor with neutralizing activity expressed in the anterior primitive endoderm of the mouse gastrula. Mech Dev 68(1–2):45–57

Bertet C, Sulak L, Lecuit T (2004) Myosin-dependent junction remodelling controls planar cell intercalation and axis elongation. Nature 429(6992):667–671

Blankenship JT, Backovic ST, Sanny JS, Weitz O, Zallen JA (2006) Multicellular rosette formation links planar cell polarity to tissue morphogenesis. Dev Cell 11(4):459–470

Brennan J, Lu CC, Norris DP, Rodriguez TA, Beddington RS, Robertson EJ (2001) Nodal signalling in the epiblast patterns the early mouse embryo. Nature 411(6840):965–969

Caneparo L, Huang YL, Staudt N, Tada M, Ahrendt R, Kazanskaya O, Niehrs C, Houart C (2007) Dickkopf-1 regulates gastrulation movements by coordinated modulation of Wnt/beta catenin and Wnt/PCP activities, through interaction with the Dally-like homolog Knypek. Genes Dev 21(4):465–480

Conti MA, Even-Ram S, Liu C, Yamada KM, Adelstein RS (2004) Defects in cell adhesion and the visceral endoderm following ablation of nonmuscle myosin heavy chain II-A in mice. J Biol Chem 279(40):41263–41266

Ding J, Yang L, Yan YT, Chen A, Desai N, Wynshaw-Boris A, Shen MM (1998) Cripto is required for correct orientation of the anterior-posterior axis in the mouse embryo. Nature 395(6703): 702–707

Gibson MC, Patel AB, Nagpal R, Perrimon N (2006) The emergence of geometric order in proliferating metazoan epithelia. Nature 442(7106):1038–1041

Kimura-Yoshida C, Nakano H, Okamura D, Nakao K, Yonemura S, Belo JA, Aizawa S, Matsui Y, Matsuo I (2005) Canonical Wnt signaling and its antagonist regulate anterior-posterior axis polarization by guiding cell migration in mouse visceral endoderm. Dev Cell 9(5):639–650

Meno C, Shimono A, Saijoh Y, Yashiro K, Mochida K, Ohishi S, Noji S, Kondoh H, Hamada H (1998) lefty-1 is required for left-right determination as a regulator of lefty-2 and nodal. Cell 94 (3):287–297

Mesnard D, Guzman-Ayala M, Constam DB (2006) Nodal specifies embryonic visceral endoderm and sustains pluripotent cells in the epiblast before overt axial patterning. Development 133 (13):2497–2505

Migeotte I, Omelchenko T, Hall A, Anderson KV (2010) Rac1-dependent collective cell migration is required for specification of the anterior-posterior body axis of the mouse. PLoS Biol 8(8): e1000442

Mukhopadhyay M, Shtrom S, Rodriguez-Esteban C, Chen L, Tsukui T, Gomer L, Dorward DW, Glinka A, Grinberg A, Huang SP, Niehrs C, Izpisua Belmonte JC, Westphal H (2001) Dickkopf1 is required for embryonic head induction and limb morphogenesis in the mouse. Dev Cell 1(3):423–434

Norris DP, Brennan J, Bikoff EK, Robertson EJ (2002) The Foxh1-dependent autoregulatory enhancer controls the level of Nodal signals in the mouse embryo. Development 129 (14):3455–3468

Perea-Gomez A, Rhinn M, Ang SL (2001) Role of the anterior visceral endoderm in restricting posterior signals in the mouse embryo. Int J Dev Biol 45(1 Spec No):311–320

Perea-Gomez A, Vella FD, Shawlot W, Oulad-Abdelghani M, Chazaud C, Meno C, Pfister V, Chen L, Robertson E, Hamada H, Behringer RR, Ang SL (2002) Nodal antagonists in the anterior visceral endoderm prevent the formation of multiple primitive streaks. Dev Cell 3(5):745–756

Piccolo S, Agius E, Leyns L, Bhattacharyya S, Grunz H, Bouwmeester T, De Robertis EM (1999) The head inducer Cerberus is a multifunctional antagonist of Nodal, BMP and Wnt signals. Nature 397(6721):707–710

Plusa B, Piliszek A, Frankenberg S, Artus J, Hadjantonakis AK (2008) Distinct sequential cell behaviours direct primitive endoderm formation in the mouse blastocyst. Development 135 (18):3081–3091

Rakeman AS, Anderson KV (2006) Axis specification and morphogenesis in the mouse embryo require Nap1, a regulator of WAVE-mediated actin branching. Development 133(16):3075–3083

Rodriguez TA, Casey ES, Harland RM, Smith JC, Beddington RS (2001) Distinct enhancer elements control Hex expression during gastrulation and early organogenesis. Dev Biol 234 (2):304–316

Rodriguez TA, Srinivas S, Clements MP, Smith JC, Beddington RS (2005) Induction and migration of the anterior visceral endoderm is regulated by the extra-embryonic ectoderm. Development 132(11):2513–2520

Rossant J, Tam PP (2009) Blastocyst lineage formation, early embryonic asymmetries and axis patterning in the mouse. Development 136(5):701–713

Sakuma R, Ohnishi Yi Y, Meno C, Fujii H, Juan H, Takeuchi J, Ogura T, Li E, Miyazono K, Hamada H (2002) Inhibition of Nodal signalling by Lefty mediated through interaction with common receptors and efficient diffusion. Genes Cells 7(4):401–412

Schlessinger K, Hall A, Tolwinski N (2009) Wnt signaling pathways meet Rho GTPases. Genes Dev 23(3):265–277

Simpson EH, Johnson DK, Hunsicker P, Suffolk R, Jordan SA, Jackson IJ (1999) The mouse Cer1 (Cerberus related or homologue) gene is not required for anterior pattern formation. Dev Biol 213(1):202–206

Srinivas S (2006) The anterior visceral endoderm-turning heads. Genesis 44(11):565–572

Srinivas S, Rodriguez T, Clements M, Smith JC, Beddington RS (2004) Active cell migration drives the unilateral movements of the anterior visceral endoderm. Development 131 (5):1157–1164

Stanley EG, Biben C, Allison J, Hartley L, Wicks IP, Campbell IK, McKinley M, Barnett L, Koentgen F, Robb L, Harvey RP (2000) Targeted insertion of a lacZ reporter gene into the mouse Cer1 locus reveals complex and dynamic expression during embryogenesis. Genesis 26 (4):259–264

Stuckey DW, Clements M, Di-Gregorio A, Senner CE, Le Tissier P, Srinivas S, Rodriguez TA (2011) Coordination of cell proliferation and anterior–posterior axis establishment in the mouse embryo. Development 138(8):1521–1530

Takaoka K, Yamamoto M, Hamada H (2011) Origin and role of distal visceral endoderm, a group of cells that determines anterior-posterior polarity of the mouse embryo. Nat Cell Biol 13(7):743–752

Takaoka K, Yamamoto M, Shiratori H, Meno C, Rossant J, Saijoh Y, Hamada H (2006) The mouse embryo autonomously acquires anterior–posterior polarity at implantation. Dev Cell 10(4):451–459

Thomas P, Beddington R (1996) Anterior primitive endoderm may be responsible for patterning the anterior neural plate in the mouse embryo. Curr Biol 6(11):1487–1496

Thomas PQ, Brown A, Beddington RS (1998) Hex: a homeobox gene revealing peri-implantation asymmetry in the mouse embryo and an early transient marker of endothelial cell precursors. Development 125(1):85–94

Trichas G, Joyce B, Crompton LA, Wilkins V, Clements M, Tada M, Rodriguez TA, Srinivas S (2011) Nodal dependent differential localisation of dishevelled-2 demarcates regions of differing cell behaviour in the visceral endoderm. PLoS Biol 9(2):e1001019

Trichas G, Smith A, White N, Wilkins V, Watanabe T, Moore A, Joyce B, Sugnaseelan J, Rodriguez TA, Kay D, Baker RE, Maini PK, Srinivas S (2012) Multi-cellular rosettes in the mouse visceral endoderm facilitate the ordered migration of AVE cells. PLoS Biol 10(2): e1001256

van Amerongen R, Nusse R (2009) Towards an integrated view of Wnt signaling in development. Development 136(19):3205–3214

Yamamoto M, Saijoh Y, Perea-Gomez A, Shawlot W, Behringer RR, Ang SL, Hamada H, Meno C (2004) Nodal antagonists regulate formation of the anteroposterior axis of the mouse embryo. Nature 428(6981):387–392

Zallen JA, Zallen R (2004) Cell-pattern disordering during convergent extension in Drosophila. J Phys Condens Matter 16(44):S5073

Chapter 13
Balancing the Dose in the Mouse

Mary E. Donohoe

Abstract Organisms that use a chromosomal basis of sex determination have a problem of gene inequality. In the mouse, this dimorphism is evident by the presence of two X-chromosomes in females, while males have a single X- and a single Y-chromosome. To balance this disparity, one of the two female X-chromosomes is transcriptionally silenced to neutralize the gene dose with the XY male. Dosage compensation in mammals is known as X-chromosome inactivation (XCI) and is a crucial early developmental process. XCI is an example of epigenetics: a phenotype resulting in changes on a chromosome without a change in nucleic acid sequence. Studies in mouse embryology and genetics have answered many questions about the process of balancing the dose. In this chapter, I highlight how the mouse dosage compensates the gene disparity between XX females and XY males in a crucial epigenetic process called X-chromosome inactivation (XCI).

13.1 The Lyon Hypothesis

In 1949 Barr and Beltram reported a dark, condensed peri-nucleolar structure present in the nuclei of female cat neurons (Barr and Bertram 1949). Later they described that other female mammalian somatic cells including mouse and man have this cytological entity now known as the "Barr body" found in somatic cells. Ten years later in 1959, Susumo Ohno suggested that the Barr body is a condensed or heteropyknotic X-chromosome (Ohno et al. 1959). That same year, Russell described that XO female mice having only one X-chromosome (instead of the normal two X-chromosomes) lack the Barr body, develop normally, and are fertile

M.E. Donohoe (✉)
The Department of Neuroscience and the Department of Cell and Developmental Biology, Burke Medical Research Institute, Weill Cornell Medical College, 785 Mamaroneck Avenue, White Plains, NY 10605, USA
e-mail: Med2008@med.cornell.edu

J.Z. Kubiak (ed.), *Mouse Development*, Results and Problems in Cell Differentiation 55,
DOI 10.1007/978-3-642-30406-4_13, © Springer-Verlag Berlin Heidelberg 2012

(Russell et al. 1959). But it was not until 1961 that a geneticist, Mary Lyon, studying X-linked coat, color variegation in the *Tabby* and *Mottled* mice proposed an explanation for the above findings. In a landmark paper, Lyon noted that XY males had a solid color, whereas heterozygous XX females had a mottled or variegated color due to gene expression differences between male and females (Lyon 1961). (Similarly, female tortoiseshell and calico cats show visible evidence of XCI with the variegation of the X-linked orange and red colors analogous to those described in the *Tabby* and *Mottled* mice.) She deduced that the condensed Barr body is the genetically inactivated X chromosome originating from either maternal or paternal origin. In addition, Lyon hypothesized that mammalian females balance X-linked gene disparity or dose with XY males by silencing one of the two X chromosomes. She posited that the inheritance of this inactive X-chromosome state is clonally propagated over cell divisions. Now over 50 years later we celebrate the golden anniversary of the discovery of XCI honoring Mary Lyon's hypothesis, "Gene action in the X-chromosome of the mouse (*Mus musculus* L)" where one of the two female X chromosomes is randomly silenced in the soma to balance the gene dosage with XY males (Lyon 1961; Lee 2011; Morey and Avner 2011). In 1961 the idea of random XCI was not widely accepted but the Lyon Hypothesis explained the non-Mendelian inheritance of many human X-linked diseases in female carriers (such as hemophilia, red–green color blindness, etc.) (Beutler et al. 1962; Linder and Gartler 1965; Gartler et al. 1972; Puck and Willard 1998). Today we know that the Lyon Hypothesis describing the balance of gene dosage on the X-chromosome is not a theory, but a fact that spawned a new field: XCI.

13.2 Two Forms of XCI: Random Versus Imprinted

There are two forms of XCI in the mouse: random and imprinted (Fig. 13.1). Both forms of XCI are essential for early development and segregate with lineage allocation in the mouse. Random and imprinted XCI offer a paradigm to studies of epigenetics in the early embryo as the XCI cycle involves rounds of inactivation and reactivation (Payer and Lee 2008). The random form occurs in all somatic cells where either of the two female X-chromosomes (maternal or paternal) may be chosen for silencing and occurs at the epiblast stage. In the extraembryonic tissues of the mouse embryo, XCI occurs selectively on the paternal X-chromosome (X^P) (Sharman 1971; Takagi and Sasaki 1975). This preferential silencing of the X^P is the imprinted form of XCI. Imprinted XCI was the first description of mammalian genomic imprinting (Sharman 1971). Genomic imprinting refers to a parent-of-origin gene expression and many examples of autosomal imprinting in the mouse have been described (McGrath and Solter 1984; Bartolomei and Ferguson-Smith 2011; Surani et al. 1984). The silencing of the X^P occurs in the male germline and shares mechanistic features with autosomal imprinting including setting and

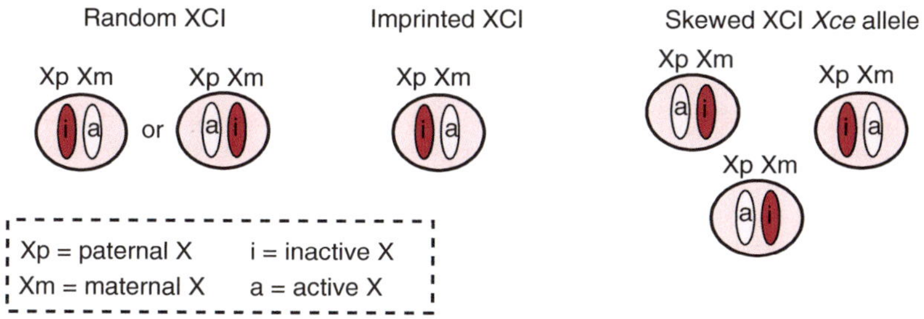

Fig. 13.1 *Two forms of X-chromosome inactivation are present in the mouse*. In the random form of XCI there is an equal chance for either the paternal or maternally inherited X-chromosome to be chosen for inactivation. In the imprinted form of XCI the paternal X is exclusively silenced. Skewing of random XCI may result by the *X-controlling element* (*Xce*) allele on particular strains of mice

erasing imprinted marks via DNA repetitive elements and methylation (Bartolomei and Ferguson-Smith 2011; Payer and Lee 2008).

13.3 XCI: Not Just for Females

XCI is not just a female-specific process. In the mouse a third form of dosage compensation occurs in the male germline and is called meiotic sex chromosome inactivation (MSCI) (Turner 2007). MSCI occurs during the first meiotic prophase of spermatogenesis when homologous pairing or synapsis takes place (Greaves et al. 2006; Lifschytz and Lindsley 1972; Turner et al. 2006). Homologous chromosomes pair during meiosis. However, the dimorphic X and Y sex chromosomes can only pair with each other at their homologous pseudoautosomal regions (PARs) situated at the tips, while the nonhomologous X- and Y-specific regions remain unpaired. The double-strand break/DNA-repair machinery initially recognizes unpaired regions. This results in the recruitment of multiple repressive marks so that the sex chromosomes form a distinct structure during the pachytene stage called the "XY" or "sex body" (Fig. 13.2). The "XY body" is heterochromatic and transcriptionally silent. The formation of the "XY body" is thought to suppress recombination across nonhomologous chromatin. The evolution of MSCI may have arisen from an ancient mechanism seen in lower organisms in which unpaired chromatin or DNA is silenced (Turner 2007). The post-meiotic sex chromatin remains transcriptionally silenced throughout spermatogenesis (Namekawa et al. 2006). As a result this partially inactivated paternal X-chromosome (X^P) is then passed onto the next generation via the mature sperm (Cooper 1971; Huynh and Lee 2003).

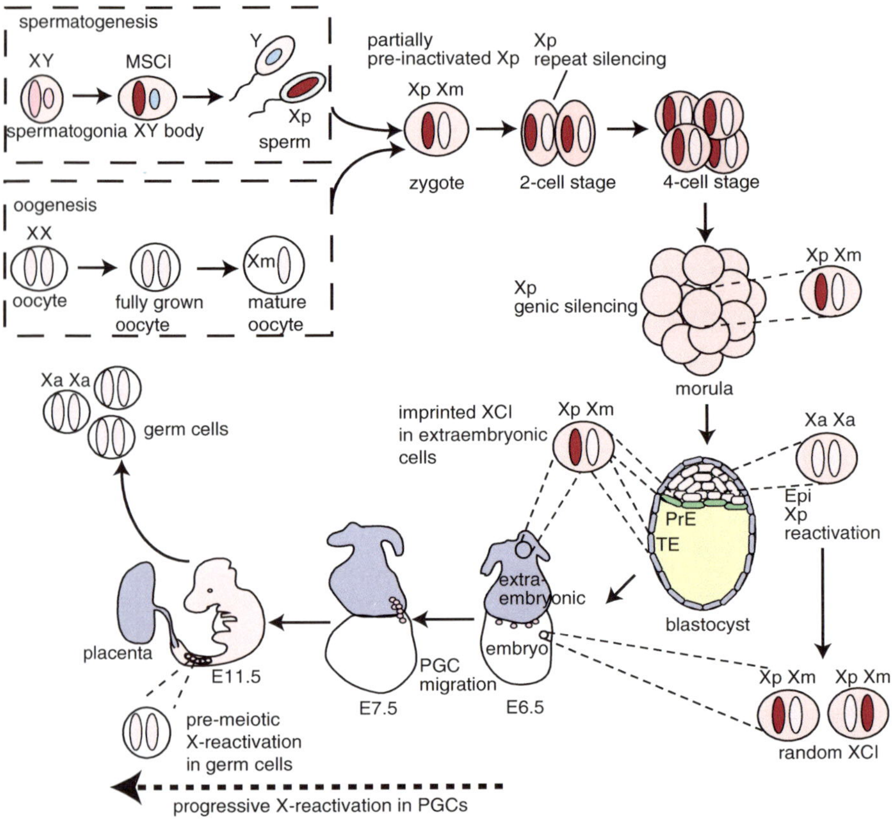

Fig. 13.2 *The cycle of X-chromosome inactivation (XCI) in mouse development*. During mouse spermatogenesis the paternal X (X^P) is partially inactivated as a result of meiotic sex chromosome inactivation (MSCI). During oogenesis the maternal X-chromosome (X^M) resists silencing. Following fertilization the female zygote has a partially pre-inactivated XP that subsequently undergoes DNA repeat element silencing followed by genic silencing. By the blastocyst stage the X^P is completely inactivated in the trophectoderm (TE) and primitive endoderm (PrE). In contrast, the epiblast (Epi) undergoes an erasure or reactivation of the X^P. Random XCI (inactivation of either the X^P or X^M) occurs shortly afterwards. As a result, the extraembryonic tissues exhibit imprinted XCI while the embryo proper shows random XCI. The Epi also gives rise to the primordial germ cells (PGCs). The PGCs undergo an X-reactivation to ensure equal meiotic segregation of the X^M and X^P

13.4 XCI and Early Lineage Allocation

At fertilization, the X^P arrives partially pre-inactivated due to MSCI. The gamete-to-embryo inactivation of the X^P occurs in two sequential steps (Namekawa et al. 2010). At the time of zygotic gene activation during the two-cell stage (Zheng and Liu 2012), the repetitive elements on the X^P are silenced (Huynh and Lee 2003; Namekawa et al. 2010). Following the two-cell stage, there is a gradual silencing of the coding genes on the X^P (Namekawa et al. 2010; Okamoto et al. 2004). The imprinted formed of XCI is completed by the blastocyst stage (Okamoto et al. 2004). The extraembryonic tissues, the trophectoderm (TE) and the primitive ectoderm (PrE), have their X^P transcriptionally silenced (Fig. 13.2). In contrast, the cells in the epiblast (Epi) undergo a remarkable reactivation of the silenced X^P (Mak et al. 2004; Martin et al. 1978; McMahon et al. 1983; Monk and Harper 1979; Rastan and Robertson 1985). Between embryonic day (E) 3.5 and E4.5, the inactive X^P is exclusively reactivated in the Epi but not in the fated extraembryonic cells of the TE and PrE. Shortly afterwards, one of the two X-chromosomes (either the maternal or paternal) is randomly chosen for silencing in the epiblast. At E6.5, the mouse extraembryonic tissues have the imprinted and the embryo proper have the random form of XCI. Following the primordial germ cells (PGCs) migration to the genital ridge (E7.0–E10.5), the PGCs undergo a progressive X-chromosome reactivation (Chuva de Sousa Lopes et al. 2008; de Napoles et al. 2007; Gartler et al. 1972; McLaren and Monk 1981; McMahon et al. 1981; Monk and McLaren 1981; Sugimoto and Abe 2007). This results in the germ cells having two active Xs prior to gametogenesis. The cycle is then continued with MSCI in the male germline and a maternal imprint that silences the *X-inactive specific transcript*, *Xist*, on the maternal X-chromosome (Tada et al. 2000).

The early mouse lineages have their own progenitor cells that are derived from and represent each of the lineages present in the blastocyst. These stem cells also reflect the form of XCI present in the lineages. The derivation and propagation of the lineage progenitor cells offer a powerful in vitro tool to study the regulation of XCI. Embryonic stem (ES) cells are derived from and represent the epiblast (Evans and Kaufman 1981). ES cells exhibit the random form of XCI (Martin et al. 1978; Rastan and Robertson 1985). Mouse ES cells faithfully recapitulate XCI as the embryo itself and are a valuable reagent to study this process. Undifferentiated ES cells are pre-XCI with both female X-chromosomes active. Upon differentiation, a random choice is made for one of the two female Xs to be silenced. This silencing is maintained over cell divisions. Trophoblast stem (TS) cells are derived from and represent the trophectoderm (Takagi and Sasaki 1975; Tanaka et al. 1998). Extra-embryonic endoderm (XEN) stem cells are derived from and represent the primitive endoderm. Both TS and XEN cells maintain the imprinted form of XCI (Kunath et al. 2005). Taken together the mouse conceptus and their lineage progenitor cells allow one to investigate the mechanisms of XCI.

13.5 Defining the Mouse X-Chromosome Inactivation Center: Counting, Choosing, and Silencing

Following Lyon's paper on XCI, research focused on cells containing more than X-chromosome per set of autosomes. It was discovered in diploid cells that all but one of the Xs was silenced as evident by the presence of the Barr body.

This suggested that each cell could "count" the number of X-chromosomes and inactivate Xs ($n - 1$) per autosome set (Gartler and Riggs 1983). The cell determines its X/autosome (X/A) ratio to maintain one active X-chromosome. Thus, cells harboring more than one X would silence all but one X-chromosome, whereas a normal XY male or an XO female shows no XCI. The XXX cells exhibit two condensed Xs or Barr bodies while XXY cells harbor one inactivated X. There is also a mutually exclusive "choice" of active versus inactive X-chromosome in somatic cells with each X having a 50:50 ratio or random chance of silencing. Nonrandom or skewing can be noted and arise from factors or genomic regions implicated in the XCI process (Payer and Lee 2008). In mice, primary XCI skewing is genetically defined by the *X-controlling element* (*Xce*) as defined by Bruce Cattanach (Cattanach and Williams 1972). They showed that certain genetic background strains might be more likely to resist XCI than others based on the *Xce* allele. For example, in the crossing of 129 and *M. castaneus* mice, XCI favors the 129 X because it carries a relatively weak *Xce* modifier. The precise genomic localization of *Xce* remains elusive and studies are actively in progress to determine how an X chooses to be inactivated.

XCI is established very early in mouse development and is initiated by a unique X-linked locus called the *X-inactivation center* (*Xic*) (Russell 1963). The *Xic* initiates the counting, choice, and cis-silencing of the 155-megabases along the X-chromosome (Payer and Lee 2008). Mouse transgenics defined this region to 450-kilobases (kb) by showing that this region of the X-chromosome can recapitulate the steps of XCI: counting, choice, and silencing (Heard et al. 1996; Lee et al. 1996). Smaller *Xic* transgenes of 100-kb are capable of XCI ectopically (Lee et al. 1999b). Thus the XCI machinery resides within a defined region on the X-chromosome within the *Xic*.

13.6 Noncoding RNAs and XCI

The molecular mechanisms for XCI have led to the discovery of numerous long, noncoding RNAs situated within the mouse *Xic* (Fig. 13.3). The search for an *XIC* candidate gene led to the discovery of a gene *XIST*, *X-inactive specific transcripts*, by Brown et al. in 1991 that was exclusively expressed from the human inactive X-chromosome (Xi) (Brown 1991). That same year the mouse *Xist* gene was described (Borsani et al. 1991; Brockdorff et al. 1991). Both the mouse and human *Xist* homologs are poorly conserved at the nucleotide level but show similar genomic organization including a long (15–17 kb), noncoding RNA whose expression is restricted to the nucleus (Brockdorff et al. 1992; Brown et al. 1992). By using fluorescence in situ hybridization (FISH) techniques, the dynamic expression of *Xist* expression can be visualized in the developing female mouse embryo and differentiating female ES cells. *Xist* RNA "paints" the entire inactive X chromosome (Barr body) (Clemson et al. 1996; Panning et al. 1997). *Xist* is essential for random and imprinted XCI (Lee et al. 1996; Marahrens et al. 1997; Penny et al. 1996;

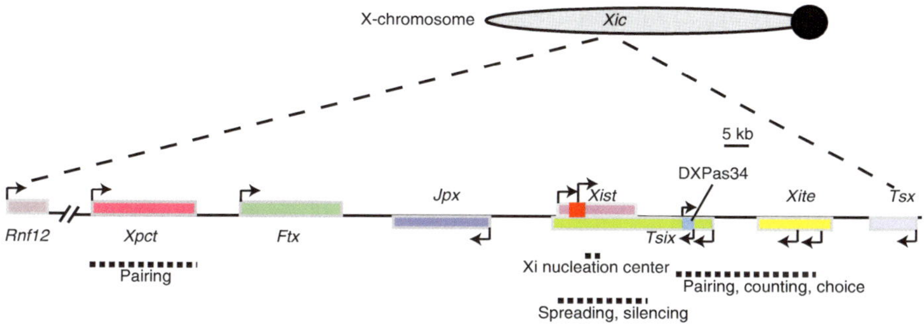

Fig. 13.3 *The mouse X-inactivation center (Xic) and surrounding regions*. The *Xic* has been defined as the minimal region necessary and sufficient to drive XCI. Noncoding RNAs are abundant at the *Xic*. A *dark dashed line* denotes the genomic regions implicated in the steps of XCI (pairing, etc.). *Rnf12* encodes a protein that is an activator of *Xist*. *Xpct* (*Slc16a2*) is an X-linked PEST-containing transporter. *Ftx* (*5′ to Xist*) is a noncoding RNA. *Jpx* (*Enox:expressed neighbor of Xist*) is an activator of *Xist*. *Xist* (*X-inactivation specific transcript*) is upregulated during cellular differentiation and its RNA coats the entire Xi. The *red box* within *Xist* is Repeat A. *Tsix* is a long, noncoding RNA antisense to *Xist* and controls its expression. The *DXPas34* repeat region is the imprint control region within *Tsix*. *Xite* harbors an enhancer for *Tsix*. *Tsx is a* testis-specific RNA and protein. The *shaded circle* depicts the centrosome

Wutz and Jaenisch 2000). However, once XCI is established, the inactive state of the X is no longer dependent on the expression of Xist (Csankovszki et al. 1999). The Xi has a battery of silencing factors such as Polycomb repressive complex 2 (PRC2) factors and trimethylated histone 3 lysine 27 marks associated with facultative heterochromatin (Okamoto et al. 2004; Plath et al. 2003, 2004; Silva et al. 2003). *Xist* has other associated proteins that localize to the Xi during and the initiation of XCI: The other polycomb group protein complex (PRC1) (de Napoles et al. 2004), the structural maintenance of chromosomes hinge domain-containing 1 protein (SmcHD1) (Blewitt et al. 2008), a histone variant macro-H2A1 (Costanzi and Pehrson 1998), a trithorax group protein Ash2I (Pullirsch et al. 2010), and a nuclear matrix heterogeneous nuclear ribonucleoprotein U (hnRNP U) (Hasegawa et al. 2010; Helbig and Fackelmayer 2003; Pullirsch et al. 2010). *Xist* RNA directly binds PRC2 via a 5′ repeat region A to target this complex to the *Xic* before spreading (Zhao et al. 2008). Although *Xist* can be spliced and polyadenylated, it is not exported to the cytoplasm but remains in the nucleus by binding hnRNP protein (Brockdorff et al. 1992; Brown et al. 1992; Hasegawa et al. 2010). *Xist* RNA is anchored to the Xi via a nucleation center within exon 1 by YY1 transcription factor (Jeon and Lee 2011).

The mouse *Xic* also harbors another long (40-kb), noncoding RNA *Tsix* discovered by Jeannie Lee (Lee et al. 1999a). *Tsix* is anti-sense to *Xist* and negatively regulates *Xist* expression early in development. During cellular differentiation, *Tsix* expression is extinguished on the future Xi and *Xist* RNA is upregulated nearly one hundred fold. *Tsix* is imprinted in the early embryo and extraembryonic tissues (Lee 2000). *Tsix* becomes randomized following X reactivation in the epiblast and is expressed exclusively from the future active X-chromosome (Xa). The importance of *Tsix* is underscored in knockout female ES cells where the female X-chromosome harboring the mutation is nearly always chosen as the Xi (Lee 2002; Lee and Lu 1999). *Tsix* is required for an X-chromosome to remain active. Forced over-expressed of *Tsix* in female ES cells suppresses *Xist* expression (Stavropoulos et al. 2001). *Tsix* contains a repeat region, *DXPas34*, which controls Tsix expression (Cohen et al. 2007; Debrand et al. 1999; Vigneau et al. 2006). The chromatin insulator Ctcf binds to *DXPas34* as well as other regions within *Tsix* (Chao et al. 2002). Ctcf has a co-factor, Yy1. *Yy1*-deficient embryos have abnormal *Tsix* and *Xist* expression and die shortly after implantation (Donohoe et al. 2007). Taken together, *Tsix* is an important epigenetic switch for XCI in early mouse development.

An additional long noncoding RNA, *Xite* (*X-inactivation intergenic transcription element*), situated upstream of *Tsix* harbors an enhancer for *Tsix* that enables its expression persistence on the future Xa (Ogawa and Lee 2003). *Xite* is *cis*-acting and has a number of transcriptional elements. Similar to *Tsix*, *Xite* is developmentally regulated functioning as a Xa choice element. Both *Xite* and *Tsix* are expressed during the pluripotent state and their expression is extinguished upon cellular differentiation.

In addition to *Xist* and *Tsix*, we now appreciate that there is a plethora of long, noncoding RNAs situated at the *Xic*. *Jpx* (also known as *Enox: expressed neighbor of Xist*), a long, noncoding RNA, is located upstream of *Xist* and is essential for its activation (Chureau et al. 2002). Unlike *Tsix* and *Xite*, *Jpx* is *trans*-acting and activates either *Xist* allele (Tian et al. 2010). *Rnf12* situated 500-kb 5′ relative to *Xist*, produces a *trans*-factor RNF12, an ubiquitin ligase. Over expression of RNF12 can induce *Xist* coating of the single X in differentiating male and both X's in female ES cells (Barakat et al. 2011; Jonkers et al. 2009). Therefore, RNF12 is an activator of *Xist*. *Ftx* (*5′ to Xist*) is another noncoding RNA situated upstream of *Jpx* (Chureau et al. 2002). Deletion of *Ftx* in male ES cells alters the chromatin structure surrounding *Xist* (Chureau et al. 2010). Taken together, long, noncoding RNAs play an important role for XCI and can provide precise regulation within the *Xic*.

13.7 Silencing Along the X-Chromosome

How does Xist spread along the Xi to transcriptionally silence nearly 155 Mb of DNA? The X-chromosome has nearly double the amount of LINE-1 (long interspersed nuclear elements) repeat elements as compared with autosomes. Mary Lyon proposed that the X-chromosome silencing marks are propagated via LINE-1 elements that act as way stations (Lyon 2000). The Heard lab has shown that LINEs create a silent nuclear compartment in female cells and a subset of young LINE-1 elements are expressed during XCI and reside near XCI escape prone regions (Chow et al. 2010). LINEs may facilitate XCI are multiple levels establishing heterochromatin and propagation of silencing.

13.8 Nuclear Dynamics of the X Chromosomes: A Developmental Dance

The conformation of the Xi as the Barr body has opened additional studies linking the compartmentalization of the silenced versus active X. Genic and nongenic elements harbored on the Xi are organized in a nonrandom fashion with chromatin conformation and looping most likely involved in gene activation versus repression. Although *Xist*, *Tsix* and *Xite* act in *cis*, a *trans*-communication between the two female X-chromosomes must take place to achieve the mutually exclusive fate of active (Xa) versus inactive X (Xi) (Bacher et al. 2006; Xu et al. 2006). Homologous X pairing is a rare example of somatic or non-meiotic pairing. This chromosomal "kissing" is mediated by sequences within *Xite* and *Tsix*, and X–X pairing is necessary for XCI to take place (Augui et al. 2007; Bacher et al. 2006; Xu et al. 2006). The transcription regulators Ctcf and Oct4 control X–X pairing (Donohoe et al. 2009; Xu et al. 2007).

13.9 XCI and the Link with Pluripotency

It is now appreciated that transcription factors are key regulators for the different mouse lineages. In a landmark study, Yamanaka and Takahasi showed that over expression of four key transcription factors, Oct4, Klf4, c-Myc, and Sox2, convert a differentiated fibroblast to an induced pluripotent stem cell (iPS) (Takahashi and Yamanaka 2006). Since this first description, iPS cells have been generated in both humans and mouse and may offer hope for future therapeutics. The coupling of the pluripotent state and reactivation of both female X-chromosomes was demonstrated in the conversion of female mouse fibroblasts to iPS cells (Maherali et al. 2007). XCI is tightly linked with cellular differentiation and Oct4, Sox2, Klf4, c-myc, and Rex transcription factors bind *Xite* and *Tsix*, the noncoding RNAs expressed during pluripotency (Navarro et al. 2008; Navarro et al. 2010; Donohoe et al. 2009). Oct4 not only activates *Tsix* but is also involved in X–X pairing, an event necessary for the mutual choice between active versus inactive X-chromosomes (Donohoe et al. 2009).

13.10 Genes That Escape XCI: Breaking the Rules

Although XCI is generally a silencing of all genes residing on the Xi, there are exceptions to the rule. Genes that escape XCI may be divided into two groups. Genes situated within the pseudoautosomal region (PAR) and genes located outside the PAR (Berletch et al. 2011). The genes harbored within the PAR do not need dosage compensation because they have equivalents on the Y-chromosome.

In humans nearly 15% of the 1,500 genes residing on the X-chromosome escape silencing (Carrel and Willard 2005). In contrast, only a few genes escape the chromosomal-wide silencing in mouse. X-chromosome escape raises many questions. Why there are fewer numbers of escaping genes in mice? What are the molecular mechanisms of XCI escape? Genes that thwart XCI are frequently organized in clusters (Carrel and Willard 2005; Li and Carrel 2008). It has been postulated that chromatin insulation by Ctcf and noncoding RNAs may shield those genes destined to escape silencing by XCI (Filippova et al. 2005; Reinius et al. 2010). Ctcf insulation on its own may not sufficient to protect from XCI (Ciavatta et al. 2006). In addition, the LINE-1 element density is low in regions where XCI escape is frequent (Chow et al. 2010). Genes that escape may reside in a different nuclear subdomain or compartment to shield silencing.

13.11 Conclusions

In the fifty years since Mary Lyon's landmark paper proposing XCI, we have learned much information on the molecular mechanism and important players in this process. Most of these findings have come from the studies using mouse embryology and genetics. These studies have opened new doors to the biology of noncoding RNAs, nuclear compartmentalization, gene regulation in epigenetics, etc. in both humans and mouse. The next 50 years will surely enlighten us and offer exciting findings.

Acknowledgements I am grateful to the members of my lab for their valuable input on this chapter.

References

Augui S, Filion GJ, Huart S, Nora E, Guggiari M, Maresca M, Stewart AF, Heard E (2007) Sensing X chromosome pairs before X inactivation via a novel X-pairing region of the Xic. Science 318:1632–1636

Bacher CP, Guggiari M, Brors B, Augui S, Clerc P, Avner P, Eils R, Heard E (2006) Transient colocalization of X-inactivation centres accompanies the initiation of X inactivation. Nat Cell Biol 8:293–299

Barakat TS, Gunhanlar N, Pardo CG, Achame EM, Ghazvini M, Boers R, Kenter A, Rentmeester E, Grootegoed JA, Gribnau J (2011) RNF12 activates Xist and is essential for X chromosome inactivation. PLoS Genet 7:e1002001

Barr ML, Bertram EG (1949) A morphological distinction between neurones of the male and female, and the behaviour of the nucleolar satellite during accelerated nucleoprotein synthesis. Nature 163:676

Bartolomei MS, Ferguson-Smith AC (2011) Mammalian genomic imprinting. Cold Spring Harb Perspect Biol 3:a002592

Berletch JB, Yang F, Xu J, Carrel L, Disteche CM (2011) Genes that escape from X inactivation. Hum Genet 130:237–245

Beutler E, Yeh M, Fairbanks VF (1962) The normal human female as a mosaic of X-chromosome activity: studies using the gene for C-6-PD-deficiency as a marker. Proc Natl Acad Sci USA 48:9–16

Blewitt ME, Gendrel AV, Pang Z, Sparrow DB, Whitelaw N, Craig JM, Apedaile A, Hilton DJ, Dunwoodie SL, Brockdorff N et al (2008) SmcHD1, containing a structural-maintenance-of-chromosomes hinge domain, has a critical role in X inactivation. Nat Genet 40:663–669

Borsani G, Tonlorenzi R, Simmler MC, Dandolo L, Arnaud D, Capra V, Grompe M, Pizzuti A, Muzny D, Lawrence C et al (1991) Characterization of a murine gene expressed from the inactive X chromosome. Nature 351:325–329

Brockdorff N, Ashworth A, Kay GF, Cooper P, Smith S, McCabe VM, Norris DP, Penny GD, Patel D, Rastan S (1991) Conservation of position and exclusive expression of mouse Xist from the inactive X chromosome. Nature 351:329–331

Brockdorff N, Ashworth A, Kay GF, McCabe VM, Norris DP, Cooper PJ, Swift S, Rastan S (1992) The product of the mouse Xist gene is a 15 kb inactive X-specific transcript containing no conserved ORF and located in the nucleus. Cell 71:515–526

Brown CJ, Hendrich BD, Rupert JL, Lafreniere RG, Xing Y, Lawrence J, Willard HF (1992) The human XIST gene: analysis of a 17 kb inactive X-specific RNA that contains conserved repeats and is highly localized within the nucleus. Cell 71:527–542

Brown SD (1991) XIST and the mapping of the X chromosome inactivation centre. Bioessays 13:607–612

Carrel L, Willard HF (2005) X-inactivation profile reveals extensive variability in X-linked gene expression in females. Nature 434:400–404

Cattanach BM, Williams CE (1972) Evidence of non-random X chromosome activity in the mouse. Genet Res 19:229–240

Chao W, Huynh KD, Spencer RJ, Davidow LS, Lee JT (2002) CTCF, a candidate trans-acting factor for X-inactivation choice. Science 295:345–347

Chow JC, Ciaudo C, Fazzari MJ, Mise N, Servant N, Glass JL, Attreed M, Avner P, Wutz A, Barillot E et al (2010) LINE-1 activity in facultative heterochromatin formation during X chromosome inactivation. Cell 141:956–969

Chureau C, Chantalat S, Romito A, Galvani A, Duret L, Avner P, Rougeulle C (2010) Ftx is a non-coding RNA which affects Xist expression and chromatin structure within the X-inactivation center region. Hum Mol Genet 20:705–718

Chureau C, Prissette M, Bourdet A, Barbe V, Cattolico L, Jones L, Eggen A, Avner P, Duret L (2002) Comparative sequence analysis of the X-inactivation center region in mouse, human, and bovine. Genome Res 12:894–908

Chuva de Sousa Lopes C, Hayashi K, Shovlin TC, Mifsud W, Surani MA, McLaren A (2008) X chromosome activity in mouse XX primordial germ cells. PLoS Genet 4:e30

Ciavatta D, Kalantry S, Magnuson T, Smithies O (2006) A DNA insulator prevents repression of a targeted X-linked transgene but not its random or imprinted X inactivation. Proc Natl Acad Sci USA 103:9958–9963

Clemson CM, McNeil JA, Willard HF, Lawrence JB (1996) XIST RNA paints the inactive X chromosome at interphase: evidence for a novel RNA involved in nuclear/chromosome structure. J Cell Biol 132:259–275

Cohen DE, Davidow LS, Erwin JA, Xu N, Warshawsky D, Lee JT (2007) The DXPas34 repeat regulates random and imprinted X inactivation. Dev Cell 12:57–71

Cooper DW (1971) Directed genetic change model for X chromosome inactivation in eutherian mammals. Nature 230:292–294

Costanzi C, Pehrson JR (1998) Histone macroH2A1 is concentrated in the inactive X chromosome of female mammals. Nature 393:599–601

Csankovszki G, Panning B, Bates B, Pehrson JR, Jaenisch R (1999) Conditional deletion of Xist disrupts histone macroH2A localization but not maintenance of X inactivation. Nat Genet 22:323–324

de Napoles M, Mermoud JE, Wakao R, Tang YA, Endoh M, Appanah R, Nesterova TB, Silva J, Otte AP, Vidal M et al (2004) Polycomb group proteins Ring1A/B link ubiquitylation of histone H2A to heritable gene silencing and X inactivation. Dev Cell 7:663–676

de Napoles M, Nesterova T, Brockdorff N (2007) Early loss of Xist RNA expression and inactive X chromosome associated chromatin modification in developing primordial germ cells. PLoS One 2:e860

Debrand E, Chureau C, Arnaud D, Avner P, Heard E (1999) Functional analysis of the DXPas34 locus, a 3′ regulator of Xist expression. Mol Cell Biol 19:8513–8525

Donohoe ME, Silva SS, Pinter SF, Xu N, Lee JT (2009) The pluripotency factor Oct4 interacts with Ctcf and also controls X-chromosome pairing and counting. Nature 460:128–132

Donohoe ME, Zhang LF, Xu N, Shi Y, Lee JT (2007) Identification of a Ctcf cofactor, Yy1, for the X chromosome binary switch. Mol Cell 25:43–56

Evans MJ, Kaufman MH (1981) Establishment in culture of pluripotential cells from mouse embryos. Nature 292:154–156

Filippova GN, Cheng MK, Moore JM, Truong JP, Hu YJ, Nguyen DK, Tsuchiya KD, Disteche CM (2005) Boundaries between chromosomal domains of X inactivation and escape bind CTCF and lack CpG methylation during early development. Dev Cell 8:31–42

Gartler SM, Liskay RM, Campbell BK, Sparkes R, Gant N (1972) Evidence for two functional X chromosomes in human oocytes. Cell Differ 1:215–218

Gartler SM, Riggs AD (1983) Mammalian X-chromosome inactivation. Annu Rev Genet 17:155–190

Greaves IK, Rangasamy D, Devoy M, Marshall Graves JA, Tremethick DJ (2006) The X and Y chromosomes assemble into H2A.Z-containing [corrected] facultative heterochromatin [corrected] following meiosis. Mol Cell Biol 26:5394–5405

Hasegawa Y, Brockdorff N, Kawano S, Tsutui K, Nakagawa S (2010) The matrix protein hnRNP U is required for chromosomal localization of Xist RNA. Dev Cell 19:469–476

Heard E, Kress C, Mongelard F, Courtier B, Rougeulle C, Ashworth A, Vourc'h C, Babinet C, Avner P (1996) Transgenic mice carrying an Xist-containing YAC. Hum Mol Genet 5:441–450

Helbig R, Fackelmayer FO (2003) Scaffold attachment factor A (SAF-A) is concentrated in inactive X chromosome territories through its RGG domain. Chromosoma 112:173–182

Huynh KD, Lee JT (2003) Inheritance of a pre-inactivated paternal X chromosome in early mouse embryos. Nature 426:857–862

Jeon Y, Lee JT (2011) YY1 tethers Xist RNA to the inactive X nucleation center. Cell 146:119–133

Jonkers I, Barakat TS, Achame EM, Monkhorst K, Kenter A, Rentmeester E, Grosveld F, Grootegoed JA, Gribnau J (2009) RNF12 is an X-Encoded dose-dependent activator of X chromosome inactivation. Cell 139:999–1011

Kunath T, Arnaud D, Uy GD, Okamoto I, Chureau C, Yamanaka Y, Heard E, Gardner RL, Avner P, Rossant J (2005) Imprinted X-inactivation in extra-embryonic endoderm cell lines from mouse blastocysts. Development 132:1649–1661

Lee JT (2000) Disruption of imprinted X inactivation by parent-of-origin effects at Tsix. Cell 103:17–27

Lee JT (2002) Homozygous Tsix mutant mice reveal a sex-ratio distortion and revert to random X-inactivation. Nat Genet 32:195–200

Lee JT (2011) Gracefully ageing at 50, X-chromosome inactivation becomes a paradigm for RNA and chromatin control. Nat Rev Mol Cell Biol 12:815–826

Lee JT, Davidow LS, Warshawsky D (1999a) Tsix, a gene antisense to Xist at the X-inactivation centre. Nat Genet 21:400–404

Lee JT, Lu N (1999) Targeted mutagenesis of Tsix leads to nonrandom X inactivation. Cell 99:47–57

Lee JT, Lu N, Han Y (1999b) Genetic analysis of the mouse X inactivation center defines an 80-kb multifunction domain. Proc Natl Acad Sci USA 96:3836–3841

Lee JT, Strauss WM, Dausman JA, Jaenisch R (1996) A 450 kb transgene displays properties of the mammalian X-inactivation center. Cell 86:83–94

Li N, Carrel L (2008) Escape from X chromosome inactivation is an intrinsic property of the Jarid1c locus. Proc Natl Acad Sci USA 105:17055–17060

Lifschytz E, Lindsley DL (1972) The role of X-chromosome inactivation during spermatogenesis (Drosophila-allocycly-chromosome evolution-male sterility-dosage compensation). Proc Natl Acad Sci USA 69:182–186

Linder D, Gartler SM (1965) Glucose-6-phosphate dehydrogenase mosaicism: utilization as a cell marker in the study of leiomyomas. Science 150:67–69

Lyon MF (1961) Gene action in the X-chromosome of the mouse (*Mus musculus* L). Nature 190:372–373

Lyon MF (2000) LINE-1 elements and X chromosome inactivation: a function for "junk" DNA? Proc Natl Acad Sci USA 97:6248–6249

Maherali N, Sridharan R, Xie W, Utikal J, Eminli S, Arnold K, Stadtfeld M, Yachechko R, Tchieu J, Jaenisch R et al (2007) Directly reprogrammed fibroblasts show global epigenetic remodeling and widespread tissue contribution. Cell Stem Cell 1:55–70

Mak W, Nesterova TB, de Napoles M, Appanah R, Yamanaka S, Otte AP, Brockdorff N (2004) Reactivation of the paternal X chromosome in early mouse embryos. Science 303:666–669

Marahrens Y, Panning B, Dausman J, Strauss W, Jaenisch R (1997) Xist-deficient mice are defective in dosage compensation but not spermatogenesis. Genes Dev 11:156–166

Martin GR, Epstein CJ, Travis B, Tucker G, Yatziv S, Martin DW Jr, Clift S, Cohen S (1978) X-chromosome inactivation during differentiation of female teratocarcinoma stem cells in vitro. Nature 271:329–333

McGrath J, Solter D (1984) Completion of mouse embryogenesis requires both the maternal and paternal genomes. Cell 37:179–183

McLaren A, Monk M (1981) X-chromosome activity in the germ cells of sex-reversed mouse embryos. J Reprod Fertil 63:533–537

McMahon A, Fosten M, Monk M (1981) Random X-chromosome inactivation in female primordial germ cells in the mouse. J Embryol Exp Morphol 64:251–258

McMahon A, Fosten M, Monk M (1983) X-chromosome inactivation mosaicism in the three germ layers and the germ line of the mouse embryo. J Embryol Exp Morphol 74:207–220

Monk M, Harper MI (1979) Sequential X chromosome inactivation coupled with cellular differentiation in early mouse embryos. Nature 281:311–313

Monk M, McLaren A (1981) X-chromosome activity in foetal germ cells of the mouse. J Embryol Exp Morphol 63:75–84

Morey C, Avner P (2011) The demoiselle of X-inactivation: 50 years old and as trendy and mesmerizing as ever. PLoS Genet 7:e1002212

Namekawa SH, Park PJ, Zhang LF, Shima JE, McCarrey JR, Griswold MD, Lee JT (2006) Postmeiotic sex chromatin in the male germline of mice. Curr Biol 16:660–667

Namekawa SH, Payer B, Huynh KD, Jaenisch R, Lee JT (2010) Two-step imprinted X inactivation: repeat versus genic silencing in the mouse. Mol Cell Biol 30:3187–3205

Navarro P, Chambers I, Karwacki-Neisius V, Chureau C, Morey C, Rougeulle C, Avner P (2008) Molecular coupling of Xist regulation and pluripotency. Science 321:1693–1695

Navarro P, Oldfield A, Legoupi J, Festuccia N, Dubois A, Attia M, Schoorlemmer J, Rougeulle C, Chambers I, Avner P (2010) Molecular coupling of Tsix regulation and pluripotency. Nature 468:457–460

Ogawa Y, Lee JT (2003) Xite, X-inactivation intergenic transcription elements that regulate the probability of choice. Mol Cell 11:731–743

Ohno S, Kaplan WD, Kinosita R (1959) Formation of the sex chromatin by a single X-chromosome in liver cells of Rattus norvegicus. Exp Cell Res 18:415–418
Okamoto I, Otte AP, Allis CD, Reinberg D, Heard E (2004) Epigenetic dynamics of imprinted X inactivation during early mouse development. Science 303:644–649
Panning B, Dausman J, Jaenisch R (1997) X chromosome inactivation is mediated by Xist RNA stabilization. Cell 90:907–916
Payer B, Lee JT (2008) X chromosome dosage compensation: how mammals keep the balance. Annu Rev Genet 42:733–772
Penny GD, Kay GF, Sheardown SA, Rastan S, Brockdorff N (1996) Requirement for Xist in X chromosome inactivation. Nature 379:131–137
Plath K, Fang J, Mlynarczyk-Evans SK, Cao R, Worringer KA, Wang H, de la Cruz CC, Otte AP, Panning B, Zhang Y (2003) Role of histone H3 lysine 27 methylation in X inactivation. Science 300:131–135
Plath K, Talbot D, Hamer KM, Otte AP, Yang TP, Jaenisch R, Panning B (2004) Developmentally regulated alterations in Polycomb repressive complex 1 proteins on the inactive X chromosome. J Cell Biol 167:1025–1035
Puck JM, Willard HF (1998) X inactivation in females with X-linked disease. N Engl J Med 338:325–328
Pullirsch D, Hartel R, Kishimoto H, Leeb M, Steiner G, Wutz A (2010) The Trithorax group protein Ash2l and Saf-A are recruited to the inactive X chromosome at the onset of stable X inactivation. Development 137:935–943
Rastan S, Robertson EJ (1985) X-chromosome deletions in embryo-derived (EK) cell lines associated with lack of X-chromosome inactivation. J Embryol Exp Morphol 90:379–388
Reinius B, Shi C, Hengshuo L, Sandhu KS, Radomska KJ, Rosen GD, Lu L, Kullander K, Williams RW, Jazin E (2010) Female-biased expression of long non-coding RNAs in domains that escape X-inactivation in mouse. BMC Genomics 11:614
Russell LB (1963) Mammalian X-chromosome action: inactivation limited in spread and region of origin. Science 140:976–978
Russell WL, Russell LB, Gower JS (1959) Exceptional inheritance of a sex-linked gene in the mouse explained on the basis that the X/O sex-chromosome constitution is female. Proc Natl Acad Sci USA 45:554–560
Sharman GB (1971) Late DNA replication in the paternally derived X chromosome of female kangaroos. Nature 230:231–232
Silva J, Mak W, Zvetkova I, Appanah R, Nesterova TB, Webster Z, Peters AH, Jenuwein T, Otte AP, Brockdorff N (2003) Establishment of histone h3 methylation on the inactive X chromosome requires transient recruitment of Eed-Enx1 polycomb group complexes. Dev Cell 4:481–495
Stavropoulos N, Lu N, Lee JT (2001) A functional role for Tsix transcription in blocking Xist RNA accumulation but not in X-chromosome choice. Proc Natl Acad Sci USA 98:10232–10237
Sugimoto M, Abe K (2007) X chromosome reactivation initiates in nascent primordial germ cells in mice. PLoS Genet 3:e116
Surani MA, Barton SC, Norris ML (1984) Development of reconstituted mouse eggs suggests imprinting of the genome during gametogenesis. Nature 308:548–550
Tada T, Obata Y, Tada M, Goto Y, Nakatsuji N, Tan S, Kono T, Takagi N (2000) Imprint switching for non-random X-chromosome inactivation during mouse oocyte growth. Development 127:3101–3105
Takagi N, Sasaki M (1975) Preferential inactivation of the paternally derived X chromosome in the extraembryonic membranes of the mouse. Nature 256:640–642
Takahashi K, Yamanaka S (2006) Induction of pluripotent stem cells from mouse embryonic and adult fibroblast cultures by defined factors. Cell 126:663–676
Tanaka S, Kunath T, Hadjantonakis AK, Nagy A, Rossant J (1998) Promotion of trophoblast stem cell proliferation by FGF4. Science 282:2072–2075

Tian D, Sun S, Lee JT (2010) The long noncoding RNA, Jpx, is a molecular switch for X chromosome inactivation. Cell 143:390–403

Turner JM (2007) Meiotic sex chromosome inactivation. Development 134:1823–1831

Turner JM, Mahadevaiah SK, Ellis PJ, Mitchell MJ, Burgoyne PS (2006) Pachytene asynapsis drives meiotic sex chromosome inactivation and leads to substantial postmeiotic repression in spermatids. Dev Cell 10:521–529

Vigneau S, Augui S, Navarro P, Avner P, Clerc P (2006) An essential role for the DXPas34 tandem repeat and Tsix transcription in the counting process of X chromosome inactivation. Proc Natl Acad Sci USA 103:7390–7395

Wutz A, Jaenisch R (2000) A shift from reversible to irreversible X inactivation is triggered during ES cell differentiation. Mol Cell 5:695–705

Xu N, Donohoe ME, Silva SS, Lee JT (2007) Evidence that homologous X-chromosome pairing requires transcription and Ctcf protein. Nat Genet 39:1390–1396

Xu N, Tsai CL, Lee JT (2006) Transient homologous chromosome pairing marks the onset of X inactivation. Science 311:1149–1152

Zhao J, Sun BK, Erwin JA, Song JJ, Lee JT (2008) Polycomb proteins targeted by a short repeat RNA to the mouse X chromosome. Science 322:750–756

Zheng W, Liu K (2012) Maternal control of mouse preimplantation development. In: Kubiak JZ (ed) Results and problems in cell differentiation. Springer-Verlag, Heidelberg

Chapter 14
Mouse Oviduct Development

C. Allison Stewart and Richard R. Behringer

Abstract The oviduct, or Fallopian tube in humans, transports oocytes and sperm, serves as the site of fertilization, and supports early embryonic development. The oviduct is essential for fertility. In the mouse, the oviduct is a coiled, complex structure that develops from the simple embryonic Müllerian duct. The oviduct consists of four segments, including the infundibulum, ampulla, isthmus, and uterotubal junction. Additionally, the mouse oviduct forms coils, develops longitudinal folds, and undergoes both mesenchymal and epithelial differentiation. Oviduct development and differentiation occurs perinatally. Several signaling pathways have been found to be involved in oviduct formation, such as Wnt, Tgfβ, microRNA processing, as well as others. Overall, the process of oviduct development is poorly understood and can be utilized to further knowledge of epithelial–mesenchymal interactions, regulation of coiling, characteristics of pseudostratified epithelia, and smooth muscle differentiation.

14.1 The Mouse Oviduct

The oviduct (Fallopian tube as it is referred to in humans) is a narrow tube through which newly ovulated oocytes travel from the ovary to the uterus. It is also the site of fertilization and regulates the number of sperm that ultimately interact with the oocytes (Pulkkinen 1995; Ikawa et al. 2010). The mature oviduct consists of four distinct regions, from anterior to posterior, the infundibulum, ampulla, isthmus, and uterotubal junction (Fig. 14.1). The segments of the oviduct are supported by the mesosalpinx, a portion of the broad ligament. In both mice and humans, some of the oviduct segments can first be distinguished prior to birth based on gross morphology

C.A. Stewart • R.R. Behringer (✉)
Department of Genetics, University of Texas M.D. Anderson Cancer Center, Houston, TX 77030, USA
e-mail: rrb@mdanderson.org

J.Z. Kubiak (ed.), *Mouse Development*, Results and Problems in Cell Differentiation 55, 247
DOI 10.1007/978-3-642-30406-4_14, © Springer-Verlag Berlin Heidelberg 2012

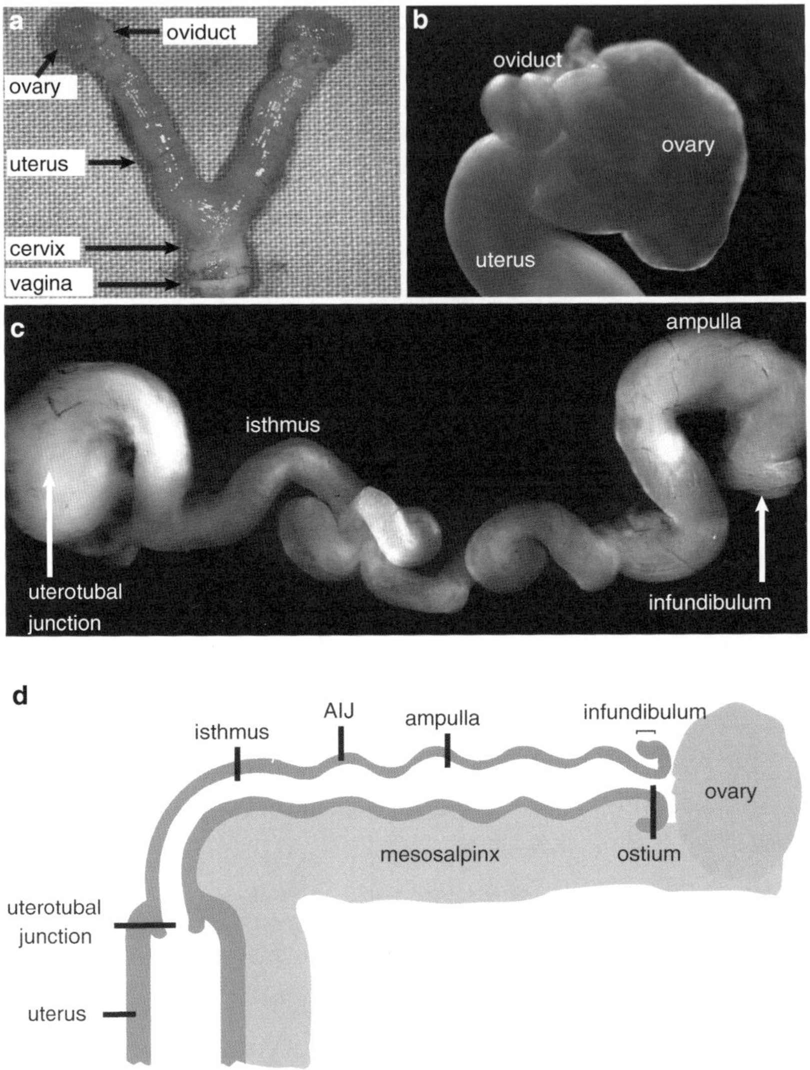

Fig. 14.1 Morphology of the mouse oviduct. (**a**) Gross morphology of the mouse female reproductive tract. (**b**) Coiling of the mouse oviduct. (**c**) A partially "uncoiled" mouse oviduct. (**d**) Schematic representation of the oviduct if it were uncoiled

(Agduhr 1927; Perlman et al. 2005). However, a significant degree of differentiation and morphogenesis occurs postnatally.

Oviduct length in mature mice is approximately 1.8 cm (Agduhr 1927) and Fallopian tube length in humans is approximately 11 cm (Blandau 1978; Eddy and Pauerstein 1980). Interestingly, oviduct length is much longer in avian species

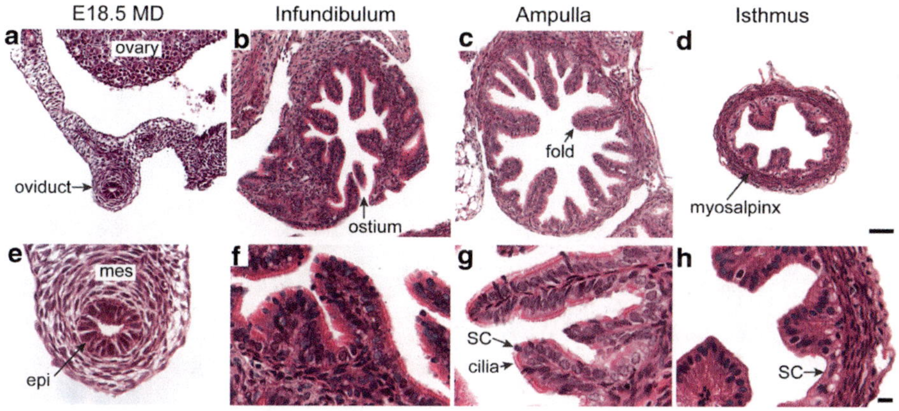

Fig. 14.2 Histology of the mouse oviduct. Hematoxylin and eosin staining of the embryonic Müllerian duct (**a**, **e**) and segments of the adult oviduct, including the infundibulum (**b**, **f**), ampulla (**c**, **g**), and isthmus (**d**, **h**). Prior to birth, the presumptive oviduct consists of an epithelium surrounded by a loose mesenchyme. In the adult, the pseudostratified epithelia consist of both ciliated and secretory cells (SC). There are more longitudinal epithelial folds within the ampulla region, while the smooth muscle layers, or myosalpinx, are thicker in the isthmus region. Scale bars represent 50 μm (**a**–**d**) and 10 μm (**e**–**h**)

and can be up to 95 cm in turkeys (Bakst 1998). The structure of the oviduct varies between species. In mice, oviducts form coiled loops containing approximately 11 turns (Agduhr 1927), whereas in humans there are no coils.

Another distinct difference between mice and humans is the presence of an ovarian bursa in mice. The ovarian bursa is a sac that encapsulates the ovary and anterior oviduct in mice and many other species. Interestingly, the ovarian bursa is absent in humans. At birth, the forming bursa does not completely surround the ovary of mice, but does so by postnatal day (P) 7.5 (Agduhr 1927).

Morphologically, the oviduct consists of a pseudostratified epithelium and a thin stromal layer surrounded by inner circular and outer longitudinal smooth muscle layers (myosalpinx). The epithelium consists of two types of cells: ciliated and secretory. The oviduct epithelium forms longitudinal folds that increase surface area (Fig. 14.2). The abundance of these longitudinal folds varies among species. The oviduct regions can be distinguished by the ratio of ciliated versus secretory epithelial cells, as well as the degree of longitudinal epithelial folding and thickness of the smooth muscle layers (Fig. 14.3).

The oviduct receives its blood supply from the ovarian and uterine arteries. The venous drainage is closely apposed to the arterial supply. In addition to the vasculature, the oviduct contains an extensive lymphatic system. Most of the available information on the lymphatic system in the oviduct comes from research in the sow. The lymphatic system is more extensive within the posterior region of the oviduct, including the isthmus and uterotubal junction. The vasculature and lymphatic drainage connect to the oviduct via the mesosalpinx (Hafez and Blandau 1969). Furthermore, both systems are sensitive to steroid hormones in mature animals.

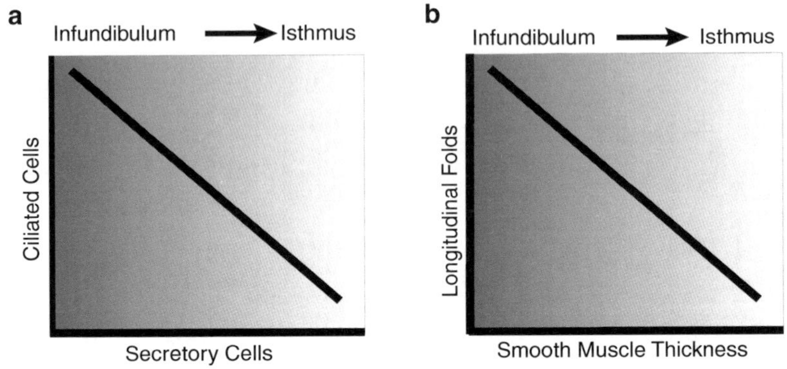

Fig. 14.3 Representation of oviduct characteristics that vary along the anterior to posterior axis. The anterior tip of the oviduct, the infundibulum, contains higher numbers of ciliated cells (**a**) and longitudinal epithelial folds (**b**). In contrast, the isthmus has more secretory epithelial cells (**a**) and a thicker smooth muscle layer (**b**). This relationship appears to be linear

14.2 The Oviduct Is Derived from the Müllerian Duct

Prior to sexual differentiation, the fetal gonads are bipotential, meaning that they are able to give rise to either ovaries or testes. They are connected to the mesonephroi that contain a pair of genital ducts. Embryos of both sexes develop Wolffian (mesonephric) and Müllerian (paramesonephric) ducts, which form the reproductive tracts in males and females, respectively. The expression of *Sry* on the Y chromosome is essential for Sertoli cell differentiation and the formation of testicular cords (Koopman et al. 1991). The Sertoli cells of the testis produce anti-Müllerian hormone (AMH), which actively induces the regression of the Müllerian ducts in males. The Wolffian ducts give rise to the epididymis, vas deferens, and seminal vesicles. In females, the absence of fetal AMH production results in the maintenance of the Müllerian ducts, which along the anterior–posterior axis, give rise to the oviducts, uterus, cervix, and anterior portion of the vagina (Kurita et al. 2001; Kobayashi and Behringer 2003; Yin and Ma 2005). The process of Müllerian duct regression in males has been well studied, but little is known of Müllerian duct differentiation into the tissues of the female reproductive tract. This review will focus primarily on the development and differentiation of the Müllerian duct into the oviduct.

The development of the male and female reproductive tracts is similar in most mammalian species, but this chapter will focus largely on mouse and human development. The urogenital system, including the kidneys, gonads, urinary tract, and both male and female reproductive tracts, is derived from intermediate mesoderm. The Wolffian ducts begin to develop at embryonic day (E) 9.5 (fertilization = E0) and grow caudally towards the urogenital sinus. A few days later, at E11.75, the Müllerian duct originates as an invagination of the mesonephric epithelium (Orvis and Behringer 2007). This invagination remains open and will form the infundibular ostium of the oviduct (Kurita et al. 2001). The Müllerian duct forms adjacent to the Wolffian duct and remains in close contact as it extends rostro-caudally to the urogenital sinus. The presence of the Wolffian duct is essential for Müllerian duct development (Gruenwald 1941; Kobayashi et al. 2005); however, the Wolffian duct epithelium does not contribute cells to the Müllerian duct (Orvis and Behringer 2007). At E13.5, embryos of both sexes have fully extended Wolffian and Müllerian ducts. The entire process, from invagination to completion, takes 30–45 h (Orvis and Behringer 2007). AMH produced by the Sertoli cells of the fetal testis bind to the AMH receptor type 2 (AMHR2) in the Müllerian duct mesenchyme to cause regression of the Müllerian duct. The absence of AMH in the fetal ovary allows for maintenance of the Müllerian duct in females. Additionally, the absence of testis hormones prevents further development of the Wolffian duct and results in its regression. Müllerian duct development and regression is similar to this point in most mammalian species.

In this chapter, the female mouse reproductive tract before birth will be referred to as the Müllerian duct and after birth by the region-specific name (i.e., oviduct, uterus, etc.). The oviduct develops from the most anterior region of the Müllerian duct. As the Müllerian duct increases in length, it remains a relatively straight tube. The anterior region of the Müllerian duct curves into a C-shape with the presence of flexures that curve the oviduct region around the ovary. The *flexura medialis* is present posterior to the ovary and is believed to delineate which regions of the Müllerian duct will give rise to the oviduct and uterus (Agduhr 1927). At later stages of gestation, the diameter of the oviduct region of the Müllerian duct is less than the diameter of the uterine region.

Morphologically, the Müllerian duct consists of a simple epithelium surrounded by mesenchyme. A few morphological differences between regions of the Müllerian duct were noted prior to birth, including more organization of the mesenchyme into smooth muscle cells within the uterine region and the presence of "club" cells, or secretory cells, within the epithelia of the oviduct region (Agduhr 1927). There are very few reports of differences in gene expression between regions of the Müllerian duct (Nielsen et al. 2000). Within days after birth the Müllerian duct differentiates into the distinct reproductive tract structures, including the oviduct, uterus, cervix, and anterior portion of the vagina, each with unique gene expression profiles.

14.3 Regions of the Oviduct

14.3.1 Infundibulum

The infundibulum is the most anterior structure of the oviduct, which forms the ostium or opening of the oviduct to receive ovulated oocytes *in cumulus* from the ovary. In mice, the infundibulum is found within the ovarian bursa. Compared to the adjacent ampulla, the infundibulum is enlarged and "inside-out", with the epithelium extending beyond the smooth muscle and stromal tissues. The exterior infundibular epithelium is continuous with the epithelial lining of the oviduct. Epithelial cells found in the infundibular region are highly ciliated. This is likely to encourage the funneling of oocytes *in cumulus* into the oviduct. Once the oocytes enter into the oviduct, they move to the ampulla. In some species, including humans and rabbits, the infundibulum is a trumpet-shaped structure that contains numerous fimbrae, which are fluted, finger-like structures. In humans, the infundibulum is approximately 1 cm in length (Eddy and Pauerstein 1980). The infundibulum is very sensitive to ovarian hormones, causing the fimbrae to extend and nearly engulf the ovary during estrus, but not during other stages of the estrous cycle (Blandau 1978; Eddy and Pauerstein 1980).

As mentioned previously, the initial invagination from the mesonephric epithelium remains open throughout development and will eventually form the infundibulum (Kurita et al. 2001). This makes it the first segment of the entire female reproductive tract to form. However, the distinct structure of the infundibulum is not morphologically evident until closer to birth. The fimbrae form in humans in the 18th week of gestation (Perlman et al. 2005). In mice, the distinct structure of the infundibulum can first be identified immediately before birth (Agduhr 1927).

14.3.2 Ampulla

The ampulla is the segment of the oviduct tube proximal to the ovary and is morphologically and histologically distinct from the isthmus segment, which is proximal to the uterus. The epithelium of the ampulla is highly folded and contains more ciliated cells than that of the isthmus. The human ampulla is 5–8 cm in length (Blandau 1978; Eddy and Pauerstein 1980). Furthermore, the smooth muscle layers are not as thick as in the isthmus. The oocytes entering the oviduct are contained within cumulus masses not as single cells. The relaxed nature of the smooth muscle layers within the ampulla allows for stretching to accommodate these relatively large clumps of cells. Fertilization occurs within the ampulla. Following fertilization, the cumulus masses are disassembled and individual zygotes will pass into the isthmus region, which contains thicker smooth muscle layers. Interestingly, ectopic embryo implantations in women occur most commonly in the ampulla (Woodruff and Paurstein 1969). The ampullary–isthmic junction (AIJ) between the two major oviduct regions can first be observed at postnatal day (P) 6 in mice (Agduhr 1927).

14.3.3 Isthmus

The isthmus is the segment of the oviduct proximal to the uterus, extending from the uterotubal junction. It joins to the ampulla at the AIJ. The isthmus is characterized by having a thick, well-developed myosalpinx and fewer longitudinal folds. The ratio of secretory to ciliated epithelial cells is higher in the isthmus than the ampulla. Very little is known about what regulates the development of the isthmus region or when it becomes distinct from the ampulla. This region is two to three cm long in the Fallopian tube (Blandau 1978; Eddy and Pauerstein 1980).

14.3.4 Uterotubal Junction

The oviduct and the uterus are joined at the uterotubal junction. In humans, this is sometimes referred to as the interstitial region. Depending on the species, the uterotubal junction is either intramural or diffuse. An intramural uterotubal junction occurs when the oviduct extends into the uterine horn and is present in humans, horses, and mice (Blandau 1978; Morris et al. 2000). In other species, such as cows, pigs, sheep, and goats, the oviduct does not extend into the uterine horn and the junction is diffuse (Agduhr 1927). The location of the uterotubal junction in cats, dogs, horses, and mice is dorsolateral to the tip of uterine horn, not in the center (Agduhr 1927). In mice, the uterotubal junction is formed as early as three days after birth (Poweirza 1912). The uterotubal junction functions to regulate passage of spermatozoa from the uterus into the oviduct, as well as embryos from the oviduct into the uterus. The ability for the uterotubal junction to allow passage between the uterus and oviduct is tightly regulated by steroid hormones during the estrous cycle and pregnancy. This can be easily observed by injecting blue dye into the uterine lumen. The dye will not enter into the oviduct when the uterotubal junction has formed properly or is functioning correctly (Newbold et al. 1983; Gonzalez and Behringer 2009). It is crucial that embryos do not remain within the oviduct. This can be a problem in species, such as humans where tubal pregnancies can be life threatening. Ectopic pregnancies account for 1–2% of pregnancies in women. This is a significant problem especially when the risks of maternal mortality can be as high as 10% (Shaw et al. 2010). An ectopic pregnancy in humans can occur when the embryo does not pass through the uterotubal junction, but remains in the Fallopian tube, where it implants and continues to develop. Ectopic pregnancies in mice are rare and have only been reported in conjunction with Chlamydia infections.

14.4 Oviduct Differentiation and Morphogenesis

Differentiation and morphogenesis of the oviduct consists of several events: oviductal coiling, mesenchymal differentiation, epithelial folding, and epithelial differentiation (Fig. 14.4). Differentiation of the mesenchyme results in a thin layer of stromal cells directly underlying the epithelia surrounded by two layers of smooth muscle cells. The simple epithelia found in the undifferentiated Müllerian duct become either ciliated or secretory cells postnatally.

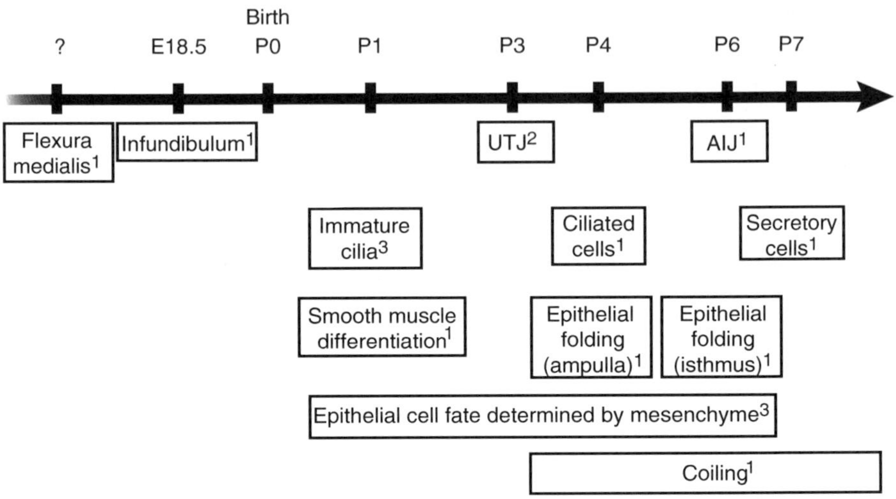

Fig. 14.4 Timeline of oviduct development in the mouse. Oviduct differentiation and development begin late during embryonic life and is completed within the first few weeks of neonatal life. It is possible that additional oviduct development occurs after puberty under the influence of ovarian hormones, as is seen in the mouse uterus. [1]Agduhr (1927), [2]Poweirza (1912), [3]Dirksen (1974), [4]Yamanouchi et al. (2010)

14.4.1 Oviductal Coiling

Oviduct coiling does not occur in all species, including humans. Interestingly, oviduct coiling is present in amphibians, including frogs, salamanders and caecilians, and birds (Wake and Dickie 1998). In mice, oviduct coiling is a very intricate system of looping that is supported by a connective tissue called the mesosalpinx. The pattern of oviduct coiling is not random, as shown by intricate tracing of mature and developing oviducts (Agduhr 1927). Distinct flexures form in the Müllerian duct and postnatal oviduct, resulting in the coiling pattern observed in mature animals (Agduhr 1927). Oviduct coiling in mice is complete by P15. However, it is currently unknown what directs the formation of these flexures. In the intestines, coiling is regulated by asymmetrical differences in the extracellular matrix, as well as adhesion molecules (Kurpios et al. 2008). It is certainly possible that a similar mechanism regulates oviductal coiling.

14.4.2 Epithelial Folding

The presence of longitudinal epithelial folds in the oviduct is evolutionarily conserved in most species. These folds are present in avian species, such as chickens and turkeys (Bakst 1998), as well as the amphibians mentioned above

(Wake and Dickie 1998). The degree of folding differs dramatically among different species. While this has not been studied extensively, there appears to be an inverse relationship between oviductal coiling and abundance of epithelial folding. Humans do not have coiled oviducts, but have extensive epithelial folding and mice are the opposite. In adult mice, seven to eight folds in the ampulla region and four to six longitudinal folds can be found in the isthmus region (Agduhr 1927).

The development of these epithelial folds has not been examined closely. Epithelial folding can first be observed three to four days after birth in the ampulla and six days after birth in the isthmus of the oviduct in mice (Agduhr 1927). Therefore, the process of epithelial folding occurs at the same time as oviduct coiling.

14.4.3 Mesenchymal Differentiation

Radial patterning of the oviduct results in the undifferentiated mesenchyme present in the Müllerian duct to form a shallow stroma and smooth muscle layers. The shallow stroma, or lamina propria, forms the scaffolding for the epithelial folds. The smooth muscle layers of the oviduct are referred to as the myosalpinx. The postnatal mesenchyme from the specific oviduct regions (i.e., ampulla, isthmus) determines epithelial cell morphology (Yamanouchi et al. 2010).

Very little information exists on the development of the myosalpinx. The morphology of the Müllerian duct mesenchyme suggests that smooth muscle differentiation occurs after birth (Agduhr 1927), which is consistent with what occurs in the uterine myometrium (Brody and Cunha 1989). The wall of the myosalpinx is thicker in the isthmus region compared to the ampulla. Longitudinal folding is more abundant in oviduct regions with thinner muscle layers. A combination of ciliary movement and smooth muscle contractions are important for oocyte transport (Jansen 1984). However, inhibition of smooth muscle contractions does not impair fertility (Halbert et al. 1989).

14.4.4 Epithelial Differentiation

The Müllerian ducts morphologically consist of simple cuboidal epithelia surrounded by mesenchyme. Following postnatal differentiation, the oviduct epithelial cells form a pseudostratified layer consisting of both secretory and ciliated epithelial cells. In general, there is an anterior to posterior gradient in which the infundibulum contains the highest abundance of ciliated cells and the isthmus contains the fewest. Conversely, the highest percentage of secretory cells can be found in the isthmus. In mice, during the first week after birth, the mesenchyme from the different oviduct segments has the ability to determine the abundance of ciliated or secretory cells (Yamanouchi et al. 2010). This suggests that oviduct epithelial cell fate is still plastic for a week after birth.

14.4.4.1 Secretory Cells

Secretory cells produce secretions important for oocyte and embryo survival. These cells can be identified based on cell morphology. The apical cell surface extends into the oviduct lumen above the beating cilia. Due to this appearance, they have also been called "club" cells (Agduhr 1927). Secretory cells are differentiated by P23 in mice (Komatsu and Fujita 1978), although they can be recognized as being devoid of cilia about a week or so after birth. These cells are highly responsive to ovarian hormones in the mature animal.

14.4.4.2 Ciliated Cells

In mature animals, a combination of ciliary movement and smooth muscle contractions is important for transport of the cumulus mass and the early embryo. The absence of Fallopian tube ciliary movement, as seen in women with Kartagener's syndrome, results in subfertility, not infertility (Afzelius et al. 1978). The cilia in the infundibulum and ampulla beat towards the uterus. However, ciliary direction in the isthmus differs among species. In women and mice, the cilia beat towards the uterus, but in rabbits the cilia beat in both directions (Blandau 1978; Shi et al. 2011).

Ciliated cells are more abundant within the fimbrae of the infundibulum than in the isthmic region of the Fallopian tube. In humans, approximately 50% of the epithelial cells in the fimbrae are ciliated and this is reduced to 35% in the isthmus (Patek et al. 1972; Lyons et al. 2006). The oviductal cilia are 10 μm in length and 0.25 μm in diameter (Satir 1992).

Ciliated cells are observed in the Fallopian tubes of humans by the 18th week of gestation (Perlman et al. 2005). Ciliated cells are first observed histologically in the ampulla region of mice on P4 (Agduhr 1927). However, taking a closer look by scanning electron microscopy reveals the presence of immature cilia as early as P1 (Dirksen 1974). These immature cilia can be found on the oviduct epithelia before P12, when only mature cilia are present.

14.5 Genes Involved in Oviduct Development

Genetic deletion studies in mice have led to the discovery of many genes involved in oviduct development. These genes can affect oviduct development at various different stages. For example, there are several genes that are known to be essential for initial Müllerian duct formation including *Wnt4* (Vainio et al. 1999), *Lhx1* (Kobayashi et al. 2004), *Pax2* (Torres et al. 1995), *Emx2* (Miyamoto et al. 1997), *Dach1*, and *Dach2* (Davis et al. 2008), and retinoic acid (RA) signaling genes (Mendelsohn et al. 1994; Kastner et al. 1997). However, there is no evidence that these genes are specifically involved in oviduct development. Oviduct development is sensitive to exposure to endocrine disruptors, such as diethylstilbestrol (DES) (Newbold et al. 1983; Yin and Ma 2005) and bisphenol A (BPA) (Suzuki et al. 2002; Newbold et al. 2009). The genes below have been linked to oviduct development and/or differentiation.

14.5.1 Müllerian Duct Differentiation/Patterning

Homeobox genes are involved in segmental patterning of many different tissues (Krumlauf 1994). In the female reproductive tract, some Abdominal B *Hox* gene family members regulate the tissue identity of the Müllerian duct into the oviduct, uterus, cervix, and anterior vagina. These *Hox* genes exhibit overlapping patterns of expression within the mesenchyme of the postnatal female reproductive tract from anterior to posterior, *Hoxa9*, *Hoxa10*, *Hoxa11*, and *Hoxa13* (Taylor et al. 1997). The reproductive tracts of *Hoxa10* null mice underwent homeotic transformation of the anterior portion of the uterus into the morphology of the oviduct. These mice also lack a uterotubal junction (Benson et al. 1996). This suggests that *Hoxa10* is required for establishing the boundary between the differentiating oviduct and uterus. A milder version of uterine homeotic transformation towards oviduct morphology can be seen in *Hoxa11* null mice (Gendron et al. 1997).

Wnt7a is expressed by the Müllerian duct epithelia and is essential for Müllerian duct regression in males, as it regulates mesenchymal *Amhr2* expression (Parr and McMahon 1998). *Wnt7a* is also involved in patterning of the female reproductive tract (Miller and Sassoon 1998; Parr and McMahon 1998). *Wnt7a* null mice exhibit posteriorization of reproductive tract structures. Thus, the oviduct exhibits uterine morphology and the uterus exhibits vaginal morphology. This further results in a loss of oviduct coiling. *Wnt7a* is expressed throughout the Müllerian duct epithelium, but becomes limited to the oviduct and uterus after birth (i.e., differentiation of the Müllerian duct into female reproductive tract structures). Furthermore, *Wnt7a* regulates uterine myometrial organization and may also regulate myosalpinx organization (Miller and Sassoon 1998).

14.5.2 Oviduct Differentiation

Several genes have been identified that play a role in mature oviduct function, including the cannabinoid receptor CB1 (*Cnr1*) (Wang et al. 2004), but the focus of this chapter is on oviduct development. Therefore, this section will address genes involved in oviduct differentiation.

Hedgehog signaling is involved in development of a number of different tissues and organs. When a dominant active allele of *smoothened* (*SmoM2*) is conditionally overexpressed in the Müllerian duct-derived mesenchyme, oviduct development is compromised. The oviducts do not coil and radial patterning is altered. These mice also exhibited increased levels of *Wnt5a* and *Hoxa13* in the anterior uterus and oviduct.

These genes regulate patterning of the posterior Müllerian-derived structures, including the cervix. Their misexpression in the oviduct may result in uterine characteristics, including a lack of coiling (Migone et al. 2012).

Beta-catenin (*Ctnnb1*) is the downstream effector of the canonical *Wnt* signaling pathway. It is required for Müllerian duct regression in males (Kobayashi et al. 2011). Conditional deletion of *Ctnnb1* in the Müllerian duct-derived mesenchyme resulted in a loss of oviduct coiling. However, no changes in *Wnt5a* or *Wnt7a* expression were detected, so the phenotype is not attributable to the loss of either *Wnt* (Deutscher and Hung-Chang Yao 2007). Interestingly, conditional deletion of *Ctnnb1* leads to a change of fate in the uterine myometrial cells from smooth muscle to adipocytes (Arango et al. 2005). It is not clear if this occurs in the myosalpinx of these mice, as well. This suggests that mesenchymal, *Ctnnb1*, and canonical *Wnt* signaling may play a role in both oviductal coiling and mesenchymal differentiation within the oviduct.

Transforming growth factor beta (TGFβ) is another large superfamily of genes important broadly during development. Conditional deletion of Tgfβ type 1 receptor (*Tgfbr1*) in Müllerian duct mesenchyme-derived tissues resulted in the development of oviductal diverticula, or fluid-filled cysts on the oviducts. These diverticula were present as early as 3 weeks of age and are similar to those observed in *Dicer1* conditional knockout mice. This similarity is not due to common regulation. Further, deletion of *Tgfbr1* decreased expression of smooth muscle markers and several microRNAs involved in smooth muscle specification at 3 weeks. However, these differences were not observed in P7 mice, suggesting that *Tgfbr1* is not essential for initial smooth muscle differentiation but rather for organization and maintenance of the smooth muscle layers (Li et al. 2011).

DICER1 is a ribonuclease that is essential for microRNA processing. Conditional deletion of *Dicer1* in the Müllerian duct mesenchyme severely impacted female reproductive tract development and function. In the oviduct, less coiling was apparent and fluid-filled cysts were present in the isthmus (Hong et al. 2008; Nagaraja et al. 2008; Gonzalez and Behringer 2009). The formation of these cysts may be due to defects in smooth muscle formation within the oviduct, causing a weakness in the oviduct. These cysts served as reservoirs for fertilized oocytes, preventing them from traveling into the uterine horns (Nagaraja et al. 2008; Gonzalez and Behringer 2009). Conditional deletion of *Dicer1* also resulted in improper formation of the uterotubal junction (Gonzalez and Behringer 2009).

The dimer of tuberous sclerosis complex (TSC) 1 and TSC2 is essential for the suppression of mammalian target of rapamycin (mTOR) to prevent tumor formation. Conditional deletion of *Tsc1* in the Müllerian duct mesenchyme resulted in an interesting oviduct phenotype. The lumen of the isthmus was occluded, but the ampulla region appeared normal (Tanaka et al. 2012). As a result, the mice were infertile. However, *Tsc1* is also important in the uterine mesenchyme, as in vitro fertilization of embryos was not able to rescue infertility. *Tsc1* in the isthmus mesenchyme directly regulates secretory epithelial cells, because in the absence of *Tsc1*, the secretory cells are the ones occluding the oviduct lumen (Tanaka et al. 2012). This mouse model genetically emphasizes the importance of the oviduct mesenchyme to direct epithelial cell fate and behavior.

FOXJ1 is a member of the forkhead box family of transcription factors. FOXJ1 is expressed by ciliated epithelia and is required for ciliogenesis throughout the body (Chen et al. 1998). In the mouse oviduct, *Foxj1* expression is induced by estrogen. Furthermore, estrogen receptor alpha (*Esr1*) expression is restricted to the ciliated cells in the infundibulum and ampulla regions of the oviduct. Postnatal estrogen treatment can accelerate ciliogenesis in the oviduct of rats. However, *Esr1* null mice undergo normal ciliogenesis (Okada et al. 2004). Thus, estrogen and *Esr1* are not essential for oviduct ciliogenesis, but they are stimulatory of *Foxj1* expression, which in turn regulates ciliogenesis.

14.6 Summary

The mammalian oviduct is a structure that is deceptively simplistic. It is derived from the embryonic Müllerian duct, which is a simple epithelial tube surrounded by mesenchyme. In a very specific perinatal time frame, the oviduct differentiates into four unique segments that are somewhat complex. The basic structure and cell types are evolutionarily conserved from fish (Borcea 1906; Wourms 1977) to amphibians (Wake and Dickie 1998), birds (Bakst 1998), rodents (Agduhr 1927) to humans (Blandau 1978), although the oviduct in egg-laying species contains significant modifications. Proper development of the oviduct is essential for fertility. Understanding how the mouse oviduct develops can give insight into the process in women. The differences between oviducts in mice and women are fewer than the similarities, and include the presence of an ovarian bursa, oviductal coiling, and a less developed infundibulum in mice.

Because most mouse oviduct differentiation occurs postnatally, it facilitates developmental studies. For example, the oviduct is an excellent model for studying epithelial–mesenchymal interactions, as the epithelia remain plastic during the first week of postnatal life. Similarly, oviductal coiling occurs postnatally and can provide insight into coiling in other tissues (i.e., gut, epididymis, etc.). Most of what we know of basic mouse oviduct development is based on a study performed in the 1920s (Agduhr 1927) or focused specifically on ciliogenesis (Dirksen 1974; Hagiwara et al. 2000). Many basic developmental questions remain about cell differentiation, how and why the oviduct coils, develops longitudinal folds, and forms the interstitial type uterotubal junction seen in both women and mice.

References

Afzelius BA, Camner P, Mossberg B (1978) On the function of cilia in the female reproductive tract. Fertil Steril 29(1):72–4

Agduhr E (1927) Studies on the structure and development of the bursa ovarica and the tuba uterina in the mouse. Acta Zoologica 8:1–133

Arango NA, Szotek PP, Manganaro TF, Oliva E, Donahoe PK, Teixeira J (2005) Conditional deletion of beta-catenin in the mesenchyme of the developing mouse uterus results in a switch to adipogenesis in the myometrium. Dev Biol 288(1):276–83

Bakst MR (1998) Structure of the avian oviduct with emphasis on sperm storage in poultry. J Exp Zool 282(4–5):618–26

Benson GV, Lim H, Paria BC, Satokata I, Dey SK, Maas RL (1996) Mechanisms of reduced fertility in Hoxa-10 mutant mice: uterine homeosis and loss of maternal Hoxa-10 expression. Development 122(9):2687–96

Blandau RJ (1978) Comparative aspects of tubal anatomy and physiology as they relate to reconstructive procedures. J Reprod Med 21(1):7–15

Borcea I (1906) Recherches sur le systeme Urogenital-genital des Elasmobranches. Arch Zool Exper et Gen 4:4,199–484

Brody JR, Cunha GR (1989) Histologic, morphometric, and immunocytochemical analysis of myometrial development in rats and mice: I. Normal development. Am J Anat 186(1):1–20

Chen J, Knowles HJ, Hebert JL, Hackett BP (1998) Mutation of the mouse hepatocyte nuclear factor/forkhead homologue 4 gene results in an absence of cilia and random left-right asymmetry. J Clin Invest 102(6):1077–82

Davis RJ, Harding M, Moayedi Y, Mardon G (2008) Mouse Dach1 and Dach2 are redundantly required for Mullerian duct development. Genesis 46(4):205–13

Deutscher E, Hung-Chang Yao H (2007) Essential roles of mesenchyme-derived beta-catenin in mouse Mullerian duct morphogenesis. Dev Biol 307(2):227–36

Dirksen ER (1974) Ciliogenesis in the mouse oviduct. A scanning electron microscope study. J Cell Biol 62(3):899–904

Eddy CA, Pauerstein CJ (1980) Anatomy and physiology of the fallopian tube. Clin Obstet Gynecol 23(4):1177–93

Gendron RL, Paradis H, Hsieh-Li HM, Lee DW, Potter SS, Markoff E (1997) Abnormal uterine stromal and glandular function associated with maternal reproductive defects in Hoxa-11 null mice. Biol Reprod 56(5):1097–105

Gonzalez G, Behringer RR (2009) Dicer is required for female reproductive tract development and fertility in the mouse. Mol Reprod Dev 76(7):678–88

Gruenwald P (1941) The relation of the growing Müllerian Duct to the Wolffian Duct and its importance for the genesis of malformations. Anat Rec 81:1–19

Hafez ESE, Blandau RJ (1969) The mammalian oviduct: comparative biology and methodology. The University of Chicago Press, Chicago

Hagiwara H, Ohwada N, Aoki T, Takata K (2000) Ciliogenesis and ciliary abnormalities. Med Electron Microsc 33(3):109–14, Official journal of the Clinical Electron Microscopy Society of Japan

Halbert SA, Becker DR, Szal SE (1989) Ovum transport in the rat oviductal ampulla in the absence of muscle contractility. Biol Reprod 40(6):1131–6

Hong X, Luense LJ, McGinnis LK, Nothnick WB, Christenson LK (2008) Dicer1 is essential for female fertility and normal development of the female reproductive system. Endocrinology 149(12):6207–12

Ikawa M, Inoue N, Benham AM, Okabe M (2010) Fertilization: a sperm's journey to and interaction with the oocyte. J Clin Invest 120(4):984–94

Jansen RP (1984) Endocrine response in the fallopian tube. Endocr Rev 5(4):525–51

Kastner P, Mark M, Ghyselinck N, Krezel W, Dupe V, Grondona JM, Chambon P (1997) Genetic evidence that the retinoid signal is transduced by heterodimeric RXR/RAR functional units during mouse development. Development 124(2):313–26

Kobayashi A, Behringer RR (2003) Developmental genetics of the female reproductive tract in mammals. Nat Rev Genet 4(12):969–80

Kobayashi A, Kwan KM, Carroll TJ, McMahon AP, Mendelsohn CL, Behringer RR (2005) Distinct and sequential tissue-specific activities of the LIM-class homeobox gene Lim1 for tubular morphogenesis during kidney development. Development 132(12):2809–23

Kobayashi A, Shawlot W, Kania A, Behringer RR (2004) Requirement of Lim1 for female reproductive tract development. Development 131(3):539–49

Kobayashi A, Stewart CA, Wang Y, Fujioka K, Thomas NC, Jamin SP, Behringer RR (2011) beta-Catenin is essential for Müllerian duct regression during male sexual differentiation. Development 138(10):1967–75

Komatsu M, Fujita H (1978) Electron-microscopic studies on the development and aging of the oviduct epithelium of mice. Anat Embryol 152(3):243–59

Koopman P, Gubbay J, Vivian N, Goodfellow P, Lovell-Badge R (1991) Male development of chromosomally female mice transgenic for Sry. Nature 351(6322):117–21

Krumlauf R (1994) Hox genes in vertebrate development. Cell 78(2):191–201

Kurita T, Cooke PS, Cunha GR (2001) Epithelial-stromal tissue interaction in paramesonephric (Müllerian) epithelial differentiation. Dev Biol 240(1):194–211

Kurpios NA, Ibanes M, Davis NM, Lui W, Katz T, Martin JF, Izpisua Belmonte JC, Tabin CJ (2008) The direction of gut looping is established by changes in the extracellular matrix and in cell:cell adhesion. Proc Natl Acad Sci USA 105(25):8499–506

Li Q, Agno JE, Edson MA, Nagaraja AK, Nagashima T, Matzuk MM (2011) Transforming growth factor beta receptor type 1 is essential for female reproductive tract integrity and function. PLoS Genet 7(10):e1002320

Lyons RA, Saridogan E, Djahanbakhch O (2006) The reproductive significance of human Fallopian tube cilia. Hum Reprod Update 12(4):363–72

Mendelsohn C, Lohnes D, Decimo D, Lufkin T, LeMeur M, Chambon P, Mark M (1994) Function of the retinoic acid receptors (RARs) during development (II). Multiple abnormalities at various stages of organogenesis in RAR double mutants. Development 120(10):2749–71

Migone FF, Ren Y, Cowan RG, Harman RM, Nikitin AY, Quirk SM (2012) Dominant activation of the hedgehog signaling pathway alters development of the female reproductive tract. Genesis 50(1):28–40

Miller C, Sassoon DA (1998) Wnt-7a maintains appropriate uterine patterning during the development of the mouse female reproductive tract. Development 125(16):3201–11

Miyamoto N, Yoshida M, Kuratani S, Matsuo I, Aizawa S (1997) Defects of urogenital development in mice lacking Emx2. Development 124(9):1653–64

Morris LH, Hunter RH, Allen WR (2000) Hysteroscopic insemination of small numbers of spermatozoa at the uterotubal junction of preovulatory mares. J Reprod Fertil 118(1):95–100

Nagaraja AK, Andreu-Vieyra C, Franco HL, Ma L, Chen R, Han DY, Zhu H, Agno JE, Gunaratne PH, DeMayo FJ et al (2008) Deletion of Dicer in somatic cells of the female reproductive tract causes sterility. Mol Endocrinol 22(10):2336–52

Newbold RR, Jefferson WN, Padilla-Banks E (2009) Prenatal exposure to bisphenol a at environmentally relevant doses adversely affects the murine female reproductive tract later in life. Environ Health Perspect 117(6):879–85

Newbold RR, Tyrey S, Haney AF, McLachlan JA (1983) Developmentally arrested oviduct: a structural and functional defect in mice following prenatal exposure to diethylstilbestrol. Teratology 27(3):417–26

Nielsen M, Bjornsdottir S, Hoyer PE, Byskov AG (2000) Ontogeny of oestrogen receptor alpha in gonads and sex ducts of fetal and newborn mice. J Reprod Fertil 118(1):195–204

Okada A, Ohta Y, Brody SL, Watanabe H, Krust A, Chambon P, Iguchi T (2004) Role of foxj1 and estrogen receptor alpha in ciliated epithelial cell differentiation of the neonatal oviduct. J Mol Endocrinol 32(3):615–25

Orvis GD, Behringer RR (2007) Cellular mechanisms of Müllerian duct formation in the mouse. Dev Biol 306(2):493–504

Parr BA, McMahon AP (1998) Sexually dimorphic development of the mammalian reproductive tract requires Wnt-7a. Nature 395(6703):707–10

Patek E, Nilsson L, Johannisson E (1972) Scanning electron microscopic study of the human fallopian tube. Report II. Fetal life, reproductive life, and postmenopause. Fertil Steril 23 (10):719–33

Perlman S, Hertweck P, Fallat ME (2005) Paratubal and tubal abnormalities. Semin Pediatr Surg 14(2):124–34
Poweirza S (1912) Über Änderungen im Bau der Ausführwege des weiblichen Geschlechts-apparates der Maus während ihres postembyonalen Lebens. Bull Int Acad Sci Krackow 349–399
Pulkkinen MO (1995) Oviductal function is critical for very early human life. Ann Med 27(3): 307–10
Satir P (1992) Mechanisms of ciliary movement: contributions from electron microscopy. Scanning Microsc 6(2):573–9
Shaw JL, Dey SK, Critchley HO, Horne AW (2010) Current knowledge of the aetiology of human tubal ectopic pregnancy. Hum Reprod Update 16(4):432–444
Shi D, Komatsu K, Uemura T, Fujimori T (2011) Analysis of ciliary beat frequency and ovum transport ability in the mouse oviduct. Genes Cells 16(3):282–90, Devoted to molecular & cellular mechanisms
Suzuki A, Sugihara A, Uchida K, Sato T, Ohta Y, Katsu Y, Watanabe H, Iguchi T (2002) Developmental effects of perinatal exposure to bisphenol-A and diethylstilbestrol on reproductive organs in female mice. Reprod Toxicol 16(2):107–16
Tanaka Y, Park JH, Tanwar PS, Kaneko-Tarui T, Mittal S, Lee HJ, Teixeira JM (2012) Deletion of tuberous sclerosis 1 in somatic cells of the murine reproductive tract causes female infertility. Endocrinology 153(1):404–16
Taylor HS, Vanden Heuvel GB, Igarashi P (1997) A conserved Hox axis in the mouse and human female reproductive system: late establishment and persistent adult expression of the Hoxa cluster genes. Biol Reprod 57(6):1338–45
Torres M, Gomez-Pardo E, Dressler GR, Gruss P (1995) Pax-2 controls multiple steps of urogenital development. Development 121(12):4057–65
Vainio S, Heikkila M, Kispert A, Chin N, McMahon AP (1999) Female development in mammals is regulated by Wnt-4 signalling. Nature 397(6718):405–9
Wake MH, Dickie R (1998) Oviduct structure and function and reproductive modes in amphibians. J Exp Zool 282(4–5):477–506
Wang H, Guo Y, Wang D, Kingsley PJ, Marnett LJ, Das SK, DuBois RN, Dey SK (2004) Aberrant cannabinoid signaling impairs oviductal transport of embryos. Nat Med 10(10):1074–80
Woodruff JD, Paurstein CJ (1969) The fallopian tube: structure, function, pathology, and management. Williams and Wilkins, Baltimore, MD
Wourms JP (1977) Reproduction and development in Chondrichthyan fishes. Am Zool 17:379–410
Yamanouchi H, Umezu T, Tomooka Y (2010) Reconstruction of oviduct and demonstration of epithelial fate determination in mice. Biol Reprod 82(3):528–33
Yin Y, Ma L (2005) Development of the mammalian female reproductive tract. J Biochem 137 (6):677–83

Chapter 15
Cell Lineages, Growth and Repair of the Mouse Heart

Fabienne Lescroart and Sigolène M. Meilhac

Abstract The formation of the heart involves diversification of lineages which differentiate into distinct cardiac cell types or contribute to different regions such as the four cardiac chambers. The heart is the first organ to form in the embryo. However, in parallel with the growth of the organism, before or after birth, the heart has to adapt its size to maintain pumping efficiency. The adult heart has only a mild regeneration potential; thus, strategies to repair the heart after injury are based on the mobilisation of resident cardiac stem cells or the transplantation of external sources of stem cells. We discuss current knowledge on these aspects and raise questions for future research.

15.1 Introduction

The function of the heart is intimately linked to its structure. Malformations of the heart which impair the connections or the confinement of cardiac chambers impinge on the circulation of the blood and its degree of oxygenation. The architecture of the heart, including trabeculations and oriented fibres of the myocardium, is essential for efficient contraction. Therefore, the study of heart morphogenesis provides important insights into the molecular and cellular mechanisms which are required to shape the heart. This is relevant to medical issues, such as congenital heart diseases and heart repair. Here, we shall deal mainly with the mouse heart, as an experimental model for mammals, unless otherwise specified.

F. Lescroart • S.M. Meilhac (✉)
Department of Developmental Biology, Institut Pasteur, Unit of Molecular Genetics of Development, 75015 Paris, France

CNRS, URA2578, 75015 Paris, France
e-mail: sigolene.meilhac@pasteur.fr

J.Z. Kubiak (ed.), *Mouse Development*, Results and Problems in Cell Differentiation 55,
DOI 10.1007/978-3-642-30406-4_15,

15.2 Lineages of Heart Cells: Origin and Clonal Segregation

Myocardium is the main component of the heart; however, many different cell types contribute to the heart. In the last decade, much progress has been made in the elucidation of the lineage relationships between them (Fig. 15.1) and between the cardiomyocytes colonising different regions of the heart (Fig. 15.2). The underlying issues include the organisation of the heart field which contains the progenitor cells in the embryo, the timing of segregation of the lineages and the identification of markers of cell lineages.

15.2.1 *Cardiogenic Mesoderm and the Layers of the Heart (Pericardium, Epicardium, Myocardium and Endocardium)*

Heart cells can be traced back to gastrulation stages. Clonal analyses have indicated that heart progenitors in the epiblast (E6.5), before gastrulation, are clonally related to the endoderm, the neuroectoderm, the extra-embryonic mesoderm and the paraxial mesoderm (Buckingham et al. 1997; Lawson and Pedersen 1987; Tzouanacou et al. 2009). Fate map experiments have shown that cardiac progenitors ingress early, at the mid-streak stage, and are located in the anterior region of the primitive streak, in close proximity with progenitors of the cranial mesoderm (Kinder et al. 1999). *Mesp1*, which encodes a bHLH transcription factor, is expressed transiently in the nascent mesoderm at gastrulation. It is the earliest marker of cardiovascular lineages but is not restricted to them (Saga et al. 2000, 1999). Mesp1 has been proposed to act as a master regulator of cardiovascular fate specification (Bondue et al. 2008).

The embryonic heart is composed of two concentric tissues: the endocardium, which is the endothelial lining of the heart, and the myocardium, which is the muscular envelope, and is enclosed in a cavity outlined by the pericardium. The progenitors of these three tissues are localised in the same region of the epiblast (Tam et al. 1997). However grafting experiments have indicated that these progenitors are not committed to a cardiac fate before they have completed their migration, as lateral plate mesoderm cells, towards the cranial region of the E7.5 embryo, where they form bilateral heart fields or cardiogenic mesoderm. The timing of the segregation between the endocardial, myocardial and pericardial lineages is still controversial (Harris and Black 2010). Distinct progenitors of the endocardium or myocardium have been described in the chick early primitive streak, following retroviral cell labelling (Wei and Mikawa 2000). However in the mouse, genetic tracing experiments have shown that the myocardium, endocardium and pericardium all derive from progenitor cells that are positive for *Nkx2-5* (Stanley et al. 2002), and that the myocardium and the endocardium also derive from progenitor cells that have expressed *Mesp1* and *Isl1* (Cai et al. 2003;

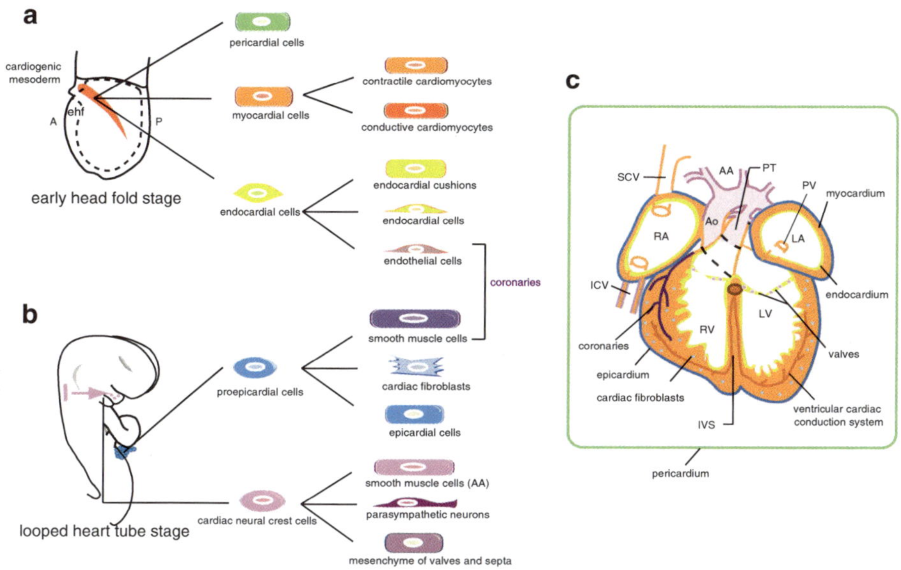

Fig. 15.1 Lineage of cardiac cell types. Schemes summarising the lineage tree of cardiac cell types, which derive from three origins: (**a**) the cardiogenic mesoderm (*in red*), at the early head fold stage, contains precursors of the endocardium (*yellow*), myocardium (*orange*) and pericardium (*green*); (**b**) the proepicardium (*in blue*) or the cardiac neural crest (*in purple*) at the looped heart tube stage. The histology of the foetal heart is represented in **c**, with colour-coded cell types. *A* anterior, *AA* arch arteries, *Ao* aorta, *ehf* early head folds, *ICV* inferior caval vein, *IVS* interventricular septum, *LA* left atrium, *LV* left ventricle, *P* posterior, *PT* pulmonary trunk, *PV* pulmonary vein, *RA* right atrium, *RV* right ventricle, *SCV* superior caval vein

Saga et al. 2000; Sun et al. 2007). As *Nkx2-5* and *Isl1*, that encode homeodomain transcription factors, are first expressed at the early head fold stage (E7.5), in undifferentiated progenitors (Cai et al. 2003; Lints et al. 1993), these genetic data indicate that endocardial, myocardial and pericardial progenitor cells still share spatial and molecular relationships in the E7.5 heart fields. Moreover, the gene *Etv2*, which encodes an ETS domain transcription factor and is expressed in endothelial cells, is essential for endocardium development and is regulated by *Nkx2-5* (Ferdous et al. 2009). *Etv2*-expressing cells give rise only to endocardial cells in the heart. However, in the absence of *Etv2*, the cells which should have expressed it differentiate into other muscle lineages including the myocardium (Rasmussen et al. 2011). This suggests that there is a tightly regulated balance between myocardial and endocardial cell lineages. In addition to their origin in the cardiogenic mesoderm, some endocardial cells may also derive from vascular endothelial cells, which have never expressed *Isl1* (Milgrom-Hoffman et al. 2011). In the cardiac crescent, the three cell types can be distinguished histologically: squamous pericardial cells are part of the somatic mesoderm, whereas myocardial cells are found in the epithelial splanchnic mesoderm and endocardial cells appear as small clusters ventrally, which delaminate from the splanchnic

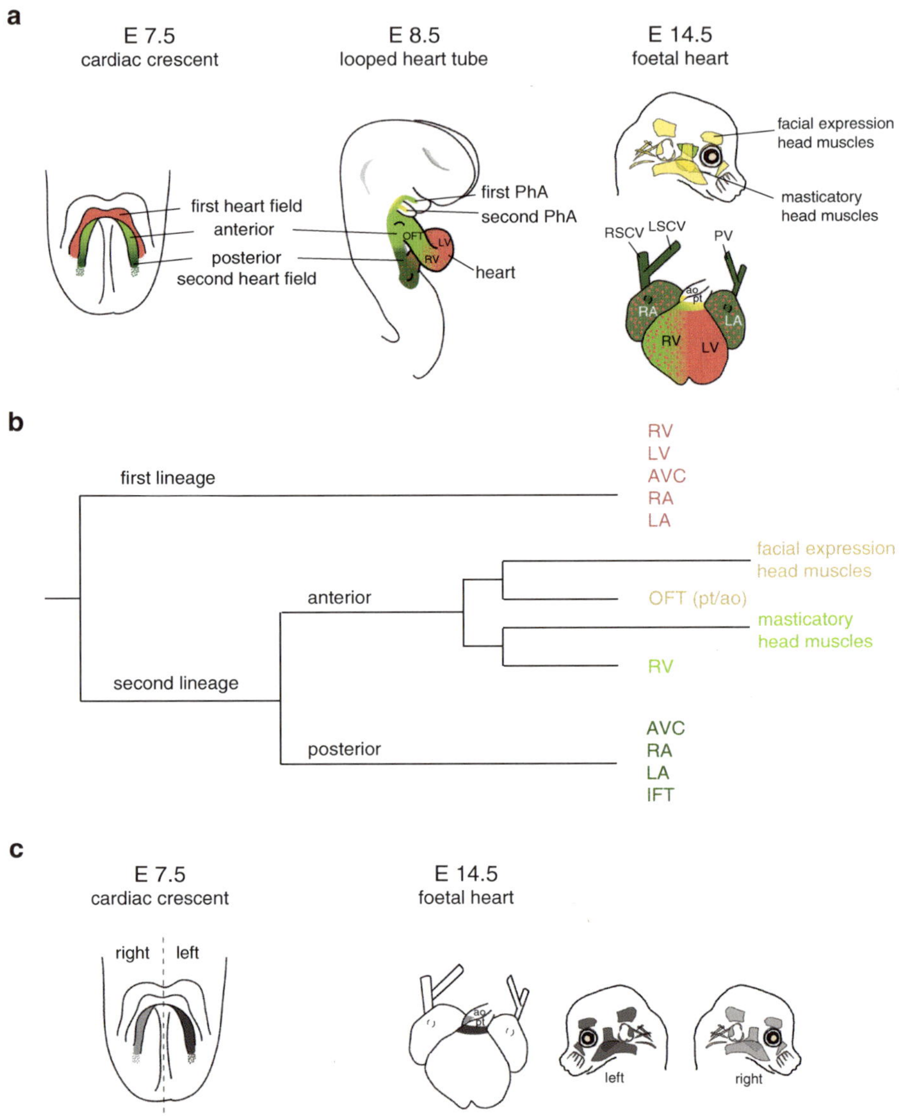

Fig. 15.2 First and second heart fields and further subdivisions of the myocardial lineage. (**a**) The myocardium of the heart forms from two cell populations: the first (*red*) and second (*green*) heart fields. The second heart field can be subdivided into an anterior (*light green* and *yellow*) and posterior (*dark green*) domain, based on molecular markers (see text). The second heart field, which extends into the pharyngeal arches (first arch in *light green*, second in *yellow*) at E8.5, is progressively added to the heart tube at both the arterial and venous poles of the heart (*arrows*). Derivatives of the heart fields are shown at E14.5, in addition to the clonally related skeletal head muscles, which derive from the pharyngeal arches. (**b**) Lineage tree of myocardial cells, as revealed by retrospective clonal analyses. (**c**) The heart fields are also patterned according to left/right signalling. The right heart field and its derivatives are shown in *light grey*, whereas the

mesoderm (Kaufman and Navaratnam 1981). It is still unclear to what extent cells of the splanchnic mesoderm are bipotent for endocardial and myocardial fates (Linask et al. 1997). As an indication of myocardial differentiation, the expression of sarcomeric proteins such as Myl7 (also known as MLC2a) is first detected at the early head fold stage, in a subset of cells of the heart fields which is more lateral compared to the *Isl1*-expressing progenitors (Cai et al. 2003). This is before the formation of the α-cardiac actin-positive cardiac crescent (Sassoon et al. 1988), in which the contractile apparatus of myocardial cells becomes functional (Nishii and Shibata 2006).

The epicardium originates from the proepicardium, a transient structure observed between E8.5 and E10.5 at the venous pole of the heart (Schulte et al. 2007). Cells detach from the proepicardium and, by E11.5, have covered most of the heart to form a single layered epithelial outer sheet. From an evolutionary point of view, the epicardium has permitted the formation of a multi-layered myocardium and thus a more efficient pumping heart (Carmona et al. 2010). The proepicardium expresses markers such as *Tbx18* and *Wt1*, encoding transcription factors which also mark the kidney. The proepicardium originates from mesoderm on the surface of the septum transversum (the primordium of the diaphragm), adjacent to the sinus venosus or inflow tract of the heart (Viragh and Challice 1981), and may be induced by signals from the neighbouring liver bud (Ishii et al. 2007). How the proepicardium relates to the other tissue layers of the heart is still unclear. Depending on the targeted allele of Cre used, controversial results have been obtained by genetic tracing experiments, regarding the origin of the epicardium from *Nkx2-5* and *Isl1* expressing progenitor cells (Lints et al. 1993; Ma et al. 2008; Sun et al. 2007; Zhou et al. 2008b). Markers such as Cited2 (Dunwoodie et al. 1998), which is traced back for earlier expression pattern, would indicate a craniolateral origin at the early head fold stage, which becomes a caudal territory upon closure of the foregut at the straight heart tube stage. This is consistent with cell labelling experiments which traced back the *Tbx18* expression domain (Mommersteeg et al. 2010). At the heart tube stage in the chick, labelling of cells in the splanchnic mesoderm of the right vitelline vein, caudal to the heart, has indicated a spatial proximity between progenitors of the inflow tract myocardium and the proepicardium (van Wijk et al. 2009). In addition to these complex movements of proepicardial cells, the epicardium in the outflow region of the chick heart has a distinct origin, from the cephalic pericardium (Perez-Pomares et al. 2003). The epicardium, as well as other tissue layers of the heart, later contributes to the diversification of cell types in the heart.

Fig. 15.2 (continued) left heart field and its derivatives are shown in *dark grey*. Skeletal head muscles, which are clonally related to the outflow tract myocardium, are also schematised. *Ao* aorta, *AVC* atrio-ventricular canal, *E* embryonic day, *IFT* inflow tract (veins), *LA* left atrium, *LSCV* left superior caval vein, *LV* left ventricle, *OFT* outflow tract, *PhA* pharyngeal arches, *pt* pulmonary trunk, *PV* pulmonary vein, *RA* right atrium, *RSCV* right superior caval vein, *RV* right ventricle

15.2.2 Derivatives of the Epicardium, Myocardium and Endocardium

Cells of the epicardium undergo an epithelial to mesenchymal transition from E11.5 and invade the subepicardial space and the myocardium. The orientation of the mitotic spindle is important for balancing the expansion of the epicardium with the invasion of the myocardium (Wu et al. 2010). Cell labelling experiments in the chick have shown that the epicardium-derived cells contribute most of the cardiac interstitial cells such as fibroblasts, as well as the smooth muscle cells of the coronary blood vessels, which irrigate the myocardium (Manner 1999; Mikawa and Gourdie 1996). Genetic tracing experiments with *Wt1-Cre* and *Tbx18-Cre* have lead to controversial conclusions on the participation of epicardium-derived cells in the myocardium, because precise knowledge on the spatiotemporal pattern of expression of the marker is required to interpret the data (Cai et al. 2008; Christoffels et al. 2009; Zhou et al. 2008a). The contribution of epicardium-derived cells to the endothelium of the coronaries has also been debated, based on discrepancy between embryological experiments in the chick, in favour of such a contribution (Manner 1999; Mikawa and Gourdie 1996), and genetic tracing in the mouse against (Cai et al. 2008; Zhou et al. 2008a). This question has been recently solved in an elegant study by Red-Horse et al. (2010), which has revealed that the endothelial lining of both the coronary veins and arteries arises initially from angiogenic sprouts of the sinus venosus at E11.5. These venous cells are not descendants of *Tbx18* expressing progenitors, and thus do not correspond to epicardium-derived cells.

Similar to the epicardium, the endocardium can also undergo an epithelial to mesenchymal transition. This occurs from E9.5 in the atrio-ventricular canal and later in the outflow tract. The endocardial cushions correspond to the accumulation of endocardium-derived mesenchymal cells, between the endocardium and myocardium of specific heart regions. These cushions later give rise to the cardiac valves and also contribute to the septation of the atria, of the ventricles and the partition of the outflow tract into the aortic and pulmonary trunks. Among the many studies which support this characterisation (see Camenisch et al. 2010) are the pioneering cell labelling experiments in the chick (De la Cruz et al. 1983) and more recent genetic tracing in the mouse using markers such as *Tie2* (de Lange et al. 2004). The endocardium has also been shown to participate in the endothelial lining of the coronaries, via endocardial blood islands in the interventricular region, although to a lesser extent than the sinus venosus derivatives (Red-Horse et al. 2010).

Derivatives of the myocardium have also been reported. The cardiac conduction system generates and transmits electrical impulses to coordinate heart contraction. It is mainly composed of modified cardiomyocytes (Christoffels et al. 2010). The timing and mode of segregation of contracting and conductive myocytes has been controversial in the chick. Progenitors of the ventricular conduction system have been postulated to arise at the crest of the interventricular septum, from an independent pool (Chan-Thomas et al. 1993; Lamers et al. 1991) or by continuous recruitment from neighbouring cardiomyocytes (Gourdie et al. 1998). A dual origin of the cardiac conduction system is supported in the mouse by genetic tracing with *Mesp1-Cre*, which indicated that 20 % of cells in the ventricular conduction system have never expressed this early marker of the cardiogenic mesoderm (Kitajima et al. 2006). Clonal analyses have characterised common progenitors of contracting and conductive myocytes of the ventricle, which are segregated by E16.5 (Miquerol et al. 2010). Cells of both lineages continue to proliferate after the segregation, although the proliferation of conductive myocytes is much lower. The origin of the central conduction system, which includes the sino-atrial and atrioventricular nodes, remains obscure. Markers of the central conduction system such as Tbx3 can potentially be traced back to the early heart tube (Hoogaars et al. 2004), although it is unclear whether they correspond to conductive cell progenitors at this early stage.

The initial cardiogenic mesoderm, or heart fields, provides most cells of the heart, after diversification of lineages. However, there is also contribution from another source.

15.2.3 Cardiac Neural Crest Cells

The cardiac neural crest, which delaminates from the neural tube, migrates through the posterior pharyngeal arches and reaches the heart from E9.5. It has been well characterised by cell labelling experiments in the chick (Kirby et al. 1983; Le Lievre and Le Douarin 1975) and genetic tracing in the mouse (Jiang et al. 2000), which have demonstrated a later contribution to the smooth muscle of the aortic and pulmonary trunks, to the smooth muscle of the pharyngeal arch arteries, to the septum of the ventricles and the septum between the aortic and pulmonary trunks, to the valves, as well as to the parasympathetic innervation of the heart.

With the identification of the derivatives of the neural crest cells and of the tissue layers of the heart, a lineage tree can be reconstructed, which explains how the different cell types, which are integrated in the heart, diverge. However, lineage analyses have also been instrumental in understanding the origin of the different regions of the heart such as the four cardiac chambers.

15.2.4 The First and Second Heart Fields of Myocardial Cells

The early (straight) heart tube, at about the six somite stage, does not contain progenitors of all regions of the heart. The myocardium of the outflow tract and right ventricle is added later by recruitment of cells from undifferentiated progenitors (Viragh and Challice 1973; Zaffran et al. 2004). This source of progenitor cells was initially located by genetic tracing with the *Mlc1v-lacZ-24* transgene, which recapitulates *Fgf10* expression (Kelly et al. 2001). It is first expressed medially to the cardiac crescent, in the *Isl1* expression domain, and later in the pharyngeal mesoderm. The perdurance of the β-galactosidase indicated contribution of cells to the outflow tract and right ventricle, which was confirmed by cell labelling experiments in the pharyngeal mesoderm. These observations show that there are two waves of cardiomyocyte differentiation, which reflect the existence of two heart fields (Buckingham et al. 2005).

A retrospective clonal analysis demonstrated the existence of two distinct myocardial cell lineages, which mainly correspond to these two heart fields. The first lineage contributes to the left ventricle, right ventricle and both atria but not to the outflow tract of the embryonic heart, whereas the second lineage contributes to the outflow tract, right ventricle and both atria, but not to the left ventricle. The two myocardial lineages, which have overlapping contributions to the heart tube, were shown to segregate early, probably at the onset of gastrulation (Meilhac et al. 2004a).

The expression of *Isl1* is more extensive than the domain of *Fgf10*. *Isl1* is expressed medially to the cardiac crescent and then dorsally to the heart tube and is not expressed in differentiated myocardial cells (Cai et al. 2003). It has been regarded as the marker of the second heart field. Genetic tracing indicates that the cells which have expressed *Isl1* have a myocardial contribution which closely resembles that of the second lineage. Further experiments, with a tamoxifen-inducible *Isl1-CreERt2*, have shown that myocardial cells have expressed *Isl1* between E6.5 and E9 (Sun et al. 2007). In *Isl1* mutants, the cells, which should have expressed it, fail to enter the heart, resulting in organ truncation (Cai et al. 2003). Therefore, Isl1 was initially considered as a marker of the second myocardial lineage. However, other studies with a new *Isl1-Cre* line (Prall et al. 2007) or a new reporter of Cre recombinase activity (Ma et al. 2008) suggest a more extensive contribution of *Isl1*-expressing cells to the heart and in particular to the left ventricle. This raises the question whether Isl1 is a bona fide marker of the second myocardial lineage. In general, such a marker of the entire second myocardial lineage is still lacking, although there are several markers of subdomains of the second heart field.

15.2.5 Subdivisions of the Second Heart Field

The second heart field is subdivided into anterior and posterior subdomains that are defined by molecular differences (see Vincent and Buckingham 2010) and by different contributions to the heart.

The anterior second heart field is characterised by the expression of *Fgf8*, *Fgf10* (Kelly et al. 2001) and *Tbx1* (Brown et al. 2004; Xu et al. 2004). An enhancer of the *Mef2c* gene is widely used to specifically drive expression in the anterior second heart field, with later derivatives in the outflow region and the right ventricle (Dodou et al. 2004; Verzi et al. 2005). Connected to this arterial pole of the heart, the pharyngeal arches are transient embryonic structures, which integrate different cell populations to generate the oro-facial–pharyngeal region. The mesodermal core of the arches can be regarded as an extension of the anterior second heart field, and expresses markers such as *Isl1* and *Fgf10* (Cai et al. 2003; Kelly et al. 2001) The second pharyngeal arch can contribute cells to the outflow tract of the heart (Kelly et al. 2001), whereas the first pharyngeal arch, the most anterior, contributes to the right ventricle (Lescroart et al. 2010). The more caudal arches (3–6) contain progenitors of endothelial cells of the pharyngeal arch arteries (Kelly et al. 2001; Sun et al. 2007). Derivatives of the pharyngeal arches also include smooth muscles of the aortic and pulmonary trunks as shown in the chick (Waldo et al. 2005; Ward et al. 2005). In proximity with these heart progenitors are the progenitors of the head muscles, which are also part of the mesodermal core of the pharyngeal arches. Retrospective clonal analysis has demonstrated a clonal relationship between cells deriving from the second heart field and head muscles, such that the outflow tract is clonally related to facial expression muscles (derivatives of the second arch), whereas the right ventricle is clonally related to the masticatory head muscles (derivatives of the first arch). A further left/right regionalisation is detected, which distinguishes the trunk of the pulmonary artery and the left facial expression muscles, from the aorta and the right facial expression muscles (Lescroart et al. 2010). This regionalisation of the outflow tract depends on the left/right patterning of the embryo. Expression of a transgene under the control of *Pitx2c*, which is a major player in left/right patterning, is consistent with a regionalisation of the outflow tract prefiguring the great arteries (Furtado et al. 2011). Other transgenic markers include the 96–16 transgene, which is inserted close to the *Semaphorin 3c* locus and is preferentially expressed in the pulmonary trunk myocardium, or the *T55* transgene, that has trapped part of the regulatory sequence of the *Hes1* locus (encoding a downstream target of the Notch pathway) and is later expressed in the base of the aorta (Bajolle et al. 2008; Rochais et al. 2009; Theveniau-Ruissy et al. 2008). This reveals differences in transcriptional regulation in these subdomains.

As distinct from the anterior second heart field, which contributes to the arterial pole of the heart, the posterior second heart field mainly gives rise to the venous pole. It is characterised as the *Isl1*-expression domain that does not express the markers of the anterior heart field (Galli et al. 2008) and positive markers include Podoplanin (Gittenberger-de Groot et al. 2007). Genetic tracing experiments suggested that the caval vein myocardium arises from a population of cells expressing *Tbx18* and not *Isl1* which is thus different from the second heart field (Christoffels et al. 2006; Mommersteeg et al. 2010). However, retrospective clonal analysis demonstrated that both the caval vein and pulmonary vein myocardium share common progenitors with atrial myocardium (FL, Mohun, Bennet and Buckingham, unpublished observations). This correlates with cell labelling experiments, which indicated a common origin for the atria and outflow tract myocardium in the posterior second heart field (expressing *Isl1* and not *Fgf10*), with an additional contribution to the atrio-ventricular canal (Domínguez, SMM, Bland, Buckingham and Brown, unpublished observations). Further left/right regionalisation was also revealed by these studies. The second heart field appears thus to be patterned both along the left–right and antero-posterior axis. Retinoic acid signalling is important for antero-posterior patterning and embryos deficient for it have an expanded anterior second heart field (Ryckebusch et al. 2008).

Downstream lie the homeogenes *Hoxb1*, *Hoxa1* and *Hoxa3*, which are expressed in a graded fashion in the second heart field (Bertrand et al. 2011).

The complexity of the myocardial lineage tree invalidates the old segmented model of heart development, which hypothesised that each region of the heart corresponds to a segment, defined as a transcriptional and lineage module, with an early origin before the formation of the heart tube (Fishman and Chien 1997). However, questions remain for the future regarding the two heart field model. For example the mechanism of the segregation between the two myocardial lineages is unknown and, if characterised, should make it possible to identify bona fide markers of the lineages. Another challenge will be to understand the precise relationship between the first/second heart fields and the first/second lineages. Indeed, the heart fields are defined by molecular markers of progenitor cells, which differentiate before or after the cardiac crescent stage, whereas the lineages are defined by the clonal history of differentiated cells. It is conceivable, for example, that some precursors of the first lineage, which differentiate after the crescent stage, are still present in the second heart field.

15.3 Growth of the Developing Heart: Hyperplasia and Hypertrophy

The previous section shows that the embryonic heart grows from the recruitment of cells from progenitor sources such as the heart fields. However, the heart also grows significantly from the proliferation of its differentiated cells during the embryonic period, or from their enlargement after birth. These mechanisms of growth are called hyperplastic and hypertrophic, respectively (Sedmera and Thompson 2011). The hyperplastic growth of the chamber myocardium results in the formation of trabeculations, i.e., finger-like protrusions pointing towards the blood cavity. In contrast, the epicardial–most myocardium remains compact and is highly proliferative. The hypertrophic growth of the myocardium is mainly based on the binucleation of cardiomyocytes, which undergo karyokinesis without cytokinesis. In the postnatal heart, cardiomyocytes no longer constitute the most abundant cell type of the heart and a significant contribution to the growth of the postnatal heart comes from the proliferation of cardiac fibroblasts. Growth of the heart is necessary to accommodate with the increased size of the organism. It is a dynamic process, which extends beyond development, which is common to all individuals, to more adaptive mechanisms such as physiological growth after exercise or pathological responses to injury. We shall give an insight in some of the many factors that have been identified to regulate developmental growth, in both rate and orientation.

Modulation of the proliferation of cardiac progenitors and of the recruitment of these cells to the heart tube will impinge on the growth of the embryonic heart. For example, the proliferation of the cardiac progenitors of the second heart field has been shown to be promoted by Fgf, hedgehog and canonical Wnt signalling. Canonical Wnt, together with Notch and Bmp signalling, also exerts an effect on the differentiation of these cells, and hence on their recruitment to the heart tube (see Vincent and Buckingham 2010).

The hyperplastic growth of the myocardium is dependent on well characterised crosstalks between the different tissue layers of the heart. Inactivation, in the epicardium, of the genes encoding receptors of retinoic acid leads to hypoplasia of the compact ventricular myocardium (Merki et al. 2005), as do mutations, like *Itga4*$^{-/-}$ or *Wt1*$^{-/-}$ (Kreidberg et al. 1993), which impair the formation of the epicardium. The epicardium requires retinoic acid and erythropoietin autocrine signalling to secrete mitogenic signals to cardiomyocytes (see Sucov et al. 2009). The nature of the secreted signal is still unclear. Direct candidates include Fgf, Pdgf, Igf2 (Li et al. 2011), or the signal may be indirect via Wnt9b secretion. The endocardium is another source of factors promoting growth, such as Notch, EphrinB2 and Neuregulin1, which favour the differentiation of trabeculations, via the activation of ErbB2/4 and Bmp10 in the myocardium (Chen et al. 2004; Grego-Bessa et al. 2007; Meyer and Birchmeier 1995).

When the myocardium becomes thicker, it is unlikely that the signals from other tissue layers can reach all myocardial cells. Factors intrinsic to the myocardium indeed intervene in the regulation of hyperplastic growth. Transcription factors such as Nkx2-5, Tbx2, Gata4 or Foxm1 control the balance between the compact and trabeculated myocardium (Bolte et al. 2011; Cai et al. 2005; Nakajima et al. 2011; Terada et al. 2011). The differential expression of transcription factors in the cardiac chambers as opposed to the rest of the cardiac tube may underlie the preferential growth of the myocardium in the chambers, as explained in the ballooning model (Christoffels et al. 2004). Another pathway active in the myocardium is the non-canonical Wnt/Planar Cell Polarity pathway, which is required for the compaction of the myocardium. Indeed, inactivation of *Scrib*, *Vangl2*, *Daam1* or *Wnt11* results in non-compaction of the ventricles (Henderson and Chaudhry 2011). A classical signalling pathway, that modulates the growth of many organs, is the Hippo pathway. It has recently been shown to be important for the growth of the myocardium, by impacting both cell proliferation and apoptosis. Heart specific deletions of members of the Hippo pathway such as *Lats2*, *Sav1* or *Mts1/2* (Heallen et al. 2011) or overexpression of *Yap1* (Xin et al. 2011) produce cardiomegaly, with a hyperplastic heart. This pathway can influence canonical Wnt signalling. As a consequence of heart specific deletions of *β-catenin*, the heart becomes hypoplastic (Klaus et al. 2007; Kwon et al. 2007; Lin et al. 2007).

Downstream effectors of the different pathways regulating myocardial growth, include K-ras, CyclinD1/2/3, Cdk2/4 and N-myc, which are directly involved in cell cycle regulation (see Sucov et al. 2009).

Growth of the heart is regulated regionally, both for its rate and orientation, which ultimately determines the shape of the heart. Hyperplastic growth of the myocardium is mainly exponential (Meilhac et al. 2003). The proliferation of heart progenitors is dispersive, whereas that of differentiated cells is more coherent (Meilhac et al. 2003). Local increased rates of cell proliferation have been documented during the embryonic period, particularly in the chick (Thompson et al. 1990; van den Berg et al. 2009), and account for the elongation of the early heart tube, the expansion of cardiac chambers and the thickening of the myocardial wall. Clonal analyses have indicated that orientation of myocardial hyperplastic growth underlies fine differential shaping of the cardiac chambers (Meilhac et al. 2004b) as well as myofibre architecture (Meilhac et al. 2003). Cardiac myofibres correspond to the alignment of individual elongated cells, connected with gap junctions, and are important for the rapid propagation of the contraction. Orientation also concerns cell hypertrophy, which has been shown to differ in dynamics between the right and left ventricles (Leu et al. 2001).

The extra-cellular matrix appears to play an important role in the growth of the heart. Disruption of *hyaluronan synthase-2,* encoding an enzyme required for the production of a glycosaminoglycan, or *versican,* encoding a chondroitin sulphate proteoglycan, abrogates ventricular trabeculation (Camenisch et al. 2000). The extra-cellular matrix is involved in direct signalling via integrin or FAK receptors, which are essential for myocardial growth (Peng et al. 2008; Shai et al. 2002). It also controls the availability of growth factors, which are trapped or even modified by its components, as exemplified by the interaction between hyaluronan and the neuregulin receptor ErbB2 (Camenisch et al. 2002). Growth factors can be released by the action of metalloproteinases. For example, the metalloproteinase ADAMTS1, which can cleave versican, is secreted by the endocardium and prevents excessive trabeculation of the myocardium (Stankunas et al. 2008).

Cardiac fibroblasts, which secrete components of the extra-cellular matrix, provide another aspect of myocardium growth control. In coculture experiments, it was demonstrated that fibroblasts secrete heparin-binding epidermal growth factor (HB-EGF), which binds to EGF or neuregulin receptors to stimulate myocardial growth. The concomitant secretion of proteins of the extra-cellular matrix, collagen and fibronectin, stimulates the mitogenic activity of HB-EGF at embryonic stages and creates a microenvironment that permits β1A-integrin-mediated proliferation (Ieda et al. 2009). In addition to being a source of paracrine growth factors, fibroblasts at least double in number during the first 2 weeks after birth, providing an important direct contribution to the growth of the postnatal heart (Snider et al. 2009). However, the balance between the number of cardiomyocytes and cardiac fibroblasts must be tightly controlled, since too many fibroblasts, a condition called fibrosis, impair myocardial stiffness and contractility (Weber and Brilla 1991).

Adult cardiac fibroblasts, as distinct from embryonic fibroblasts, induce more cell hypertrophy, but less proliferation (Ieda et al. 2009). Hypertrophy corresponds to the enlargement of cells, by accumulation of sarcomeric proteins, either longitudinally (eccentric hypertrophy) or radially (concentric hypertrophy). It is not confined to postnatal stages, as it has been shown in the mouse that the length of cardiomyocytes increases threefold between E8.5 and birth (Hirschy et al. 2006), whereas cardiomyocyte volume raises up to 15-fold between birth and adult stages (Leu et al. 2001). Direct transcriptional regulation of sarcomere integrity, as for instance by the homeodomain transcription factor Prox1, is required for foetal cardiomyocyte hypertrophy (Risebro et al. 2009). When the organism grows to adult stages, there is a volume overload, which stimulates cell hypertrophy, via stretch-induced gene expression. An important determinant of physiological hypertrophy is the Igf1-PI3K-Akt pathway. Overexpression of the Igf1 receptor in transgenic mice causes cardiac hypertrophy, with an increased size of cardiomyocytes (McMullen et al. 2004). Physiological hypertrophy (see Catalucci et al. 2008) is molecularly distinct from pathological hypertrophy (see Barry et al. 2008), and is beneficial for cardiac function. In pathological conditions of the adult heart, stimulation of myocardial hyperplastic growth is essential to regenerate tissue.

15.4 Cardiac Stem Cells: A Tool for Heart Repair

Myocardial infarction is the leading cause of death. Because of an interruption of blood supply in the myocardium, a massive loss of cardiomyocytes occurs. In response, scar tissue is formed to prevent leakage. Cardiomyocytes undergo hypertrophy to compensate for the impairment of contractility. However the scar, which is mainly composed of extra-cellular matrix, fibroblasts and myofibroblasts, has a poor contractility and thus induces cardiac remodelling, that corresponds to a change of shape (dilation) of the heart. This compromises in the longer term the sustained function of the heart.

Strategies for heart repair (Fig. 15.3) address several challenges, including the short-term minimisation of the damage and, in the longer term, the supply of new cardiomyocytes to restore a sufficient pumping capacity. We shall consider the intrinsic regeneration potential of the mouse heart, survey the characterised sources of resident cardiac stem cells and discuss the possibilities of transplanting external sources of stem cells.

15.4.1 Regeneration Potential of the Heart

The postnatal mammalian heart is no longer considered as a fully post-mitotic organ. During the first week of life, hyperplastic growth of the mouse heart is still important (Fernandez et al. 2001; Soonpaa et al. 1996), such that the heart, when injured the first day after birth, is able to regenerate within 21 days and produce new cardiomyocytes (Porrello et al. 2011). This proliferative response is preceded by partial dedifferentiation to less mature cardiomyocytes. There is a limited time window during which regeneration can occur, as opposed to the situation in the fish heart, which conserves a regeneration potential through adulthood (Poss et al. 2002).

Cardiomyocyte proliferation decreases after this neonatal period, although some turnover persists throughout life. For example, analysis of the integration of ^{14}C, delivered by nuclear bomb tests during the cold war, lead to an estimated turnover rate of 1 % per year in adult humans (Bergmann et al. 2009). However, depending on the technique used, the absolute figures remain controversial, and may vary with the age and between individuals (see Mummery et al. 2010). After myocardial infarction proliferation can be further activated (Beltrami et al. 2001). In a pulse-chase

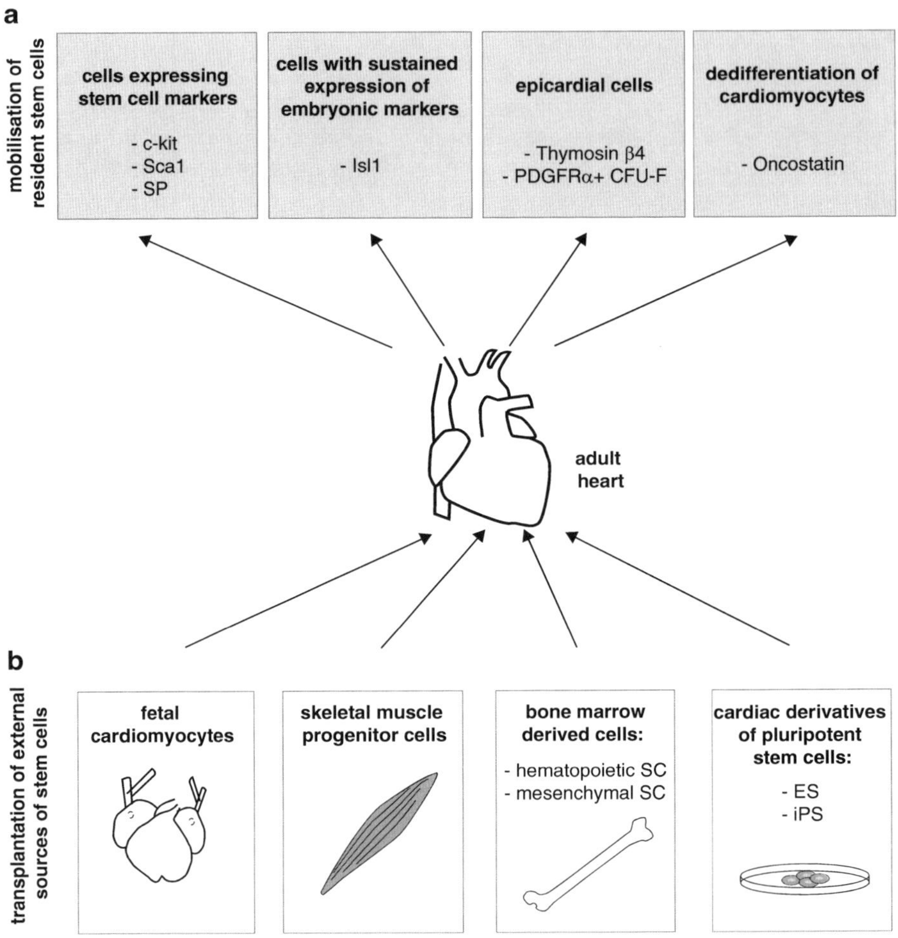

Fig. 15.3 Cardiac stem cells for heart repair. Different strategies are considered for repairing the injured heart and can be classified as mobilisation of resident stem cells of the adult heart (**a**) or transplantation of external sources of cardiac stem cells (**b**). *CFU-F* colony-forming units-fibroblasts, *ES* embryonic stem cells, *iPS* induced pluripotent stem cells, *SC* stem cells, *SP* side population

experiment in the adult mouse heart, based on inducible genetic tracing (Hsieh et al. 2007), all cardiomyocytes which express *Myh6* (also known as *αMHC*) were labelled irreversibly by a pulse of tamoxifen. After a chase period during normal ageing or following injury, the percentage of labelled cardiomyocytes was found unchanged or decreased, respectively. The decreased percentage indicates that new cardiomyocytes in the injured heart were formed from non-myocardial cells, such as stem or progenitor cells. In contrast, stem cells during normal ageing did not contribute significantly new cardiomyocytes, which may rather derive from the division of other cardiomyocytes. This study however did not characterise the progenitor cells which regenerate myocardium, nor their origin.

These data indicate that the mouse heart conserves a regeneration potential after birth, but it is insufficient in adulthood to repair an extensive injury after myocardial infarction. A challenge for heart repair is therefore to stimulate this potential or to complement it by other sources of cardiomyocytes.

15.4.2 Mobilisation of Resident Cardiac Stem Cells

The existence of resident stem cells in the adult heart is suggested by cell surface markers. This is exemplified by the classical stem cell marker c-kit, which was initially identified in hematopoietic stem cells as a receptor tyrosine kinase important for cell survival, proliferation and differentiation. In the rat heart, c-kit positive cells, which are negative for blood lineage markers, are able to self-renew and differentiate into cardiomyocytes, as well as smooth muscles and endothelial cells, both in vitro and in vivo (Beltrami et al. 2003). These cells can be found at a low frequency of 1 out of 10^4 cardiomyocytes. Another classical surface receptor of stem cells is Sca1 (Ly6a), which facilitates cell signalling by generating lipid rafts. In the heart, Sca1-positive cells are small interstitial cells, able to differentiate into cardiomyocytes in vitro and expressing the telomerase reverse transcriptase, which is associated with self-renewal (Matsuura et al. 2004). Transplantations in infarcted hearts showed that these cells have homing abilities, such that they can find their cardiac niche when injected intravenously. They survive and contribute to the heart. However, cardiac contribution is mainly due to fusion of the candidate stem cells with pre-existing cardiomyocytes (Oh et al. 2003). Stem cells are also frequently identified based on their capacity to pump out drugs, such as the dye Hoechst, via transporters of the ATP-binding cassette family. In the heart, this population, referred to as Side Population (SP), is present at a low frequency of 1 cell per 3×10^4 cardiac cells (Martin et al. 2004). In vitro these cells are able to differentiate into hematopoietic and cardiac lineages, but their in vivo potential is unknown. Although three markers associated with stem cells are found in the adult heart, labelling by one of these is not sufficient to conclude on stemness. A perivascular niche of cardiac stem cells was revealed in the ventricles of juvenile 4-week old mice. These cardiac mesoangioblasts express endothelial (CD31) and pericyte (NG2) markers, as well as the stem cell markers Sca1 and c-kit and early cardiac transcription factors (Nkx2.5). They can generate cardiomyocytes locally, when transplanted in an injured heart (Galvez et al. 2008). In general, the lineage origin of the resident cardiac stem cells remains uncertain, for example whether they derive from cardiogenic mesoderm or rather correspond to genuine hematopoietic stem cells, which have left their bone marrow niche.

Another indication of resident cardiac stem cells is the persistence of embryonic cardiac progenitor cell markers. For example, Isl1-positive cells were found in the postnatal heart at a level of 500–600 cells per heart in the 5-day old rat (Laugwitz et al. 2005). These cells can differentiate in vitro and in the embryo into cardiomyocytes, as well as into smooth muscle and endothelial cells (Moretti et al. 2006; Sun et al. 2007).

Their potential for producing new cardiomyocytes after infarction has not been determined so far. A pending question is to understand whether cells of the first myocardial lineage, which contribute exclusively to the embryonic left ventricle, would be better in repairing the adult left ventricle. For the purification of such cells, the molecular signature of the first and second lineages would be useful knowledge (Domian et al. 2009), although no bona fide marker of myocardial lineages is currently available to distinguish the two populations.

Cardiac stem cells may not necessarily be localised in the myocardium, as indicated by studies on the epicardium. Under physiological conditions, there seems to be no differentiation of epicardial cells into cardiac muscle (see Sect. 15.2.2). Proliferation of the epicardium ceases by 4 days after birth, accompanied by repression of the embryonic epicardial genes. Upon infarction, re-activation of these genes, such as *Wt1* and *Tbx18*, is observed. Treatment with thymosin β4, an actin regulating peptide, primes the epicardium and can lead to the recruitment of cells, which have re-activated *Wt1,* into the myocardial lineage (Smart et al. 2011). Additional benefits include the stimulation of vascularisation, as reflected by the natural potential of epicardial cells to give rise to coronary vessels. Epicardium-derived cells have also been identified as a new source of resident stem cells, which are isolated as colony-forming units-fibroblasts (CFU-F) expressing the receptor PDGFRα and resembling mesenchymal stem cells with a perivascular origin (Chong et al. 2011).

An emerging theme is the dedifferentiation of cardiomyocytes upon injury, which is no longer a peculiarity of the zebrafish model (Jopling et al. 2010). In mouse, the cytokine oncostatin was identified as a factor which promotes cardiac dedifferentiation, as indicated by the re-expression of embryonic cardiac markers and the activation of stem-cell markers (Kubin et al. 2011). Higher efficiency was provided by a combined treatment with Fgf2. Activation of the c-kit marker after treatment raises the possibility that cells previously identified as resident cardiac stem cells, in fact correspond to dedifferentiated cardiomyocytes. The difficulty of such stimulation of dedifferentiation lies in the control of the ratio between stem cells and differentiated cells, to provide sufficient repair without compromising immediate contractility. A way to ensure supply of cardiomyocytes is to graft them from external sources.

15.4.3 Transplantation of External Sources of Cardiac Stem Cells

External sources would favour control over the number and the state of differentiation of cardiac stem cells, but may raise problems of ethics, immunoreaction or tumorigenicity. Foetal cardiomyocytes derived from transgenic embryos and transplanted in adult mouse hearts were shown to differentiate into mature cardiomyocytes in vivo and to be maintained in the myocardium up to 2 months after transplantation (Soonpaa et al. 1994). Such transplantation after injury is beneficial, as indicated by improvement of ventricular function as well as better survival of endogenous cells (Roell et al. 2002). These cells integrate well into the host myocardium, as electrical coupling between host and grafted cardiomyocytes is established (Rubart et al. 2003).

Because such a strategy of heart repair is difficult to implement in human, other sources of muscular cells were explored, such as satellite cells or myoblasts, which can potentially be used in autologous grafts. The C2C12 mouse cell line, which was initially derived from perinatal satellite cells, has been transplanted into an adult mouse heart. The grafted cells survived and differentiated but were never coupled electrically with host myocardial cells (Koh et al. 1993). They can fuse with host myocardial cells (Reinecke et al. 2004; Rubart et al. 2004). However this is a very low frequent event, which does not provide an increase in the number of cardiac cells. Clinical trials involving the transplantation of skeletal myoblasts in injured hearts showed a positive effect on left ventricular function. Nonetheless, the trials were interrupted because of the appearance of deleterious arrhythmias, which correspond to the different contraction and electrical properties of skeletal and cardiac muscle cells (Menasche et al. 2003; Smits et al. 2003).

C-kit-positive, lin-negative bone marrow cells, which are known to have potential for other than haematopoietic lineages, were an attractive source of cells for heart repair. The fate of hematopoietic stem cells after transplantation in an injured heart has been controversial: initial report of transdifferentiation into cardiomyocytes, endothelial cells and smooth muscles (Orlic et al. 2001) has not been reproduced and other studies have rather demonstrated fusion with host cardiomyocytes (Balsam et al. 2004; Murry et al. 2004). Using the pulse-chase technique, a recent study has elegantly shown that new cardiomyocytes are formed after transplantation of bone marrow cells. However none of the newly cardiomyocytes derived from the transplanted cells (Loffredo et al. 2011). The consensus now is that benefits of bone marrow cell treatment result from paracrine effects, which protect myocardial cells from the ischemic damage and stimulate the recruitment of endogenous progenitor cells in the myocardial lineage. Clinical trials showed only modest long-term improvements (see Laflamme and Murry 2005). Another population of the bone marrow is that of mesenchymal stem cells, which has the advantage of escaping the host immune system and thus being compatible with allogeneic transplantations. These cells secrete growth factors, cytokines, components of the extra-cellular matrix and matrix modulating enzymes (see Choi et al. 2011). However, these cells have been reported to be less competent than hematopoietic stem cells to promote the production of new cardiomyocytes in the injured heart (Loffredo et al. 2011). Another potential source of mesenchymal stem cells is the perivascular cell population called mesoangioblasts. When transplanted into the infarcted heart, they may be recruited to the smooth muscle lineage and mainly promote short-term improvement by a paracrine effect on angiogenesis and cardiomyocyte survival (Galli et al. 2005).

To increase the number of contractile cells after injury, which is required for long-term improvements of ventricular function, pluripotent stem cells which are committed to the cardiac lineage are appealing candidates, with the additional benefit of providing vessels as well. These pluripotent cells are either Embryonic Stem (ES) cells derived from blastocyst embryos, or induced Pluripotent Stem (iPS) cells, which derive from reprogrammation of differentiated cells such as fibroblasts.

A challenge is to trigger cardiac differentiation. Studies of mouse ES cells have shown that the transient activation of *Mesp1*, described as a master gene of cardiac differentiation (see Sect. 15.2.1), can enhance cardiac specification (Bondue et al. 2008; Lindsley et al. 2008). Other studies with human ES cells have characterised alternative protocols, such as coculture with mouse visceral endoderm cells (Mummery et al. 2003), depletion of insulin (Freund et al. 2008), prostaglandin (Xu et al. 2008), or a combination of growth factors including Bmp4/activinA (Laflamme et al. 2007) or Bmp2/Wnt3A/Fgf inhibitor (Blin et al. 2010). These protocols are working well to initiate the differentiation of cardiomyocytes, but fail to result in more mature phenotypes. Another issue is the reprogrammation of differentiated cells into iPS cells and cardiomyocytes. Mouse mesodermal cells exposed ectopically to Smarcd3, Tbx5 and Gata4 are directed to differentiate into cardiomyocytes (Takeuchi and Bruneau 2009). Cardiac fibroblasts can be induced to transdifferentiate into cardiomyocytes via the transfection of *Mef2c*, *Gata4* and *Tbx5* (Ieda et al. 2010). Under these conditions, 20 % of cardiomyocytes can be formed, based on the expression of cardiac markers, and only a sub-fraction has cardiac physiological properties. The transduction of *Oct4*, *Sox2* and *Klf4* in mouse embryonic or tail fibroblasts led to a complete reprogrammation to a pluripotent state (Efe et al. 2011) from which cardiomyocytes can be derived following an ES-cell like protocol. To purify the population of differentiated cardiomyocytes, identification of cell surface markers is progressing (Dubois et al. 2011).

Transplantation of mouse ES cells that were committed to the cardiac lineage improved cardiac function after injury in a mouse model. This effect is greater with transplantation of cells overexpressing VEGF to promote vascularisation (Yang et al. 2002). Cardiomyocytes derived from human ES cells can survive up to 9 months and facilitate short- and long-term recovery, with increased myocardial vascularisation. However, there is no alignment of fibres and no coupling between human and mouse cells, due to the species difference in the contraction rate (van Laake et al. 2009). iPS treatment can restore contractile performance and electric stability without a long-term contribution to the myocardium. It halts progression of the pathological remodelling in infarcted hearts and provides new cardiac, smooth muscle and endothelial tissue (Nelson et al. 2009). The use of iPS cells has the advantage of being suitable for autologous transplantation, but is associated with a risk of tumour formation, which will need to be assessed in the long term.

For the future, questions remain about the protocol of delivery of cardiac stem cells such as the appropriate number of cells or time of delivery after the injury to maximise their survival and integration. An important parameter of the survival in the infracted zone is the resistance of cells to hypoxia, which is more a feature of stem cells than differentiated cells. A new possibility for improving the integration of cells would be to graft tissue grown in vitro, rather than isolated cells. Such bioengineered myocardium would include supportive cells, as for example fibroblasts and vessels which provide trophic factors, and would be more mature, with an appropriate architecture of oriented fibres, which could rapidly increase contraction, if it can be synchronised with the heart. This strategy, which is already being explored, will require integration of our knowledge on the lineage of heart cells, to control the cell types of the graft, on the growth of the myocardium, to generate a mature and functional tissue, and on heart repair, to optimise transplantation.

Acknowledgements We thank M. Buckingham and D. Montarras for insightful discussions. SMM is an INSERM research scientist and FL is supported by the CardioCell (FP7 - HEALTH-2007-2.4.2-5) EU project to M. Buckingham.

References

Bajolle F, Zaffran S, Meilhac SM, Dandonneau M, Chang T, Kelly RG, Buckingham ME (2008) Myocardium at the base of the aorta and pulmonary trunk is prefigured in the outflow tract of the heart and in subdomains of the second heart field. Dev Biol 313:25–34

Balsam LB, Wagers AJ, Christensen JL, Kofidis T, Weissman IL, Robbins RC (2004) Haematopoietic stem cells adopt mature haematopoietic fates in ischaemic myocardium. Nature 428:668–673

Barry SP, Davidson SM, Townsend PA (2008) Molecular regulation of cardiac hypertrophy. Int J Biochem Cell Biol 40:2023–2039

Beltrami AP, Barlucchi L, Torella D, Baker M, Limana F, Chimenti S, Kasahara H, Rota M, Musso E, Urbanek K et al (2003) Adult cardiac stem cells are multipotent and support myocardial regeneration. Cell 114:763–776

Beltrami AP, Urbanek K, Kajstura J, Yan SM, Finato N, Bussani R, Nadal-Ginard B, Silvestri F, Leri A, Beltrami CA et al (2001) Evidence that human cardiac myocytes divide after myocardial infarction. N Engl J Med 344:1750–1757

Bergmann O, Bhardwaj RD, Bernard S, Zdunek S, Barnabe-Heider F, Walsh S, Zupicich J, Alkass K, Buchholz BA, Druid H et al (2009) Evidence for cardiomyocyte renewal in humans. Science 324:98–102

Bertrand N, Roux M, Ryckebusch L, Niederreither K, Dolle P, Moon A, Capecchi M, Zaffran S (2011) Hox genes define distinct progenitor sub-domains within the second heart field. Dev Biol 353:266–274

Blin G, Nury D, Stefanovic S, Neri T, Guillevic O, Brinon B, Bellamy V, Rucker-Martin C, Barbry P, Bel A et al (2010) A purified population of multipotent cardiovascular progenitors derived from primate pluripotent stem cells engrafts in postmyocardial infarcted nonhuman primates. J Clin Invest 120:1125–1139

Bolte C, Zhang Y, Wang IC, Kalin TV, Molkentin JD, Kalinichenko VV (2011) Expression of Foxm1 transcription factor in cardiomyocytes is required for myocardial development. PLoS One 6:e22217

Bondue A, Lapouge G, Paulissen C, Semeraro C, Iacovino M, Kyba M, Blanpain C (2008) Mesp1 acts as a master regulator of multipotent cardiovascular progenitor specification. Cell Stem Cell 3:69–84

Brown CB, Wenning JM, Lu MM, Epstein DJ, Meyers EN, Epstein JA (2004) Cre-mediated excision of Fgf8 in the Tbx1 expression domain reveals a critical role for Fgf8 in cardiovascular development in the mouse. Dev Biol 267:190–202

Buckingham M, Biben C, Lawson KA (1997) Fate mapping of pre-cardiac cells in the developing mouse. In: Olson EN, Harvey RP, Schultz RA, Altman JS (eds) Genetic control of heart development. HSFP, Strasbourg, pp 31–33

Buckingham M, Meilhac S, Zaffran S (2005) Building the mammalian heart from two sources of myocardial cells. Nat Rev Genet 6:826–835

Cai CL, Liang X, Shi Y, Chu PH, Pfaff SL, Chen J, Evans S (2003) Isl1 identifies a cardiac progenitor population that proliferates prior to differentiation and contributes a majority of cells to the heart. Dev Cell 5:877–889

Cai CL, Martin JC, Sun Y, Cui L, Wang L, Ouyang K, Yang L, Bu L, Liang X, Zhang X et al (2008) A myocardial lineage derives from Tbx18 epicardial cells. Nature 454:104–108

Cai CL, Zhou W, Yang L, Bu L, Qyang Y, Zhang X, Li X, Rosenfeld MG, Chen J, Evans S (2005) T-box genes coordinate regional rates of proliferation and regional specification during cardiogenesis. Development 132:2475–2487

Camenisch T, Runyan R, Markwald RR (2010) Molecular regulation of cushion morphogenesis. In: Rosenthal N, Harvey R (eds) Heart development and regeneration. Elsevier, Amsterdam, pp 363–387

Camenisch TD, Schroeder JA, Bradley J, Klewer SE, McDonald JA (2002) Heart-valve mesenchyme formation is dependent on hyaluronan-augmented activation of ErbB2-ErbB3 receptors. Nat Med 8:850–855

Camenisch TD, Spicer AP, Brehm-Gibson T, Biesterfeldt J, Augustine ML, Calabro A Jr, Kubalak S, Klewer SE, McDonald JA (2000) Disruption of hyaluronan synthase-2 abrogates normal cardiac morphogenesis and hyaluronan-mediated transformation of epithelium to mesenchyme. J Clin Invest 106:349–360

Carmona R, Guadix JA, Cano E, Ruiz-Villalba A, Portillo-Sanchez V, Perez-Pomares JM, Munoz-Chapuli R (2010) The embryonic epicardium: an essential element of cardiac development. J Cell Mol Med 14:2066–2072

Catalucci D, Latronico MV, Ellingsen O, Condorelli G (2008) Physiological myocardial hypertrophy: how and why? Front Biosci 13:312–324

Chan-Thomas PS, Thompson RP, Robert B, Yacoub MH, Barton PJ (1993) Expression of homeobox genes Msx-1 (Hox-7) and Msx-2 (Hox-8) during cardiac development in the chick. Dev Dyn 197:203–216

Chen H, Shi S, Acosta L, Li W, Lu J, Bao S, Chen Z, Yang Z, Schneider MD, Chien KR et al (2004) BMP10 is essential for maintaining cardiac growth during murine cardiogenesis. Development 131:2219–2231

Choi YH, Kurtz A, Stamm C (2011) Mesenchymal stem cells for cardiac cell therapy. Hum Gene Ther 22:3–17

Chong JJ, Chandrakanthan V, Xaymardan M, Asli NS, Li J, Ahmed I, Heffernan C, Menon MK, Scarlett CJ, Rashidianfar A et al (2011) Adult cardiac-resident MSC-like stem cells with a proepicardial origin. Cell Stem Cell 9:527–540

Christoffels V, Hoogaars WM, Moorman A (2010) Patterning and development of the conduction system of the heart: origins of the conduction system in development. In: Rosenthal N, Harvey R (eds) Heart development and regeneration. Elsevier, Amsterdam, pp 171–192

Christoffels VM, Grieskamp T, Norden J, Mommersteeg MT, Rudat C, Kispert A (2009) Tbx18 and the fate of epicardial progenitors. Nature 458:E8–9, discussion E9–10

Christoffels VM, Hoogaars WM, Tessari A, Clout DE, Moorman AF, Campione M (2004) T-box transcription factor Tbx2 represses differentiation and formation of the cardiac chambers. Dev Dyn 229:763–770

Christoffels VM, Mommersteeg MT, Trowe MO, Prall OW, de Gier-de Vries C, Soufan AT, Bussen M, Schuster-Gossler K, Harvey RP, Moorman AF et al (2006) Formation of the venous pole of the heart from an Nkx2-5-negative precursor population requires Tbx18. Circ Res 98:1555–1563

De la Cruz MV, Gimenez-Ribotta M, Saravalli O, Cayre R (1983) The contribution of the inferior endocardial cushion of the atrioventricular canal to cardiac septation and to the development of the atrioventricular valves: study in the chick embryo. Am J Anat 166:63–72

de Lange FJ, Moorman AF, Anderson RH, Manner J, Soufan AT, de Gier-de Vries C, Schneider MD, Webb S, van den Hoff MJ, Christoffels VM (2004) Lineage and morphogenetic analysis of the cardiac valves. Circ Res 95:645–654

Dodou E, Verzi MP, Anderson JP, Xu SM, Black BL (2004) Mef2c is a direct transcriptional target of ISL1 and GATA factors in the anterior heart field during mouse embryonic development. Development 131:3931–3942

Domian IJ, Chiravuri M, van der Meer P, Feinberg AW, Shi X, Shao Y, Wu SM, Parker KK, Chien KR (2009) Generation of functional ventricular heart muscle from mouse ventricular progenitor cells. Science 326:426–429

Dubois NC, Craft AM, Sharma P, Elliott DA, Stanley EG, Elefanty AG, Gramolini A, Keller G (2011) SIRPA is a specific cell-surface marker for isolating cardiomyocytes derived from human pluripotent stem cells. Nat Biotechnol 29:1011–1018

Dunwoodie SL, Rodriguez TA, Beddington RS (1998) Msg1 and Mrg1, founding members of a gene family, show distinct patterns of gene expression during mouse embryogenesis. Mech Dev 72:27–40

Efe JA, Hilcove S, Kim J, Zhou H, Ouyang K, Wang G, Chen J, Ding S (2011) Conversion of mouse fibroblasts into cardiomyocytes using a direct reprogramming strategy. Nat Cell Biol 13:215–222

Ferdous A, Caprioli A, Iacovino M, Martin CM, Morris J, Richardson JA, Latif S, Hammer RE, Harvey RP, Olson EN et al (2009) Nkx2-5 transactivates the Ets-related protein 71 gene and specifies an endothelial/endocardial fate in the developing embryo. Proc Natl Acad Sci USA 106:814–819

Fernandez E, Siddiquee Z, Shohet RV (2001) Apoptosis and proliferation in the neonatal murine heart. Dev Dyn 221:302–310

Fishman MC, Chien KR (1997) Fashioning the vertebrate heart: earliest embryonic decisions. Development 124:2099–2117

Freund C, Ward-van Oostwaard D, Monshouwer-Kloots J, van den Brink S, van Rooijen M, Xu X, Zweigerdt R, Mummery C, Passier R (2008) Insulin redirects differentiation from cardiogenic mesoderm and endoderm to neuroectoderm in differentiating human embryonic stem cells. Stem Cells 26:724–733

Furtado MB, Biben C, Shiratori H, Hamada H, Harvey RP (2011) Characterization of Pitx2c expression in the mouse heart using a reporter transgene. Dev Dyn 240:195–203

Galli D, Dominguez JN, Zaffran S, Munk A, Brown NA, Buckingham ME (2008) Atrial myocardium derives from the posterior region of the second heart field, which acquires left-right identity as Pitx2c is expressed. Development 135:1157–1167

Galli D, Innocenzi A, Staszewsky L, Zanetta L, Sampaolesi M, Bai A, Martinoli E, Carlo E, Balconi G, Fiordaliso F et al (2005) Mesoangioblasts, vessel-associated multipotent stem cells, repair the infarcted heart by multiple cellular mechanisms: a comparison with bone marrow progenitors, fibroblasts, and endothelial cells. Arterioscler Thromb Vasc Biol 25:692–697

Galvez BG, Sampaolesi M, Barbuti A, Crespi A, Covarello D, Brunelli S, Dellavalle A, Crippa S, Balconi G, Cuccovillo I et al (2008) Cardiac mesoangioblasts are committed, self-renewable progenitors, associated with small vessels of juvenile mouse ventricle. Cell Death Differ 15:1417–1428

Gittenberger-de Groot AC, Mahtab EA, Hahurij ND, Wisse LJ, Deruiter MC, Wijffels MC, Poelmann RE (2007) Nkx2.5-negative myocardium of the posterior heart field and its correlation with podoplanin expression in cells from the developing cardiac pacemaking and conduction system. Anat Rec (Hoboken) 290:115–122

Gourdie RG, Wei Y, Kim D, Klatt SC, Mikawa T (1998) Endothelin-induced conversion of embryonic heart muscle cells into impulse-conducting Purkinje fibers. Proc Natl Acad Sci USA 95:6815–6818

Grego-Bessa J, Luna-Zurita L, del Monte G, Bolos V, Melgar P, Arandilla A, Garratt AN, Zang H, Mukouyama YS, Chen H et al (2007) Notch signaling is essential for ventricular chamber development. Dev Cell 12:415–429

Harris IS, Black BL (2010) Development of the endocardium. Pediatr Cardiol 31:391–399

Heallen T, Zhang M, Wang J, Bonilla-Claudio M, Klysik E, Johnson RL, Martin JF (2011) Hippo pathway inhibits Wnt signaling to restrain cardiomyocyte proliferation and heart size. Science 332:458–461

Henderson DJ, Chaudhry B (2011) Getting to the heart of planar cell polarity signaling. Birth Defects Res A Clin Mol Teratol 91:460–467

Hirschy A, Schatzmann F, Ehler E, Perriard JC (2006) Establishment of cardiac cytoarchitecture in the developing mouse heart. Dev Biol 289:430–441

Hoogaars WM, Tessari A, Moorman AF, de Boer PA, Hagoort J, Soufan AT, Campione M, Christoffels VM (2004) The transcriptional repressor Tbx3 delineates the developing central conduction system of the heart. Cardiovasc Res 62:489–499

Hsieh PC, Segers VF, Davis ME, MacGillivray C, Gannon J, Molkentin JD, Robbins J, Lee RT (2007) Evidence from a genetic fate-mapping study that stem cells refresh adult mammalian cardiomyocytes after injury. Nat Med 13:970–974

Ieda M, Fu JD, Delgado-Olguin P, Vedantham V, Hayashi Y, Bruneau BG, Srivastava D (2010) Direct reprogramming of fibroblasts into functional cardiomyocytes by defined factors. Cell 142:375–386

Ieda M, Tsuchihashi T, Ivey KN, Ross RS, Hong TT, Shaw RM, Srivastava D (2009) Cardiac fibroblasts regulate myocardial proliferation through beta1 integrin signaling. Dev Cell 16:233–244

Ishii Y, Langberg JD, Hurtado R, Lee S, Mikawa T (2007) Induction of proepicardial marker gene expression by the liver bud. Development 134:3627–3637

Jiang X, Rowitch DH, Soriano P, McMahon AP, Sucov HM (2000) Fate of the mammalian cardiac neural crest. Development 127:1607–1616

Jopling C, Sleep E, Raya M, Marti M, Raya A, Belmonte JC (2010) Zebrafish heart regeneration occurs by cardiomyocyte dedifferentiation and proliferation. Nature 464:606–609

Kaufman MH, Navaratnam V (1981) Early differentiation of the heart in mouse embryos. J Anat 133:235–246

Kelly RG, Brown NA, Buckingham ME (2001) The arterial pole of the mouse heart forms from Fgf10-expressing cells in pharyngeal mesoderm. Dev Cell 1:435–440

Kinder SJ, Tsang TE, Quinlan GA, Hadjantonakis AK, Nagy A, Tam PP (1999) The orderly allocation of mesodermal cells to the extraembryonic structures and the anteroposterior axis during gastrulation of the mouse embryo. Development 126:4691–4701

Kirby ML, Gale TF, Stewart DE (1983) Neural crest cells contribute to normal aorticopulmonary septation. Science 220:1059–1061

Kitajima S, Miyagawa-Tomita S, Inoue T, Kanno J, Saga Y (2006) Mesp1-nonexpressing cells contribute to the ventricular cardiac conduction system. Dev Dyn 235:395–402

Klaus A, Saga Y, Taketo MM, Tzahor E, Birchmeier W (2007) Distinct roles of Wnt/beta-catenin and Bmp signaling during early cardiogenesis. Proc Natl Acad Sci USA 104:18531–18536

Koh GY, Klug MG, Soonpaa MH, Field LJ (1993) Differentiation and long-term survival of C2C12 myoblast grafts in heart. J Clin Invest 92:1548–1554

Kreidberg JA, Sariola H, Loring JM, Maeda M, Pelletier J, Housman D, Jaenisch R (1993) WT-1 is required for early kidney development. Cell 74:679–691

Kubin T, Poling J, Kostin S, Gajawada P, Hein S, Rees W, Wietelmann A, Tanaka M, Lorchner H, Schimanski S et al (2011) Oncostatin m is a major mediator of cardiomyocyte dedifferentiation and remodeling. Cell Stem Cell 9:420–432

Kwon C, Arnold J, Hsiao EC, Taketo MM, Conklin BR, Srivastava D (2007) Canonical Wnt signaling is a positive regulator of mammalian cardiac progenitors. Proc Natl Acad Sci USA 104:10894–10899

Laflamme MA, Chen KY, Naumova AV, Muskheli V, Fugate JA, Dupras SK, Reinecke H, Xu C, Hassanipour M, Police S et al (2007) Cardiomyocytes derived from human embryonic stem cells in pro-survival factors enhance function of infarcted rat hearts. Nat Biotechnol 25:1015–1024

Laflamme MA, Murry CE (2005) Regenerating the heart. Nat Biotechnol 23:845–856

Lamers WH, De Jong F, De Groot IJ, Moorman AF (1991) The development of the avian conduction system, a review. Eur J Morphol 29:233–253

Laugwitz KL, Moretti A, Lam J, Gruber P, Chen Y, Woodard S, Lin LZ, Cai CL, Lu MM, Reth M et al (2005) Postnatal isl1+ cardioblasts enter fully differentiated cardiomyocyte lineages. Nature 433:647–653

Lawson KA, Pedersen RA (1987) Cell fate, morphogenetic movement and population kinetics of embryonic endoderm at the time of germ layer formation in the mouse. Development 101:627–652

Le Lievre CS, Le Douarin NM (1975) Mesenchymal derivatives of the neural crest: analysis of chimaeric quail and chick embryos. J Embryol Exp Morphol 34:125–154

Lescroart F, Kelly RG, Le Garrec JF, Nicolas JF, Meilhac SM, Buckingham M (2010) Clonal analysis reveals common lineage relationships between head muscles and second heart field derivatives in the mouse embryo. Development 137:3269–3279

Leu M, Ehler E, Perriard JC (2001) Characterisation of postnatal growth of the murine heart. Anat Embryol (Berl) 204:217–224

Li P, Cavallero S, Gu Y, Chen TH, Hughes J, Hassan AB, Bruning JC, Pashmforoush M, Sucov HM (2011) IGF signaling directs ventricular cardiomyocyte proliferation during embryonic heart development. Development 138:1795–1805

Lin L, Cui L, Zhou W, Dufort D, Zhang X, Cai CL, Bu L, Yang L, Martin J, Kemler R et al (2007) Beta-catenin directly regulates Islet1 expression in cardiovascular progenitors and is required for multiple aspects of cardiogenesis. Proc Natl Acad Sci USA 104:9313–9318

Linask KK, Knudsen KA, Gui YH (1997) N-cadherin-catenin interaction: necessary component of cardiac cell compartmentalization during early vertebrate heart development. Dev Biol 185:148–164

Lindsley RC, Gill JG, Murphy TL, Langer EM, Cai M, Mashayekhi M, Wang W, Niwa N, Nerbonne JM, Kyba M et al (2008) Mesp1 coordinately regulates cardiovascular fate restriction and epithelial-mesenchymal transition in differentiating ESCs. Cell Stem Cell 3:55–68

Lints TJ, Parsons LM, Hartley L, Lyons I, Harvey RP (1993) Nkx-2.5: a novel murine homeobox gene expressed in early heart progenitor cells and their myogenic descendants. Development 119:969

Loffredo FS, Steinhauser ML, Gannon J, Lee RT (2011) Bone marrow-derived cell therapy stimulates endogenous cardiomyocyte progenitors and promotes cardiac repair. Cell Stem Cell 8:389–398

Ma Q, Zhou B, Pu WT (2008) Reassessment of Isl1 and Nkx2-5 cardiac fate maps using a Gata4-based reporter of Cre activity. Dev Biol 323:98–104

Manner J (1999) Does the subepicardial mesenchyme contribute myocardioblasts to the myocardium of the chick embryo heart? A quail-chick chimera study tracing the fate of the epicardial primordium. Anat Rec 255:212–226

Martin CM, Meeson AP, Robertson SM, Hawke TJ, Richardson JA, Bates S, Goetsch SC, Gallardo TD, Garry DJ (2004) Persistent expression of the ATP-binding cassette transporter, Abcg2, identifies cardiac SP cells in the developing and adult heart. Dev Biol 265:262–275

Matsuura K, Nagai T, Nishigaki N, Oyama T, Nishi J, Wada H, Sano M, Toko H, Akazawa H, Sato T et al (2004) Adult cardiac Sca-1-positive cells differentiate into beating cardiomyocytes. J Biol Chem 279:11384–11391

McMullen JR, Shioi T, Huang WY, Zhang L, Tarnavski O, Bisping E, Schinke M, Kong S, Sherwood MC, Brown J et al (2004) The insulin-like growth factor 1 receptor induces physiological heart growth via the phosphoinositide 3-kinase(p110alpha) pathway. J Biol Chem 279:4782–4793

Meilhac SM, Esner M, Kelly RG, Nicolas JF, Buckingham ME (2004a) The clonal origin of myocardial cells in different regions of the embryonic mouse heart. Dev Cell 6:685–698

Meilhac SM, Esner M, Kerszberg M, Moss JE, Buckingham ME (2004b) Oriented clonal cell growth in the developing mouse myocardium underlies cardiac morphogenesis. J Cell Biol 164:97–109

Meilhac SM, Kelly RG, Rocancourt D, Eloy-Trinquet S, Nicolas JF, Buckingham ME (2003) A retrospective clonal analysis of the myocardium reveals two phases of clonal growth in the developing mouse heart. Development 130:3877–3889

Menasche P, Hagege AA, Vilquin JT, Desnos M, Abergel E, Pouzet B, Bel A, Sarateanu S, Scorsin M, Schwartz K et al (2003) Autologous skeletal myoblast transplantation for severe postinfarction left ventricular dysfunction. J Am Coll Cardiol 41:1078–1083

Merki E, Zamora M, Raya A, Kawakami Y, Wang J, Zhang X, Burch J, Kubalak SW, Kaliman P, Belmonte JC et al (2005) Epicardial retinoid X receptor alpha is required for myocardial growth and coronary artery formation. Proc Natl Acad Sci USA 102:18455–18460

Meyer D, Birchmeier C (1995) Multiple essential functions of neuregulin in development. Nature 378:386–390

Mikawa T, Gourdie RG (1996) Pericardial mesoderm generates a population of coronary smooth muscle cells migrating into the heart along with ingrowth of the epicardial organ. Dev Biol 174:221–232

Milgrom-Hoffman M, Harrelson Z, Ferrara N, Zelzer E, Evans SM, Tzahor E (2011) The heart endocardium is derived from vascular endothelial progenitors. Development 138:4777–4787

Miquerol L, Moreno-Rascon N, Beyer S, Dupays L, Meilhac SM, Buckingham ME, Franco D, Kelly RG (2010) Biphasic development of the mammalian ventricular conduction system. Circ Res 107:153–161

Mommersteeg MT, Dominguez JN, Wiese C, Norden J, de Gier-de Vries C, Burch JB, Kispert A, Brown NA, Moorman AF, Christoffels VM (2010) The sinus venosus progenitors separate and diversify from the first and second heart fields early in development. Cardiovasc Res 87:92–101

Moretti A, Caron L, Nakano A, Lam JT, Bernshausen A, Chen Y, Qyang Y, Bu L, Sasaki M, Martin-Puig S et al (2006) Multipotent embryonic isl1+ progenitor cells lead to cardiac, smooth muscle, and endothelial cell diversification. Cell 127:1151–1165

Mummery C, Ward-van Oostwaard D, Doevendans P, Spijker R, van den Brink S, Hassink R, van der Heyden M, Opthof T, Pera M, de la Riviere AB et al (2003) Differentiation of human embryonic stem cells to cardiomyocytes: role of coculture with visceral endoderm-like cells. Circulation 107:2733–2740

Mummery CL, Davis RP, Krieger JE (2010) Challenges in using stem cells for cardiac repair. Sci Transl Med 2:27ps17

Murry CE, Soonpaa MH, Reinecke H, Nakajima H, Nakajima HO, Rubart M, Pasumarthi KB, Virag JI, Bartelmez SH, Poppa V et al (2004) Haematopoietic stem cells do not transdifferentiate into cardiac myocytes in myocardial infarcts. Nature 428:664–668

Nakajima K, Inagawa M, Uchida C, Okada K, Tane S, Kojima M, Kubota M, Noda M, Ogawa S, Shirato H et al (2011) Coordinated regulation of differentiation and proliferation of embryonic cardiomyocytes by a jumonji (Jarid2)-cyclin D1 pathway. Development 138:1771–1782

Nelson TJ, Martinez-Fernandez A, Yamada S, Perez-Terzic C, Ikeda Y, Terzic A (2009) Repair of acute myocardial infarction by human stemness factors induced pluripotent stem cells. Circulation 120:408–416

Nishii K, Shibata Y (2006) Mode and determination of the initial contraction stage in the mouse embryo heart. Anat Embryol (Berl) 211:95–100

Oh H, Bradfute SB, Gallardo TD, Nakamura T, Gaussin V, Mishina Y, Pocius J, Michael LH, Behringer RR, Garry DJ et al (2003) Cardiac progenitor cells from adult myocardium: homing, differentiation, and fusion after infarction. Proc Natl Acad Sci USA 100:12313–12318

Orlic D, Kajstura J, Chimenti S, Jakoniuk I, Anderson SM, Li B, Pickel J, McKay R, Nadal-Ginard B, Bodine DM et al (2001) Bone marrow cells regenerate infarcted myocardium. Nature 410:701–705

Peng X, Wu X, Druso JE, Wei H, Park AY, Kraus MS, Alcaraz A, Chen J, Chien S, Cerione RA et al (2008) Cardiac developmental defects and eccentric right ventricular hypertrophy in cardiomyocyte focal adhesion kinase (FAK) conditional knockout mice. Proc Natl Acad Sci USA 105:6638–6643

Perez-Pomares JM, Phelps A, Sedmerova M, Wessels A (2003) Epicardial-like cells on the distal arterial end of the cardiac outflow tract do not derive from the proepicardium but are derivatives of the cephalic pericardium. Dev Dyn 227:56–68

Porrello ER, Mahmoud AI, Simpson E, Hill JA, Richardson JA, Olson EN, Sadek HA (2011) Transient regenerative potential of the neonatal mouse heart. Science 331:1078–1080

Poss KD, Wilson LG, Keating MT (2002) Heart regeneration in zebrafish. Science 298:2188–2190

Prall OW, Menon MK, Solloway MJ, Watanabe Y, Zaffran S, Bajolle F, Biben C, McBride JJ, Robertson BR, Chaulet H et al (2007) An Nkx2-5/Bmp2/Smad1 negative feedback loop controls heart progenitor specification and proliferation. Cell 128:947–959

Rasmussen TL, Kweon J, Diekmann MA, Belema-Bedada F, Song Q, Bowlin K, Shi X, Ferdous A, Li T, Kyba M et al (2011) ER71 directs mesodermal fate decisions during embryogenesis. Development 138:4801–4812

Red-Horse K, Ueno H, Weissman IL, Krasnow MA (2010) Coronary arteries form by developmental reprogramming of venous cells. Nature 464:549–553

Reinecke H, Minami E, Poppa V, Murry CE (2004) Evidence for fusion between cardiac and skeletal muscle cells. Circ Res 94:e56–60

Risebro CA, Searles RG, Melville AA, Ehler E, Jina N, Shah S, Pallas J, Hubank M, Dillard M, Harvey NL et al (2009) Prox1 maintains muscle structure and growth in the developing heart. Development 136:495–505

Rochais F, Dandonneau M, Mesbah K, Jarry T, Mattei MG, Kelly RG (2009) Hes1 is expressed in the second heart field and is required for outflow tract development. PLoS One 4:e6267

Roell W, Lu ZJ, Bloch W, Siedner S, Tiemann K, Xia Y, Stoecker E, Fleischmann M, Bohlen H, Stehle R et al (2002) Cellular cardiomyoplasty improves survival after myocardial injury. Circulation 105:2435–2441

Rubart M, Pasumarthi KB, Nakajima H, Soonpaa MH, Nakajima HO, Field LJ (2003) Physiological coupling of donor and host cardiomyocytes after cellular transplantation. Circ Res 92:1217–1224

Rubart M, Soonpaa MH, Nakajima H, Field LJ (2004) Spontaneous and evoked intracellular calcium transients in donor-derived myocytes following intracardiac myoblast transplantation. J Clin Invest 114:775–783

Ryckebusch L, Wang Z, Bertrand N, Lin SC, Chi X, Schwartz R, Zaffran S, Niederreither K (2008) Retinoic acid deficiency alters second heart field formation. Proc Natl Acad Sci USA 105:2913–2918

Saga Y, Kitajima S, Miyagawa-Tomita S (2000) Mesp1 expression is the earliest sign of cardiovascular development. Trends Cardiovasc Med 10:345–352

Saga Y, Miyagawa-Tomita S, Takagi A, Kitajima S, Miyazaki J, Inoue T (1999) MesP1 is expressed in the heart precursor cells and required for the formation of a single heart tube. Development 126:3437–3447

Sassoon DA, Garner I, Buckingham M (1988) Transcripts of alpha-cardiac and alpha-skeletal actins are early markers for myogenesis in the mouse embryo. Development 104:155–164

Schulte I, Schlueter J, Abu-Issa R, Brand T, Manner J (2007) Morphological and molecular left-right asymmetries in the development of the proepicardium: a comparative analysis on mouse and chick embryos. Dev Dyn 236:684–695

Sedmera D, Thompson RP (2011) Myocyte proliferation in the developing heart. Dev Dyn 240:1322–1334

Shai SY, Harpf AE, Babbitt CJ, Jordan MC, Fishbein MC, Chen J, Omura M, Leil TA, Becker KD, Jiang M et al (2002) Cardiac myocyte-specific excision of the beta1 integrin gene results in myocardial fibrosis and cardiac failure. Circ Res 90:458–464

Smart N, Bollini S, Dube KN, Vieira JM, Zhou B, Davidson S, Yellon D, Riegler J, Price AN, Lythgoe MF et al (2011) De novo cardiomyocytes from within the activated adult heart after injury. Nature 474:640–644

Smits PC, van Geuns RJ, Poldermans D, Bountioukos M, Onderwater EE, Lee CH, Maat AP, Serruys PW (2003) Catheter-based intramyocardial injection of autologous skeletal myoblasts as a primary treatment of ischemic heart failure: clinical experience with six-month follow-up. J Am Coll Cardiol 42:2063–2069

Snider P, Standley KN, Wang J, Azhar M, Doetschman T, Conway SJ (2009) Origin of cardiac fibroblasts and the role of periostin. Circ Res 105:934–947

Soonpaa MH, Kim KK, Pajak L, Franklin M, Field LJ (1996) Cardiomyocyte DNA synthesis and binucleation during murine development. Am J Physiol 271:H2183–2189

Soonpaa MH, Koh GY, Klug MG, Field LJ (1994) Formation of nascent intercalated disks between grafted fetal cardiomyocytes and host myocardium. Science 264:98–101

Stankunas K, Hang CT, Tsun ZY, Chen H, Lee NV, Wu JI, Shang C, Bayle JH, Shou W, Iruela-Arispe ML et al (2008) Endocardial Brg1 represses ADAMTS1 to maintain the microenvironment for myocardial morphogenesis. Dev Cell 14:298–311

Stanley EG, Biben C, Elefanty A, Barnett L, Koentgen F, Robb L, Harvey RP (2002) Efficient Cre-mediated deletion in cardiac progenitor cells conferred by a 3′UTR-ires-Cre allele of the homeobox gene Nkx2-5. Int J Dev Biol 46:431–439

Sucov HM, Gu Y, Thomas S, Li P, Pashmforoush M (2009) Epicardial control of myocardial proliferation and morphogenesis. Pediatr Cardiol 30:617–625

Sun Y, Liang X, Najafi N, Cass M, Lin L, Cai CL, Chen J, Evans SM (2007) Islet 1 is expressed in distinct cardiovascular lineages, including pacemaker and coronary vascular cells. Dev Biol 304:286–296

Takeuchi JK, Bruneau BG (2009) Directed transdifferentiation of mouse mesoderm to heart tissue by defined factors. Nature 459:708–711

Tam PP, Parameswaran M, Kinder SJ, Weinberger RP (1997) The allocation of epiblast cells to the embryonic heart and other mesodermal lineages: the role of ingression and tissue movement during gastrulation. Development 124:1631–1642

Terada R, Warren S, Lu JT, Chien KR, Wessels A, Kasahara H (2011) Ablation of Nkx2-5 at mid-embryonic stage results in premature lethality and cardiac malformation. Cardiovasc Res 91:289–299

Theveniau-Ruissy M, Dandonneau M, Mesbah K, Ghez O, Mattei MG, Miquerol L, Kelly RG (2008) The del22q11.2 candidate gene Tbx1 controls regional outflow tract identity and coronary artery patterning. Circ Res 103:142–148

Thompson RP, Lindroth JR, Wong M (1990) Regional differences in DNA-synthetic activity in the preseptation myocardium of the chick. In: Clark EB, Takao A (eds) Developmental cardiology: morphogenesis and function. PFuturama Publishing Co, NY, pp 219–234

Tzouanacou E, Wegener A, Wymeersch FJ, Wilson V, Nicolas JF (2009) Redefining the progression of lineage segregations during mammalian embryogenesis by clonal analysis. Dev Cell 17:365–376

van den Berg G, Abu-Issa R, de Boer BA, Hutson MR, de Boer PA, Soufan AT, Ruijter JM, Kirby ML, van den Hoff MJ, Moorman AF (2009) A caudal proliferating growth center contributes to both poles of the forming heart tube. Circ Res 104:179–188

van Laake LW, Passier R, den Ouden K, Schreurs C, Monshouwer-Kloots J, Ward-van Oostwaard D, van Echteld CJ, Doevendans PA, Mummery CL (2009) Improvement of mouse cardiac function by hESC-derived cardiomyocytes correlates with vascularity but not graft size. Stem Cell Res 3:106–112

van Wijk B, van den Berg G, Abu-Issa R, Barnett P, van der Velden S, Schmidt M, Ruijter JM, Kirby ML, Moorman AF, van den Hoff MJ (2009) Epicardium and myocardium separate from a common precursor pool by crosstalk between bone morphogenetic protein- and fibroblast growth factor-signaling pathways. Circ Res 105:431–441

Verzi MP, McCulley DJ, De Val S, Dodou E, Black BL (2005) The right ventricle, outflow tract, and ventricular septum comprise a restricted expression domain within the secondary/anterior heart field. Dev Biol 287:134–145

Vincent SD, Buckingham ME (2010) How to make a heart: the origin and regulation of cardiac progenitor cells. Curr Top Dev Biol 90:1–41

Viragh S, Challice CE (1973) Origin and differentiation of cardiac muscle cells in the mouse. J Ultrastruct Res 42:1–24

Viragh S, Challice CE (1981) The origin of the epicardium and the embryonic myocardial circulation in the mouse. Anat Rec 201:157–168

Waldo KL, Hutson MR, Ward CC, Zdanowicz M, Stadt HA, Kumiski D, Abu-Issa R, Kirby ML (2005) Secondary heart field contributes myocardium and smooth muscle to the arterial pole of the developing heart. Dev Biol 281:78–90

Ward C, Stadt H, Hutson M, Kirby ML (2005) Ablation of the secondary heart field leads to tetralogy of Fallot and pulmonary atresia. Dev Biol 284:72–83

Weber KT, Brilla CG (1991) Pathological hypertrophy and cardiac interstitium. Fibrosis and renin-angiotensin-aldosterone system. Circulation 83:1849–1865

Wei Y, Mikawa T (2000) Fate diversity of primitive streak cells during heart field formation in ovo. Dev Dyn 219:505–513

Wu M, Smith CL, Hall JA, Lee I, Luby-Phelps K, Tallquist MD (2010) Epicardial spindle orientation controls cell entry into the myocardium. Dev Cell 19:114–125

Xin M, Kim Y, Sutherland LB, Qi X, McAnally J, Schwartz RJ, Richardson JA, Bassel-Duby R, Olson EN (2011) Regulation of insulin-like growth factor signaling by yap governs cardiomyocyte proliferation and embryonic heart size. Sci Signal 4:ra70

Xu H, Morishima M, Wylie JN, Schwartz RJ, Bruneau BG, Lindsay EA, Baldini A (2004) Tbx1 has a dual role in the morphogenesis of the cardiac outflow tract. Development 131:3217–3227

Xu XQ, Graichen R, Soo SY, Balakrishnan T, Rahmat SN, Sieh S, Tham SC, Freund C, Moore J, Mummery C et al (2008) Chemically defined medium supporting cardiomyocyte differentiation of human embryonic stem cells. Differentiation 76:958–970

Yang Y, Min JY, Rana JS, Ke Q, Cai J, Chen Y, Morgan JP, Xiao YF (2002) VEGF enhances functional improvement of postinfarcted hearts by transplantation of ESC-differentiated cells. J Appl Physiol 93:1140–1151

Zaffran S, Kelly RG, Meilhac SM, Buckingham ME, Brown NA (2004) Right ventricular myocardium derives from the anterior heart field. Circ Res 95:261–268

Zhou B, Ma Q, Rajagopal S, Wu SM, Domian I, Rivera-Feliciano J, Jiang D, von Gise A, Ikeda S, Chien KR et al (2008a) Epicardial progenitors contribute to the cardiomyocyte lineage in the developing heart. Nature 454:109–113

Zhou B, von Gise A, Ma Q, Rivera-Feliciano J, Pu WT (2008b) Nkx2-5- and Isl1-expressing cardiac progenitors contribute to proepicardium. Biochem Biophys Res Commun 375:450–453

Chapter 16
Cellular Reprogramming During Mouse Development

Zoë D. Burke, Gabriela Miron-Buchacra, and David Tosh

Abstract States of terminal cell differentiation are often considered to be fixed. There are examples, however, in which cells of one type can be converted to a completely different cell type. The process whereby one cell type can be converted to another is referred to as cellular reprogramming. Cellular reprogramming is also referred to in the literature as transdifferentiation (or the direct conversion of one cell type to another without dedifferentiation to an intermediate cell type). Where the conversion between cell types occurs in the developing embryo, the process is referred to as transdetermination. Herein we examine some well-defined examples of transdetermination. Defining the molecular and cellular basis of transdetermination will help us to understand the normal developmental biology of the cells that interconvert, as well as identifying key regulatory transcription factors (master switch genes) that may be important for the reprogramming of stem cells. Harnessing the therapeutic potential of reprogramming and master genes is an important goal in regenerative medicine.

16.1 Definitions

States of terminal cell differentiation were often considered to be fixed, but we now know that this is not the case. A number of examples of the reprogramming of one cell type to another exist in the literature. Such changes usually arise in situations of chronic tissue damage and associated regeneration or under defined experimental conditions. Some examples of reprogramming may be indirect, occurring through an intervening stem cell, whereas others may be direct transformations, referred to as transdifferentiation. The term transdifferentiation was originally coined by

Z.D. Burke (✉) • G. Miron-Buchacra • D. Tosh
Department of Biology and Biochemistry, Centre for Regenerative Medicine, University of Bath, Claverton Down, Bath BA2 7AY, UK
e-mail: Z.D.Burke@bath.ac.uk; D.Tosh@bath.ac.uk

J.Z. Kubiak (ed.), *Mouse Development*, Results and Problems in Cell Differentiation 55,
DOI 10.1007/978-3-642-30406-4_16,

Selman and Kafatos to describe the conversion of the cuticle-producing cells to salt-secreting cells in the silk moth during metamorphosis from the larval to the adult moth (Selman and Kafatos 1974). The issue of whether reprogramming necessarily involves dedifferentiation depends on the timescale of the switch in phenotype. Inevitably, one set of genes needs to be switched off and a new set needs to be turned on. If this happens quickly then the products of the first may still persist as those of the second accumulate. In this scenario, there will be no dedifferentiated intermediate stage between the original and end states.

Reprogramming of one cell type is important in a pathological context. Some cell type conversions are known to predispose to cancer. In this case the switch in cellular phenotype is known as metaplasia. One such example is Barrett's metaplasia. Barrett's metaplasia is a pathological condition in which the distal oesophageal epithelium switches from stratified squamous to intestinal-type columnar epithelium (Colleypriest et al. 2009a). Barrett's metaplasia is important clinically as it predisposes to oesophageal adenocarcinoma. The condition is the consequence of long-term gastro-oesophageal reflux disease (Colleypriest et al. 2009b).

One of the best known examples of transdetermination comes from *Drosophila* developmental biology. The imaginal discs of *Drosophila* are larval structures that metamorphose to the main epidermal body parts during pupation. While the discs are not visibly differentiated in the larva, the molecular biology of imaginal discs during development is well understood (Cohen 1993). In a now classical series of experiments, Hadorn cultured disc fragments in the abdomens of adult female flies (Hadorn 1965). During the culture, the disc type (i.e., leg, wing) was usually conserved, but in the longer term it sometimes switched to another disc type. This switch in phenotype was called transdetermination (Hadorn 1965). In terms of the molecular biology of transdetermination, it has been found that ectopic expression of the transcription factor *Vestigial* can bring about the conversion of the leg to a wing (Maves and Schubiger 1995). Vestigial is not normally expressed in the leg, but is expressed in the wing, and experimental activation of the Wingless (Wnt) pathway in the leg disc will turn on vestigial and cause a transdetermination to wing.

16.2 Examples of Cell-Type Conversions in the Gastrointestinal Tract

The gastrointestinal tract provides some excellent examples of reprogramming events that occur both during normal development as well as following genetic manipulation.

During development, cephalo-caudal and lateral folding of the embryo creates an endodermally lined cavity, the primitive gut tube. The primitive gut tube extends the entire length of the body and gives rise to both the gastrointestinal and respiratory tracts. Between embryonic days 11 and 12 (E11–E12) in the mouse embryo, an outgrowth in the anterior region of the primitive gut tube emerges and generates two distinct tissues: the respiratory and digestive tracts. The respiratory tube forms the trachea which bifurcates into two bronchi and the lungs while the digestive tube becomes regionalised to give rise to the oesophagus, stomach and intestine. Furthermore, buds arising in the foregut region of the digestive tube give rise to the liver, pancreas and gallbladder (Wells and Melton 1999).

16.3 Transdetermination in Normal Mouse Development

16.3.1 Development of the Oesophagus

The oesophagus is the region of the gastrointestinal tract that connects the pharynx to the stomach, serving primarily to allow the passage of food into the gut. In the adult mouse the oesophagus is lined by a stratified squamous epithelium which exhibits a distinct basal layer surrounded by the spinous, granulated and cornified suprabasal layers. The inner epithelial layer is surrounded by two layers of muscle; the longitudinal and circular layers (Raymond et al. 1991; Yu et al. 2005). In the developing mouse (E11–E12) the early oesophagus and more caudal regions of the gastrointestinal tract are lined by a simple columnar epithelium surrounded by undifferentiated mesenchyme. Between E13.5 and E17.5 the epithelial lining of the oesophagus remains columnar but becomes multilayered while the surrounding mesenchyme separates into zones that will later contribute to the submucosal and smooth muscle layers. The cornified layer of the stratified squamous epithelium typically found in the adult oesophagus appears perinatally while the basal layer becomes distinct only in the adult oesophagus. Therefore, the simple columnar epithelium of the early oesophagus is replaced by the stratified squamous epithelium in the adult and represents a transdetermination of epithelial type since it begins during embryonic development (Fig. 16.1). This observation led to the proposal that the basal layer of the stratified squamous epithelium is derived from the simple columnar epithelium of the early oesophagus (Thorey et al. 1993). A model has been developed for studying oesophageal development in vitro which fully recapitulates the in vivo development of the oesophagus and provides the direct evidence for this conversion (Yu et al. 2005).

First, Yu and colleagues characterised the profile of expression of intermediate filament proteins through immunohistochemical analysis of sections taken from developing mouse embryos. The intermediate filament protein cytokeratin 14 (CK14) is a marker for the basal layer of stratified squamous tissue while cytokeratins 8 (CK8) and 18 (CK18) are typically found in columnar epithelium (Moll et al. 1982). The cells of the columnar epithelium lining the embryonic oesophagus at E11.5 express CK8 but not CK14. Expression of CK8 persists as the oesophagus becomes multilayered but diminishes until it is almost absent from the basal layer at E18.5 and disappears postnatally from the oesophageal epithelium

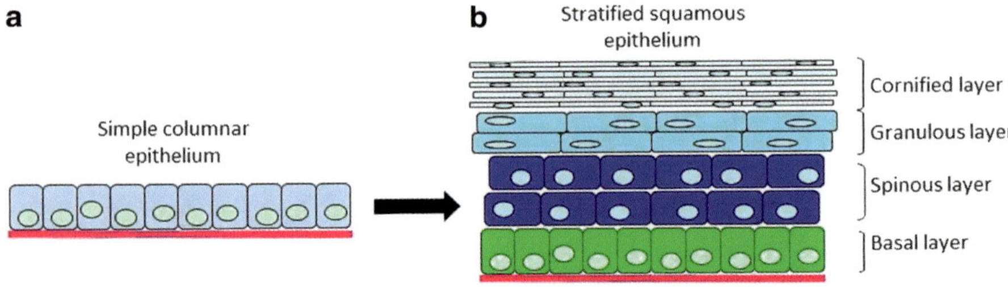

Fig. 16.1 Illustration depicting the epithelial organisation of the oesophagus. During development the oesophagus is lined by a simple columnar epithelium (**a**) which is replaced postnatally by a fully differentiated stratified squamous epithelium (**b**)

completely. However, the expression of CK14 is quite different. CK14 is first detected at around E17.5 in the basal cells which no longer express CK8 and later it is localised to all of the basal cells in the adult oesophagus. The change in expression of CK8 and CK14 in the basal layer correlates with the switch in organisation of the epithelium from a columnar to stratified squamous type.

Next, Yu and colleagues developed a novel in vitro method for the culture of embryonic oesophagus in order to study the cellular basis of the cell-type conversion (Yu et al. 2005). Oesophageal tissue dissected from E11.5 embryos and cultured on fibronectin-coated coverslips differentiates as stratified squamous epithelium and exhibits the same switch in expression from CK8 to CK14 in the basal cells as seen in vivo. When oesophageal explants were co-stained for CK8 and K14 it was revealed that some cells expressed both markers suggesting that CK14-positive basal cells arose directly from the CK8-positive columnar cells (Yu et al. 2005). In order to determine the exact origin of the basal cells of the squamous epithelium in the adult oesophagus, a lineage-tracing experiment was performed. A reporter plasmid expressing GFP under the control of the CK14 promoter was electroporated into oesophageal tissue isolated from E15.5 embryos prior to culture, thus cells in which the CK14 promoter is active express the reporter gene GFP. Subsequent co-staining of the cultures for GFP and CK8 revealed that some of the GFP positive cells also expressed CK8 indicating that the basal squamous cells arise through direct conversion from CK8-positive columnar cells. The embryonic oesophageal explant model described by Yu and colleagues provided the first evidence of a direct conversion from columnar to stratified squamous cells (Yu et al. 2005). Interestingly in Barrett's Metaplasia the lining of the adult oesophagus undergoes metaplastic transformation from normal stratified squamous epithelium to columnar-lined epithelium. In patients with Barrett's metaplasia abnormal oesophageal tissue adopts an epithelial organisation typically found in embryonic oesophagus and other parts of the gastrointestinal tract and is often characterised by intestinal differentiation (Barbera and Fitzgerald 2010). Therefore understanding the molecular basis of the conversion of one epithelial type to another during development will provide insights into the mechanisms driving the reverse conversion in Barrett's metaplasia.

16.4 Transdetermination in the Developing Lung

The alveolar epithelium consists of two types of cells alveolar "Type I" and alveolar "Type II" epithelial cells. Type I cells provide the gas exchange surface whereas Type II cells function to produce the surfactant proteins A, B, C and D. In vitro models have shown that adult Type II cells cultured on plastic over a period of 4–5 days gradually lose their phenotype and transdifferentiate into Type I cells (Chen et al. 2005; Dobbs et al. 1988; Rock and Hogan 2011). Therefore, Type II cells are considered to act as progenitor cells during restoration of the alveolar epithelium following injury, either by giving rise to new Type II cells or transdifferentiating into Type I cells (Qiao et al. 2008) . The molecular basis for the switch from Type II to Type I mouse alveolar epithelial cells has recently been elucidated. FoxM1, a member of the mammalian forkhead box family of transcription factors, plays a crucial role mediating morphogenesis and differentiation of the embryonic lung. During tissue repair, FoxM1 plays an important role in restoring the alveolar epithelial barrier. FoxM1 is highly expressed in Type II alveolar epithelial cells. Intriguingly, when FoxM1 is deleted in mouse Type II cells, the cells are no longer capable of proliferation or transdifferentiation into Type I cells, resulting in alveolar barrier dysfunction and suggesting that Type II cells are the progenitors for Type I alveolar epithelial cells (Liu et al. 2011). It would be of interest to determine whether the conversion of Type II epithelial cells to Type I is a direct or indirect transdifferentiation event.

To date there is limited evidence for the transdifferentiation of adult human alveolar epithelium (Fuchs et al. 2003), and until recently there were no previous descriptions of transdetermination in the human foetal lung. The differentiated phenotype of Type II alveolar epithelial cells isolated from human foetal lung can be maintained in vitro following culture with dexamethasone, cAMP and isobutylmethylxanthine (referred to as DCI treatment) (Gonzales et al. 2001). These culture conditions are also able to induce the differentiation of Type II cells from human foetal lung epithelium (Gonzales et al. 2002). Removing the DCI from the culture medium caused the human foetal Type II alveolar cells to transdetermine towards Type I-like cells (Foster et al. 2007). Transdetermination was associated with a change in cell morphology, reduced expression of Type II alveolar markers and induction of Type I cell markers. Since models for Type I cells are lacking, the transdetermination of Type II human foetal alveolar cells to Type I could potentially be used as a tool for studying human alveolar Type I cell biology. However, the maturation status of the Type I cells must be determined to provide unequivocal evidence of a switch to a more mature phenotype.

16.5 Transdetermination of Liver to Pancreas in the Hes 1 Knockout Mouse

During normal development the liver and ventral pancreas arise as evaginations of the ventral foregut endoderm in response to specifying signals, such as bone morphogenetic proteins (BMPs) and fibroblast growth factors (FGFs), secreted by the cardiac and septum transversum mesenchyme, respectively (Jung et al. 1999; Rossi et al. 2001). Outgrowth of the cranial region of the foregut endoderm generates the hepatocytes of the liver parenchyma and the intrahepatic biliary system while the caudal foregut endoderm gives rise to extrahepatic biliary structures such as the gallbladder, common hepatic duct, cystic duct and common bile duct. The pancreas is constructed from the ventral pancreatic diverticula, located adjacent to the hepatic foregut region, and a second outgrowth arising from the dorsal foregut endoderm. At around E12.5 of mouse development the dorsal and ventral pancreatic outgrowths fuse and continued differentiation generates the pancreatic ducts and the hormone-producing cells of the islets of Langerhans (Jorgensen et al. 2007).

Many transcription factors have now been shown to be essential for the specification and differentiation of the liver and pancreas. One pathway of particular interest is the Notch signalling pathway, first discovered in *Drosophila*, which controls developmental decisions made by cells according to the state of their immediate neighbours. In *Drosophila* the transmembrane glycoprotein receptor Notch is activated by two ligands, Delta and Serrate (Portin 2002). In mammals there are four Notch receptors (Notch 1–Notch 4) and five ligands, two serrate-like (Jagged 1 and Jagged 2) and three delta-like (Delta-like 1, Delta-like 3 and Delta-like 4). Binding of the ligand on one cell to the receptor on an adjacent cell causes the cleavage of the intracellular domain of Notch (ICN) which then translocates to the nucleus where it mediates the upregulation of the transcription factor Hairy Enhancer of Split 1 (HES1). A role for the Notch signalling pathway in fate determination of pancreatic progenitors and maintenance of progenitor cell populations has been well documented (Apelqvist et al. 1999; Jensen et al. 2000; Murtaugh and Melton 2003; Murtaugh et al. 2003; Norgaard et al. 2003). Hes1 is initially expressed throughout the entire pancreatic progenitor population where it strongly suppresses expression of differentiation factors, thereby allowing the progenitor state to be maintained. Hes1 then directs some progenitor cells towards an endocrine cell fate through lateral inhibition and co-ordinated regulation of the bHLH transcription factors, Ngn3 and NeuroD1. Ngn3 expression is restricted to endocrine cells while Hes1, which is a strong repressor of Ngn3, is expressed in the exocrine cell population. Although a role for Notch signalling in liver development is less well understood there are two pieces of evidence to suggest that the Notch pathway may be important for the development of the biliary system. The first comes from studies in mice which have shown that all four Notch receptors are expressed in the liver. However, there is a rise in Notch 2 expression around E15.5d of liver development which coincides with the formation of the ductal plate and the appearance of ductal cell precursors (Loomes et al. 2002; Shiojiri 1997). The second piece of evidence comes from the observation that mutations in the gene encoding Jagged 1 is the main cause of the human disorder Alagille syndrome, a condition characterised by abnormalities in liver development including a lack of intrahepatic bile ducts (Alagille et al. 1987; Li et al. 1997; Nijjar et al. 2002).

Remarkably, it has been demonstrated that Hes1 deficiency in mice results in the conversion of the biliary system to ectopic pancreatic tissue (Sumazaki et al. 2004). Despite the initial budding phase of bile duct development occurring as normal, continued elongation and growth of the extrahepatic and common bile duct is severely impaired. The truncated remnant of the duct in these animals contains aggregations of epithelial cells that resemble pancreatic exocrine acini. In addition to exocrine tissue expressing amylase and carboxypeptidase, the ectopic tissue also contains endocrine pancreatic cell types expressing glucagon (α-cells), insulin (β-cells), somatostatin (δ-cells) and pancreatic polypeptide (PP cells) (Fig. 16.2). The conversion of the common bile duct to pancreatic tissue represents an example of transdetermination and demonstrates a role for Hes1 in the maintenance of the biliary phenotype. Interestingly, it is only the extrahepatic biliary system, and not the intrahepatic component of the liver, that undergoes transdetermination to a pancreatic fate. While Hes1 is initially localised to the region of the foregut endoderm that gives rise to the intra- and extrahepatic biliary systems, its expression persists only in the extrahepatic ductal tissue and is lost from the liver by E12.5. Thus the first stage of intrahepatic biliary differentiation from periportal hepatoblasts occurs independently of the Notch signalling pathway as removal of Hes1 has no effect on early development of the intrahepatic bile ducts. However, the intrahepatic biliary system continues to develop postnatally. It has been shown that Hes1 expression is detectable in the neonatal liver; since Hes1 null mice die at E18.5 it remains to be determined whether Hes1 contributes to the later stages of development of the intrahepatic biliary system.

Another interesting observation in the Hes1 null mice is that normal pancreatic development is also perturbed. Due to depletion of the pancreatic precursor population through accelerated endocrine cell differentiation and apoptosis, the $Hes^{-/-}$ pancreas is hypoplastic and underdeveloped. However, the ectopic pancreatic tissue in the biliary system appears to develop like a wild-type pancreas containing clusters of insulin- and glucagon-expressing cells migrating away from the biliary epithelium. The conversion of extrahepatic biliary components to pancreatic tissue led the authors to conclude that Hes1 not only maintains the biliary phenotype but also represses differentiation along the pancreatic pathway (Sumazaki et al. 2004). The potential for the extrahepatic biliary system to give rise to pancreatic tissue is supported by the following observations. The common bile duct, which is the site of ectopic pancreatic tissue, develops from the caudal hepatic endoderm immediately adjacent to the pancreatic endoderm; therefore, both tissues are very similar with respect to the signals they encounter during development (Li et al. 1997). During human foetal liver development, hepatocytes and the primitive ducts in the hilar region, where the intrahepatic ducts meet the common hepatic duct, express pancreatic amylase (Terada et al. 1998). Finally, the endocrine pancreas in the sea lamprey arises through transdifferentiation of the common bile duct (Elliott and Youson 1993).

16.6 Transdifferentiation of Intestine to Forestomach in the Cdx2 Knockout Mouse

The endodermal lining of the gut tube extends the full length of the body, and when the anterior and posterior ends form the surrounding splancnic mesenchyme becomes regionally specified. As the developing gut tube interacts with regionally

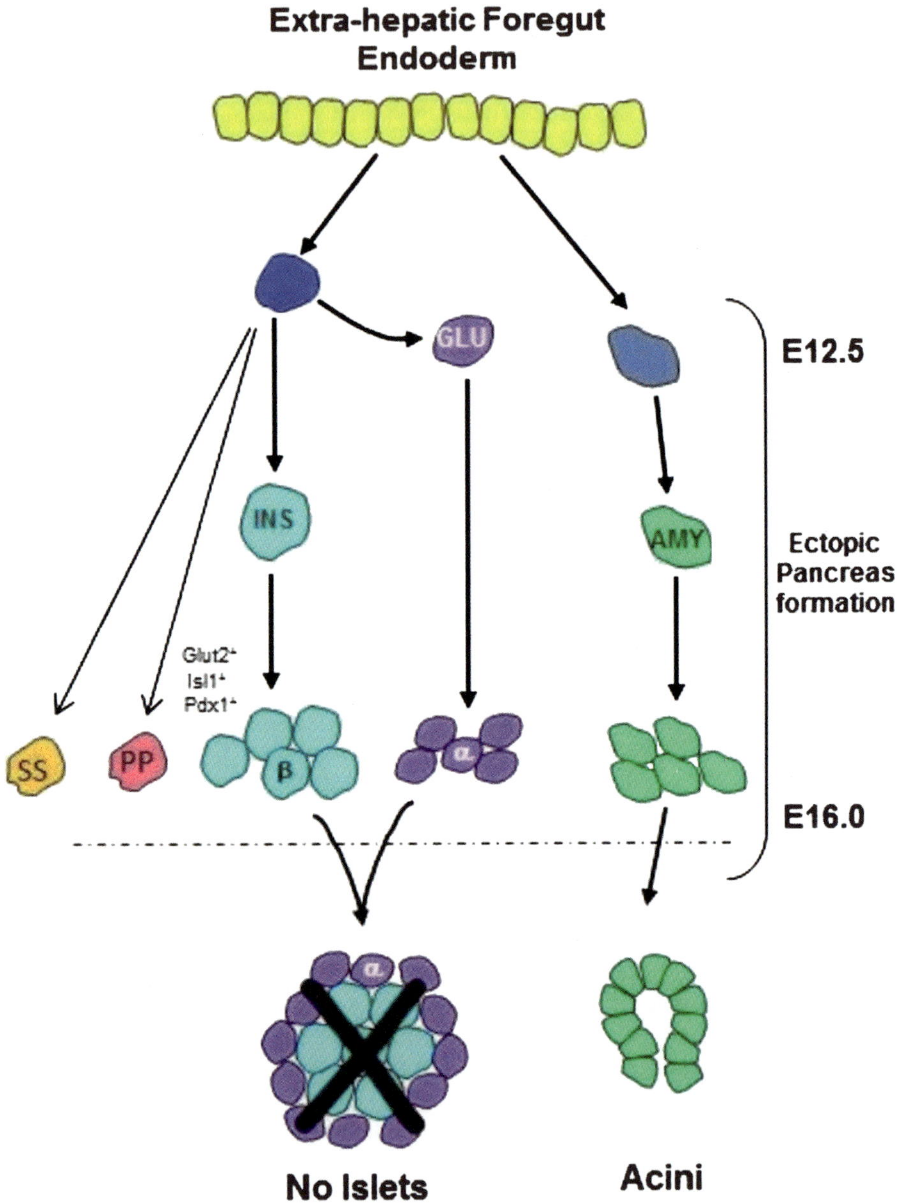

Fig. 16.2 Representation of the generation of ectopic pancreas in Hes1$^{-/-}$ extrahepatic endoderm. The common bile duct primordia of the extrahepatic biliary system in Hes1 mutant mice is completely replaced by pancreatic tissue possessing both well-differentiated β cells and exocrine acinar structures. Prior to the premature death of these animals mature islets were not identified

specified mesenchyme, it too becomes regionalised differentiating to form the oesophagus, stomach and intestine. During early development the entire gastrointestinal tract is lined by a simple columnar epithelium which, with the exception of the oesophagus, is maintained into adulthood. The lower oesophageal sphincter marks the junction between the oesophagus and the non-glandular forestomach, both of which are lined by stratified squamous epithelium. The columnar epithelium of the gastric mucosa is divided into three parts based on the different type of glands present; the cardia, the corpus and the distal region of the stomach. The pylorus contains the cardiac, gastric and pyloric glands. In more posterior regions of the gastrointestinal tract the lining epithelia remains columnar in nature. The small intestinal region is comprised of an epithelium organised into many highly specialised villi containing absorptive, goblet, Paneth and enteroendocrine cells. The epithelium lining the large intestine contains the same cell types found in the small intestine, with the exception of Paneth cells, but is smooth in appearance due to the absence of villi.

Specification of different regions along the anterior–posterior axis of the gut tube is mediated by the Sonic hedgehog (Shh) signalling pathway (Ramalho-Santos et al. 2000). Initially Shh is expressed in endoderm of the pharynx and posterior hindgut, but the domain of expression expands throughout the endoderm of the entire gut tube as the anterior and posterior ends extend towards the centre of the embryo (Roberts et al. 1998). Shh secreted from the hindgut region binds to the Patched receptor expressed by the splancnic mesenchyme surrounding the gut tube. In doing so the mesenchyme is induced to express a set of 'posterior' Hox genes the anterior boundaries of which delineate the morphological boundaries of different gastrointestinal tissues (Kawazoe et al. 2002; Roberts et al. 1995). The Hox genes contain a conserved sequence called the homeobox domain. They were first identified in *Drosophila* and are known to confer positional information to tissue during development. For example *Drosophila* mutants for the homeobox gene *cad* exhibit shortening of the body axis and loss of posterior segments (Mlodzik et al. 1985). Mouse homologues of the *Caudal* (*Cad*) gene, *Cdx1*, *Cdx2*, and *Cdx4* have been identified and are widely expressed throughout development. Interestingly, *Cdx1* and *Cdx2* expression is restricted to the endoderm of the posterior hindgut later in development and after birth (Silberg et al. 2000).

In a landmark study Beck and collaborators demonstrated that *Cdx2* is essential for the development of the posterior gastrointestinal tract (Beck et al. 1999). *Cdx2* homozygous-null mice die due to a failure to undergo implantation. *Cdx2* heterozygous mice however survive into adulthood but exhibit many abnormalities including a tendency to develop intestinal polyp-like lesions in the proximal colon. The colon is typically lined by a simple columnar epithelium containing non-branching mucus-producing glands. The polyp-like lesions arising in the colon of *Cdx2* heterozygous mice were found to contain epithelial tissue type typically found in more anterior regions of the gastrointestinal tract. Detailed histological analysis revealed that the polyps regularly contained stratified squamous epithelium similar to that found in the forestomach, gastric epithelia reminiscent of the cardia, corpus and pyloric regions of the stomach as well as small-intestinal type epithelium (Beck et al. 1999). Histologically normal colonic tissue surrounding the polyps expressed *Cdx2* while the epithelia within the polyps themselves lacked *Cdx2* expression. The authors proposed that the levels of *Cdx2* in most cells are sufficient to allow normal morphogenesis of the proximal colon; however, the levels of *Cdx2* are lower in some cells possibly due to epigenetic factors. Since the gut is a rapidly dividing tissue, cells expressing a lower level of *Cdx2* could very quickly give rise to a clone of cells in which the positional information that normally allows them to form proximal colon is insufficient. Consequently the cells in that region will differentiate towards a more anterior fate specified by the lower levels or absence of the colon differentiating factor *Cdx2*.

16.7 Summary and Future Perspectives

Although there is accumulating evidence for the transdetermination or conversion of one cell type to another during embryonic development, several challenges remain to be addressed. The first challenge is to identify the molecular basis for the switch in phenotype and whether the gene(s) responsible can be used to distinguish the two cell types during development. The second concerns testing whether ectopic expression of the gene responsible can induce the reprogramming of the cell types in vitro. Although some in vitro models may need to be developed, this approach might provide direct proof of the nature of the master switch genes. Identifying the molecular basis for transdetermination will be an important step in harnessing the therapeutic potential for cell therapy.

References

Alagille D, Estrada A, Hadchouel M, Gautier M, Odievre M, Dommergues JP (1987) Syndromic paucity of interlobular bile ducts (Alagille syndrome or arteriohepatic dysplasia): review of 80 cases. J Pediatr 110:195–200

Apelqvist A, Li H, Sommer L, Beatus P, Anderson DJ, Honjo T, Hrabe de Angelis M, Lendahl U, Edlund H (1999) Notch signalling controls pancreatic cell differentiation. Nature 400:877–881

Barbera M, Fitzgerald RC (2010) Cellular origin of Barrett's metaplasia and oesophageal stem cells. Biochem Soc Trans 38:370–373

Beck F, Chawengsaksophak K, Waring P, Playford RJ, Furness JB (1999) Reprogramming of intestinal differentiation and intercalary regeneration in Cdx2 mutant mice. Proc Natl Acad Sci USA 96:7318–7323

Chen SP, Zhou B, Willis BC, Sandoval AJ, Liebler JM, Kim KJ, Ann DK, Crandall ED, Borok Z (2005) Effects of transdifferentiation and EGF on claudin isoform expression in alveolar epithelial cells. J Appl Physiol 98:322–328

Cohen SM (1993) Imaginal disc development. In: Martinez-Arias A (ed) The Drosophila Melanogaster. Cold Spring Harbor Press, Cold Spring Harbor

Colleypriest BJ, Palmer RM, Ward SG, Tosh D (2009a) Cdx genes, inflammation and the pathogenesis of Barrett's metaplasia. Trends Mol Med 15:313–322

Colleypriest BJ, Ward SG, Tosh D (2009b) How does inflammation cause Barrett's metaplasia? Curr Opin Pharmacol 9:721–726

Dobbs LG, Williams MC, Gonzalez R (1988) Monoclonal antibodies specific to apical surfaces of rat alveolar type I cells bind to surfaces of cultured, but not freshly isolated, type II cells. Biochim Biophys Acta 970:146–156

Elliott WM, Youson JH (1993) Development of the adult endocrine pancreas during metamorphosis in the sea lamprey, Petromyzon marinus L. II. Electron microscopy and immunocytochemistry. Anat Rec 237:271–290

Foster CD, Varghese LS, Skalina RB, Gonzales LW, Guttentag SH (2007) In vitro transdifferentiation of human fetal type II cells toward a type I-like cell. Pediatr Res 61:404–409

Fuchs S, Hollins AJ, Laue M, Schaefer UF, Roemer K, Gumbleton M, Lehr CM (2003) Differentiation of human alveolar epithelial cells in primary culture: morphological characterization and synthesis of caveolin-1 and surfactant protein-C. Cell Tissue Res 311:31–45

Gonzales LW, Angampalli S, Guttentag SH, Beers MF, Feinstein SI, Matlapudi A, Ballard PL (2001) Maintenance of differentiated function of the surfactant system in human fetal lung type II epithelial cells cultured on plastic. Pediatr Pathol Mol Med 20:387–412

Gonzales LW, Guttentag SH, Wade KC, Postle AD, Ballard PL (2002) Differentiation of human pulmonary type II cells in vitro by glucocorticoid plus cAMP. Am J Physiol Lung Cell Mol Physiol 283:L940–951

Hadorn E (1965) Problems of determination and transdetermination. Brookhaven Sym Biol, 148–161

Jensen J, Pedersen EE, Galante P, Hald J, Heller RS, Ishibashi M, Kageyama R, Guillemot F, Serup P, Madsen OD (2000) Control of endodermal endocrine development by Hes-1. Nat Genet 24:36–44

Jorgensen MC, Ahnfelt-Ronne J, Hald J, Madsen OD, Serup P, Hecksher-Sorensen J (2007) An illustrated review of early pancreas development in the mouse. Endocr Rev 28:685–705

Jung J, Zheng M, Goldfarb M, Zaret KS (1999) Initiation of mammalian liver development from endoderm by fibroblast growth factors. Science 284:1998–2003

Kawazoe Y, Sekimoto T, Araki M, Takagi K, Araki K, Yamamura K (2002) Region-specific gastrointestinal Hox code during murine embryonal gut development. Dev Growth Differ 44:77–84

Li L, Krantz ID, Deng Y, Genin A, Banta AB, Collins CC, Qi M, Trask BJ, Kuo WL, Cochran J, Costa T, Pierpont ME, Rand EB, Piccoli DA, Hood L, Spinner NB (1997) Alagille syndrome is caused by mutations in human Jagged1, which encodes a ligand for Notch1. Nat Genet 16:243–251

Liu Y, Sadikot RT, Adami GR, Kalinichenko VV, Pendyala S, Natarajan V, Zhao YY, Malik AB (2011) FoxM1 mediates the progenitor function of type II epithelial cells in repairing alveolar injury induced by Pseudomonas aeruginosa. J Exp Med 208:1473–1484

Loomes KM, Taichman DB, Glover CL, Williams PT, Markowitz JE, Piccoli DA, Baldwin HS, Oakey RJ (2002) Characterization of Notch receptor expression in the developing mammalian heart and liver. Am J Med Genet 112:181–189

Maves L, Schubiger G (1995) Wingless induces transdetermination in developing Drosophila imaginal discs. Development 121:1263–1272

Mlodzik M, Fjose A, Gehring WJ (1985) Isolation of caudal, a Drosophila homeo box-containing gene with maternal expression, whose transcripts form a concentration gradient at the pre-blastoderm stage. EMBO J 4:2961–2969

Moll R, Franke WW, Schiller DL, Geiger B, Krepler R (1982) The catalog of human cytokeratins: patterns of expression in normal epithelia, tumors and cultured cells. Cell 31:11–24

Murtaugh LC, Melton DA (2003) Genes, signals, and lineages in pancreas development. Annu Rev Cell Dev Biol 19:71–89

Murtaugh LC, Stanger BZ, Kwan KM, Melton DA (2003) Notch signaling controls multiple steps of pancreatic differentiation. Proc Natl Acad Sci USA 100:14920–14925

Nijjar SS, Wallace L, Crosby HA, Hubscher SG, Strain AJ (2002) Altered Notch ligand expression in human liver disease: further evidence for a role of the Notch signaling pathway in hepatic neovascularization and biliary ductular defects. Am J Pathol 160:1695–1703

Norgaard GA, Jensen JN, Jensen J (2003) FGF10 signaling maintains the pancreatic progenitor cell state revealing a novel role of Notch in organ development. Dev Biol 264:323–338

Portin P (2002) General outlines of the molecular genetics of the Notch signalling pathway in Drosophila melanogaster: a review. Hereditas 136:89–96

Qiao R, Yan W, Clavijo C, Mehrian-Shai R, Zhong Q, Kim KJ, Ann D, Crandall ED, Borok Z (2008) Effects of KGF on alveolar epithelial cell transdifferentiation are mediated by JNK signaling. Am J Respir Cell Mol Biol 38:239–246

Ramalho-Santos M, Melton DA, McMahon AP (2000) Hedgehog signals regulate multiple aspects of gastrointestinal development. Development 127:2763–2772

Raymond C, Anne V, Millane G (1991) Development of esophageal epithelium in the fetal and neonatal mouse. Anat Rec 230:225–234

Roberts DJ, Johnson RL, Burke AC, Nelson CE, Morgan BA, Tabin C (1995) Sonic hedgehog is an endodermal signal inducing Bmp-4 and Hox genes during induction and regionalization of the chick hindgut. Development 121:3163–3174

Roberts DJ, Smith DM, Goff DJ, Tabin CJ (1998) Epithelial-mesenchymal signaling during the regionalization of the chick gut. Development 125:2791–2801

Rock JR, Hogan BL (2011) Epithelial progenitor cells in lung development, maintenance, repair, and disease. Annu Rev Cell Dev Biol 27:493–512

Rossi JM, Dunn NR, Hogan BL, Zaret KS (2001) Distinct mesodermal signals, including BMPs from the septum transversum mesenchyme, are required in combination for hepatogenesis from the endoderm. Genes Dev 15:1998–2009

Selman K, Kafatos FC (1974) Transdifferentiation in the labial gland of silk moths: is DNA required for cellular metamorphosis? Cell Differ 3:81–94

Shiojiri N (1997) Development and differentiation of bile ducts in the mammalian liver. Microsc Res Tech 39:328–335

Silberg DG, Swain GP, Suh ER, Traber PG (2000) Cdx1 and cdx2 expression during intestinal development. Gastroenterology 119:961–971

Sumazaki R, Shiojiri N, Isoyama S, Masu M, Keino-Masu K, Osawa M, Nakauchi H, Kageyama R, Matsui A (2004) Conversion of biliary system to pancreatic tissue in Hes1-deficient mice. Nat Genet 36:83–87

Terada T, Kato M, Horie S, Endo K, Kitamura Y (1998) Expression of pancreatic alpha-amylase protein and messenger RNA in hilar primitive bile ducts and hepatocytes during human fetal liver organogenesis: an immunohistochemical and in situ hybridization study. Liver 18:313–319

Thorey IS, Meneses JJ, Neznanov N, Kulesh DA, Pedersen RA, Oshima RG (1993) Embryonic expression of human keratin 18 and K18-beta-galactosidase fusion genes in transgenic mice. Dev Biol 160:519–534

Wells JM, Melton DA (1999) Vertebrate endoderm development. Annu Rev Cell Dev Biol 15:393–410

Yu WY, Slack JM, Tosh D (2005) Conversion of columnar to stratified squamous epithelium in the developing mouse oesophagus. Dev Biol 284:157–170

Chapter 17
Differentiation of Definitive Endoderm from Mouse Embryonic Stem Cells

Peter T.W. Kim and Christopher J. Ong

Abstract Efficient production of definitive endoderm from embryonic stem (ES) cells opens doors to the possibilities of differentiation of endoderm-derived tissues such as the intestines, pancreas, and liver that could address the needs of people with chronic diseases involving these organs. The lessons learned from developmental biology have contributed significantly to in vitro differentiation of definitive endoderm. Gastrulation, a process that results in the production of all three embryonic germ cell layers, definitive endoderm, mesoderm, and ectoderm, is an important step in embryonic development. Gastrulation occurs as a result of the events that are orchestrated by the signaling pathways involving Nodal, FGF, Wnt, and BMP. Understanding these signaling pathways has led to the introduction of key ingredients such as Activin A, FGF, Wnt, and BMP to the differentiation protocols that have been able to produce definitive endoderm from ES cells. Efficient production of definitive endoderm needs to meet the specific criteria that include (a) increase in the production of markers of definitive endoderm such as Sox 17, FOXA2, GSC, and Mixl1; (b) decrease in the production of markers of primitive/visceral/parietal endoderm, Sox 7 and OCT4; and (c) decrease in the mesoderm markers (Brachyury, MEOX) and ectoderm markers (Sox1 and ZIC1).

P.T.W. Kim • C.J. Ong (✉)
Department of Surgery, University of British Columbia, 2660 Oak Street, Vancouver, BC, Canada V6H 3Z6
e-mail: chris.ong@ubc.ca

J.Z. Kubiak (ed.), *Mouse Development*, Results and Problems in Cell Differentiation 55, DOI 10.1007/978-3-642-30406-4_17,

17.1 Mouse Embryonic Development

17.1.1 Early Development

In mice, by day 3.5 after fertilization, the zygote develops into a blastocyst. The mammalian blastocyst structurally resembles a vertebrate blastula, which is a hollow sphere of cells that is comprised of undifferentiated cells called blastomeres that form an epithelial layer (blastoderm) enclosing a fluid-filled cavity (blastocoel). At preimplantation stages, the early blastocyst of the mouse is a fluid-filled blastula-like sphere comprised of the trophectoderm, inner cell mass, and blastocyst cavity that is formed through a process called blastulation. Despite their morphological similarity and semantic relatedness, the blastocyst is not homologous to and should not be confused with the blastula, and the blastocyst cavity to the blastocoel of amphibian blastula. The latter is the primary cavity of the embryo, while the blastocyst cavity is extraembryonic. The real homologue of the blastula stage during mouse development can be distinguished at the egg cylinder stage (6.5 dpc) right before gastrulation (see the next section) (O'Farrell et al. 2004). The outer cells of the blastocyst (trophoblast) form the trophectoderm that gives rise to extraembryonic tissues such as placenta. The inner cell mass is composed of a cluster of pluripotent stem cells located at one end of the blastocyst that gives rise to the embryo proper as well as the amnion, yolk sac, and allantois and is the cell source from which embryonic stem cells are derived.

By day 4.5, late blastocyst contains three cell types: epiblast (EPI), trophectoderm (TE), and primitive endoderm (PE). Trophectoderm gives rise to the extraembryonic ectoderm (ExE) and all the trophoblast lineages that eventually form the majority of the fetal part of the placenta (Rossant 2004). The PE gives rise to the visceral (VE) and parietal endoderm. The PE and VE are called "endoderm," but they are different from the definitive endoderm in that they do not contribute to the embryo but are extraembryonic. Due to the fact that they share the term endoderm, this has led to some confusion. The EPI gives rise to all of the cells that make up the entire fetus, as well as the allantois and the extraembryonic mesoderm cells that line the visceral yolk sac.

While the trophoblasts (outer cells of the preimplanted embryo) are invading the uterine wall, the inner cell mass develops an epithelial structure distally. This structure has been variously named: primitive endoderm (PE), extraembryonic, primary endoderm, or hypoblast. The primitive endoderm gives rise to two structures: parietal endoderm and the visceral endoderm (Verheijen and Defize 1999). Parietal endoderm gives rise to the parietal yolk sac and this is critical for absorption of nutrients during early development and it protects the embryo from mechanical damage. The visceral endoderm later in gestation, together with extraembryonic mesoderm, gives rise to the visceral yolk sac (Verheijen and Defize 1999). These extraembryonic tissues are precursors of the placenta and serve multiple functions; they are required for invasion of the uterine lining, they protect the embryo, and facilitate the embryo–maternal interactions and are responsible for nutrient and waste exchange between mother and the fetus.

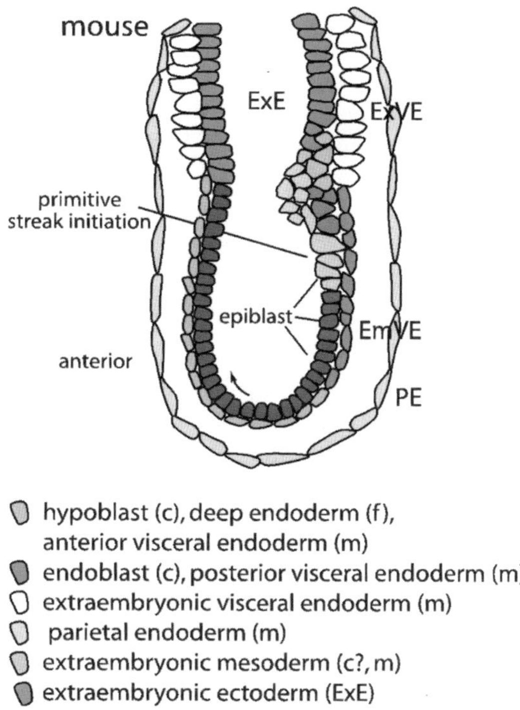

Fig. 17.1 Mouse embryo at gastrulation illustrating the location and organization of the extraembryonic structures and the precursor of definitive endoderm (epiblasts). Adapted from Evolution of the mechanisms and molecular control of endoderm formation by Grapin-Botton and Constam. Mechanism of Development 2007; 124:253–278

17.1.2 Gastrulation

Gastrulation in mice, as in humans, is a set of morphogenetic processes that occur during development that transforms an unstructured embryo into a gastrula with three germ layers: endoderm, mesoderm, and ectoderm. Coined by Haeckel, it is derived from the Greek word, "gaster" which means a stomach or a gut (Solnica-Krezel 2005). In mice, the early gastrula looks like a cup and undergoes four morphogenetic movements: internalization, epiboly, convergence, and extension (Solnica-Krezel 2005). Internalization brings the cells of the prospective mesoderm and endoderm beneath the prospective ectoderm via an opening in the blastula proper stage embryo called the primitive streak (PS) also known as the blastopore. Epiboly movements spread and thin germ layers. Convergence and extension movements narrow the germ layers mediolaterally and elongate the embryo from head to tail. These movements are preceded and accompanied by processes that are largely in part controlled by the Spemman-Mangold organizer (SMO) located in the dorsal or axial aspect of the blastopore (Spemann 1938).

The formation of the primitive streak marks the beginning of gastrulation. As outlined in Fig. 17.1, the PS forms in the region of the embryo that will eventually form the posterior end (Tam and Behringer 1997). The EPI cells then undergo an epithelial-to-mesenchymal transition (EMT) and exit from PS to form mesoderm or endoderm. The distribution of the distinct population of mesoderm and endoderm

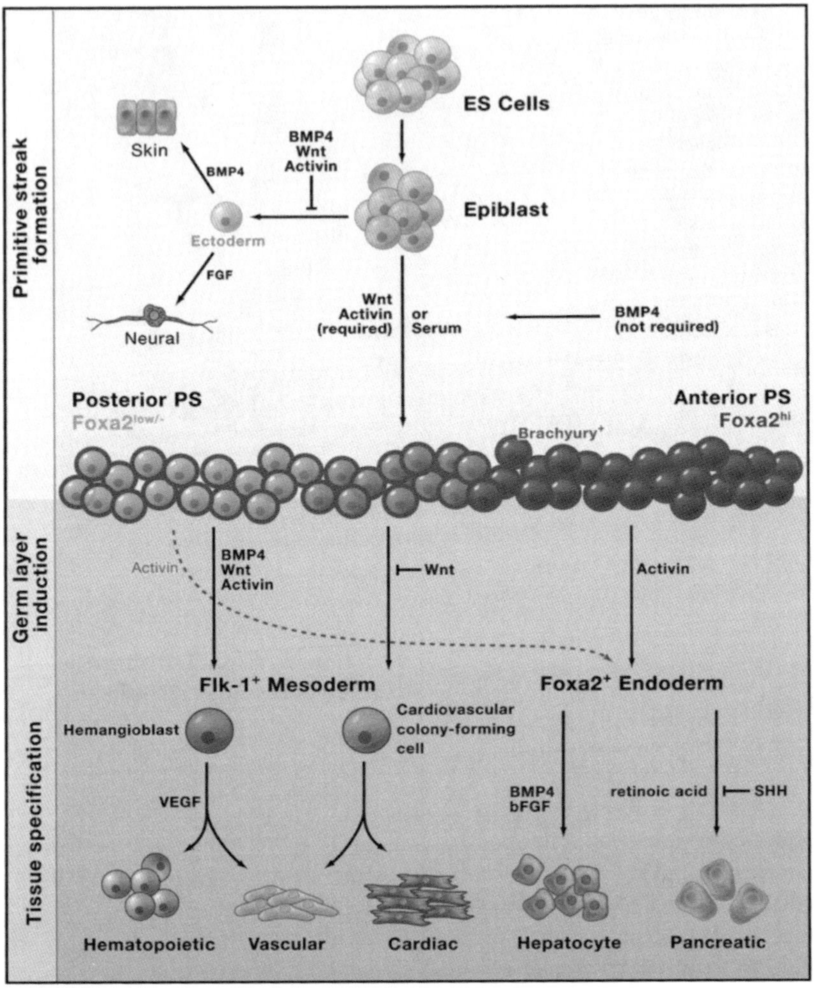

Fig. 17.2 A description of the production of the three germ cell layers and the regulatory factors. Adapted from Differentiation of embryonic stem cells to clinically relevant populations: lessons from embryonic development by Murry and Keller. Cell 132:661–680. 2008

are not random but they follow a specific sequence. The mobilized population of EPI cells traverses the posterior PS and gives rise to the extraembryonic mesoderm that forms the allantois, amnion, and the hematopoietic, endothelial, and vascular smooth muscle cells of the yolk sac (Kinder et al. 1999). As illustrated in Fig. 17.2, the EPI cells migrate further toward the more anterior parts of the PS and form cranial and cardiac mesoderm (Murry and Keller 2008). The definitive endoderm arises from the EPI cells that travel to the most anterior region of the PS. In contrast to the mesoderm and endoderm, ectoderm arises from the anterior part of the epibast that does not enter the PS. The ectoderm will eventually give rise to epidermis and the neural tissues.

The definitive endoderm is subdivided into medial and lateral anterior and posterior endoderm. The first EPI cells that exit the PS migrate toward anterior giving rise to medial and lateral anterior endoderm and axial mesoderm (Kimura et al., 2006; Lawson and Schoenwolf, 2003; Lawson and Pedersen, 1987). Endoderm that exits the PS later contributes to the posterior endoderm.

The genetic expression patterns in the embryo support the spatial distribution of different population of EPI cells. As the different type of EPI cells occupy different parts of the PS, the gene expression patterns also have a spatial distribution. The genes such as Brachyury (Hart et al. 2002; Kispert and Herrmann 1994) and Mixl1 (Hart et al. 2002) are expressed throughout the PS. In contrast, Foxa2 and Goosecoid are found predominantly in the anterior regions (Kinder et al. 2001; Sasaki and Hogan 1993). HoxB1 and Evx1 are found in the posterior regions (Dush and Martin 1992; Forlani et al. 2003).

17.2 Definitive Endoderm

The definitive endoderm gives rise to an array of epithelial cell types that line the respiratory and gastrointestinal tract and contributes to associated essential organs such as the thymus, thyroid, lungs, liver, bile ducts, and the pancreas. There is some evidence that suggest the endoderm and the mesoderm arise from a transient common precursor tissue called the mesendoderm (Gouon-Evans et al. 2006; Tada et al. 2005). After gastrulation, the endoderm is transformed into the primitive gut tube and assumes the specific orientation and becomes regionalized along the dorsal–ventral (D-V) and anterior–posterior (A-P) axes into foregut, midgut, and hindgut (Zorn and Wells 2009). The foregut gives rise to the esophagus, trachea, stomach, lungs, and the first two portions of the duodenum. Thyroid, lungs, liver, bile ducts, and the pancreas develop as outpouchings of the foregut. The midgut gives rise to the small intestine, and the hindgut forms the large intestine.

17.2.1 Endoderm Formation

Nodal, Wnt, FGF, and BMP signaling pathways play important roles in lineage allocation and germ layer specification during early embryogenesis (Tam and Loebel 2007).

17.2.1.1 TGFβ Signaling

The TGFβ superfamily of ligands includes Nodal, Activin A, BMPs, and TGFβ. As illustrated in Fig. 17.3, these ligands bind to the type II and type I serine–threonine kinase receptors and signal through the Smad2/Smad3 branch (Payne et al. 2011;

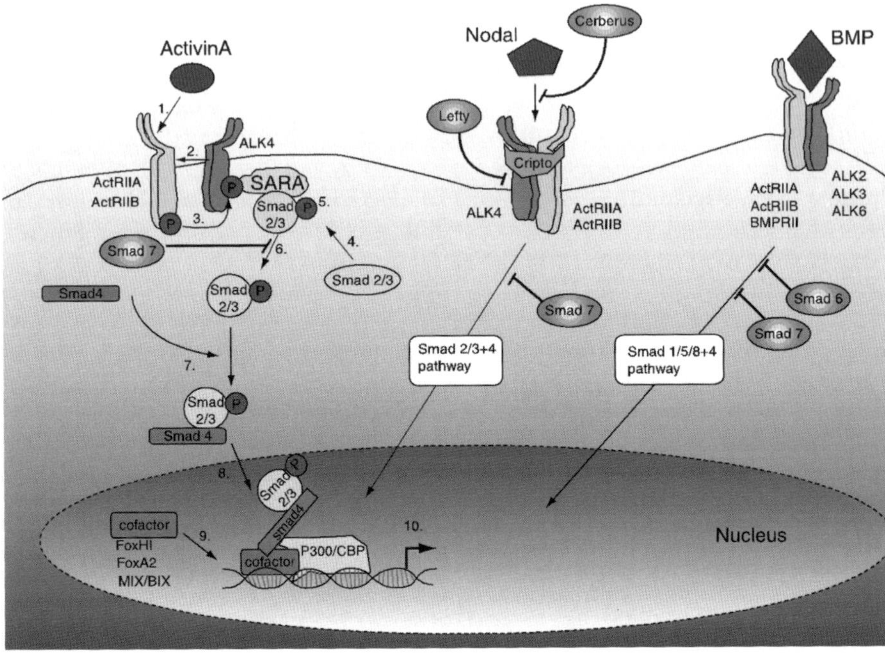

Fig. 17.3 Smad-mediated pathway of Activin/Nodal and BMP signaling. Adapted from The role of activin/nodal and Wnt signaling in endoderm formation. Vitamins and Hormones 85:207–216. 2011

Schier 2003; Whitman 2001). Nodal/Activin A bind to the Activin receptor type 2 (ActRIIA or ActRIIB) dimer resulting in phosphorylation of the Activin receptor type 1B (ALK4). The Smad 2/3 complex is recruited to ALK4 by SARA. Phosphorylation of Smad 2/3 allows it to bind to Smad 4. The Smad 2/3/4 complex enters the nucleus and initiates gene expression along with cofactors (Fork head box h1 (FoxH1), FoxA2, MIX/BIX) and P300/CBP. For Nodal signaling, a member of the EGF-CFC coreceptor family, such as Cripto and Criptic in mouse, needs to bind to ALK4 to stabilize the tetrametric ALK-4–AcRII receptor complex. Once activated, intracellular signaling is the same as the Activin A signaling. Nodal signaling can be inhibited extracellularly by Lefty and Cerberus.

BMP signaling also requires an assembly of a tetrameric receptor complex, but there are more options for assembly since there is an additional type II receptor (BMPRII) and three type I receptors (ALK2, ALK3, and ALK6). Contrary to Activin A/Nodal, BMP signaling uses the Smad1/5/8/4 pathway and affects gene transcription (Feng and Derynck 2005; Heldin et al. 1997; Itoh and ten Dijke 2007).

The Nodal signaling pathway is necessary for endoderm formation. In vitro, Activin A binds to the same receptor as Nodal and mimics Nodal signaling (Gadue et al. 2006). This response is dose dependent. High levels of Nodal signaling promote endoderm development, whereas lower doses promote mesoderm development (Kubo et al. 2004; D'Amour et al. 2005). Loss of function in any core

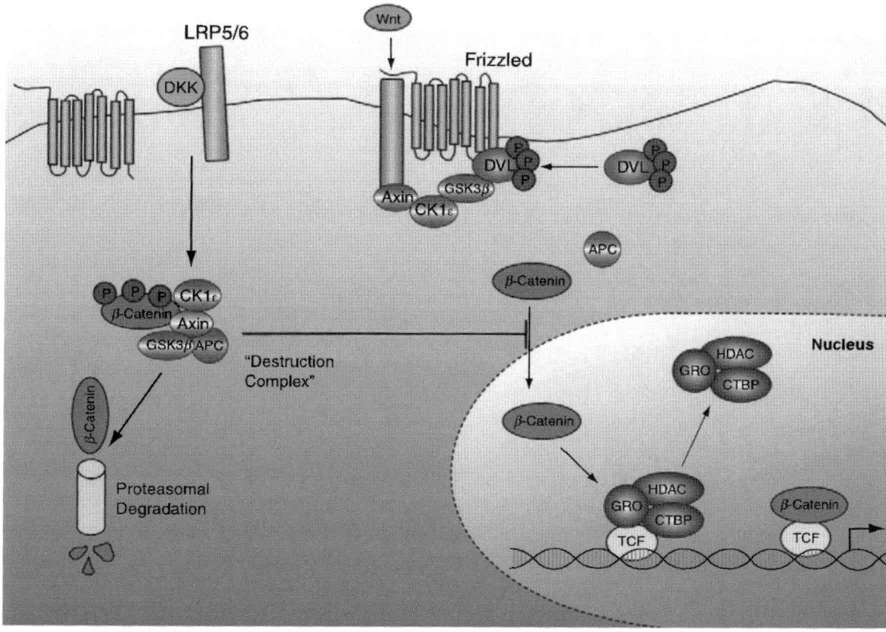

Fig. 17.4 Wnt/β-catenin signaling pathway. Adapted from The role of activin/nodal and Wnt signaling in endoderm formation. Vitamins and Hormones 85:207–216. 2011

pathway component results in a compromised mesendoderm development (Shen 2007). In a recent study, inhibition of endogenous BMP4 signaling has been shown to promote production of definitive endoderm in mouse (Li et al. 2011)

17.2.1.2 Wnt/β-Catenin Signaling

As illustrated in Fig. 17.4, the Wnt signaling pathway is broadly divided into canonical and non-canonical pathways. The canonical pathway involves binding of Wnt to its cell surface receptor Frizzled and its coreceptor low-density lipoprotein receptor protein (LRP5/6). This results in the activation of the Dishevelled (DSH) family proteins which recruits axin to the plasma membrane and inhibits the assembly of the "destruction complex" made of Axin, GSK3β, APC, and CKI (Logan and Nusse 2004; Moon et al. 1997). The absence of destruction complex results in β-catenin stability allowing it to travel to the nucleus. In the absence of Wnt signaling, β-catenin is phosphorylated and ubiquitinated by the "destruction complex" promoting its degradation (Clevers 2006). In the nucleus, β-catenin interacts with the T-cell factor/lymphocyte enhancing binding factor (TCF/LEF) family of transcription factors and results in the displacement of transcriptional repressor Groucho and HDAC and eventual activation of transcription (Arce et al. 2006; Clevers 2006; Moon 2005).

The non-canonical Wnt pathway does not involve the frizzled receptor and β-catenin. This pathway involves activation of protein kinase C resulting in an increase in intracellular calcium or activation of small Rho GTPases (e.g., c-Jun N terminus kinase (Jnk)) (Kuhl et al. 2000; Nateri et al. 2005; Topol et al. 2003; Westfall et al. 2003).

The canonical Wnt signaling plays an important role in formation of PS and endoderm. This has been shown in knockout mouse studies where mouse embryos lacking β-catenin fail to form PS, mesoderm, or endoderm (Haegel et al. 1995). In addition, Wnt3 has been shown to be important in PS formation (Liu et al. 1999) and Wnt signaling has been shown to be important in human liver development (Hay et al. 2008).

A complex regulatory circuit involving Nodal, Wnt3, and BMP4 occurs in mouse embryo. To stimulate the formation of primitive streak, an unprocessed Nodal ligand can signal from the epiblast to the adjacent extraembryonic ectoderm to induce expression of Furin and Pace4 proprotein convertases and BMP4 (Ben-Haim et al. 2006). Then BMP4 signals back to the epiblast to activate Wnt3 expression which can upregulate Nodal and Cripto expression in the epiblast through the canonical Wnt pathway. Gdf3 is likely involved in this feedback loop (Chen et al. 2006).

The definitive endoderm can further be differentiated to hepatic fate with FGF and BMP-4 (Gouon-Evans et al. 2006). Retinoic acid with inhibition of the sonic hedgehog (SHH) results in the production of cells of pancreatic fate (D'Amour et al. 2006).

17.2.1.3 Endoderm Transcription Factors

Markers of Definitive Endoderm

There are a number of transcription factors that are expressed in definitive endoderm. These include Sox17 (Sry (sex-determining region Y)-related HMG box), FOXA2, Goosecoid (GSC), and Mixl1.

Originally identified as a stage-specific transcriptional activator during spermatogenesis, based on a study involving the Sox 17 null mutant mouse embryo, Sox 17 is expressed in the definitive endoderm as well as the primitive endoderm (Kanai-Azuma et al. 2002).

FOXA2 (HNF3β) is one of forkhead domain (FD)-containing genes. It is expressed in the definitive endoderm as well as anterior part of the primitive streak, notochord, mesoderm, and primitive endoderm. Co-expression of Sox17 and FOXA2 imply production of definitive endoderm (Ang et al. 1993; Monaghan et al. 1993; Sasaki and Hogan 1993).

Goosecoid (GSC) is a homeobox gene expressed during gastrulation, but it is also expressed in primitive endoderm (Blum et al. 1992). Mixl1, a homeobox gene, is a primitive streak marker, and it is also expressed in the visceral endoderm (Hart et al. 2002; Pearce and Evans 1999).

Because most markers of the definitive endoderm are also commonly expressed in the primitive/visceral/parietal endoderm as well, additional markers are required to distinguish between extraembryonic endoderm versus definitive endoderm. Markers that have restricted expression in extraembryonic endoderm such as Sox7 (Futaki et al. 2004) and pluripotency marker OCT4 (POU5F1) (Pesce and Scholer 2001; Rosner et al. 1990) are used to help delineate whether an increase in endoderm marker expression is truly from production of definitive endoderm. For example, a decrease in extraembryonic endoderm markers, Sox7 and OCT4, in combination with an increase in the endoderm markers Sox17, FOXA2, Mixl1, and GSC implies production of definitive endoderm.

To ensure that there is enrichment and efficient production of definitive endoderm rather than chaotic differentiation of all three germ layers, the levels of the mesoderm markers such as Brachyury and MEOX1 are monitored (Candia et al. 1992; Candia and Wright 1996; Kubo et al. 2004; Wilkinson et al. 1990) as well as the ectoderm markers SOX1 and ZIC1 (Nagai et al. 1997; Pevny et al. 1998).

CXCR4 is a cell surface chemokine receptor specific to stromal-derived factor-1 (SDF-1) expressed in definitive endoderm and mesoderm and not expressed in visceral/primitive endoderm (McGrath et al. 1999). It has been used as a tool for isolating definitive endodermal cells (D'Amour et al. 2005).

17.3 Production of Definitive Endoderm from Embryonic Stem (ES) Cells

17.3.1 Embryoid Body Formation

The capacity of ES cells to differentiate into definitive endoderm was first demonstrated by Keller et al. who showed that Brachyury expressing definitive endoderm cells were contained within embryoid bodies (EBs) (Kubo et al. 2004). Embryoid bodies are three-dimensional, multicellular aggregates of differentiated and undifferentiated ES cells that develop following culture of ES cells under conditions that promote cell aggregation and differentiation (Kubo et al. 2004). Descendants of all three germ layers can be generated from ES cells following EB formation. Under these conditions, a wide spectrum of cell types can be produced from ES cells including hemopoietic, cardiac, pancreatic, hepatic, neural, and even germ cells. However, as opposed to the highly organized and coordinated spatial and temporal development of the normal embryo in vivo, cell differentiation within embryoid bodies in vitro occurs in a relatively disorganized and chaotic fashion. However, features of organized differentiation have been observed within embryoid bodies including the establishment of anterior–posterior polarity, the formation of primitive streak-like regions, epithelial-to-mesenchymal transition-generating mesoendodermal progenitor cells and execution of a coordinated gastrulation-like process (ten Berge et al. 2008).

Variability in EB size can greatly influence cell differentiation. For example, during EB-induced gastrulation-like differentiation, cells within the vicinity of the original inducing cell are recruited to the polarizing center, whereas an inhibitory signal acting over a longer range prevents formation of multiple polarizing centers. Small EBs typically have single polarizing centers; however, as the size of EBs increased, multiple polarizing centers were often observed suggesting that the size of the EBs grew beyond the range of action of key morphogens. Thus, regulating the size of EBs and the timing of their formation is crucial for controlling the differentiation of ES cells within EBs. Current conventional methods typically result in EBs that are heterogeneous in size and shape, leading to inefficient and uncontrolled differentiation. Hence, strategies for generation of uniform-sized embryoid bodies are currently being developed (Jeong et al. 2011; Mohr et al. 2010; Torisawa et al. 2007; Xu et al. 2011a).

Directed cell differentiation within the embryoid bodies is difficult to control due to the stochastic nature of differentiation within embryoid bodies, the chaotic organization of diverse cell types, and the intrinsic variability of cell-to-cell interactions and exposure to variable levels of external signals within embryoid bodies. Furthermore, the three-dimensional cellular structure of EBs restricts the ability of exogenously added growth factors to diffuse more than a limited distance through the outer shell of the embryoid body (Sachlos and Auguste 2008). Recent studies have shown that both human and mouse ES cells can be induced to differentiate as monolayers in the absence of EB formation providing a platform for more controlled directed differentiation (D'Amour et al. 2005; Hansson et al. 2009; Kim et al. 2010; Shi et al. 2005; Xu et al. 2011b).

17.3.2 Role of Serum

Fetal bovine serum (FBS) is commonly used in conventional methods for embryoid body differentiation of ES cells. FBS provides the key factors that influence differentiation of ES cells toward diverse lineages including endoderm. For example, an important role for serum in triggering gastrulation-like processes during EB differentiation is to provide Bmp activity that activates Wnt signaling that is needed to initiate primitive streak formation (ten Berge et al. 2008).

The yield of endoderm cells within EBs is typically low and highly variable. Variability in endoderm differentiation within EBs is highly influenced by different batches of FBS and/or by addition of exogenous factors. FBS contains complex mixtures of undefined growth and differentiation factors. Different batches of FBS often contain highly variable levels of endoderm-inducing factors. In addition, FBS may also contain inhibitory factors that actively suppress endoderm differentiation, and/or contain high levels of competing morphogens that drive differentiation toward alternate lineages. This concept is supported by the fact that the production of definitive endoderm from ES cells is decreased by culture in high serum concentrations, whereas limiting exposure to serum increases production of definitive endoderm (D'Amour et al. 2005; Kubo et al. 2004; Sherwood et al. 2011). Furthermore, batch selection of FBS was found to be critical for efficient definitive endoderm differentiation (Kim et al. 2010).

Since the addition of serum which consists of an undefined mixture of growth factors and inhibitors is a major cause of variability in regulating ES cell differentiation, attempts to replace the serum-containing medium with purified factors in chemically defined serum-free media are actively being pursued to improve consistency and reproducibility of differentiation cultures. Nusse and coworkers showed that serum-containing media can be replaced with purified growth factors such as Wnt3a, BMP4, or Activin A in chemically defined medium to induce gastrulation-like process in EBs (ten Berge et al. 2008). Furthermore, Kubo and coworkers (2004) found that the presence of Activin A in serum-free conditions could replace the need for serum to induce formation of definitive endoderm in EBs.

17.3.3 Effects of Exogenous Agents

17.3.3.1 Activin A

Activin A, a member of the TGF-β superfamily, has been used in the initial studies in mouse to produce definitive endoderm from embryonic stem cells using embryoid bodies (Kubo et al. 2004; Tada et al. 2005). Activin A has been a key agent in the development of endoderm in human embryonic stem cells as well (D'Amour et al. 2005; Xu et al. 2011b). While Activin A is an essential component in formation of definitive endoderm from ES cells, a number of factors can modulate the ability of Activin A to induce production of definitive endoderm.

High doses of Activin A have been required for efficient production of definitive endoderm. Sufficient concentration appears to be between 30 and 100 ng/ml (D'Amour et al. 2005; Kubo et al. 2004; Shim et al. 2007). Lower concentrations such as 10 ng/ml seem to be acceptable when the ES cells are cultured in monolayers (Tada et al. 2005) but higher concentrations are required for EBs. In mouse ES cells, the duration of exposure to Activin A has been dependent on the concentration and the nature of the cells (EBs vs. monolayer). When higher concentration was used in EBs, the duration of Activin exposure was shorter (1–3 days) (Gouon-Evans et al. 2006; Shi et al. 2005), whereas, when lower concentration was used in monolayers, a longer duration was used (4–6 days) (Tada et al. 2005; Yasunaga et al. 2005).

17.3.3.2 Wnt

Wnt signaling has been shown to be required for endoderm formation from ES cells (D'Amour et al. 2006; Gadue et al. 2006; Lindsley et al., 2006). Activin-induced endoderm differentiation of mES cells is potentiated by Wnt signaling while antagonizing Wnt signaling impairs endoderm yield (Sherwood et al. 2011). Exogenous Wnt3a protein posteriorizes the embryoid body, resulting in predominantly mesoendodermal differentiation (ten Berge et al. 2008). Conversely, Wnt antagonists promoted anterior character and neuroectodermal differentiation (ten Berge et al. 2008).

17.3.3.3 BMPs

Previous studies on the role of BMP on the production of definitive endoderm have yielded conflicting results in human and mouse embryonic stem cells. Addition of BMP4 to human ES in culture results in the production of trophectoderm, and native BMP signaling is necessary for the production of extraembryonic endoderm (Pera et al. 2004; Xu et al. 2002). Introduction of BMP4 and FGF inhibition decreased the production of primitive streak markers even in the presence of high-dose Activin A (D'Amour et al. 2005). However, in mouse, BMP4 alone, Activin A alone or both combined induced primitive streak markers (Jackson et al. 2010; Nostro et al. 2008). Other studies have suggested that the combination of factors BMP, Activin A, and bFGF was necessary to induce mesendoderm differentiation and that BMP4 combined with the other factors does not adversely affect DE production (Vallier et al. 2009; Xu et al. 2011b). On the other hand, inhibition of BMP4 signaling by Noggin promoted Activin A-induced differentiation of mES cells to definitive endoderm (Li et al. 2011). Inhibiting BMP4 with dorsomorphin promoted endoderm differentiation of mES cells, whereas addition of BMP4 inhibited endoderm differentiation (Sherwood et al. 2011).

17.3.3.4 Fibroblast Growth Factors

Gene-targeting experiments have shown that FGF plays an important role in endoderm formation. Treatment of mouse embryos with FGF4 resulted in a dose-dependent expression of anterior endoderm (NeuroD) and posterior endoderm (Somatostatin) markers (Niswander and Martin 1992). FGF2 with Activin A resulted in an additive effect of cells expressing Sox 17 and PDX1 (Shiraki et al. 2008).

17.3.3.5 Chemical Inducers of Endoderm Differentiation

Cell permeable small molecules can alter cellular processes by modulating signal transduction pathways, gene expression, or cell metabolism. A number of small molecules have been effectively used to promote ESC differentiation toward definitive endoderm.

Lithium chloride (LiCl) treatment in combination with Activin A and Noggin improved production of definitive endoderm from mESC (Li et al. 2011). LiCl treatment triggers downstream effects that mimic Wnt pathway activation such as stabilization and accumulation of β-catenin. LiCl is presumed to mimic Wnt signaling by inhibiting GSK3β—a key regulator of the β-catenin degradation complex. Interestingly, GSK3β inhibitors, CHIR99021 and m1, have been shown to promote definitive endoderm production from hIPS and hESC, respectively (Bone et al. 2011; Kunisada et al. 2012).

Inducer of definitive endoderm (IDE)-1 and IDE-2 are two small molecules that can efficiently induce definitive endoderm from mouse and human ESCs that were identified through screening of a small-molecule library (Borowiak et al. 2009). IDE1 and IDE2 activate SMAD2 phosphorylation by upregulating Nodal signaling. IDE1 and IDE2 promote definitive endoderm formation through activation of canonical Activin/Nodal signaling pathways.

From a high content imaging screen, stauprimide was identified as a small molecule that increased the efficiency of Activin A-induced differentiation of mouse and human ESCs toward definitive endoderm (Zhu et al. 2009). Stauprimide binds to NME2 and inhibits activation of c-Myc transcription. c-Myc plays an important role in the maintenance of the proliferative, self-renewing ES state; its downregulation reduces this proliferative capacity.

17.4 Conclusions

The lessons learned from developmental biology have formed the basis for in vitro differentiation of definitive endoderm from embryonic stem cells. The common thread in the culture protocols described in the literature includes Activin A, FGF, and serum-free media. Although there are areas of improvement in the field, efficient production of definitive endoderm is now possible. Ability to form specific population of definitive endoderm forms a foundation on which specific tissues such as liver and pancreas that could be produced in vitro.

References

Ang SL, Wierda A, Wong D, Stevens KA, Cascio S, Rossant J, Zaret KS (1993) The formation and maintenance of the definitive endoderm lineage in the mouse: involvement of HNF3/forkhead proteins. Development 119:1301–1315

Arce L, Yokoyama NN, Waterman ML (2006) Diversity of LEF/TCF action in development and disease. Oncogene 25:7492–7504

Ben-Haim N, Lu C, Guzman-Ayala M, Pescatore L, Mesnard D, Bischofberger M, Naef F, Robertson EJ, Constam DB (2006) The nodal precursor acting via activin receptors induces mesoderm by maintaining a source of its convertases and BMP4. Dev Cell 11:313–323

Blum M, Gaunt SJ, Cho KW, Steinbeisser H, Blumberg B, Bittner D, De Robertis EM (1992) Gastrulation in the mouse: the role of the homeobox gene goosecoid. Cell 69:1097–1106

Bone HK, Nelson AS, Goldring CE, Tosh D, Welham MJ (2011) A novel chemically directed route for the generation of definitive endoderm from human embryonic stem cells based on inhibition of GSK-3. J Cell Sci 124:1992–2000

Borowiak M, Maehr R, Chen S, Chen AE, Tang W, Fox JL, Schreiber SL, Melton DA (2009) Small molecules efficiently direct endodermal differentiation of mouse and human embryonic stem cells. Cell Stem Cell 4:348–358
Candia AF, Wright CV (1996) Differential localization of Mox-1 and Mox-2 proteins indicates distinct roles during development. Int J Dev Biol 40:1179–1184
Candia AF, Hu J, Crosby J, Lalley PA, Noden D, Nadeau JH, Wright CV (1992) Mox-1 and Mox-2 define a novel homeobox gene subfamily and are differentially expressed during early mesodermal patterning in mouse embryos. Development 116:1123–1136
Chen C, Ware SM, Sato A, Houston-Hawkins DE, Habas R, Matzuk MM, Shen MM, Brown CW (2006) The Vg1-related protein Gdf3 acts in a Nodal signaling pathway in the pre-gastrulation mouse embryo. Development 133:319–329
Clevers H (2006) Wnt/beta-catenin signaling in development and disease. Cell 127:469–480
D'Amour KA, Agulnick AD, Eliazer S, Kelly OG, Kroon E, Baetge EE (2005) Efficient differentiation of human embryonic stem cells to definitive endoderm. Nat Biotechnol 23:1534–1541
D'Amour KA, Bang AG, Eliazer S, Kelly OG, Agulnick AD, Smart NG, Moorman MA, Kroon E, Carpenter MK, Baetge EE (2006) Production of pancreatic hormone-expressing endocrine cells from human embryonic stem cells. Nat Biotechnol 24:1392–1401
Dush MK, Martin GR (1992) Analysis of mouse Evx genes: Evx-1 displays graded expression in the primitive streak. Dev Biol 151:273–287
Feng XH, Derynck R (2005) Specificity and versatility in tgf-beta signaling through Smads. Annu Rev Cell Dev Biol 21:659–693
Forlani S, Lawson KA, Deschamps J (2003) Acquisition of Hox codes during gastrulation and axial elongation in the mouse embryo. Development 130:3807–3819
Futaki S, Hayashi Y, Emoto T, Weber CN, Sekiguchi K (2004) Sox7 plays crucial roles in parietal endoderm differentiation in F9 embryonal carcinoma cells through regulating Gata-4 and Gata-6 expression. Mol Cell Biol 24:10492–10503
Gadue P, Huber TL, Paddison PJ, Keller GM (2006) Wnt and TGF-beta signaling are required for the induction of an in vitro model of primitive streak formation using embryonic stem cells. Proc Natl Acad Sci U S A 103:16806–16811
Gouon-Evans V, Boussemart L, Gadue P, Nierhoff D, Koehler CI, Kubo A, Shafritz DA, Keller G (2006) BMP-4 is required for hepatic specification of mouse embryonic stem cell-derived definitive endoderm. Nat Biotechnol 24:1402–1411
Haegel H, Larue L, Ohsugi M, Fedorov L, Herrenknecht K, Kemler R (1995) Lack of beta-catenin affects mouse development at gastrulation. Development 121:3529–3537
Hansson M, Olesen DR, Peterslund JM, Engberg N, Kahn M, Winzi M, Klein T, Maddox-Hyttel P, Serup P (2009) A late requirement for Wnt and FGF signaling during activin-induced formation of foregut endoderm from mouse embryonic stem cells. Dev Biol 330:286–304
Hart AH, Hartley L, Sourris K, Stadler ES, Li R, Stanley EG, Tam PP, Elefanty AG, Robb L (2002) Mixl1 is required for axial mesendoderm morphogenesis and patterning in the murine embryo. Development 129:3597–3608
Hay DC, Fletcher J, Payne C, Terrace JD, Gallagher RC, Snoeys J, Black JR, Wojtacha D, Samuel K, Hannoun Z et al (2008) Highly efficient differentiation of hESCs to functional hepatic endoderm requires ActivinA and Wnt3a signaling. Proc Natl Acad Sci U S A 105:12301–12306
Heldin CH, Miyazono K, ten Dijke P (1997) TGF-beta signalling from cell membrane to nucleus through SMAD proteins. Nature 390:465–471
Itoh S, ten Dijke P (2007) Negative regulation of TGF-beta receptor/Smad signal transduction. Curr Opin Cell Biol 19:176–184
Jackson SA, Schiesser J, Stanley EG, Elefanty AG (2010) Differentiating embryonic stem cells pass through 'temporal windows' that mark responsiveness to exogenous and paracrine mesendoderm inducing signals. PLoS One 5:e10706
Jeong GS, Jun Y, Song JH, Shin SH, Lee SH (2011) Meniscus induced self organization of multiple deep concave wells in a microchannel for embryoid bodies generation. Lab Chip 12:159–166

Kanai-Azuma M, Kanai Y, Gad JM, Tajima Y, Taya C, Kurohmaru M, Sanai Y, Yonekawa H, Yazaki K, Tam PP et al (2002) Depletion of definitive gut endoderm in Sox17-null mutant mice. Development 129:2367–2379

Kim PT, Hoffman BG, Plesner A, Helgason CD, Verchere CB, Chung SW, Warnock GL, Mui AL, Ong CJ (2010) Differentiation of mouse embryonic stem cells into endoderm without embryoid body formation. PLoS One 5:e14146

Kinder SJ, Tsang TE, Quinlan GA, Hadjantonakis AK, Nagy A, Tam PP (1999) The orderly allocation of mesodermal cells to the extraembryonic structures and the anteroposterior axis during gastrulation of the mouse embryo. Development 126:4691–4701

Kinder SJ, Tsang TE, Wakamiya M, Sasaki H, Behringer RR, Nagy A, Tam PP (2001) The organizer of the mouse gastrula is composed of a dynamic population of progenitor cells for the axial mesoderm. Development 128:3623–3634

Kispert A, Herrmann BG (1994) Immunohistochemical analysis of the Brachyury protein in wild-type and mutant mouse embryos. Dev Biol 161:179–193

Kubo A, Shinozaki K, Shannon JM, Kouskoff V, Kennedy M, Woo S, Fehling HJ, Keller G (2004) Development of definitive endoderm from embryonic stem cells in culture. Development 131:1651–1662

Kuhl M, Sheldahl LC, Park M, Miller JR, Moon RT (2000) The Wnt/Ca2+ pathway: a new vertebrate Wnt signaling pathway takes shape. Trends Genet 16:279–283

Kimura W et al (2006) Fate and plasticity of the endoderm in the early chick embryo. Dev Biol 289(2):283–295

Kunisada Y, Tsubooka-Yamazoe N, Shoji M, Hosoya M (2012) Small molecules induce efficient differentiation into insulin-producing cells from human induced pluripotent stem cells. Stem Cell Res 8(2):274–284

Lawson KA, Pedersen RA (1987) Cell fate, morphogenetic movement and population kinetics of embryonic endoderm at the time of germ layer formation in the mouse. Development 101 (3):627–652

Lawson A, Schoenwolf GC (2003) Epiblast and primitive-streak origins of the endoderm in the gastrulating chick embryo. Development 130(15):3491–3501

Li F, He Z, Li Y, Liu P, Chen F, Wang M, Zhu H, Ding X, Wangensteen KJ, Hu Y et al (2011) Combined activin A/LiCl/Noggin treatment improves production of mouse embryonic stem cell-derived definitive endoderm cells. J Cell Biochem 112:1022–1034

Lindsley RC, Gill JG, Kyba M, Murphy TL, Murphy KM (2006) Canonical Wnt signaling is required for development of embryonic stem cell-derived mesoderm. Development 133:3787–3796

Liu P, Wakamiya M, Shea MJ, Albrecht U, Behringer RR, Bradley A (1999) Requirement for Wnt3 in vertebrate axis formation. Nat Genet 22:361–365

Logan CY, Nusse R (2004) The Wnt signaling pathway in development and disease. Annu Rev Cell Dev Biol 20:781–810

McGrath KE, Koniski AD, Maltby KM, McGann JK, Palis J (1999) Embryonic expression and function of the chemokine SDF-1 and its receptor, CXCR4. Dev Biol 213:442–456

Mohr JC, Zhang J, Azarin SM, Soerens AG, de Pablo JJ, Thomson JA, Lyons GE, Palecek SP, Kamp TJ (2010) The microwell control of embryoid body size in order to regulate cardiac differentiation of human embryonic stem cells. Biomaterials 31:1885–1893

Monaghan AP, Kaestner KH, Grau E, Schutz G (1993) Postimplantation expression patterns indicate a role for the mouse forkhead/HNF-3 alpha, beta and gamma genes in determination of the definitive endoderm, chordamesoderm and neuroectoderm. Development 119:567–578

Moon RT (2005) Wnt/beta-catenin pathway. Sci STKE 2005(217):cm1

Moon RT, Brown JD, Torres M (1997) WNTs modulate cell fate and behavior during vertebrate development. Trends Genet 13:157–162

Murry CE, Keller G (2008) Differentiation of embryonic stem cells to clinically relevant populations: lessons from embryonic development. Cell 132:661–680

Nagai T, Aruga J, Takada S, Gunther T, Sporle R, Schughart K, Mikoshiba K (1997) The expression of the mouse Zic1, Zic2, and Zic3 gene suggests an essential role for Zic genes in body pattern formation. Dev Biol 182:299–313

Nateri AS, Spencer-Dene B, Behrens A (2005) Interaction of phosphorylated c-Jun with TCF4 regulates intestinal cancer development. Nature 437:281–285

Niswander L, Martin GR (1992) Fgf-4 expression during gastrulation, myogenesis, limb and tooth development in the mouse. Development 114:755–768

Nostro MC, Cheng X, Keller GM, Gadue P (2008) Wnt, activin, and BMP signaling regulate distinct stages in the developmental pathway from embryonic stem cells to blood. Cell Stem Cell 2:60–71

O'Farrell PH, Stumpff J, Su TT (2004) Embryonic cleavage cycles: how is a mouse like a fly? Curr Biol 14:R35–R45

Payne C, King J, Hay D (2011) The role of activin/nodal and Wnt signaling in endoderm formation. Vitam Horm 85:207–216

Pearce JJ, Evans MJ (1999) Mml, a mouse Mix-like gene expressed in the primitive streak. Mech Dev 87:189–192

Pera MF, Andrade J, Houssami S, Reubinoff B, Trounson A, Stanley EG, Ward-van Oostwaard D, Mummery C (2004) Regulation of human embryonic stem cell differentiation by BMP-2 and its antagonist noggin. J Cell Sci 117:1269–1280

Pesce M, Scholer HR (2001) Oct-4: gatekeeper in the beginnings of mammalian development. Stem Cells 19:271–278

Pevny LH, Sockanathan S, Placzek M, Lovell-Badge R (1998) A role for SOX1 in neural determination. Development 125:1967–1978

Rosner MH, Vigano MA, Ozato K, Timmons PM, Poirier F, Rigby PW, Staudt LM (1990) A POU-domain transcription factor in early stem cells and germ cells of the mammalian embryo. Nature 345:686–692

Rossant J (2004) Lineage development and polar asymmetries in the peri-implantation mouse blastocyst. Semin Cell Dev Biol 15:573–581

Sachlos E, Auguste DT (2008) Embryoid body morphology influences diffusive transport of inductive biochemicals: a strategy for stem cell differentiation. Biomaterials 29:4471–4480

Sasaki H, Hogan BL (1993) Differential expression of multiple fork head related genes during gastrulation and axial pattern formation in the mouse embryo. Development 118:47–59

Schier AF (2003) Nodal signaling in vertebrate development. Annu Rev Cell Dev Biol 19:589–621

Shen MM (2007) Nodal signaling: developmental roles and regulation. Development 134:1023–1034

Sherwood RI, Maehr R, Mazzoni EO, Melton DA (2011) Wnt signaling specifies and patterns intestinal endoderm. Mech Dev 128:387–400

Shi Y, Hou L, Tang F, Jiang W, Wang P, Ding M, Deng H (2005) Inducing embryonic stem cells to differentiate into pancreatic beta cells by a novel three-step approach with activin A and all-trans retinoic acid. Stem Cells 23:656–662

Shim JH, Kim SE, Woo DH, Kim SK, Oh CH, McKay R, Kim JH (2007) Directed differentiation of human embryonic stem cells towards a pancreatic cell fate. Diabetologia 50:1228–1238

Shiraki N, Yoshida T, Araki K, Umezawa A, Higuchi Y, Goto H, Kume K, Kume S (2008) Guided differentiation of embryonic stem cells into Pdx1-expressing regional-specific definitive endoderm. Stem Cells 26:874–885

Solnica-Krezel L (2005) Conserved patterns of cell movements during vertebrate gastrulation. Curr Biol 15:R213–R228

Spemann H (1938) Embryonic Development and Induction. Yale University Press, New Haven, CT

Tada S, Era T, Furusawa C, Sakurai H, Nishikawa S, Kinoshita M, Nakao K, Chiba T (2005) Characterization of mesendoderm: a diverging point of the definitive endoderm and mesoderm in embryonic stem cell differentiation culture. Development 132:4363–4374

Tam PP, Behringer RR (1997) Mouse gastrulation: the formation of a mammalian body plan. Mech Dev 68:3–25

Tam PP, Loebel DA (2007) Gene function in mouse embryogenesis: get set for gastrulation. Nat Rev Genet 8:368–381

ten Berge D, Koole W, Fuerer C, Fish M, Eroglu E, Nusse R (2008) Wnt signaling mediates self-organization and axis formation in embryoid bodies. Cell Stem Cell 3:508–518

Topol L, Jiang X, Choi H, Garrett-Beal L, Carolan PJ, Yang Y (2003) Wnt-5a inhibits the canonical Wnt pathway by promoting GSK-3-independent beta-catenin degradation. J Cell Biol 162:899–908

Torisawa YS, Chueh BH, Huh D, Ramamurthy P, Roth TM, Barald KF, Takayama S (2007) Efficient formation of uniform-sized embryoid bodies using a compartmentalized microchannel device. Lab Chip 7:770–776

Vallier L, Touboul T, Chng Z, Brimpari M, Hannan N, Millan E, Smithers LE, Trotter M, Rugg-Gunn P, Weber A et al (2009) Early cell fate decisions of human embryonic stem cells and mouse epiblast stem cells are controlled by the same signalling pathways. PLoS One 4:e6082

Verheijen MH, Defize LH (1999) Signals governing extraembryonic endoderm formation in the mouse: involvement of the type 1 parathyroid hormone-related peptide (PTHrP) receptor, p21Ras and cell adhesion molecules. Int J Dev Biol 43:711–721

Westfall TA, Brimeyer R, Twedt J, Gladon J, Olberding A, Furutani-Seiki M, Slusarski DC (2003) Wnt-5/pipetail functions in vertebrate axis formation as a negative regulator of Wnt/beta-catenin activity. J Cell Biol 162:889–898

Whitman M (2001) Nodal signaling in early vertebrate embryos: themes and variations. Dev Cell 1:605–617

Wilkinson DG, Bhatt S, Herrmann BG (1990) Expression pattern of the mouse T gene and its role in mesoderm formation. Nature 343:657–659

Xu RH, Chen X, Li DS, Li R, Addicks GC, Glennon C, Zwaka TP, Thomson JA (2002) BMP4 initiates human embryonic stem cell differentiation to trophoblast. Nat Biotechnol 20:1261–1264

Xu F, Sridharan B, Wang S, Gurkan UA, Syverud B, Demirci U (2011a) Embryonic stem cell bioprinting for uniform and controlled size embryoid body formation. Biomicrofluidics 5:22207

Xu X, Browning VL, Odorico JS (2011b) Activin, BMP and FGF pathways cooperate to promote endoderm and pancreatic lineage cell differentiation from human embryonic stem cells. Mech Dev 128(7–10):412–427

Yasunaga M, Tada S, Torikai-Nishikawa S, Nakano Y, Okada M, Jakt LM, Nishikawa S, Chiba T, Era T (2005) Induction and monitoring of definitive and visceral endoderm differentiation of mouse ES cells. Nat Biotechnol 23:1542–1550

Zhu S, Wurdak H, Wang J, Lyssiotis CA, Peters EC, Cho CY, Wu X, Schultz PG (2009) A small molecule primes embryonic stem cells for differentiation. Cell Stem Cell 4:416–426

Zorn AM, Wells JM (2009) Vertebrate endoderm development and organ formation. Annu Rev Cell Dev Biol 25:221–251

Chapter 18
Mouse and Human Pluripotent Stem Cells and the Means of Their Myogenic Differentiation

I. Grabowska, K. Archacka, A.M. Czerwinska, M. Krupa, and M.A. Ciemerych

Abstract Pluripotent stem cells, such as embryonic stem cells and induced pluripotent stem cells, are an important tool in the studies focusing at the differentiation of various cell types, including skeletal myoblasts. They are also considered as a source of the cells that due to their pluripotent character and availability could be turned into any required tissue and then used in future in regenerative medicine. However, the methods of the derivation of some of cell types from pluripotent cells still need to be perfected. This chapter summarizes the history and current advancements in the derivation and testing of pluripotent stem cells-derived skeletal myoblasts. It focuses at the in vitro methods allowing the differentiation of stem cells grown in monolayer or propagated as embryoid bodies, and also at in vivo tests allowing the verification of the functionality of obtained skeletal myoblasts.

18.1 Introduction

Skeletal muscles possess a great capacity to regenerate in response to injury or disease. The key role in this process is played by satellite cells (SCs), i.e., muscle stem cells (Collins et al. 2005). However, repetitive injuries may lead to the exhaustion of SC population and, in consequence, to the impairment of regeneration and muscle dysfunction. Such scenario is characteristic for the progression of muscular dystrophies, a group of incurable genetic diseases affecting proper structure and functionality of muscles (Goyenvalle et al. 2011). Transplantation of cells that could replenish SC population and support muscle regeneration could be considered as possible therapy of such diseases. The low number of SCs within muscle as well as difficulties with their isolation and in vitro culture limit their use

I. Grabowska • K. Archacka • A.M. Czerwinska • M. Krupa • M.A. Ciemerych (✉)
Department of Cytology, Institute of Zoology, University of Warsaw, Miecznikowa 1, 02-096 Warsaw, Poland
e-mail: ciemerych@biol.uw.edu.pl

J.Z. Kubiak (ed.), *Mouse Development*, Results and Problems in Cell Differentiation 55, DOI 10.1007/978-3-642-30406-4_18,

in cell therapies. Therefore, other sources of cells for transplantation are sought. Due to their unique properties, stem cells, as well as myogenic cells derived from them, could serve as such sources. Many types of cells have been tested so far, including undifferentiated pluripotent stem cells or their more differentiated progeny. Such studies revealed several problems that have to be solved before cells application in the clinic. First of all it is necessary to fully characterize the potential of pluripotent cells both in vitro and in vivo. Such issues as progression of their myogenic differentiation, assessment of functionality and integration after transplantation into injured tissue, as well as tumorigeneity and immunogeneity must be addressed before pluripotent stem cells can be used for muscle diseases treatment. In this chapter, we summarize current knowledge on the myogenic pluripotent stem cells differentiation in vitro and in vivo after their transplantation into ectopic sites, e.g., subcutaneously, or into the injured skeletal muscle. But at first we will start with a brief history of these stem cells and their characteristics.

18.2 Pluripotent Stem Cells

18.2.1 The Origin of Pluripotent Stem Cells

Thirty years ago first papers describing the derivation of embryonic stem cells (ESCs) have been published by Gail Martin (1981) as well as Martin Evans and Matthew Kaufman (1981). Since then, mouse ESCs were subjected to various tests and analyses that allowed characterization of the molecular basis of their pluripotent state and also molecular changes associated with their differentiation into various cell lines. Properly stimulated ESCs differentiate into ecto-, endo-, mesoderm, and also germ cells. However, when cultured under suitable conditions, they are also able to self-renew and retain their pluripotent character. The knowledge accumulated during the studies on mouse ESCs led to the derivation of ESCs from human (Thomson et al. 1998), monkey (Thomson et al. 1996; Sasaki et al. 2005), and rat embryos (Buehr et al. 2008; Li et al. 2008, a; Ueda et al. 2008). Finally, it was creatively used in the experiments leading to the reprogramming of various mammalian somatic cells into induced pluripotent stem cells (iPSCs) (e.g., Takahashi and Yamanaka 2006; Takahashi et al. 2007; Maherali et al. 2008).

Cellular pluripotency, however, was discussed much earlier. In 1941, Jackson and Brues published the results of their studies on embryoma that developed in the female of C3H mice (Jackson and Brues 1941). Repeated transplantations of the small fragments of this tumor resulted in the development of as many as 250 tumors containing various tissues. Analysis of their histology led the authors to speculate that "the pluripotential nature of the embryonal cells in this tumor seems probable." Similar idea was presented later by Fekete and Ferrigno who, describing teratomas developing in the same strain of mice, wrote that "undifferentiated embryonic tissues of teratomas are pluripotent and are able to give rise to cells that differentiate into the diverse components of these complex tumors" (Fekete and Ferrigno 1952). Next, Stevens and Little presented the paper describing testicular teratomas in males of 129 strain of mice and wrote that "pluripotent embryonic cells appear to give rise to both rapidly differentiating cells and others which, like themselves, remain undifferentiated" (Stevens and Little 1954). Since then, their work on 129 and also LT/Sv strain of mice is widely considered as the initial step in the history of the isolation of pluripotent stem cells (e.g., Ciemerych et al. 2009; Evans 2011). Teratomas can be characterized as mature (benign), i.e., containing well-differentiated tissues and somatic structures, or as immature (malignant), i.e., containing masses of undifferentiated embryonic cells together with differentiated tissues. Teratomas containing undifferentiated cells were described as "teratocarcinomas". Following studies on the cells isolated from teratomas or teratocarcinomas revealed that even single cell has the potency to reproduce tumor composed of multiple tissues (Kleinsmith and Pierce 1964; Stevens 1967). In vitro culture of these cells led to the formation of spherical structures, so-called embryoid bodies (EBs), and also resulted in the derivation of the first pluripotent cell lines, i.e., embryonal carcinoma cells (ECCs) characterized by high potential to proliferate (Kahan and Ephrussi 1970; Rosenthal et al. 1970; Evans 1972; Martin and Evans 1974), as well as to differentiate in vitro (Martin and Evans 1975). ECCs were also able to form teratomas and to participate in the development of chimeras (Brinster 1974; Mintz and Illmensee 1975; Papaioannou et al. 1975; Bradley et al. 1984). Simultaneously, the studies involving the transplantation of cells isolated from the inner cell mass (ICM) of mouse blastocysts into the cavities of host blastocysts were carried out (Gardner 1968; Rossant et al. 1978). These experiments, together with those ones involving transplantation of ICM cells into ectopic sites such as kidney capsule, established the basis for the future in vivo testing of stem cells, i.e., generating chimeras and teratomas (Evans 2011, Fig. 18.1). Finally, first ESC lines were derived (Evans and Kaufman 1981; Martin 1981) and soon became the model for the studies on the pluripotency and differentiation. ESCs also became indispensable, thanks to the development of techniques allowing genetic modifications of their genomes, as a tool for the creation of genetically modified mouse (Mak 2007).

18.2.2 Basic Molecular Networks Regulating Pluripotency

Among the factors that have been shown to play a crucial role in the securing self-renewal and pluripotency, i.e., preventing differentiation, were the proteins playing a crucial role during early mammalian development. The core pluripotency network relies on the Oct4, Nanog, and Sox2 transcription factors (e.g., Nichols et al. 1998; Avilion et al. 2003; Chambers et al. 2003; Mitsui et al. 2003); however, many others were shown to be involved [for review, see Czyz et al. (2003), Suwinska and Ciemerych (2011)]. Importantly two of these factors, i.e., Oct4 and Sox2, together with Klf4 that influences the expression of Nanog, and c-Myc oncogene, were used to achieve first mouse and human fibroblasts conversion into iPSCs (Takahashi and Yamanaka 2006; Takahashi et al. 2007).

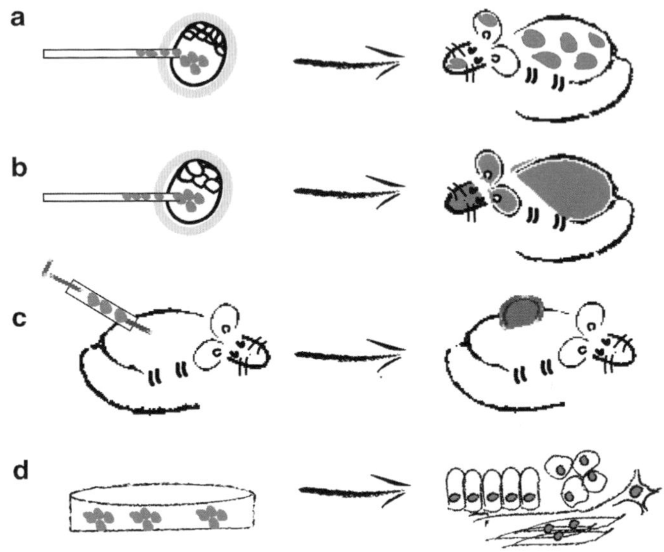

Fig. 18.1 *Testing of the pluripotency and developmental abilities of ESC and iPSC involves*: (**a**) Production of mouse chimeras via microinjection of tested cells into cavity of diploid blastocysts; (**b**) production of non-chimeric mouse via microinjection of tested cells into the cavity of tetraploid blastocyst [so-called tetraploid complementation technique (Eggan et al. 2001)]. In such case tetraploid host cells will contribute to extraembryonic tissues only when tested diploid stem cells will give rise to the embryo proper; (**c**) transplantation of animal or human stem cells under the skin of immunodeficient mice leading to the development of teratomas; (**d**) in vitro differentiation of animal or human stem cells

Importantly, the cells hidden behind the term "pluripotent" can differ not only depending on the technique of their derivation and culture, or the species of their origin. Comparisons of morphology, clonogenicity, signaling pathways, and gene expression patterns of most extensively studied mouse, rat, and human ESCs revealed several important differences (e.g., Czyz et al. 2003). In order to remain pluripotent, mouse ESCs require to be cultured in the presence of LIF (leukemia inhibitory factor) (Rathjen et al. 1990) and BMP4 (bone morphogenetic protein 4) (Ying et al. 2003). LIF interacts with LIFR-gp130 receptor activating signaling pathways impacting at the activation of STAT3 transcription factor that controls the expression of crucial pluripotency factors such as Klf4, Sox2, and Oct4 (e.g., Niwa et al. 1998, 2009; Burdon et al. 1999; Bourillot et al. 2009) [reviewed in Suwinska and Ciemerych (2011)]. Interestingly, involvement of such LIF-STAT3-dependent mechanism is questioned in rat (Buehr et al. 2008) and in human ESCs (Daheron et al. 2004; Humphrey et al. 2004). Activation of LIFR-gp130 receptor is a double-edged sword since it may also induce ESC differentiation via induction of MAPK ERK1/2 signaling pathway. Moreover, MAPK activation can be also driven by the interaction of FGF4 with FGFR. Thus, blocking MAPK activating kinase MEK1 prevents ESC differentiation and improves their self-renewal capacity (Kunath et al. 2007; Ying et al. 2008; Nichols et al. 2009). Importantly, preventing the activation of MAPK

ERK1/2 appeared to be sine qua non condition in experiments leading to the derivation of rat ESCs (Buehr et al. 2008). Surprisingly, blocking of MAPK signaling was shown to result in the decrease of self-renewal of human ESCs (Li et al. 2007). Activation of BMP-dependent signaling suppresses differentiation of mouse but not of human ESCs (Xu et al. 2002; Ying et al. 2003; Pera et al. 2004). Except that also other differences of cellular signaling can be pinpointed [for review, see Suwinska and Ciemerych (2011)]. Interestingly, even the cell lines of the same species, e.g., mouse, manifest evident differences. For example, the success rate of the derivation of mouse ESCs varies between various strains. From embryos of some strains, such as BALB/c, ESCs have been obtained with the frequency as low as 5% while the success rate of derivation of ESCs from 129/Sv mouse strain was 61% (Schoonjans et al. 2003). The major improvement in the derivation of ESCs from "difficult" strains of mice has been made after the inhibitors of MAPK, GSK3, or TGFbeta have been introduced (Lorthongpanich et al. 2008; Lee et al. 2011; Hassani et al. 2012) suggesting that the variability in the activity of the "differentiation" pathways could have been the cause of the observed phenomenon. Moreover, even the ESC lines derived from the embryos of the same strain of mice are not the same, and their ability to differentiate into given cell type might not be comparable. It was noted, for example, by Rohwedel and coworkers who described the ESC line used in their study as failing to form cardiomyocytes in in vitro culture (Rohwedel et al. 1994).

All of the primarily derived ESC and iPSC lines were carefully in vivo tested by chimera or teratoma formation and were proven to be pluripotent (Fig. 18.1). But over the time numerous other ESC lines were derived and not all of them were subjected to such stringent tests. In 2001, Rudolf Yaenisch's group proved that development of the embryos cloned from different ESC lines may be significantly different. Embryos which were produced from ESCs generated from inbred embryos had lower developmental potential than those originating from ESCs of mixed genetic background (Eggan et al. 2001). Analysis of subclones originating from V6.5 ESC line and of mice cloned from these ESCs revealed different methylation status of *H19* and *Igf2* genes, strongly suggesting that ESCs are genetically unstable (Humpherys et al. 2001). Moreover, it is possible that even the characteristics of the one that has been carefully checked might change with time, i.e., as a result of in vitro culture and passaging. Conditions of ESC derivation, culture, or the number of passages can greatly impact at the transcriptome, proteome, or epigenome of ESCs (Humpherys et al. 2001; Newman and Cooper 2010; Ooi et al. 2010; Bock et al. 2011). Same was shown to be true for iPSCs (Newman and Cooper 2010; Ooi et al. 2010; Bock et al. 2011). Meta-analysis comparing human ESCs and iPSCs generated in the several labs vividly documented variability in the gene expression pattern, described as the "laboratory signature" (Newman and Cooper 2010). To make the story even more complicated, iPSCs that were not passaged enough times can retain the molecular "memory" of the type of cell of their origin (Polo et al. 2010; Ohi et al. 2011).

18.2.3 Insight Into the Differentiation of Pluripotent Cells

As already mentioned pluripotent cells can differ from each other. However, despite described differences these cells should be able to differentiate into all types of cells within the body, including germ cells (Fig. 18.1). However, not all types of cells are formed from pluripotent cells with the same frequency. Among first cells that arise almost "spontaneously," both in vitro and in vivo, are cardiomyocytes (Doetschman et al. 1985; Wobus et al. 1991; Boheler et al. 2002). Modification of culture conditions, such as addition of activin A and BMP4 increased the rate of cardiac differentiation of human ESCs up to 90% (Laflamme et al. 2007). This example can be considered as an exceptional one, since other types of cells, such as osteoblasts or myoblasts, can be generated from pluripotent cells with much lower frequency. Thus, derivation of many cell types in vitro is usually inefficient and requires many additional actions. Despite many difficulties a broad spectrum of cell types have been generated from pluripotent cells so far, including various types of neurons, pancreatic β-cells, and many other [see Grivennikov (2008), Murry and Keller (2008)]. Nevertheless, identity and functionality of many cell types derived from ESCs or iPSCs are still debatable. Especially that numerous of the early studies describing derivation of particular cell type from pluripotent cells limited the characteristic of obtained cells only to the description of their "in vitro phenotype." Changes of cell morphology or expression of tissue-specific markers were usually taken as the solid indicators of differentiation. However, to prove that obtained cells are functional such analyses must be complemented by the one proving that the cell can be functional in vivo. For example, the key feature of myoblasts that is crucial for proper development and functionality of skeletal muscle is their ability to fuse and form multinucleated myotubes, and then myofibers. As long as myoblasts obtained from pluripotent cells do not possess capacity for fusion and contraction they cannot be treated as "real" ones, even if they express muscle-specific genes, such as myogenic regulatory factors (MRFs). Moreover, identity and functionality of ESC- or iPSC-derived cells must be verified in vivo, for example by their transplantation into animals, e.g., those one that serve as animal "models" of human diseases (see Sect. 4.1).

Formation of functional, specialized cells requires global rearrangement of cellular machinery leading to the epigenetic changes, new gene expression, as well as protein synthesis profile, cell cycle modifications, and many other cellular effects. Despite the great progress in our understanding of these processes, many of such mechanisms associated with myogenic differentiation of pluripotent stem cells remain not fully understood. It seems, however, that the central role in the myogenic differentiation, similarly as it happens during embryogenesis, is played by MyoD, i.e., the transcription factor of MRFs family. MyoD interacts with both epigenetic modifiers, such as histone acetylases, chromatin remodeling complexes (SWI/SNF), and cell cycle regulators, such as Rb or CDK4 [reviewed in Ciemerych et al. (2011)]. MyoD influences also the expression of other muscle-specific genes, such as *Myogenin*, *M-cadherin*, or *Myosin Heavy Chains* (*MyHC*) (Berkes and

Tapscott 2005). Overexpression of *MyoD* in P19 ECCs or ESCs led to their myogenic differentiation, however, only if they were cultured under suitable conditions, i.e., propagated as EBs in low-mitogen medium (Dekel et al. 1992; Shani et al. 1992; Armour et al. 1999; Gianakopoulos et al. 2011). Although overexpression of *MyoD* or other muscle-specific genes, such as *Pax3* or *Pax7,* leads to the formation of myogenic cells (Darabi et al. 2008, 2011b) (see Sect. 3.1 and 4.1); due to genetic manipulations such cells cannot be considered as clinically relevant. Apart from the methods involving overexpression of selected factors, a variety of other approaches have been proposed to derive myogenic cells from pluripotent cells. Nevertheless, this process is still inefficient.

During embryonic development, induction of myogenic differentiation requires precisely controlled co-operation of many factors, such as Sonic Hedgehog, Wnts, or TGFbeta family proteins (Tajbakhsh 2009). It is still unknown which factors should be combined to create "the cocktail" that would sufficiently increase the efficiency of myogenic cells generation. Thus, it is absolutely necessary to assess and fully understand the influence of different chemical substances or culture conditions modifications on the fate of pluripotent cells. In the next chapters, we will review the approaches and solutions proposed so far by different research groups, indicating both their advantages and limitations.

18.3 In Vitro Derivation of Skeletal Myoblasts

18.3.1 Embryoid Bodies

18.3.1.1 Embryoid Bodies as a Model of Periimplantation Development

One of the methods used to differentiate pluripotent cells into cells of various lineages is to allow them to form embryoid bodies, three-dimensional structures in which cells spontaneously differentiate in a manner resembling periimplantation embryonic development (Czyz and Wobus 2001). Thus, ectoderm forms inner mass of cells, endodermal monolayer is positioned at the surface of EB, and mesodermal cells can be found between ecto- and endoderm. First described EBs, containing differentiating cells, were obtained from mouse teratocarcinoma cells established as an ascitic tumor (Pierce and Dixon 1959; Stevens 1960). Also mouse ECCs when cultured without feeder layer formed tight round aggregates (Kleinsmith and Pierce 1964; Martin and Evans 1975; Ducibella et al. 1982). Importantly, cells forming outer layer of these aggregates exhibited endodermal features, as they produced characteristic mucopolysaccharide layer, i.e., Reichert's membrane, which during murine embryonic development surrounds the egg cylinder. Next, reattachment of such EBs to the dish surface resulted in the formation of the outgrowths and differentiation into many cell types, such as fibroblasts, neurons, cartilage, and keratinizing epithelium (Martin and Evans 1975).

Similarly to ECCs, murine ESCs form solid EBs (Evans and Kaufman 1981; Martin 1981). However, when cultured for prolonged time such EBs may convert into so-called cystic EBs characterized by the presence of the central cavity (Doetschman et al. 1985). Cystic EBs consist of two layers: endoderm and columnar ectoderm-like cells making the structure similar to egg-cylinder stage mouse embryo. Further culture of cystic EBs resulted in their expansion and formation of extraembryonic tissues such as yolk sac-like structures. Moreover, the presence of blood islands and contracting cardiomyocytes also proved that the cells within such EBs were able to differentiate into mesodermal derivatives (Doetschman et al. 1985). Analyses of the sequence of endoderm-specific genes expression, such as *α-fetoprotein*, or *Transthyretin*, during the formation of yolk sac-like structures revealed that it reflected the changes observed during embryonic development in vivo (Abe et al. 1996). To further investigate the similarity between EBs and early postimplantation embryos, Leahy and coworkers compared changes in the expression of subset of genes both in in vitro formed EBs and in vivo developing mouse embryos. By means of in situ hybridization they followed the expression of mRNAs encoding markers of germ layers: *Oct4*, *Fgf5* for embryonic ectoderm, *Gata4* for primitive endoderm, *Brachury T*, *Nodal* for mesoderm, and also additional genes expressed in the variety of more "advanced" cell lineages, e.g., *Flk1, Nkx2.3, Eklf,* and *Msx3*. These analyses showed that temporal and spatial relations in gene expression occurring in vivo are also maintained in EBs. During first 3 days of in vitro culture, EBs were shown to be equivalent to pregastrulation mouse embryos (4.5–6.5 dpc, *days post-coitum*), while EBs analyzed from 3rd to 5th day of in vitro culture resembled gastrulating mouse embryo (6.5–7.5 dpc). Afterwards changes in genes expression characteristic for early organogenesis were observed (Leahy et al. 1999). Resemblance of mouse ESC differentiation in EBs to differentiation process occurring during embryonic development was also confirmed by Karbanova and Mokry, who showed that at 26th day of culture in suspension expression of *Pax7* can be detected in EBs, suggesting the initiation of myogenic differentiation (Karbanova and Mokry 2002). Analysis of EBs formed by human ESCs also showed sequential expression of embryonic markers specific for different cellular lineages, i.e., ectoderm (neurofilament 68 kDa), endoderm (*α-FETO-PROTEIN*), mesoderm (*α-CARDIAC ACTIN*), and mesodermal hematopoietic lineage (*γ-GLOBIN*) (Itskovitz-Eldor et al. 2000).

Similarity of EBs to postimplantation embryos makes them a suitable model to examine differentiation processes occurring during embryogenesis. It is also one of the best in vitro tests used to prove pluripotency of murine (Evans and Kaufman 1981; Martin 1981), human (Itskovitz-Eldor et al. 2000; Cowan et al. 2004), or rette ESCs (Li et al. 2009a; Demers et al. 2011), as well as murine, porcine, bovine, rhesus monkey, or human iPSCs (e.g., Takahashi and Yamanaka 2006; Takahashi et al. 2007; Liu et al. 2008; Ezashi et al. 2009; Li et al. 2009b; Sumer et al. 2011). Apart from pluripotent stem cells also other types of cells, such as mesenchymal ones were tested using this system (e.g., Fong et al. 2007). Importantly, EBs give an opportunity to examine functions of different genes during early development, for example by analyzing fate of ESCs in that certain gene has been silenced by RNAi, knocked out, or replaced by its modified version (e.g., Weitzer et al. 1995; Desbaillets et al. 2000; Myer et al. 2001). One of these techniques, i.e., RNAi, was used in the study investigating the role of *Pax3* and *Pax7* during myogenic differentiation of mouse ESCs. Analysis of EBs formed by such modified cells showed that silencing the *Pax3* or *Pax7* expression led to decreased expression of *MyoD*, *Myf5,* and *Desmin* (Bhagavati et al. 2007). EBs system can also be used for the assessment of effects of genes overexpression, like in the studies documenting the delay in myogenic differentiation of ESCs in that the expression of *M-twist* was increased (Rohwedel et al. 1995), or acceleration of myogenic differentiation when *IGF2 (*Insulin Growth Factor 2) was overexpressed (Prelle et al. 2000). Finally, EBs can serve during the analyses of the influence of various drugs or growth factors on differentiation processes (e.g., Edwards et al. 1983; Wiles and Keller 1991; Schuldiner et al. 2000) or used in embryotoxicity tests (Schmidt et al. 2001).

18.3.1.2 Methods of Embryoid Bodies Generation

Traditional method of generating EBs, initially used by Martin and Evans for ECCs (Martin and Evans 1975), and then for ESCs (Evans and Kaufman 1981; Martin 1981), requires the removal of factors maintaining pluripotency, such as LIF in case of mouse ESCs (Shen and Leder 1992), or bFGF in case of human ESCs (Itskovitz-Eldor et al. 2000), and culture without feeder cells. Under such culture conditions, ECCs or ESCs spontaneously formed three-dimensional aggregates and then differentiated asynchronously (Martin and Evans 1975; Edwards et al. 1983; Doetschman et al. 1985). In order to synchronize differentiation, i.e., to induce simultaneous formation of EBs, ESCs need to be cultured in the suspension preventing them from incidental adhesion to the dish. To achieve this goal several approaches were proposed, among them culture of cells on methylcellulose (Wiles and Keller 1991), in nonadhesive bacterial-grade dishes or in culture flasks placed on rotary shaker (Doetschman et al. 1985). Unfortunately, EBs formed in such conditions still differed in size, meaning that each of them can differently advance in the differentiation processes. To obtain more uniform EBs so-called hanging drops method was proposed. Initially this method was used for culture of mouse embryos to study blastocysts development (Morris et al. 1982; Potter and Morris 1985) and then successfully adapted for the generation of EBs from P19 ECCs (van den Eijnden-van Raaij et al. 1991), and ESCs (Wobus et al. 1991; Maltsev et al. 1993; Rohwedel et al. 1994). ESCs that were cultured on feeder layer should be at first purified from feeder cells, e.g., using preplating method (Sanchez et al. 1991). Next, ESCs are transferred into the droplets of medium placed on the lid of Petri dish (Fig. 18.2). The number of the cells transferred to each drop may differ depending on the protocol used, but usually varies between 400 and 800 (Rohwedel et al. 1994, 1998; Rose et al. 1994; Weitzer et al. 1995; Myer et al. 2001; Schmidt et al. 2001). During the first 24 h of culture in hanging drops, ESCs form aggregates and start differentiation (Fig. 18.2). Since volume of the drop is limited, after several days of culture EBs should be transferred to low-adhesive cell culture or nonadhesive bacterial dishes and cultured

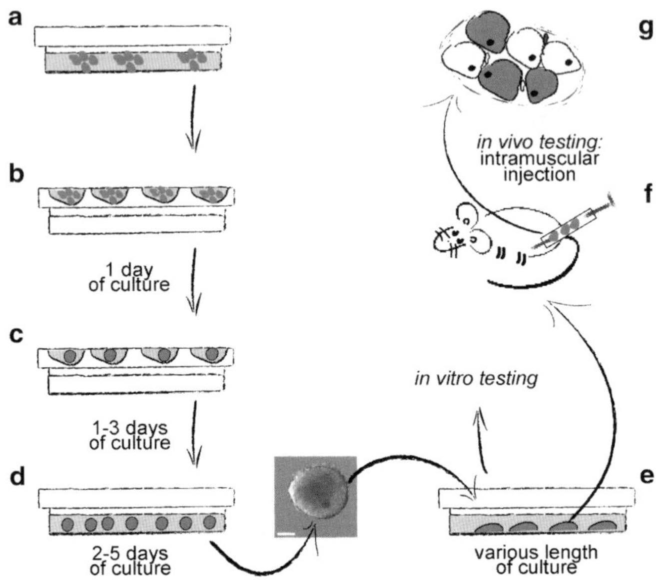

Fig. 18.2 *In vitro differentiation of pluripotent stem cells using embryoid bodies formation technique*. (**a**) Stem cells are cultured in monolayer in medium containing LIF; (**b**) 400–800 stem cells are transferred into medium drops positioned at the cover of the culture dish. The dish itself is filled with PBS which prevents the drying of the medium drops. (**c**) 1–3 day long culture allows the formation of EBs; (**d**) EBs are transferred into the dish filled with medium and cultured in suspension for additional 2–5 days. The photograph shows the EBs after 2 days culture in suspension bar 100 μm; (**e**) EBs are plated in the culture dish allowing their attachments and formation of outgrowth. Next, depending on the protocol, the outgrowths are cultured for different periods and then harvested either for molecular (RT-PCR) or cytological analyses (in situ hybridization, immunocytochemistry, FACS), or cells are FACS sorted according to the expression of selected markers; (**f**) selected cells are used for transplantation into injured or dysfunctional muscles; (**g**) analyses of regenerated skeletal muscle including the analysis of localization of myofibers and satellite cells originating from ESCs or iPSCs

in the abundant amount of medium. Several modifications of EBs derivation methods have been proposed to ensure better reproducibility. Among them are the use of so-called multiwell chip (Kim et al. 2007), culture in conical tubes (Kurosawa et al. 2003), in 96-well round-bottom plates that are either low adhesive (Ezekiel et al. 2007) or coated with polymers, like pluronic (Dang et al. 2002). To produce high number of EBs, spinner flasks or stirred bioreactors were recommended (Schroeder et al. 2005; Cameron et al. 2006). It has to be noted, however, that some of these systems may lead to EBs aggregation, which can significantly change experimental conditions (Dang et al. 2002).

Formation and differentiation of cells within EBs may differ due to the composition of medium used in experiments. Standard medium used for differentiation of both murine and human ESCs in EBs consists of DMEM or IMDM (Iscove's modification of DMEM) supplemented with fetal calf serum (FCS), sodium pyruvate, nonessential amino acids, antibiotics, L-glutamine, and β-mercaptoethanol, and also other supplements, such as bovine serum albumin (BSA), selenium, or transferrin [(Wobus et al. 1994), see also Sect. 3.2]. Interestingly, the concentration of glucose in the medium influences the differentiation of pluripotent cells in EBs. Culture of EBs in high glucose medium (25 mM) resulted in the differentiation into multiple cell lineages, manifested by the expression of markers of cardiac muscle, endodermal cells, and neurons (Mochizuki et al. 2011). On the other hand, low concentration of glucose (5.5 mM) preferentially induces differentiation of pluripotent cells into neuronal lineage (Mochizuki et al. 2011). The quality of the serum is another crucial factor influencing differentiation of cells, and it is widely accepted that some batches are more "effective" than others (Rohwedel et al. 1994; Abe et al. 1996). Some sort of serum "standardization" can be achieved by the removal of growth factors and hormones by dextran-coated charcoal treatment (e.g., Edwards et al. 1983) or by the use of defined serum replacement (SR). However, SR is designed to maintain ESCs' pluripotency; thus, it does not facilitate their differentiation (Cheng et al. 2004; Koivisto et al. 2004).

18.3.1.3 Myogenic Differentiation of Pluripotent Stem Cells Within Embryoid Bodies

EBs formation is one of the most frequently used methods in the studies attempting to derive myogenic precursor cells in vitro. However, no single, simple, and effective protocol was designed so far. Instead, a variety of experimental designs were tested with variable success rate (Table 18.1). In general, the initial step of experiments aiming at the induction of myogenic differentiation of pluripotent cells often starts with the production of EBs using hanging drops technique. Then after 4–7 days of culture in suspension EBs are plated onto gelatin- or Matrigel-coated dishes, and cultured for several days or weeks, depending on the protocol. Using such experimental design myogenic differentiation was induced in EBs formed by ECCs (Edwards et al. 1983; Edwards and McBurney 1983) and then also by ESCs. Myogenic differentiation of pluripotent cells can be enhanced by culturing them in the medium temporarily supplemented with ascorbic acid and activin A (Tian et al. 2008), retinoic acid (RA) (Wobus et al. 1994), or either in the medium supplemented with insulin, transferrin, and selenium (ITS), as well as dexamethasone and epidermal growth factor (EGF), or in medium supplemented with horse serum (HS) (Zheng et al. 2006).

However, the most efficient methods, so far, for induction of myogenic differentiation using EBs system, involve genetic modification of ESCs (Table 18.1). In trailblazing experiments, expression of *MyoD* in monolayer-cultured mouse ESCs resulted in the synthesis of Myogenin and Myosin light chain 2 but was not sufficient to promote myotube formation (Dekel et al. 1992). Myotubes were formed only if cells overexpressing *MyoD* were differentiated in EBs, indicating that apart from endogenous factors also environmental factors, such as signals coming from surrounding cells are necessary for myogenic differentiation of pluripotent cells (Dekel et al. 1992). Microarray analyses of gene expression

Table 18.1 Synopsis of myogenic differentiation of pluripotent stem cells in vitro

Cells (line name)	Experimental conditions	Markers of myogenic differentiation analyzed (technique applied)	Ability to form myotubes	Efficiency of myoblast derivation/myotube formation	References
mESCs (D3)	EBs	embryonic *MyHC* (Northern blot)	ND	ND, but estimated as extremely low	Robbins et al. (1990)
mESCs (BLC6)	EBs' outgrowths	*M-cadherin* (RT-PCR, IF)	Yes	myoblasts detected in up to 100% of EBs	Rose et al. (1994)
mECCs (P19)	outgrowths from DMSO-treated EBs	morphology, MyHC (IF)	Yes	myotubes detected in ~90% of EBs, up to 12% MyHC + cells in each EB	Edwards et al. (1983)
mECCs (P19)	outgrowths from RA-treated EBs	morphology, MyHC (IF)	Yes	myotubes detected in ~40% of EBs	Edwards and McBurney (1983)
mESCs (BLC6)	EBs' outgrowths, culture medium containing BSA and transferrin	morphology, *Myf5*, *MyoD*, *Myogenin*, *Mrf4* (RT-PCR); desmin, MyHC (IF), analysis of ionic currents	Yes	myoblasts detected in up to 100% of EBs	Rohwedel et al. (1994)
mESCs (D3)	outgrowths from RA-treated EBs, culture medium containing selenite, transferrin, BSA	Myogenin (IF), analysis of ionic currents	ND	myoblasts detected in ~50% of EBs	Wobus et al. (1994)
mESCs	EBs' outgrowths, protocol involving several various media, ascorbic acid and activin	*Myf5*, *MyoD*, *Myogenin* (RT-PCR), desmin, MyHC (IF)	Yes	1-3 contracting myotubes per culture plate present in 95% of plates	Tian et al. (2008)
mESCs (D3, R1)	outgrowths cultured in medium containing HS	morphology, MyHC (IF)	Yes	myotubes detected in ~30% of EBs	Hirsch et al. (1998)
hESCs	EBs cultured in custom medium enriched with BMP4, PDGFαR-positive cells were FACS sorted	PDGFαR (FACS)	ND	55-61% of cells positive for PDGFαR	Ng et al. (2008)

mESCs (D3)	EBs' outgrowths, culture medium containing HS; SM/C-2.6-positive myogenic precursors were FACS sorted	SM/C-2.6 (FACS), *Pax7*, *Myf5*, *MyoD*, *Myogenin* (RT-PCR, IF)	yes	16% of cells positive for SM/C-2.6; 11% of these cells formed myotubes	Chang et al. (2009)
mESCs (D3)	EBs' outgrowths, culture medium containing HS; SM/C-2.6-positive myogenic precursors were FACS sorted	SM/C-2.6 (FACS), MyHC (IF), *Pax3*, *Pax7*, *Myf5*, *MyoD*, *Miogenin* (RT-PCR, IF)	Yes	32% of cells positive for SM/C-2.6; 22% of these cells formed myotubes	Mizuno et al. (2010)
miPSCs	EBs' outgrowths, culture medium containing HS; SM/C-2.6-positive myogenic precursors were FACS sorted	SM/C-2.6 (FACS), MyHC (IF), *Pax3*, *Pax7*, *Myf5*, *MyoD*, *Miogenin* (RT-PCR, IF)	Yes	32% of cells positive for SM/C-2.6+; 15% of these cells formed myotubes	Mizuno et al. (2010)
hESCs (H1, SH39)	EBs incubated for 24 h with 5-aza-C; EBs' outgrowths cultured in medium containing HS	Pax7 and MyoD (IF), desmin and MyHC not detected (IF)	No	50% of cells were Pax7+; 57% of cells were MyoD+	Zheng et al. (2006)
hPGCs	EBs-derived cells cultured in endothelial basal medium supplemented with galectin-1	*DESMIN*, *MyHC*, *DYSTROPHIN* (RT-PCR, IF), *MYF5*, *MYOD*, *MYOGENIN*, *MRF4* (RT-PCR)	Yes	8-14% of cells expressed myogenic markers	Shao et al. (2009)
mESCs	EBs' outgrowths formed by MyoD-overexpressing ESCs, cultured in medium containing HS and insulin	morphology	Yes	20-50% of cells in outgrowths	Dekel et al. (1992), Shani et al. (1992)
mESCs (R1)	EBs' outgrowths formed by *IGF2*-overexpressing ESCs	morphology, *Myf5*, *Myogenin*, *MyoD* (RT-PCR), M-cadherin, titin , MyHC (IF)	Yes	myotubes detected in up to 80% of EBs	Prelle et al. (2000)
mESCs (D3)	EBs' outgrowths formed from *Pax3*- or *MyoD*-overexpressing ESCs	upregulation of desmin expression (microarray)	ND	ND	Craft et al. (2008)
mESCs	EBs' outgrowths formed from *Pax3*-overexpressing ESCs (expression induced at 2nd day of culture); PDGFαR-positive and Flk-negative cells were FACS sorted and cultured in medium containing HS	PDGFαR, M-cadherin (FACS), MRFs, M-cadherin, MyHC (IF)	Yes	55-61% of cells positive for PDGFαR and negative for Flk1, 78% of sorted cells formed myotubes	Darabi et al. (2008)

(continued)

Table 18.1 (continued)

Cells (line name)	Experimental conditions	Markers of myogenic differentiation analyzed (technique applied)	Ability to form myotubes	Efficiency of myoblast derivation/myotube formation	References
mESCs	EBs' outgrowths formed from *Pax7*-overexpressing ESCs (expression induced at 2nd day of culture); PDGFαR-positive and Flk-negative cells FACS sorted and cultured in medium for further culture in medium containing bFGF and HS	PDGFαR, M-cadherin (FACS), Pax7, Myf5, MyoD, MyHC (IF)	Yes	84% of cells positive for PDGFαR and negative for Flk1 formed myotubes	Darabi et al. (2011b)
hESCs	ESCs cultured in monolayer, protocol involving several various media, ITS, FACS-sorting of CD73 and N-CAM-positive cells	CD73, NCAM (FACS), *MYOD*, *MYOGENIN*, *MyHC* (RT-PCR, IF)	Yes	46% of cells positive for MYOGENIN	Barberi et al. (2007)

m murine, *h* human, *ND* not determined, *RA* retinoic acid, *DMSO* dimethyl sulfoxide, *IF* immunofluorescence

profiles in EBs produced from ESCs overexpressing *MyoD* or *Pax3* documented upregulation of the whole set of genes crucial for myogenic differentiation, such as *Desmin* (Craft et al. 2008). Moreover, as it was shown in elegant experiments performed by Darabi and coworkers, overexpression of *Pax3* or *Pax7* is sufficient to produce functional myoblasts (Darabi et al. 2008, 2011b). EBs were formed from ESCs carrying doxycycline-inducible *Pax3* gene. Next, after 2 days of culture in suspension *Pax3* expression was stimulated with doxycycline. Such manipulation resulted in the robust myogenic differentiation within outgrowths obtained after EBs plating. However, the transplantation of such cells into injured skeletal muscle resulted in the formation of teratomas documenting that at least some of ESCs did not differentiate, i.e., remained pluripotent. To select pure population of myoblasts precursors EBs were disaggregated and obtained cells were FACS sorted on the basis of the presence of Platelet-Derived Growth Factor-α Receptor (PDGF-αR), i.e., the marker of paraxial mesoderm which during embryonic development is a source of myogenic precursors, and the absence of Flk1, i.e., marker of lateral plate mesoderm. When cells synthesizing PDGF-αR and lacking Flk1 were transplanted into injured muscle no teratoma formation was observed, confirming the adequacy of such experimental approach [(Darabi et al. 2008), Fig. 18.3]. Following the same scheme PDGF-αR-positive/Flk1-negative cells overexpressing *Pax7* were tested and also shown to undergo robust myogenic differentiation (Darabi et al. 2011b). It has to be stressed that *Pax3* or *Pax7* overexpressing cells actively participated in muscle regeneration and, thus, were functional. Although the above approach resulted in the efficient generation of functional myoblasts, such cells could not be used in the therapy as they were obtained using DNA integrating vectors.

FACS sorting was also used in the studies in that EBs were formed from mouse ESCs (Chang et al. 2009) and iPSCs (Mizuno et al. 2010) that were not subjected to any genetic modifications. After the initial induction of EBs formation, myogenic differentiation was stimulated by culturing them in the presence of HS, and then plating in the Matrigel-coated culture dishes. Cells resulting from such outgrowths were dissociated, labeled with SM/C-2.6 antibody recognizing yet unknown surface antigen characteristic for SCs (Fukada et al. 2004), and FACS sorted (Chang et al. 2009; Mizuno et al. 2010). Selected population of SM/C-2.6.-positive cells was significantly enriched for cells expressing markers of SCs, such as M-cadherin and Pax7, when compared to SM/C-2.6-negative cells. During in vitro culture cells expressing SM/C-2.6 formed MyHC-positive myotubes while cells negative for SM/C-2.6 were not able to undergo myogenic differentiation (Chang et al. 2009; Mizuno et al. 2010, Table 18.1).

Myogenic differentiation of pluripotent cells forming EBs could be also induced and enhanced by signals released by in vitro differentiating myoblasts. Such strategy turned out to be efficient, for example, in promoting differentiation of mesenchymal stem cells (MSCs) (Shi et al. 2004) or neural stem cells (Galli et al. 2000). Such protocol was also used in experiments in that EBs initially grown in suspension were then co-cultured with myoblasts (Bhagavati and Xu 2005). Interactions between both types of cells probably induced differentiation of ESCs as they participated in the formation of new myofibers after introduction into skeletal muscle (Bhagavati and Xu 2005, for details see Sect. 4.1 and Fig. 18.3).

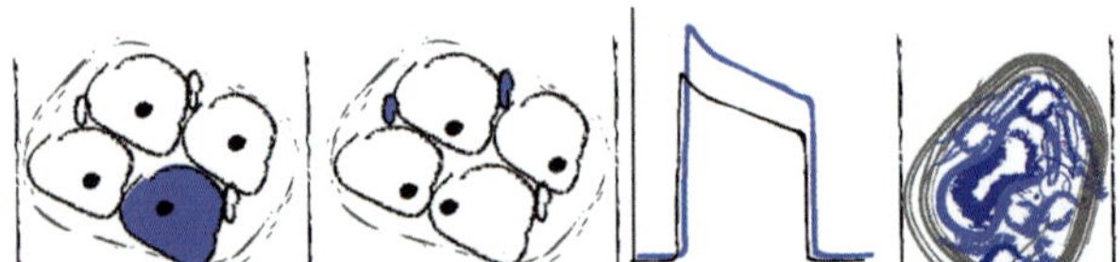

pluripotent cells transplanted into muscle	muscle engraftment	satellite cells niche localization	functional improvement	teratoma formation	references
EB-derived cells					
PDGFαR+ FACS sorted	✓ 11-20%	✓ <1%	✓	✗	Darabi et al., 2008 Darabi et al., 2009 Darabi et al., 2011a Darabi et al., 2011b Sakurai et al., 2008
SM/C-2.6+ FACS sorted	✓ <1%	✓	ND	✗	Chang et a., 2009 Mizuno et al., 2010
EGF and ITS treatment	✓ 28%	✓	ND	✗	Zheng et al., 2006
co-culture with muscle cells	✓	ND	ND	✗	Bhagavati and Xu, 2005
cells cultured in monolayer					
CD73+ N-CAM+ FACS sorted	✓ 7%	✗	ND	✗	Barberi et al., 2007
IGF2 over-expression	✓	ND	✓	✗	Kamochi et al., 2006
not treated *in isogeneic mouse*	✗	ND	ND	✓	Tian et al., 2008
not treated *in allogeneic mouse*	✗	ND	ND	✗	Tian et al., 2008

Fig. 18.3 Synopsis of myogenic differentiation of pluripotent stem cells in vivo

Spatiotemporal relationship between differentiating cells, that to some extend is reproduced within EBs, facilitates formation of mesoderm and its derivatives, such as myoblasts (Czyz and Wobus 2001). However, the efficiency of this process is very low (Table 18.1) and various supplementary strategies aiming to increase it have been proposed.

18.3.2 Differentiation of Pluripotent Stem Cells in the Presence of Insulin, Transferrin, and Selenium

Amid supplements that was shown to increase efficiency of mesoderm formation and also myogenic differentiation is ITS (i.e., insulin, transferrin, selenium). For example, it was shown that ITS promoted myogenic differentiation of mouse myoblasts C2C12 or 23A2 (Wyzykowski et al. 2002), human primary myoblasts (Pawlikowski et al. 2009), or *MyoD*-null and *Myf5*-null mouse embryonic fibroblasts (Di Padova et al. 2007). Also the outgrowths obtained from human EBs were exposed to the medium supplemented with ITS, as well as dexamethasone and epidermal growth factor (EGF), or to medium supplemented with HS (Zheng et al. 2006). Two and 4 weeks after EBs plating cells expressing myogenic markers, such as MYOD and PAX7 were detected. Much higher number of such cells were identified when plated EBs were treated with hemimethylating agent, 5-azacytidine for 24 h (see also Sect. 3.3). However, despite the induction of the expression of myogenic regulators, under none of these culture conditions myotubes formation occurred, indicating that tested cells did not convert into functional myoblasts. Thus, the cells neither changed their morphology nor synthesized skeletal muscle structural proteins, such as MyHC, DESMIN, DYSTROPHIN, or TROPONIN (Zheng et al. 2006).

Differentiation of ESCs into mesodermal cells that can further become myogenic precursors, and finally myoblasts fusing into myotubes, is also possible in monolayer, i.e., omitting EBs formation (Barberi et al. 2007; Stavropoulos et al. 2009). Medium containing BMP4 and ITS was shown to induce the expression of PDGF-αR in as many as 60% of cells within EBs (Ng et al. 2008). Furthermore, Barberi and coworkers described protocols requiring ITS in order to obtain population of myogenic precursors (Barberi et al. 2007; Stavropoulos et al. 2009). This method involves several steps of cell culture in that various media with or without ITS are used at selected points of the experiment. Briefly, human ESCs were at first cultured in the medium supplemented with ITS for 3 weeks, and then for 1 week in the absence of ITS. At this point CD73-positive cells, i.e., mesodermal precursors were FACS sorted. Such subpopulation of cells was cultured for additional 2–4 days, and then sorted, this time to select NCAM-bright cells. After NCAM-bright cells reached confluence, culture medium was shifted again to the one supplemented with ITS, as well as progesterone and putrescine (Stavropoulos et al. 2009). As a result 46% of NCAM-positive cells were shown to express *MYOGENIN* already 24 h after sorting and culture in ITS-containing medium (Table 18.1). These cells were also able to fuse and form MyHC-expressing contracting myotubes and to participate in muscle regeneration [(Barberi et al. 2007), see Sect. 4.1]. Although the protocol was shown to enhance myogenic differentiation of pluripotent cells, it is rather time-consuming and still not very efficient.

18.3.3 Azacytidine Analogues

Among the factors used to induce myogenic differentiation of pluripotent cells are 5-azacytidine (5-aza-C) and 5-aza-2′-deoxycytidine (5-aza-dC). They were primarily described as cytotoxic agents (Sorm et al. 1964), but soon it was shown that they are potent inhibitors of DNA methylation able to change gene expression and induce differentiation of in vitro cultured cells (Constantinides et al. 1977; Jones and Taylor

1980). Within the cell both analogues replace cytosine during DNA replication (Constantinides et al. 1978). Their impact at the epigenetic modifications was shown to be connected to the interference with the function of DNA methylases. DNA methylation, which along various histone modifications is the major player in epigenesis, is executed by the methylation of C5 position of cytosine residues resulting in the formation of 5-methylcytosine (Holliday and Pugh 1975). CpG islands, that are DNA regions of more than 500 bp with GC content greater than 55%, were shown to be located within the promoter regions of about 40% of mammalian genes. Their methylation maintained by a number of DNA methyltransferases (DNMTs) causes stable heritable transcriptional silencing (Takai and Jones 2002). The molecular mechanism of 5-aza-C and 5-aza-dC action most probably relies on their permanent covalent binding to methylases which depletes enzyme pool and prevents correct DNA methylation (Jones and Taylor 1980; Juttermann et al. 1994). Such change in the methylation pattern may impact at the embryonic development since changes in the expression of various genes decide about cell fate (Constantinides et al. 1978; Taylor and Jones 1982; Konieczny and Emerson 1984; Li et al. 1992; Okano et al. 1999).

The cytotoxicity of 5-aza-dC is tenfold stronger than this of 5-aza-C (Constantinides et al. 1978; Wakitani et al. 2003), although the cytotoxic effect of both factors depends also on the sensitivity of treated cells. For example, incubation of 10 T1/2 fibroblasts with 1 μM solution of 5-aza-C resulted in the survival of only 10% of cells (Constantinides et al. 1978). Under the same culture conditions over 20% of MSCs survived (Wakitani et al. 2003). Cytotoxicity of azacytidine analogues seems to be only partially connected with hypomethylation mechanism. In fact, decreased levels of methylating enzymes make cells more resistant to azacytidine analogues (Juttermann et al. 1994). Thus, trapped methylases that act as protein adducts to DNA strand may be the main reason for cytotoxic effect. Indeed, firm position of methylases causes replication fork to collapse, resulting in the formation of DNA double strand breaks (Michel et al. 2001; McGlynn and Lloyd 2002) and eventually in the induction of apoptosis (Kiziltepe et al. 2007).

5-azacytidine and 5-aza-2′-deoxycytidine were first used to induce myogenic differentiation in the 70s of XX century. The 24 h culture of mouse fibroblasts in the medium containing various concentrations of 5-aza-C or 5-aza-dC (from 0.03 μM of 5-aza-dC to 10 μM of 5-aza-C) resulted in the formation of multinucleated, contracting myotubes (Constantinides et al. 1977, 1978; Taylor and Jones 1979). Detailed analysis of these cells revealed that increasing amount of azacytidine analogues induced the formation of more multinucleated myotubes, although due to 5-aza-C or 5-aza-dC toxicity only few of them survived (Constantinides et al. 1977, 1978; Taylor and Jones 1979). Following studies, in that MSCs isolated from rat bone marrow were treated with 5-aza-C or 5-aza-dC, resulted in the low rate of myogenic differentiation, i.e., none or few myotubes were observed. In fact, myotubes were found only when MSCs were cultured in the medium containing high concentrations, i.e., 1 μM of 5-aza-C; however, only 20% of cultured cells survived (Wakitani et al. 1995). De Coppi and coworkers tested human amniotic fluid stem cells (AFSCs) expressing c-kit, and showed that these cells, treated with 3 μM of 5-aza-dC and then cultured in the presence of HS and chicken embryo extract initiated the expression of muscle specific factors, such as *MRF4*, *MYOD*, and *DESMIN*. Similar data were obtained using mouse AFSCs (De Coppi et al. 2007). However, Bossolasco and coworkers were not able to induce myogenic differentiation of cells isolated from human amniotic fluid using the same concentration of 5-aza-dC in medium containing also FCS, EGF, and PDGF-BB (Bossolasco et al. 2006). Observed differences most probably resulted from the different culture conditions and experimental design applied in both studies, and also from the fact that De Coppi and coworkers analyzed c-kit expressing cells only, while Bossolasco and coworkers studied the whole population of cells isolated from amniotic fluid (Bossolasco et al. 2006; De Coppi et al. 2007).

Except that azacytidine analogues change the DNA methylation status they can also impact at the stem cells differentiation via different mechanism. In the case of ESCs and iPSCs efficient differentiation occurs when pluripotency factors such as Oct4 or Nanog are downregulated. Degradation of these factors is driven by caspases 3 and 7 which are activated in the response to double strand breaks caused by the azacytidine analogues (Musch et al. 2010). Since Oct4 negatively regulates the expression of *MyoD* gene (Watanabe et al. 2011) its degradation leads to the expression of *MyoD* [reviewed in Ciemerych et al. (2011)]. Thus, the use of azacytidine analogues may enhance myogenic differentiation by lowering the expression of pluripotency markers. For this reason application of 5-aza-C or 5-aza-dC in the experiments in that ESCs or iPSCs differentiate within EBs could be considered to facilitate myogenic differentiation. However, Tian and coworkers claimed that this approach does not result in myogenic differentiation of pluripotent cells (Tian et al. 2008). This stays in contradiction with results of Zheng and coworkers who showed that treatment of human ESCs with 10 mM 5-aza-C resulted in the changes of both cell morphology and gene expression. Four weeks after 5-aza-C treatment about 50% of cells were elongated and synthesized such myogenic factors as Pax7 and MyoD. However, these cells did not synthesize Myf5, Desmin, Dystrophin, MyHC, Troponin I, M-cadherin, or NCAM, and never formed myotubes or myofibers (Zheng et al. 2006). Thus, their myogenic differentiation was not completed.

Although knowledge concerning myogenic differentiation of pluripotent stem cells is accumulating, presented results of in vitro studies indicate that this process is far more complex and difficult to reproduce in vitro than expected. It also has to be remembered that full recognition of properties of obtained cells is possible only if the proper in vivo studies are performed. Only under such condition issues such as integration and functionality of cells transplanted into muscle as well as their tumorigeneity and immunogeneity can be addressed.

18.4 Studies on Myogenic Potential of Pluripotent Stem Cells In Vivo

18.4.1 Pluripotent Stem Cells and Regeneration of Skeletal Muscle

The first experiments testing the myogenic potential of pluripotent cells in vivo were conducted by Bhagavati and Xu (2005). Mouse ESCs expressing GFP were first propagated as EBs (for 1–5 days) and then co-cultured for another 4 days with cells isolated from skeletal muscle of *mdx* (murine X-linked dystrophy) mice, which lack dystrophin, i.e., one of the structural protein necessary for proper function of skeletal muscle (Sicinski et al. 1989). Mixture of $5–10 \times 10^4$ co-cultured cells was transplanted into muscle of *mdx* mice, and 2 weeks later both dystrophin and GFP were detected within injected muscle. However, it should be noted that GFP-positive fibers were detected in the muscles of only 2 out of 8 analyzed mice, and one of these mice had as few as 30 of such fibers (Bhagavati and Xu 2005). Interestingly, no dystrophin-positive fibers were detected when EBs were co-cultured with C2C12 myoblasts, suggesting that they can release different signals that primary myoblasts isolated directly from the muscle. Myogenic potential of pluripotent stem cells in vivo was also investigated by other research groups. Both undifferentiated (Tian et al. 2008) and induced to differentiate mouse ESCs (e.g., Kamochi et al. 2006; Chang et al. 2009) were transplanted into skeletal muscles. In other studies participation of differentiating human ESCs was examined in the regeneration of skeletal muscle (Zheng et al. 2006; Barberi et al. 2007). Not all of these experiments addressed, however, such crucial issues as the survival rate of transplanted cells or their ability to proliferate and migrate within the muscle. Moreover, successful cell therapy for dysfunctional muscles requires both efficient contribution of transplanted cells to myofibers formation and to the population of SCs. Barberi and coworkers showed that ESC-derived cells, selected on the basis of CD73 and NCAM expression, were able to express markers of SCs – M-cadherin, syndecan-4, and CD34. However, after their transplantation into skeletal muscle none of these cells were detected in the niche of SCs [(Barberi et al. 2007), Fig. 18.3]. In other studies cells obtained from ESCs pretreated with EGF, ITS, and FCS, or from ESCs overexpressing *Pax3* or *Pax7* were shown to synthesize markers of SCs and to occupy their position when introduced into muscle (Darabi et al. 2008). However, in all of these studies such cells accounted for only few percent of all cells localized within SC niche (Fig. 18.3), what is much below the threshold (20–30%) that gives a chance to obtain therapeutic effect [see Phelps et al. (1995), Wells et al. (1995) and discussion in Chan et al. (2007)].

The influence of transplanted pluripotent cells or their progeny on the muscle weight, strength, or fatigue resistance was shown only in few studies so far (Fig. 18.3). In mice, recovery of motor function as well as improvement of contractile properties of muscles was observed one month after transplantation of cells derived from ESCs overexpressing either *IGF2* (Kamochi et al. 2006), *Pax3* (Darabi et al. 2008), or *Pax7* (Darabi et al. 2011b). However, use of genetically modified cells, as it was mentioned in the previous chapters, limits clinical translation of proposed protocols. It should be also realized that in many of the described studies pluripotent cells or their progeny were transplanted to immunodeficient animals what diminished the risk of their rejection (Barberi et al. 2007; Darabi et al. 2008; Sakurai et al. 2008, Fig. 18.3). Although such approach allows transplanting cells regardless their immunological profile, it is distant from clinical practice during which allogeneic transplantations prevail. For many years ESCs were considered to be immunologically privileged as they express low levels of MHC I and lack MHC II (Drukker et al. 2002; Li et al. 2004). However, recently it became clear that it is not the case. First, undifferentiated ESCs were shown to be removed by NK cells that recognize and kill cells expressing low level of MHC I (Ma et al. 2010). Second, expression of MHC I increases during differentiation of ESCs, and thus, leads to the development of immune response of allogeneic host (Swijnenburg et al. 2005; Nussbaum et al. 2007). Our own results indicate that transplantation of undifferentiated mouse ESCs into injured mouse skeletal muscle increase inflammatory response, and in consequence perturbs regeneration (Archacka et al. unpublished data). Administration of immunosuppressive agent, Tacrolimus®, increased the number of surviving cells. However, such treatment led to further impairment of regeneration as this drug inhibits calcineurin/NFAT pathway that is crucial for proper fusion and formation of right-sized myotubes and myofibers (Horsley and Pavlath 2004). Therefore, in both cases introduction of undifferentiated ESCs into injured muscle of allogeneic host led to extensive fibrosis and impaired muscle regeneration (Archacka et al. in preparation).

Application of myogenic cells derived from genetically matched iPSCs could solve the problem of immunological mismatch between donor and host. Two reports addressed the issue of myogenic potential of iPSCs so far. Mizuno and coworkers showed that mouse iPSCs cultured as EBs and then plated on Matrigel-coated dishes start to express myogenic markers such as *Pax3*, *Pax7*, and MRFs (Mizuno et al. 2010). After 27 days of culture nearly 50% of iPSC-derived cells synthesized protein recognized by SM/C-2.6 antibody (Fukada et al. 2004; Mizuno et al. 2010). Subpopulation of these cells also synthesized M-cadherin, CD34, and integrin α7, characteristic for SCs, and underwent final myogenic differentiation, forming myotubes in vitro. However, when transplanted into injured muscle of *mdx* mice SM/C-2.6-positive cells engrafted as little as 1% of all fibers [(Mizuno et al. 2010); Fig. 18.3]. Much higher efficiency of muscle engraftment was reported by Darabi and coworkers who obtained myogenic precursors from mouse iPSCs that overexpressed *Pax7* transcription factor (Darabi et al. 2011a). When introduced into dystrophic muscle iPSC-derived cells contributed to formation of 15–20% of all fibers, leading to the functional improvement of injected muscles (Fig. 18.3). Transplanted cells were also found in the SC niche, but the frequency of such events was not determined (Darabi et al. 2011a). Although these results are promising, both low efficiency of iPSC derivation technique and lack of efficient, nongenetic method of their myogenic differentiation, do not allow considering this solution as "therapeutically" realistic, so far. However, it cannot be excluded that due to the rapid advances in iPSC technology as well as the progressive comprehension of mechanisms regulating myogenic differentiation it will happen one day.

Last but not least, a great attention must be paid to the precise selection of differentiated cells derived from pluripotent cells from the cells that are not fully committed to myogenic lineage. Transplantation of such cells may lead to the formation of teratomas as it was shown in the experiments during which undifferentiated or partially differentiated ESCs were introduced into immunodeficient or immunosuppressed mice (Darabi et al. 2008; Tian et al. 2008). On the other hand, the transplantation of properly selected, differentiated cells seems to be safe as no teratomas were detected within mice injected systemically or locally with such myogenic precursors (Kamochi et al. 2006; Barberi et al. 2007; Darabi et al. 2008, 2011a, b). As it was mentioned earlier, teratomas development is considered as a serious risk associated with the transplantation of ESC- or iPSC-derived cells that are not sufficiently differentiated. On the other hand, detailed studies on teratoma formation could be beneficial for deciphering processes that drive myogenic differentiation. Thus, apart from its role as a valid and commonly used test of the pluripotency, teratomas are valuable, but still underestimated model for investigating formation of skeletal muscle in vivo.

18.4.2 Myogenic Differentiation of Pluripotent Stem Cells During Teratoma Formation

Although chimera formation is the most stringent in vivo test for pluripotency, it is limited by the requirement for special equipment allowing micromanipulation of preimplantation embryos, and more importantly, it cannot be applied to human cells. Therefore, generation of teratomas is widely used as a valid and crucial method for in vivo evaluation of differentiation ability of the given cell lines, also those of human origin. The term *teratoma* comes from Greek word "*teratos*" (monster) and reflects the composition of these structures, i.e., disordered mixture of mature tissues and deformed organs. German pathologist, Rudolf Virchow introduced this term in 1863 in his monograph on tumors "Die krankhaften Geschwülste" (Virchow 1863). However, documented cases of human or even mouse teratomas were certainly known much earlier. At the end of 40s and at 50s of the past century teratomas were described as a rare mouse tumor consisting of several decipherable embryonic tissues at various levels of differentiation (e.g., Fawcett 1950; Fekete and Ferrigno 1952). Results of one of the most influential teratoma-focused studies were published by Stevens and collaborators (Stevens and Little 1954; Stevens 1962; Stevens and Varnum 1974; reviewed in Ciemerych et al. 2009). Histological analysis showed that these ovarian teratomas originated from spontaneously activated oocytes which begun parthenogenetic development. The "intra-ovarian" parthenogenetic embryos developed until blastocyst stage and during further development became disorganized and formed tumors containing several tissue types and also undifferentiated cells (Stevens and Varnum 1974). Other studies documented that such teratomas can develop from preimplantation embryos transferred into extra-uterine sites, e.g., subcutaneously or under kidney capsule (Stevens 1968, 1970; Solter et al. 1970).

Analysis of the contribution of various tissues within teratomas involves standard histological methods, i.e., different stainings allowing identification of skeletal muscles, epithelia, connective tissue, neurons, pigmented cells, and many other. However, as such methods are a bit crude and in many cases do not allow to reveal the true identity of differentiated cells and tissues, currently in situ hybridization, immunolocalization of selected tissue-specific markers, RT-PCR, as well as Western blotting are commonly used. One has also to remember that teratoma develops within the host organism and it is possible that host cells can contribute to its formation. Analysis performed by Gertow and coworkers, involving FISH detecting total human genomic DNA, undoubtedly showed that mouse host cells can contribute to some structures of teratoma developing from human ESCs. Within the teratomas studied more than 90% of all tissues were almost exclusively human. However, mouse cells were present mainly within blood vessel endothelium (Bloch et al. 1997; Gertow et al. 2004).

Since the derivation of ECC and ESC teratomas became indispensable as a "tool" allowing characterization of differentiation potential of stem cell, such as mouse (Kleinsmith and Pierce 1964) and human ECCs (Przyborski et al. 2004); mouse, human, and rat ESCs (Thomson et al. 1998; Cao et al. 2007; Buehr et al. 2008); mouse embryonic germ cells (Kanatsu-Shinohara et al. 2004); multipotent adult germline stem cell (Guan et al. 2006); and finally mouse and human iPSCs (Takahashi and Yamanaka 2006; Takahashi et al. 2007). However, it should be remembered that progression of in vivo differentiation and variety of formed cells and tissues may depend on various factors. Thus, among the issues that can influence formation and composition of teratomas are (1) the type and the characteristics, e.g., species of origin, derivation technique, culture conditions of the pluripotent cells tested (ESCs versus iPSCs); (2) transplantation site, e.g., subcutaneously, under kidney capsule, into the selected organ or tissue; (3) number of cells transplanted; (4) type of the host, i.e., syngeneic or allogeneic one.

Pluripotent cells such as ESC or iPSC lines can differ in their ability to form teratomas. Gutierrez-Aranda and coworkers compared the efficiency of the formation, latency, and composition of teratomas resulting from seven human ESC and four iPSC lines (Gutierrez-Aranda et al. 2010). They did not find any differences in the histological composition of obtained teratomas. However, the rate of their formation was 81–94% for ESCs and 100% for iPSCs. More importantly, iPSCs were more "aggressive" in vivo and after transplantation they started to grow faster (latency was shortened from 59 to 31 days after subcutaneous injection, and from 66 days to 31 days after intratesticular injection) than human ESCs analyzed (Gutierrez-Aranda et al. 2010). Interestingly, meta-analysis comparing gene expression patterns in various human ESCs and iPSCs generated in several laboratories documented certain differences that can cause differences in cells potency to differentiate into various tissues, also within the teratomas (Newman and Cooper 2010).

Except the strain of mice used, also the site of transplantation affects cell growth and differentiation. Among the most "popular" transplantation sites are intracapsular space of the kidney (Heins et al. 2004), testis subcapsular compartment (Sundstrom et al. 1999; Gertow et al. 2004; Mikkola et al. 2006; Valbuena et al. 2006), and subcutaneous transplantation (Martin 1981; Cao et al. 2007). Moreover, pluripotent cells injected into the heart muscle (Nussbaum et al. 2007), liver (Cooke et al. 2006), or skeletal muscle (Thompson et al. 1998; Lee et al. 2009) also resulted in the development of teratomas. However, depending on the tissue in that they were localized their growth and composition differed. For example, ECCs or ESCs transplanted into the liver produced large tumors containing predominantly differentiated tissues and immature cells expressing pluripotent cells marker SSEA3 (Cooke et al. 2006). Tumors obtained from cells transplanted subcutaneously were five to eight times smaller even if they were allowed to grow much longer than the ones developing in the liver, but contained well-differentiated tissues (Cooke et al. 2006). The site of the transplantation was also shown to impact at the frequency of teratoma formation. Subcutaneous transplantations of human ESCs into NOD/SCID mice gave variable results, i.e., tumors were formed in 25–100% of cases. Moreover, injections of these cells under the kidney capsule resulted in 100% success rate, while intratesticular in 60%, and intramuscular in only 12.5% (Prokhorova et al. 2009). The rate of subcutaneous teratoma formation increased to 80–100% when human ESCs were transplanted together with Matrigel (Prokhorova et al. 2009). The differentiation of ESCs within teratomas can also be determined by the site of the injection. Introduced into the knee joint mouse ESCs preferably form cartilage (Wakitani et al. 2003), transplanted into heart form cardiomyocytes, however, with low frequency (Singla et al. 2006; Cao et al. 2007). The efficiency of teratoma formation can also be influenced by the number of transplanted cells. Even as few as only two or ten cells can initiate teratoma formation if transplanted with the additional support of other nondividing or non-neoplastic cells and Matrigel (Lawrenz et al. 2004; Cao et al. 2007). However, the efficiency of the procedure is much lower when only few cells are transplanted (Lawrenz et al. 2004). Despite the fact that there is no evidence that the number of transplanted cells has any effect on the differentiation processes in teratomas, in most studies 10^3–10^6 cells is needed to ensure teratoma formation.

Among the factors that may influence in vivo differentiation of tested cells is immunological mismatch between donor and host. Initially, formation of teratomas was induced after the transplantation of tested cells into syngeneic mice (e.g., Jackson and Brues 1941; Kleinsmith and Pierce 1964). Mouse ESCs injected into the both: allogeneic and syngeneic mice were shown to form teratomas. However, in allogeneic mice these teratomas can be almost completely annihilated due to immune reaction (Swijnenburg et al. 2005, 2008). For example, teratomas formed from mouse D3 ESCs (derived from 129/Sv mice) transplanted into the skeletal muscles of BALB/c mice were shown to be infiltrated by host CD45+ inflammatory cells and to gradually decrease in size (Swijnenburg et al. 2008). Currently, SCID or nude mice, that accept xenogeneic transplants, are the "first of choice" in pluripotency analysis of human cells and teratoma formation assays (e.g., Tveit et al. 1980; Wirth et al. 1983; Oakley et al. 1993).

As mentioned before, except the general analyses testing the pluripotency of ESCs, iPSCs, or other cells, formation of the teratomas may be used in the studies focusing at the differentiation of only one tissue, such as skeletal muscle. The presence of skeletal muscle tissues within the teratomas was reported in multiple reports analyzing pluripotency of ESCs or iPSCs. However, only few of the projects were using teratomas to study skeletal muscle formation in detail. One of such studies analyzing myogenic differentiation within testicular teratomas was reported in 1990 (Muntener et al. 1990). In their study, Muntener and coworkers investigated the relationship between nervous and muscular tissues in teratomas of 129/Sv mice. In such tumor skeletal muscle tissue was present and restricted to one-third of its volume in approximately half of the teratomas. In contrast, nervous tissue was dispersed over the entire tumor. The authors also reported their unpublished data documenting that parts of this testicular teratomas were able to contract. Striated muscle fibers in the teratomas were arranged in bundles of variable size and contained up to 400 fibers. In most cases these fibers were parallel and uniform in type of myofibrillar ATPase activity, suggesting that they were composed of only one fiber type. However, myofibers varied in the reaction for cytochrome c oxidase activity in given fiber bundle, being generally higher in the small myofibers what suggests existence of metabolically different fiber types, characteristic for normal adult muscles. Sarcoplasmic activity of acetylcholinesterase (AChE) was uniformly very low. Müntener and coworkers also reported that probably each muscle fiber contained one endplate. In the teratoma sections stained for AChE and motor nerve terminals, neuromuscular contacts were visible (Muntener et al. 1990). Thus, in mouse teratomas, development and innervation of the muscle fibers largely mimics normal developmental processes, although fiber type differentiation remains partially incomplete.

Teratomas can also serve as easily available assay allowing the in vivo testing of differentiation potency of pluripotent cells carrying various mutations or disease-specific iPSCs. Studies, involving generation of iPSCs from patients suffering from various genetic diseases, were initiated as early as in 2008 (Park et al. 2008). Among the first patient-specific iPSCs generated were cells from those suffering from Duchenne's and Becker's muscular dystrophies (Park et al. 2008). Teratomas were used to analyze the differentiation potency of iPSCs derived from *mdx* mice in which the mutation in dystrophin gene was annihilated by the introduction of human artificial chromosome carrying functional gene (Kazuki et al. 2010). This system was also used to prove that iPSCs derived from mesangioblasts, i.e., muscle-derived cells associated with blood vessels, are more prone to differentiate into skeletal muscle tissue than into other types of cells (Quattrocelli et al. 2011). In teratomas formed from such iPSCs striated muscle tissue accounted for up to 70% of tumor mass.

18.5 Concluding Remarks

Techniques promoting the differentiation of pluripotent stem cells give the opportunity of creating the precursor cells that can be transplanted into damaged or dysfunctional tissues or organs. However, in order to apply such methods in the clinics mechanisms governing differentiation impacting at the quality of obtained cells have to be understood. Despite the vast knowledge that has been accumulated since the first ESC lines were derived we still fail to efficiently reproduce the environment necessary to produce many functional cells, including skeletal myoblasts. However, rapid progress in uncovering mechanisms regulating both pluripotency and differentiation of cells gives a chance to reach this aim, eventually.

Acknowledgments The authors regret the omission of any relevant literature, which happened either because of the space limitation or oversight. This work was supported by the grant from National Science Center N N303 548139 to IG and N N303 320537 to KA.

References

Abe K, Niwa H, Iwase K, Takiguchi M, Mori M, Abe SI, Yamamura KI (1996) Endoderm-specific gene expression in embryonic stem cells differentiated to embryoid bodies. Exp Cell Res 229:27–34

Armour C, Garson K, McBurney MW (1999) Cell-cell interaction modulates myod-induced skeletal myogenesis of pluripotent p19 cells in vitro. Exp Cell Res 251:79–91

Avilion AA, Nicolis SK, Pevny LH, Perez L, Vivian N, Lovell-Badge R (2003) Multipotent cell lineages in early mouse development depend on sox2 function. Genes Dev 17:126–140

Barberi T, Bradbury M, Dincer Z, Panagiotakos G, Socci ND, Studer L (2007) Derivation of engraftable skeletal myoblasts from human embryonic stem cells. Nat Med 13:642–648

Berkes CA, Tapscott SJ (2005) Myod and the transcriptional control of myogenesis. Semin Cell Dev Biol 16:585–595

Bhagavati S, Song X, Siddiqui MA (2007) Rnai inhibition of pax3/7 expression leads to markedly decreased expression of muscle determination genes. Mol Cell Biochem 302:257–262

Bhagavati S, Xu W (2005) Generation of skeletal muscle from transplanted embryonic stem cells in dystrophic mice. Biochem Biophys Res Commun 333:644–649

Bloch W, Forsberg E, Lentini S, Brakebusch C, Martin K, Krell HW, Weidle UH, Addicks K, Fassler R (1997) Beta 1 integrin is essential for teratoma growth and angiogenesis. J Cell Biol 139:265–278

Bock C, Kiskinis E, Verstappen G, Gu H, Boulting G, Smith ZD, Ziller M, Croft GF, Amoroso MW, Oakley DH, Gnirke A, Eggan K, Meissner A (2011) Reference maps of human es and ips cell variation enable high-throughput characterization of pluripotent cell lines. Cell 144:439–452

Boheler KR, Czyz J, Tweedie D, Yang HT, Anisimov SV, Wobus AM (2002) Differentiation of pluripotent embryonic stem cells into cardiomyocytes. Circ Res 91:189–201

Bossolasco P, Montemurro T, Cova L, Zangrossi S, Calzarossa C, Buiatiotis S, Soligo D, Bosari S, Silani V, Deliliers GL, Rebulla P, Lazzari L (2006) Molecular and phenotypic characterization of human amniotic fluid cells and their differentiation potential. Cell Res 16:329–336

Bourillot PY, Aksoy I, Schreiber V, Wianny F, Schulz H, Hummel O, Hubner N, Savatier P (2009) Novel stat3 target genes exert distinct roles in the inhibition of mesoderm and endoderm differentiation in cooperation with nanog. Stem Cells 27:1760–1771

Bradley A, Evans M, Kaufman MH, Robertson E (1984) Formation of germ-line chimaeras from embryo-derived teratocarcinoma cell lines. Nature 309:255–256

Brinster RL (1974) The effect of cells transferred into the mouse blastocyst on subsequent development. J Exp Med 140:1049–1056

Buehr M, Meek S, Blair K, Yang J, Ure J, Silva J, McLay R, Hall J, Ying QL, Smith A (2008) Capture of authentic embryonic stem cells from rat blastocysts. Cell 135:1287–1298

Burdon T, Stracey C, Chambers I, Nichols J, Smith A (1999) Suppression of shp-2 and erk signalling promotes self-renewal of mouse embryonic stem cells. Dev Biol 210:30–43

Cameron CM, Hu WS, Kaufman DS (2006) Improved development of human embryonic stem cell-derived embryoid bodies by stirred vessel cultivation. Biotechnol Bioeng 94:938–948

Cao F, van der Bogt KE, Sadrzadeh A, Xie X, Sheikh AY, Wang H, Connolly AJ, Robbins RC, Wu JC (2007) Spatial and temporal kinetics of teratoma formation from murine embryonic stem cell transplantation. Stem Cells Dev 16:883–891

Chambers I, Colby D, Robertson M, Nichols J, Lee S, Tweedie S, Smith A (2003) Functional expression cloning of nanog, a pluripotency sustaining factor in embryonic stem cells. Cell 113:643–655

Chan J, Waddington SN, O'Donoghue K, Kurata H, Guillot PV, Gotherstrom C, Themis M, Morgan JE, Fisk NM (2007) Widespread distribution and muscle differentiation of human fetal mesenchymal stem cells after intrauterine transplantation in dystrophic mdx mouse. Stem Cells 25:875–884

Chang H, Yoshimoto M, Umeda K, Iwasa T, Mizuno Y, Fukada S, Yamamoto H, Motohashi N, Miyagoe-Suzuki Y, Takeda S, Heike T, Nakahata T (2009) Generation of transplantable, functional satellite-like cells from mouse embryonic stem cells. FASEB J 23:1907–1919

Cheng J, Dutra A, Takesono A, Garrett-Beal L, Schwartzberg PL (2004) Improved generation of c57bl/6j mouse embryonic stem cells in a defined serum-free media. Genesis 39:100–104

Ciemerych MA, Archacka K, Grabowska I, Przewozniak M (2011) Cell cycle regulation during proliferation and differentiation of mammalian muscle precursor cells. Results Probl Cell Differ 53:473–527

Ciemerych MA, Archacka K, Polanski Z, Kubiak JZ (2009) Cell cycle modifications during oocyte maturation and early development of lt/sv mice: normal development or teratoma formation? In: Kubiak JZ, Ciemerych MA, Richard-Parpaillon L (eds) Cell cycle and development in vertebrates. Research Signpost, Kerala, pp 181–202

Collins CA, Olsen I, Zammit PS, Heslop L, Petrie A, Partridge TA, Morgan JE (2005) Stem cell function, self-renewal, and behavioral heterogeneity of cells from the adult muscle satellite cell niche. Cell 122:289–301

Constantinides PG, Jones PA, Gevers W (1977) Functional striated muscle cells from non-myoblast precursors following 5-azacytidine treatment. Nature 267:364–366

Constantinides PG, Taylor SM, Jones PA (1978) Phenotypic conversion of cultured mouse embryo cells by aza pyrimidine nucleosides. Dev Biol 66:57–71

Cooke MJ, Stojkovic M, Przyborski SA (2006) Growth of teratomas derived from human pluripotent stem cells is influenced by the graft site. Stem Cells Dev 15:254–259

Cowan CA, Klimanskaya I, McMahon J, Atienza J, Witmyer J, Zucker JP, Wang S, Morton CC, McMahon AP, Powers D, Melton DA (2004) Derivation of embryonic stem-cell lines from human blastocysts. N Engl J Med 350:1353–1356

Craft AM, Krisky DM, Wechuck JB, Lobenhofer EK, Jiang Y, Wolfe DP, Glorioso JC (2008) Herpes simplex virus-mediated expression of pax3 and myod in embryoid bodies results in lineage-related alterations in gene expression profiles. Stem Cells 26:3119–3129

Czyz J, Wiese C, Rolletschek A, Blyszczuk P, Cross M, Wobus AM (2003) Potential of embryonic and adult stem cells in vitro. Biol Chem 384:1391–1409

Czyz J, Wobus A (2001) Embryonic stem cell differentiation: the role of extracellular factors. Differentiation 68:167–174

Daheron L, Opitz SL, Zaehres H, Lensch MW, Andrews PW, Itskovitz-Eldor J, Daley GQ (2004) Lif/stat3 signaling fails to maintain self-renewal of human embryonic stem cells. Stem Cells 22:770–778

Dang SM, Kyba M, Perlingeiro R, Daley GQ, Zandstra PW (2002) Efficiency of embryoid body formation and hematopoietic development from embryonic stem cells in different culture systems. Biotechnol Bioeng 78:442–453

Darabi R, Baik J, Clee M, Kyba M, Tupler R, Perlingeiro RC (2009) Engraftment of embryonic stem cell-derived myogenic progenitors in a dominant model of muscular dystrophy. Exp Neurol 220:212–216

Darabi R, Gehlbach K, Bachoo RM, Kamath S, Osawa M, Kamm KE, Kyba M, Perlingeiro RC (2008) Functional skeletal muscle regeneration from differentiating embryonic stem cells. Nat Med 14:134–143

Darabi R, Pan W, Bosnakovski D, Baik J, Kyba M, Perlingeiro RC (2011a) Functional myogenic engraftment from mouse ips cells. Stem Cell Rev 7:948–957

Darabi R, Santos FN, Filareto A, Pan W, Koene R, Rudnicki MA, Kyba M, Perlingeiro RC (2011b) Assessment of the myogenic stem cell compartment following transplantation of pax3/pax7-induced embryonic stem cell-derived progenitors. Stem Cells 29:777–790

De Coppi P, Bartsch G Jr, Siddiqui MM, Xu T, Santos CC, Perin L, Mostoslavsky G, Serre AC, Snyder EY, Yoo JJ, Furth ME, Soker S, Atala A (2007) Isolation of amniotic stem cell lines with potential for therapy. Nat Biotechnol 25:100–106

Dekel I, Magal Y, Pearson-White S, Emerson CP, Shani M (1992) Conditional conversion of es cells to skeletal muscle by an exogenous myod1 gene. New Biol 4:217–224

Demers SP, Desmarais JA, Vincent P, Smith LC (2011) Rat blastocyst-derived stem cells are precursors of embryonic and extraembryonic lineages. Biol Reprod 84:1128–1138

Desbaillets I, Ziegler U, Groscurth P, Gassmann M (2000) Embryoid bodies: an in vitro model of mouse embryogenesis. Exp Physiol 85:645–651

Di Padova M, Caretti G, Zhao P, Hoffman EP, Sartorelli V (2007) Myod acetylation influences temporal patterns of skeletal muscle gene expression. J Biol Chem 282:37650–37659

Doetschman TC, Eistetter H, Katz M, Schmidt W, Kemler R (1985) The in vitro development of blastocyst-derived embryonic stem cell lines: formation of visceral yolk sac, blood islands and myocardium. J Embryol Exp Morphol 87:27–45

Drukker M, Katz G, Urbach A, Schuldiner M, Markel G, Itskovitz-Eldor J, Reubinoff B, Mandelboim O, Benvenisty N (2002) Characterization of the expression of mhc proteins in human embryonic stem cells. Proc Natl Acad Sci USA 99:9864–9869

Ducibella T, Anderson D, Aalberg J, DeWolf WC (1982) Cell surface polarization, tight junctions, and eccentric inner cells characterize human teratocarcinoma embryoid bodies. Dev Biol 94:197–205

Edwards MK, Harris JF, McBurney MW (1983) Induced muscle differentiation in an embryonal carcinoma cell line. Mol Cell Biol 3:2280–2286

Edwards MK, McBurney MW (1983) The concentration of retinoic acid determines the differentiated cell types formed by a teratocarcinoma cell line. Dev Biol 98:187–191

Eggan K, Akutsu H, Loring J, Jackson-Grusby L, Klemm M, Rideout WM 3rd, Yanagimachi R, Jaenisch R (2001) Hybrid vigor, fetal overgrowth, and viability of mice derived by nuclear cloning and tetraploid embryo complementation. Proc Natl Acad Sci USA 98:6209–6214

Evans M (2011) Discovering pluripotency: 30 years of mouse embryonic stem cells. Nat Rev Mol Cell Biol 12:680–686

Evans MJ (1972) The isolation and properties of a clonal tissue culture strain of pluripotent mouse teratoma cells. J Embryol Exp Morphol 28:163–176

Evans MJ, Kaufman MH (1981) Establishment in culture of pluripotential cells from mouse embryos. Nature 292:154–156

Ezashi T, Telugu BP, Alexenko AP, Sachdev S, Sinha S, Roberts RM (2009) Derivation of induced pluripotent stem cells from pig somatic cells. Proc Natl Acad Sci USA 106:10993–10998

Ezekiel UR, Muthuchamy M, Ryerse JS, Heuertz RM (2007) Single embryoid body formation in a multi-well plate. Electron J Biotechnol 10:328–335

Fawcett DW (1950) Bilateral ovarian teratomas in a mouse. Cancer Res 10:705–707

Fekete E, Ferrigno MA (1952) Studies on a transplantable teratoma of the mouse. Cancer Res 12:438–440

Fong CY, Richards M, Manasi N, Biswas A, Bongso A (2007) Comparative growth behaviour and characterization of stem cells from human wharton's jelly. Reprod Biomed Online 15:708–718
Fukada S, Higuchi S, Segawa M, Koda K, Yamamoto Y, Tsujikawa K, Kohama Y, Uezumi A, Imamura M, Miyagoe-Suzuki Y, Takeda S, Yamamoto H (2004) Purification and cell-surface marker characterization of quiescent satellite cells from murine skeletal muscle by a novel monoclonal antibody. Exp Cell Res 296:245–255
Galli R, Borello U, Gritti A, Minasi MG, Bjornson C, Coletta M, Mora M, De Angelis MG, Fiocco R, Cossu G, Vescovi AL (2000) Skeletal myogenic potential of human and mouse neural stem cells. Nat Neurosci 3:986–991
Gardner RL (1968) Mouse chimeras obtained by the injection of cells into the blastocyst. Nature 220:596–597
Gertow K, Wolbank S, Rozell B, Sugars R, Andang M, Parish CL, Imreh MP, Wendel M, Ahrlund-Richter L (2004) Organized development from human embryonic stem cells after injection into immunodeficient mice. Stem Cells Dev 13:421–435
Gianakopoulos PJ, Mehta V, Voronova A, Cao Y, Yao Z, Coutu J, Wang X, Waddington MS, Tapscott SJ, Skerjanc IS (2011) Myod directly up-regulates premyogenic mesoderm factors during induction of skeletal myogenesis in stem cells. J Biol Chem 286:2517–2525
Goyenvalle A, Seto JT, Davies KE, Chamberlain J (2011) Therapeutic approaches to muscular dystrophy. Hum Mol Genet 20:R69–78
Grivennikov IA (2008) Embryonic stem cells and the problem of directed differentiation. Biochemistry (Mosc) 73:1438–1452
Guan K, Nayernia K, Maier LS, Wagner S, Dressel R, Lee JH, Nolte J, Wolf F, Li M, Engel W, Hasenfuss G (2006) Pluripotency of spermatogonial stem cells from adult mouse testis. Nature 440:1199–1203
Gutierrez-Aranda I, Ramos-Mejia V, Bueno C, Munoz-Lopez M, Real PJ, Macia A, Sanchez L, Ligero G, Garcia-Parez JL, Menendez P (2010) Human induced pluripotent stem cells develop teratoma more efficiently and faster than human embryonic stem cells regardless the site of injection. Stem Cells 28:1568–1570
Hassani SN, Totonchi M, Farrokhi A, Taei A, Larijani MR, Gourabi H, Baharvand H (2012) Simultaneous suppression of tgf-beta and erk signaling contributes to the highly efficient and reproducible generation of mouse embryonic stem cells from previously considered refractory and non-permissive strains. Stem Cell Rev 8:472–481
Heins N, Englund MC, Sjoblom C, Dahl U, Tonning A, Bergh C, Lindahl A, Hanson C, Semb H (2004) Derivation, characterization, and differentiation of human embryonic stem cells. Stem Cells 22:367–376
Hirsch E, Lohikangas L, Gullberg D, Johansson S, Fassler R (1998) Mouse myoblasts can fuse and form a normal sarcomere in the absence of beta1 integrin expression. J Cell Sci 111(Pt 16):2397–2409
Holliday R, Pugh JE (1975) DNA modification mechanisms and gene activity during development. Science 187:226–232
Horsley V, Pavlath GK (2004) Forming a multinucleated cell: molecules that regulate myoblast fusion. Cells Tissues Organs 176:67–78
Humpherys D, Eggan K, Akutsu H, Hochedlinger K, Rideout WM 3rd, Biniszkiewicz D, Yanagimachi R, Jaenisch R (2001) Epigenetic instability in es cells and cloned mice. Science 293:95–97
Humphrey RK, Beattie GM, Lopez AD, Bucay N, King CC, Firpo MT, Rose-John S, Hayek A (2004) Maintenance of pluripotency in human embryonic stem cells is stat3 independent. Stem Cells 22:522–530
Itskovitz-Eldor J, Schuldiner M, Karsenti D, Eden A, Yanuka O, Amit M, Soreq H, Benvenisty N (2000) Differentiation of human embryonic stem cells into embryoid bodies compromising the three embryonic germ layers. Mol Med 6:88–95
Jackson EB, Brues AM (1941) Studies on transplantable embryoma of the mouse. Cancer Res 1:494–498

Jones PA, Taylor SM (1980) Cellular differentiation, cytidine analogs and DNA methylation. Cell 20:85–93

Juttermann R, Li E, Jaenisch R (1994) Toxicity of 5-aza-2′-deoxycytidine to mammalian cells is mediated primarily by covalent trapping of DNA methyltransferase rather than DNA demethylation. Proc Natl Acad Sci USA 91:11797–11801

Kahan BW, Ephrussi B (1970) Developmental potentialities of clonal in vitro cultures of mouse testicular teratoma. J Natl Cancer Inst 44:1015–1036

Kamochi H, Kurokawa MS, Yoshikawa H, Ueda Y, Masuda C, Takada E, Watanabe K, Sakakibara M, Natuki Y, Kimura K, Beppu M, Aoki H, Suzuki N (2006) Transplantation of myocyte precursors derived from embryonic stem cells transfected with igfii gene in a mouse model of muscle injury. Transplantation 82:516–526

Kanatsu-Shinohara M, Inoue K, Lee J, Yoshimoto M, Ogonuki N, Miki H, Baba S, Kato T, Kazuki Y, Toyokuni S, Toyoshima M, Niwa O, Oshimura M, Heike T, Nakahata T, Ishino F, Ogura A, Shinohara T (2004) Generation of pluripotent stem cells from neonatal mouse testis. Cell 119:1001–1012

Karbanova J, Mokry J (2002) Histological and histochemical analysis of embryoid bodies. Acta Histochem 104:361–365

Kazuki Y, Hiratsuka M, Takiguchi M, Osaki M, Kajitani N, Hoshiya H, Hiramatsu K, Yoshino T, Kazuki K, Ishihara C, Takehara S, Higaki K, Nakagawa M, Takahashi K, Yamanaka S, Oshimura M (2010) Complete genetic correction of ips cells from duchenne muscular dystrophy. Mol Ther 18:386–393

Kim C, Lee IH, Lee K, Ryu SS, Lee SH, Lee KJ, Lee J, Kang JY, Kim TS (2007) Multi-well chip for forming a uniform embryoid body in a tiny droplet with mouse embryonic stem cells. Biosci Biotechnol Biochem 71:2985–2991

Kiziltepe T, Hideshima T, Catley L, Raje N, Yasui H, Shiraishi N, Okawa Y, Ikeda H, Vallet S, Pozzi S, Ishitsuka K, Ocio EM, Chauhan D, Anderson KC (2007) 5-azacytidine, a DNA methyltransferase inhibitor, induces atr-mediated DNA double-strand break responses, apoptosis, and synergistic cytotoxicity with doxorubicin and bortezomib against multiple myeloma cells. Mol Cancer Ther 6:1718–1727

Kleinsmith LJ, Pierce GB Jr (1964) Multipotentiality of single embryonal carcinoma cells. Cancer Res 24:1544–1551

Koivisto H, Hyvarinen M, Stromberg AM, Inzunza J, Matilainen E, Mikkola M, Hovatta O, Teerijoki H (2004) Cultures of human embryonic stem cells: Serum replacement medium or serum-containing media and the effect of basic fibroblast growth factor. Reprod Biomed Online 9:330–337

Konieczny SF, Emerson CP Jr (1984) 5-azacytidine induction of stable mesodermal stem cell lineages from 10 t1/2 cells: evidence for regulatory genes controlling determination. Cell 38:791–800

Kunath T, Saba-El-Leil MK, Almousailleakh M, Wray J, Meloche S, Smith A (2007) Fgf stimulation of the erk1/2 signalling cascade triggers transition of pluripotent embryonic stem cells from self-renewal to lineage commitment. Development 134:2895–2902

Kurosawa H, Imamura T, Koike M, Sasaki K, Amano Y (2003) A simple method for forming embryoid body from mouse embryonic stem cells. J Biosci Bioeng 96:409–411

Laflamme MA, Chen KY, Naumova AV, Muskheli V, Fugate JA, Dupras SK, Reinecke H, Xu C, Hassanipour M, Police S, O'Sullivan C, Collins L, Chen Y, Minami E, Gill EA, Ueno S, Yuan C, Gold J, Murry CE (2007) Cardiomyocytes derived from human embryonic stem cells in pro-survival factors enhance function of infarcted rat hearts. Nat Biotechnol 25:1015–1024

Lawrenz B, Schiller H, Willbold E, Ruediger M, Muhs A, Esser S (2004) Highly sensitive biosafety model for stem-cell-derived grafts. Cytotherapy 6:212–222

Leahy A, Xiong JW, Kuhnert F, Stuhlmann H (1999) Use of developmental marker genes to define temporal and spatial patterns of differentiation during embryoid body formation. J Exp Zool 284:67–81

Lee AS, Tang C, Cao F, Xie X, van der Bogt K, Hwang A, Connolly AJ, Robbins RC, Wu JC (2009) Effects of cell number on teratoma formation by human embryonic stem cells. Cell Cycle 8:2608–2612

Lee KH, Chuang CK, Guo SF, Tu CF (2011) Simple and efficient derivation of mouse embryonic stem cell lines using differentiation inhibitors or proliferation stimulators. Stem Cells Dev 21:373–383

Li C, Yang Y, Gu J, Ma Y, Jin Y (2009a) Derivation and transcriptional profiling analysis of pluripotent stem cell lines from rat blastocysts. Cell Res 19:173–186

Li C, Yu H, Ma Y, Shi G, Jiang J, Gu J, Yang Y, Jin S, Wei Z, Jiang H, Li J, Jin Y (2009b) Germline-competent mouse-induced pluripotent stem cell lines generated on human fibroblasts without exogenous leukemia inhibitory factor. PLoS One 4:e6724

Li E, Bestor TH, Jaenisch R (1992) Targeted mutation of the DNA methyltransferase gene results in embryonic lethality. Cell 69:915–926

Li J, Wang G, Wang C, Zhao Y, Zhang H, Tan Z, Song Z, Ding M, Deng H (2007) Mek/erk signaling contributes to the maintenance of human embryonic stem cell self-renewal. Differentiation 75:299–307

Li L, Baroja ML, Majumdar A, Chadwick K, Rouleau A, Gallacher L, Ferber I, Lebkowski J, Martin T, Madrenas J, Bhatia M (2004) Human embryonic stem cells possess immune-privileged properties. Stem Cells 22:448–456

Li P, Tong C, Mehrian-Shai R, Jia L, Wu N, Yan Y, Maxson RE, Schulze EN, Song H, Hsieh CL, Pera MF, Ying QL (2008) Germline competent embryonic stem cells derived from rat blastocysts. Cell 135:1299–1310

Liu H, Zhu F, Yong J, Zhang P, Hou P, Li H, Jiang W, Cai J, Liu M, Cui K, Qu X, Xiang T, Lu D, Chi X, Gao G, Ji W, Ding M, Deng H (2008) Generation of induced pluripotent stem cells from adult rhesus monkey fibroblasts. Cell Stem Cell 3:587–590

Lorthongpanich C, Yang SH, Piotrowska-Nitsche K, Parnpai R, Chan AW (2008) Chemical enhancement in embryo development and stem cell derivation from single blastomeres. Cloning Stem Cells 10:503–512

Ma M, Ding S, Lundqvist A, San H, Fang F, Konoplyannikov M, Berry C, Beltran LE, Chen G, Kovacic JC, Boehm M (2010) Major histocompatibility complex-i expression on embryonic stem cell-derived vascular progenitor cells is critical for syngeneic transplant survival. Stem Cells 28:1465–1475

Maherali N, Ahfeldt T, Rigamonti A, Utikal J, Cowan C, Hochedlinger K (2008) A high-efficiency system for the generation and study of human induced pluripotent stem cells. Cell Stem Cell 3:340–345

Mak TW (2007) Gene targeting in embryonic stem cells scores a knockout in stockholm. Cell 131:1027–1031

Maltsev VA, Rohwedel J, Hescheler J, Wobus AM (1993) Embryonic stem cells differentiate in vitro into cardiomyocytes representing sinusnodal, atrial and ventricular cell types. Mech Dev 44:41–50

Martin GR (1981) Isolation of a pluripotent cell line from early mouse embryos cultured in medium conditioned by teratocarcinoma stem cells. Proc Natl Acad Sci USA 78:7634–7638

Martin GR, Evans MJ (1974) The morphology and growth of a pluripotent teratocarcinoma cell line and its derivatives in tissue culture. Cell 2:163–172

Martin GR, Evans MJ (1975) Differentiation of clonal lines of teratocarcinoma cells: formation of embryoid bodies in vitro. Proc Natl Acad Sci USA 72:1441–1445

McGlynn P, Lloyd RG (2002) Recombinational repair and restart of damaged replication forks. Nat Rev Mol Cell Biol 3:859–870

Michel B, Flores MJ, Viguera E, Grompone G, Seigneur M, Bidnenko V (2001) Rescue of arrested replication forks by homologous recombination. Proc Natl Acad Sci USA 98:8181–8188

Mikkola M, Olsson C, Palgi J, Ustinov J, Palomaki T, Horelli-Kuitunen N, Knuutila S, Lundin K, Otonkoski T, Tuuri T (2006) Distinct differentiation characteristics of individual human embryonic stem cell lines. BMC Dev Biol 6:40

Mintz B, Illmensee K (1975) Normal genetically mosaic mice produced from malignant teratocarcinoma cells. Proc Natl Acad Sci USA 72:3585–3589

Mitsui K, Tokuzawa Y, Itoh H, Segawa K, Murakami M, Takahashi K, Maruyama M, Maeda M, Yamanaka S (2003) The homeoprotein nanog is required for maintenance of pluripotency in mouse epiblast and es cells. Cell 113:631–642

Mizuno Y, Chang H, Umeda K, Niwa A, Iwasa T, Awaya T, Fukada S, Yamamoto H, Yamanaka S, Nakahata T, Heike T (2010) Generation of skeletal muscle stem/progenitor cells from murine induced pluripotent stem cells. FASEB J 24:2245–2253

Mochizuki H, Ohnuki Y, Kurosawa H (2011) Effect of glucose concentration during embryoid body (eb) formation from mouse embryonic stem cells on eb growth and cell differentiation. J Biosci Bioeng 111:92–97

Morris JE, Potter SW, Buckley PM (1982) Mouse embryos and uterine epithelia show adhesive interactions in culture. J Exp Zool 222:195–198

Muntener M, Kagi U, Stevens LC, Walt H (1990) Innervation and maturation of muscular tissue in testicular teratomas in strain 129/sv-ter mice. Virchows Arch B Cell Pathol Incl Mol Pathol 59:223–229

Murry CE, Keller G (2008) Differentiation of embryonic stem cells to clinically relevant populations: lessons from embryonic development. Cell 132:661–680

Musch T, Oz Y, Lyko F, Breiling A (2010) Nucleoside drugs induce cellular differentiation by caspase-dependent degradation of stem cell factors. PLoS One 5:e10726

Myer A, Olson EN, Klein WH (2001) Myod cannot compensate for the absence of myogenin during skeletal muscle differentiation in murine embryonic stem cells. Dev Biol 229:340–350

Newman AM, Cooper JB (2010) Lab-specific gene expression signatures in pluripotent stem cells. Cell Stem Cell 7:258–262

Ng ES, Davis R, Stanley EG, Elefanty AG (2008) A protocol describing the use of a recombinant protein-based, animal product-free medium (apel) for human embryonic stem cell differentiation as spin embryoid bodies. Nat Protoc 3:768–776

Nichols J, Silva J, Roode M, Smith A (2009) Suppression of erk signalling promotes ground state pluripotency in the mouse embryo. Development 136:3215–3222

Nichols J, Zevnik B, Anastassiadis K, Niwa H, Klewe-Nebenius D, Chambers I, Scholer H, Smith A (1998) Formation of pluripotent stem cells in the mammalian embryo depends on the pou transcription factor oct4. Cell 95:379–391

Niwa H, Burdon T, Chambers I, Smith A (1998) Self-renewal of pluripotent embryonic stem cells is mediated via activation of stat3. Genes Dev 12:2048–2060

Niwa H, Ogawa K, Shimosato D, Adachi K (2009) A parallel circuit of lif signalling pathways maintains pluripotency of mouse es cells. Nature 460:118–122

Nussbaum J, Minami E, Laflamme MA, Virag JA, Ware CB, Masino A, Muskheli V, Pabon L, Reinecke H, Murry CE (2007) Transplantation of undifferentiated murine embryonic stem cells in the heart: teratoma formation and immune response. FASEB J 21:1345–1357

Oakley CS, Welsch MA, Zhai YF, Chang CC, Gould MN, Welsch CW (1993) Comparative abilities of athymic nude mice and severe combined immune deficient (scid) mice to accept transplants of induced rat mammary carcinomas: enhanced transplantation efficiency of those rat mammary carcinomas that have elevated expression of neu oncogene. Int J Cancer 53:1002–1007

Ohi Y, Qin H, Hong C, Blouin L, Polo JM, Guo T, Qi Z, Downey SL, Manos PD, Rossi DJ, Yu J, Hebrok M, Hochedlinger K, Costello JF, Song JS, Ramalho-Santos M (2011) Incomplete DNA methylation underlies a transcriptional memory of somatic cells in human iPS cells. Nat Cell Biol 13:541–549

Okabe S, Forsberg-Nilsson K, Spiro AC, Segal M, McKay RD (1996) Development of neuronal precursor cells and functional postmitotic neurons from embryonic stem cells in vitro. Mech Dev 59:89–102

Okano M, Bell DW, Haber DA, Li E (1999) DNA methyltransferases dnmt3a and dnmt3b are essential for de novo methylation and mammalian development. Cell 99:247–257

Ooi SK, Wolf D, Hartung O, Agarwal S, Daley GQ, Goff SP, Bestor TH (2010) Dynamic instability of genomic methylation patterns in pluripotent stem cells. Epigenetics Chromatin 3:17

Papaioannou VE, McBurney MW, Gardner RL, Evans MJ (1975) Fate of teratocarcinoma cells injected into early mouse embryos. Nature 258:70–73

Park IH, Zhao R, West JA, Yabuuchi A, Huo H, Ince TA, Lerou PH, Lensch MW, Daley GQ (2008) Reprogramming of human somatic cells to pluripotency with defined factors. Nature 451:141–146

Pawlikowski B, Lee L, Zuo J, Kramer RH (2009) Analysis of human muscle stem cells reveals a differentiation-resistant progenitor cell population expressing pax7 capable of self-renewal. Dev Dyn 238:138–149

Pera MF, Andrade J, Houssami S, Reubinoff B, Trounson A, Stanley EG, Ward-van Oostwaard D, Mummery C (2004) Regulation of human embryonic stem cell differentiation by bmp-2 and its antagonist noggin. J Cell Sci 117:1269–1280

Phelps SF, Hauser MA, Cole NM, Rafael JA, Hinkle RT, Faulkner JA, Chamberlain JS (1995) Expression of full-length and truncated dystrophin mini-genes in transgenic mdx mice. Hum Mol Genet 4:1251–1258

Pierce GB, Dixon FJ Jr (1959) Testicular teratomas. Ii. Teratocarcinoma as an ascitic tumor. Cancer 12:584–589

Polo JM, Liu S, Figueroa ME, Kulalert W, Eminli S, Tan KY, Apostolou E, Stadtfeld M, Li Y, Shioda T, Natesan S, Wagers AJ, Melnick A, Evans T, Hochedlinger K (2010) Cell type of origin influences the molecular and functional properties of mouse induced pluripotent stem cells. Nat Biotechnol 28:848–855

Potter SW, Morris JE (1985) Development of mouse embryos in hanging drop culture. Anat Rec 211:48–56

Prelle K, Wobus AM, Krebs O, Blum WF, Wolf E (2000) Overexpression of insulin-like growth factor-ii in mouse embryonic stem cells promotes myogenic differentiation. Biochem Biophys Res Commun 277:631–638

Prokhorova TA, Harkness LM, Frandsen U, Ditzel N, Schroder HD, Burns JS, Kassem M (2009) Teratoma formation by human embryonic stem cells is site dependent and enhanced by the presence of matrigel. Stem Cells Dev 18:47–54

Przyborski SA, Christie VB, Hayman MW, Stewart R, Horrocks GM (2004) Human embryonal carcinoma stem cells: models of embryonic development in humans. Stem Cells Dev 13:400–408

Quattrocelli M, Palazzolo G, Floris G, Schoffski P, Anastasia L, Orlacchio A, Vandendriessche T, Chuah MK, Cossu G, Verfaillie C, Sampaolesi M (2011) Intrinsic cell memory reinforces myogenic commitment of pericyte-derived ipscs. J Pathol 223:593–603

Rathjen PD, Nichols J, Toth S, Edwards DR, Heath JK, Smith AG (1990) Developmentally programmed induction of differentiation inhibiting activity and the control of stem cell populations. Genes Dev 4:2308–2318

Robbins J, Gulick J, Sanchez A, Howles P, Doetschman T (1990) Mouse embryonic stem cells express the cardiac myosin heavy chain genes during development in vitro. J Biol Chem 265:11905–11909

Rohwedel J, Guan K, Zuschratter W, Jin S, Ahnert-Hilger G, Furst D, Fassler R, Wobus AM (1998) Loss of beta1 integrin function results in a retardation of myogenic, but an acceleration of neuronal, differentiation of embryonic stem cells in vitro. Dev Biol 201:167–184

Rohwedel J, Horak V, Hebrok M, Fuchtbauer EM, Wobus AM (1995) M-twist expression inhibits mouse embryonic stem cell-derived myogenic differentiation in vitro. Exp Cell Res 220:92–100

Rohwedel J, Maltsev V, Bober E, Arnold HH, Hescheler J, Wobus AM (1994) Muscle cell differentiation of embryonic stem cells reflects myogenesis in vivo: developmentally regulated expression of myogenic determination genes and functional expression of ionic currents. Dev Biol 164:87–101

Rose O, Rohwedel J, Reinhardt S, Bachmann M, Cramer M, Rotter M, Wobus A, Starzinski-Powitz A (1994) Expression of m-cadherin protein in myogenic cells during prenatal mouse development and differentiation of embryonic stem cells in culture. Dev Dyn 201:245–259

Rosenthal MD, Wishnow RM, Sato GH (1970) In vitro growth and differetiation of clonal populations of multipotential mouse clls derived from a transplantable testicular teratocarcinoma. J Natl Cancer Inst 44:1001–1014

Rossant J, Gardner RL, Alexandre HL (1978) Investigation of the potency of cells from the postimplantation mouse embryo by blastocyst injection: a preliminary report. J Embryol Exp Morphol 48:239–247

Sakurai H, Okawa Y, Inami Y, Nishio N, Isobe K (2008) Paraxial mesodermal progenitors derived from mouse embryonic stem cells contribute to muscle regeneration via differentiation into muscle satellite cells. Stem Cells 26:1865–1873

Sanchez A, Jones WK, Gulick J, Doetschman T, Robbins J (1991) Myosin heavy chain gene expression in mouse embryoid bodies. An in vitro developmental study. J Biol Chem 266:22419–22426

Sasaki E, Hanazawa K, Kurita R, Akatsuka A, Yoshizaki T, Ishii H, Tanioka Y, Ohnishi Y, Suemizu H, Sugawara A, Tamaoki N, Izawa K, Nakazaki Y, Hamada H, Suemori H, Asano S, Nakatsuji N, Okano H, Tani K (2005) Establishment of novel embryonic stem cell lines derived from the common marmoset (callithrix jacchus). Stem Cells 23:1304–1313

Schmidt MM, Guan K, Wobus AM (2001) Lithium influences differentiation and tissue-specific gene expression of mouse embryonic stem (es) cells in vitro. Int J Dev Biol 45:421–429

Schoonjans L, Kreemers V, Danloy S, Moreadith RW, Laroche Y, Collen D (2003) Improved generation of germline-competent embryonic stem cell lines from inbred mouse strains. Stem Cells 21:90–97

Schroeder M, Niebruegge S, Werner A, Willbold E, Burg M, Ruediger M, Field LJ, Lehmann J, Zweigerdt R (2005) Differentiation and lineage selection of mouse embryonic stem cells in a stirred bench scale bioreactor with automated process control. Biotechnol Bioeng 92:920–933

Schuldiner M, Yanuka O, Itskovitz-Eldor J, Melton DA, Benvenisty N (2000) Effects of eight growth factors on the differentiation of cells derived from human embryonic stem cells. Proc Natl Acad Sci USA 97:11307–11312

Shani M, Faerman A, Emerson CP, Pearson-White S, Dekel I, Magal Y (1992) The consequences of a constitutive expression of myod1 in es cells and mouse embryos. Symp Soc Exp Biol 46:19–36

Shao H, Chen B, Tao M (2009) Skeletal myogenesis by human primordial germ cell-derived progenitors. Biochem Biophys Res Commun 378:750–754

Shen MM, Leder P (1992) Leukemia inhibitory factor is expressed by the preimplantation uterus and selectively blocks primitive ectoderm formation in vitro. Proc Natl Acad Sci USA 89:8240–8244

Shi D, Reinecke H, Murry CE, Torok-Storb B (2004) Myogenic fusion of human bone marrow stromal cells, but not hematopoietic cells. Blood 104:290–294

Sicinski P, Geng Y, Ryder-Cook AS, Barnard EA, Darlison MG, Barnard PJ (1989) The molecular basis of muscular dystrophy in the mdx mouse: a point mutation. Science 244:1578–1580

Singla DK, Hacker TA, Ma L, Douglas PS, Sullivan R, Lyons GE, Kamp TJ (2006) Transplantation of embryonic stem cells into the infarcted mouse heart: formation of multiple cell types. J Mol Cell Cardiol 40:195–200

Solter D, Skreb N, Damjanov I (1970) Extrauterine growth of mouse egg-cylinders results in malignant teratoma. Nature 227:503–504

Sorm F, Piskala A, Cihak A, Vesely J (1964) 5-azacytidine, a new, highly effective cancerostatic. Experientia 20:202–203

Stavropoulos ME, Mengarelli I, Barberi T (2009) Differentiation of multipotent mesenchymal precursors and skeletal myoblasts from human embryonic stem cells. Curr Protoc Stem Cell Biol Chapter 1:Unit 1F 8

Stevens LC (1960) Embryonic potency of embryoid bodies derived from a transplantable testicular teratoma of the mouse. Dev Biol 2:285–297

Stevens LC (1962) Testicular teratomas in fetal mice. J Natl Cancer Inst 28:247–267

Stevens LC (1967) The biology of teratomas. Adv Morphog 6:1–31

Stevens LC (1968) The development of teratomas from intratesticular grafts of tubal mouse eggs. J Embryol Exp Morphol 20:329–341

Stevens LC (1970) The development of transplantable teratocarcinomas from intratesticular grafts of pre- and postimplantation mouse embryos. Dev Biol 21:364–382

Stevens LC, Little CC (1954) Spontaneous testicular teratomas in an inbred strain of mice. Proc Natl Acad Sci USA 40:1080–1087

Stevens LC, Varnum DS (1974) The development of teratomas from parthenogenetically activated ovarian mouse eggs. Dev Biol 37:369–380

Sumer H, Liu J, Malaver-Ortega LF, Lim ML, Khodadadi K, Verma PJ (2011) Nanog is a key factor for induction of pluripotency in bovine adult fibroblasts. J Anim Sci 89:2708–2716

Sundstrom J, Verajnkorva E, Salminen E, Pelliniemi LJ, Pollanen P (1999) Experimental testicular teratoma promotes formation of humoral immune responses in the host testis. J Reprod Immunol 42:107–126

Suwinska A, Ciemerych MA (2011) Factors regulating pluripotency and differentiation in early mammalian embryos and embryo-derived stem cells. Vitam Horm 87:1–38

Swijnenburg RJ, Schrepfer S, Cao F, Pearl JI, Xie X, Connolly AJ, Robbins RC, Wu JC (2008) In vivo imaging of embryonic stem cells reveals patterns of survival and immune rejection following transplantation. Stem Cells Dev 17:1023–1029

Swijnenburg RJ, Tanaka M, Vogel H, Baker J, Kofidis T, Gunawan F, Lebl DR, Caffarelli AD, de Bruin JL, Fedoseyeva EV, Robbins RC (2005) Embryonic stem cell immunogenicity increases upon differentiation after transplantation into ischemic myocardium. Circulation 112: I166–172

Tajbakhsh S (2009) Skeletal muscle stem cells in developmental versus regenerative myogenesis. J Intern Med 266:372–389

Takahashi K, Tanabe K, Ohnuki M, Narita M, Ichisaka T, Tomoda K, Yamanaka S (2007) Induction of pluripotent stem cells from adult human fibroblasts by defined factors. Cell 131:861–872

Takahashi K, Yamanaka S (2006) Induction of pluripotent stem cells from mouse embryonic and adult fibroblast cultures by defined factors. Cell 126:663–676

Takai D, Jones PA (2002) Comprehensive analysis of cpg islands in human chromosomes 21 and 22. Proc Natl Acad Sci USA 99:3740–3745

Taylor SM, Jones PA (1979) Multiple new phenotypes induced in 10 t1/2 and 3 t3 cells treated with 5-azacytidine. Cell 17:771–779

Taylor SM, Jones PA (1982) Changes in phenotypic expression in embryonic and adult cells treated with 5-azacytidine. J Cell Physiol 111:187–194

Thompson EM, Legouy E, Renard JP (1998) Mouse embryos do not wait for the mbt: chromatin and rna polymerase remodeling in genome activation at the onset of development. Dev Genet 22:31–42

Thomson JA, Itskovitz-Eldor J, Shapiro SS, Waknitz MA, Swiergiel JJ, Marshall VS, Jones JM (1998) Embryonic stem cell lines derived from human blastocysts. Science 282:1145–1147

Thomson JA, Kalishman J, Golos TG, Durning M, Harris CP, Hearn JP (1996) Pluripotent cell lines derived from common marmoset (callithrix jacchus) blastocysts. Biol Reprod 55:254–259

Tian C, Lu Y, Gilbert R, Karpati G (2008) Differentiation of murine embryonic stem cells in skeletal muscles of mice. Cell Transplant 17:325–335

Tveit KM, Fodstad O, Brogger A, Olsnes S (1980) Human embryonal carcinoma grown in athymic mice and in vitro. Cancer Res 40:949–953

Ueda S, Kawamata M, Teratani T, Shimizu T, Tamai Y, Ogawa H, Hayashi K, Tsuda H, Ochiya T (2008) Establishment of rat embryonic stem cells and making of chimera rats. PLoS One 3: e2800

Valbuena D, Galan A, Sanchez E, Poo ME, Gomez E, Sanchez-Luengo S, Melguizo D, Garcia A, Ruiz V, Moreno R, Pellicer A, Simon C (2006) Derivation and characterization of three new spanish human embryonic stem cell lines (val -3–4 -5) on human feeder and in serum-free conditions. Reprod Biomed Online 13:875–886

van den Eijnden-van Raaij AJ, van Achterberg TA, van der Kruijssen CM, Piersma AH, Huylebroeck D, de Laat SW, Mummery CL (1991) Differentiation of aggregated murine p19 embryonal carcinoma cells is induced by a novel visceral endoderm-specific fgf-like factor and inhibited by activin a. Mech Dev 33:157–165

Virchow R (1863) Die krankhaften geschwülste. Dreissig vorlesungen, gehalten während des wintersemesters 1862–1863. Berlin

Wakitani S, Saito T, Caplan AI (1995) Myogenic cells derived from rat bone marrow mesenchymal stem cells exposed to 5-azacytidine. Muscle Nerve 18:1417–1426

Wakitani S, Takaoka K, Hattori T, Miyazawa N, Iwanaga T, Takeda S, Watanabe TK, Tanigami A (2003) Embryonic stem cells injected into the mouse knee joint form teratomas and subsequently destroy the joint. Rheumatology (Oxford) 42:162–165

Watanabe S, Hirai H, Asakura Y, Tastad C, Verma M, Keller C, Dutton JR, Asakura A (2011) Myod gene suppression by oct4 is required for reprogramming in myoblasts to produce induced pluripotent stem cells. Stem Cells 29:505–516

Weitzer G, Milner DJ, Kim JU, Bradley A, Capetanaki Y (1995) Cytoskeletal control of myogenesis: a desmin null mutation blocks the myogenic pathway during embryonic stem cell differentiation. Dev Biol 172:422–439

Wells DJ, Wells KE, Asante EA, Turner G, Sunada Y, Campbell KP, Walsh FS, Dickson G (1995) Expression of human full-length and minidystrophin in transgenic mdx mice: implications for gene therapy of duchenne muscular dystrophy. Hum Mol Genet 4:1245–1250

Wiles MV, Keller G (1991) Multiple hematopoietic lineages develop from embryonic stem (es) cells in culture. Development 111:259–267

Wirth M, Inglin B, Romen W, Ackermann R (1983) Human embryonal cell carcinoma in nude mice. Cancer Res 43:5526–5532

Wobus A, Rohwedel J, Maltsev V, Hescheler J (1994) In vitro differentiation of embryonic stem cells into cardiomyocytes or skeletal muscle cells is specifically modulated by retinoic acid. Roux's Arch Dev Biol 204:36–45

Wobus AM, Wallukat G, Hescheler J (1991) Pluripotent mouse embryonic stem cells are able to differentiate into cardiomyocytes expressing chronotropic responses to adrenergic and cholinergic agents and ca^{2+} channel blockers. Differentiation 48:173–182

Wyzykowski JC, Winata TI, Mitin N, Taparowsky EJ, Konieczny SF (2002) Identification of novel myod gene targets in proliferating myogenic stem cells. Mol Cell Biol 22:6199–6208

Xu RH, Chen X, Li DS, Li R, Addicks GC, Glennon C, Zwaka TP, Thomson JA (2002) Bmp4 initiates human embryonic stem cell differentiation to trophoblast. Nat Biotechnol 20:1261–1264

Ying QL, Nichols J, Chambers I, Smith A (2003) Bmp induction of id proteins suppresses differentiation and sustains embryonic stem cell self-renewal in collaboration with stat3. Cell 115:281–292

Ying QL, Wray J, Nichols J, Batlle-Morera L, Doble B, Woodgett J, Cohen P, Smith A (2008) The ground state of embryonic stem cell self-renewal. Nature 453:519–523

Zheng JK, Wang Y, Karandikar A, Wang Q, Gai H, Liu AL, Peng C, Sheng HZ (2006) Skeletal myogenesis by human embryonic stem cells. Cell Res 16:713–722

Chapter 19
Stem Cells and Corneal Epithelial Maintenance: Insights from the Mouse and Other Animal Models

Richard L. Mort, Panagiotis Douvaras, Steven D. Morley, Natalie Dorà, Robert E. Hill, J. Martin Collinson, and John D. West

Abstract Maintenance of the corneal epithelium is essential for vision and is a dynamic process incorporating constant cell production, movement and loss. Although cell-based therapies involving the transplantation of putative stem cells are well advanced for the treatment of human corneal defects, the scientific understanding of these interventions is poor. No definitive marker that discriminates stem cells that maintain the corneal epithelium from the surrounding tissue has been discovered and the identity of these elusive cells is, therefore, hotly debated. The key elements of corneal epithelial maintenance have long been recognised but it is still not known how this dynamic balance is co-ordinated during normal homeostasis to ensure the corneal epithelium is maintained at a uniform thickness. Most indirect experimental evidence supports the limbal epithelial stem cell (LESC) hypothesis, which proposes that the adult corneal epithelium is maintained by stem cells located in the limbus at the corneal periphery. However, this has been challenged recently by the corneal epithelial stem cell (CESC) hypothesis, which

R.L. Mort • R.E. Hill
Medical and Developmental Genetics Section, MRC Human Genetics Unit, MRC IGMM, University of Edinburgh, Western General Hospital, Crewe Road, Edinburgh EH4 2XU, UK

P. Douvaras
The New York Stem Cell Foundation, 3960 Broadway, 4th Floor, Suite 440B, New York, NY 10032, USA

S.D. Morley
Clinical Biochemistry Section, Division of Health Sciences, Chancellor's Building, University of Edinburgh Medical School, 49 Little France Crescent, Edinburgh EH16 1SB, UK

N. Dorà • J.D. West (✉)
Genes and Development Group, Centre for Integrative Physiology, School of Clinical Sciences, University of Edinburgh, Hugh Robson Building, George Square, Edinburgh EH8 9XD, UK
e-mail: John.West@ed.ac.uk

J.M. Collinson
School of Medical Sciences, Institute of Medical Sciences, University of Aberdeen, Foresterhill, Aberdeen AB25 2ZD, UK

J.Z. Kubiak (ed.), *Mouse Development*, Results and Problems in Cell Differentiation 55, DOI 10.1007/978-3-642-30406-4_19, © Springer-Verlag Berlin Heidelberg 2012

proposes that during normal homeostasis the mouse corneal epithelium is maintained by stem cells located throughout the basal corneal epithelium with LESCs only contributing during wound healing. In this chapter we review experimental studies, mostly based on animal work, that provide insights into how stem cells maintain the normal corneal epithelium and consider the merits of the alternative LESC and CESC hypotheses. Finally, we highlight some recent research on other stem cell systems and consider how this could influence future research directions for identifying the stem cells that maintain the corneal epithelium.

19.1 Introduction

19.1.1 Introduction to the Cornea

The transparent adult cornea has rightly been called our window on the world. Its unique properties allow it to maintain transparency, refract light and form a protective, impermeable barrier. The cornea comprises an outer squamous, non-keratinised epithelium of keratinocytes, which is about 5–6 cells thick, a thick stroma of flattened keratocytes embedded in collagen and the corneal endothelium, comprising a single inner cell layer (Fig. 19.1). In addition, an acellular, collagenous basement membrane (Descemet's membrane) separates the corneal stroma and endothelium, and in humans and other primates there is also a distinct acellular Bowman's layer (anterior limiting lamina) between the stroma and corneal epithelium. This is rudimentary and indistinct in mice but visible by electron microscopy (Haustein 1983). The cornea is avascular and absorbs oxygen and nutrients from the tear film and aqueous humour but it is innervated and the nerves provide additional trophic support. Mouse corneal anatomy is described in detail in Smith et al. (2002).

The corneal epithelium develops from the head surface ectoderm and both the stromal keratocytes and corneal endothelium are produced by mesenchyme (Haustein 1983), which in mice is derived predominantly from neural crest cells with an additional contribution from cranial mesoderm (Gage et al. 2005). During development, nerves grow into the stroma from the limbus and form a nerve plexus beneath the epithelium which projects fine nerves through the epithelium to the ocular surface (McKenna and Lwigale 2011).

The corneal epithelium has more cell layers than the neighbouring conjunctival epithelium, which is distinguished by the presence of goblet cells and blood vessels, both of which are incompatible with transparency and absent from the corneal epithelium (Smith et al. 2002). Mitosis is restricted to the basal layer in both the corneal and conjunctival epithelia. The basal corneal epithelial cells are cuboidal while the suprabasal cells are progressively more flattened towards the anterior. These comprise 2–3 layers of polyhedral 'wing cells' and 1–3 layers of superficial squamous cells with flattened nuclei (Fig. 19.1), which are held together by tight junctions to form an effective barrier. Corneal epithelial cells are continuously

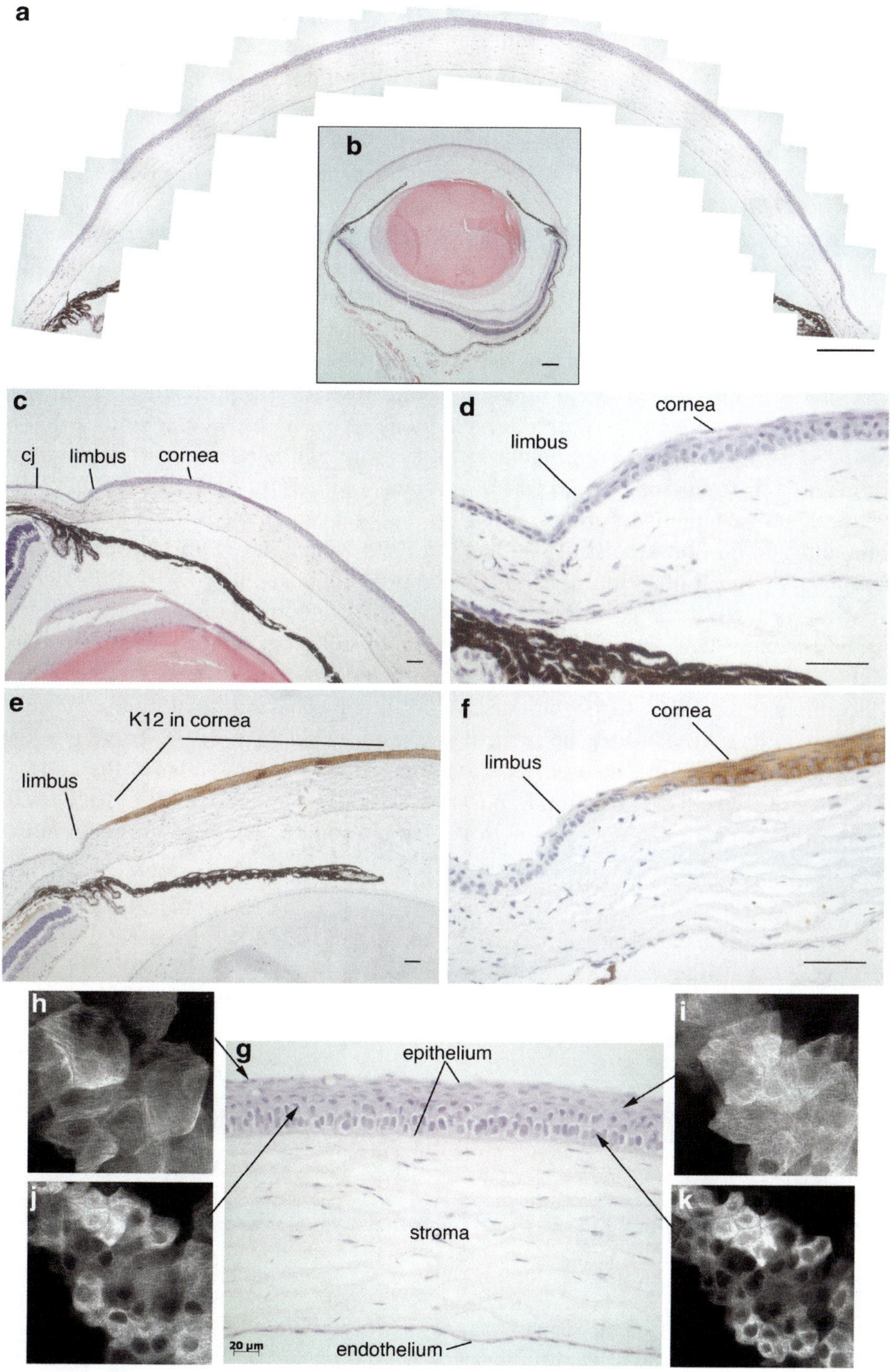

Fig. 19.1 Mouse cornea and limbus. (**a**) Montage of mouse cornea. (**b**) Whole eye used to produce montage shown in (**a**). (**c**) Peripheral cornea, limbus and part of the conjunctiva.

being shed (desquamated) from the superficial layer and replenished, yet the tissue maintains a uniform structure and thickness, so transparency is not compromised.

In the adult, neither the corneal stromal nor endothelial cells divide unless injured; endothelial cells are arrested in G1 and show contact inhibition (Joyce 2003) whereas stromal keratocytes exit the cell cycle around the time the eyes open in mice, at postnatal days (P) 12–14, and remain quiescent in G0 (Zieske 2004; Zieske et al. 2004). The corneal endothelium consists of a single layer of cells that is critical for maintaining correct hydration of the corneal stroma via metabolic pumps that actively transport fluid out of the stroma and into the anterior chamber. The corneal stroma is less hydrated than the neighbouring sclera and if the cornea becomes too hydrated it swells and becomes opaque.

Laterally, the corneal stroma merges with the sclera and forms a region known as the limbus at the corneoscleral junction. The limbus is less pronounced in mouse than humans but it forms a morphological 'dent' in the mouse ocular surface that is not always apparent in histological sections. The epithelial layer of the limbus forms a transition zone between the corneal epithelium and the conjunctival epithelium. The mouse corneo-limbal epithelial boundary can be identified by immunostaining for keratin 15 (K15) or K19, which are both present in the conjunctiva and limbus but not the cornea, or immunostaining for K12, which is specific for the corneal epithelium (Fig. 19.1e,f). The corneal epithelium is thinner at the periphery and, unlike the human, the mouse limbal epithelium is also thin. According to the conventional limbal epithelial stem cell (LESC) hypothesis (Schermer et al. 1986; Cotsarelis et al. 1989), some basal limbal epithelial cells are stem cells that maintain the corneal epithelium (see Sect. 19.3). There is some evidence that there may also be separate populations of stem cells for the stromal keratocytes (Du et al. 2005; Funderburgh et al. 2005) and endothelium (McGowan et al. 2007) that reside in or close to the limbal region (but below the epithelial layer).

Fig. 19.1 (continued) (**d**) Higher magnification of the limbal region shown in (**c**). (**e**) Peripheral cornea, limbus and part of the conjunctiva immunostained for keratin 12 (K12; *dark staining*) to show the border between the corneal epithelium (K12 positive) and limbal epithelium (K12 negative). (**f**) Higher magnification of dark K12 immunostaining shown in (**e**) to demonstrate border between corneal and limbal epithelia. (**g**) Central cornea showing corneal epithelium, stroma and endothelium. (**h–k**) Confocal images of corneal epithelium from tau-GFP transgenic mouse (TgTP6.3; Pratt et al. 2000) showing optical sections of different epithelial layers, as indicated in (**g**), from flattened squamous cells in the superficial layer at the corneal surface (**h**) to intermediate 'wing cells' in the suprabasal layers and compact basal epithelial cells (**k**). *Scale bars*: (**a, b**) 200 μm; (**c–f**), 50 μm; (**g**), 20 μm. Abbreviation in (c): *cj* conjunctiva

19.1.2 *Scope of This Review*

The focus of this review is how adult stem cells maintain a healthy cornea. Maintenance involves a balance of cell production, movement and loss, orchestrated by stem cells. The conventional LESC hypothesis has recently been challenged. The alternative view proposes that stem cells scattered throughout the mouse corneal epithelium are responsible for homeostatic maintenance of the tissue and that LESCs are only active during wound healing (Majo et al. 2008). Although human limbal epithelium is used successfully as a source of stem cells for treating human corneal epithelial conditions that are thought to involve a stem cell deficiency (Rama et al. 2010), this does not prove that limbal epithelial cells maintain the cornea during normal homeostasis. Such controversies highlight the difficulty in defining a stem cell niche; however, this is a matter crucial to understanding the basis of corneal stem cell biology. The availability of genetic and transgenic resources makes the mouse the experimental animal of choice for many investigations of the cornea. Mouse models of human corneal abnormalities provide insights into many aspects of normal and abnormal corneal epithelial maintenance. The aim of this review, therefore, is to discuss our present understanding of the basic cell biology of stem cells that maintain the corneal epithelium, to focus on the mouse to examine its contribution and to discuss its relevance as an effective model for human corneal epithelial maintenance.

19.2 Maintenance of Stratified Squamous Epithelia by Adult Stem Cells

Adult stratified squamous epithelia are composed of layers of flattened epithelial cells lying on top of a basement membrane and are often found in tissues that are subjected to constant abrasive forces such as the skin, the oral mucosa and the corneal epithelium. These tissues are highly dynamic; cells are constantly lost from the tissue surface and therefore must be replaced in order to maintain a constant cellular mass. Their inherent self-renewing capacity is driven by an adult stem cell (SC) population that resides in the basal layer of the epithelium. SCs divide to produce transient (or transit) amplifying cells (TACs) that proliferate in the basal layer before differentiating, moving through the suprabasal layers to the outer stratified layers and finally being shed from the tissue surface (Kruse 1994; Ren and Wilson 1996). Although, overall, net asymmetric division is required to maintain the SC pool and produce TACs, in some systems SCs need not always divide asymmetrically but may also divide symmetrically to produce two TACs or two SCs, providing a stochastic mechanism whereby some SC lineages expand and others are lost (Nakagawa et al. 2007; Snippert et al. 2010; Lopez-Garcia et al. 2010; Klein and Simons 2011). The adult stem cells that maintain these epithelia divide relatively infrequently; however, they have the potential to divide indefinitely (Braun and Watt 2004). Identification of the stem cells has, therefore, exploited these features using incorporation of tritiated thymidine (^{3}H-TdR) or the thymidine analogue bromodeoxyuridine (BrdU) into the DNA of dividing cells during S-phase of the cell cycle (Potten and Loeffler 1990). Although the stem cells divide infrequently they can be labelled by prolonged exposure and the label is retained over time but lost from the more rapidly dividing daughter cells. However, these traditional label-retaining methods may preferentially identify a subset of functional stem cells. For example, SCs may alternate between active and quiescent periods or separate populations of active and quiescent SC populations may exist (see Sect. 19.8). The more quiescent SCs, though less likely to incorporate label during the period of exposure, are more likely to retain that label during the chase period (Li and Clevers 2010). The use of this approach to identify stem cells that maintain the corneal epithelium is discussed in Sect. 19.3.2. In addition to cell production by stem cells and proliferation of TACs, the mechanism of corneal epithelial maintenance involves centripetal migration of cells from the periphery (as discussed in Sect. 19.4.2).

Label-retaining stem cells in the mouse epidermis have also been identified by an elegant transgenic system whereby chromatin is labelled with GFP when a transgene producing a histone-2B (H2B)-GFP fusion protein is expressed early in development and then switched off for the chase period via a Tet-Off switch, activated by continuous doxycycline treatment (Tumbar et al. 2004). This GFP label-retaining cell approach is more powerful and easier to interpret than the conventional BrdU label-retaining cell method because (1) all the potential stem cells can be labelled with H2B-GFP before the chase period, (2) live H2B-GFP-positive cells can be FACS sorted and used for transcriptional profiling, (3) H2B-GFP labelling can be restricted to specific tissues, so simplifying analysis of label-retaining cells. It is likely that this and similar transgenic approaches will be used to identify putative stem cells as GFP label-retaining cells in other tissues, including the ocular surface.

19.3 Stem Cells That Maintain the Corneal Epithelium

19.3.1 The Limbal Epithelial Stem Cell Hypothesis

The burden of evidence (discussed in Sect. 19.3.2) suggests that during adult corneal epithelial homeostasis cell production depends on a population of adult stem cells at the periphery of the cornea. These are known as LESCs because they are thought to reside in the basal layer of the limbal epithelium (limbus) at the corneoscleral junction (Schermer et al. 1986; Cotsarelis et al. 1989; Lehrer et al. 1998; Li et al. 2007); Fig. 19.1. LESCs cycle slowly unless stimulated to proliferate by corneal insult. Daughter TACs move into the corneal epithelium, where they undergo several further rounds of division to maintain the epithelium, move centripetally and leave the basal layer of the epithelium at variable times as pairs of cells (Beebe and Masters 1996). TAC populations near the periphery have a greater replicative potential than their more centrally located counterparts (Lehrer et al. 1998). Once in the suprabasal layers, the cells differentiate, become post-mitotic, move vertically to the superficial layer and are subsequently lost by desquamation at the corneal surface. In order to maintain a uniform corneal epithelial thickness, some TACs must leave the basal layer at the periphery of the cornea and others must move on towards the centre, but it is not known how this process is regulated.

19.3.2 Indirect Evidence That Stem Cells in the Limbus Maintain the Corneal Epithelium

The hypothesis that the corneal epithelium is maintained by SCs in the limbus was proposed by Schermer et al. (1986), based on the conclusion that the basal limbal epithelium was less differentiated than the suprabasal limbal epithelium and both basal and suprabasal corneal epithelia in the rabbit. More recent studies also indicate that basal limbal epithelial cells are morphologically distinct from basal corneal epithelial cells, being smaller and euchromatin-rich with a high nucleus to cytoplasm ratio (Romano et al. 2003; Chen et al. 2004). These are properties that are believed to be typical of stem cells in a variety of tissues (Barrandon and Green 1985; Tani et al. 2000; Gaspar-Maia et al. 2009). Evidence that cells move centripetally from the limbal region was also emerging at the time that the LESC hypothesis was proposed (Kinoshita et al. 1981; Buck 1985) and this was confirmed later (see Sect. 19.4.2).

Subsequent label retaining experiments in the mouse identified a population of putative stem cells in the basal layer of the limbal epithelium (Cotsarelis et al. 1989; Lehrer et al. 1998). The early label retaining studies relied on wounding of the central cornea in order to stimulate LESCs to divide during an initial pulse of tritiated thymidine (^{3}H-TdR) label. Such methods may not recapitulate the unwounded homeostatic mechanisms accurately (Cotsarelis et al. 1989). However, the results were confirmed using BrdU perfusion by osmotic minipump, which avoided the need for wounding because of the much-increased length of the pulse period (Lehrer et al. 1998).

SCs may divide infrequently in vivo but their proliferative potential can be unmasked by culturing them in vitro. Barrandon and Green classified human epidermal cultures as holoclones, paraclones or meroclones, depending on their proliferation characteristics and the morphology of the colonies produced (Barrandon and Green 1987). Holoclones have the greatest proliferative potential and are thought to be produced from SCs. Paraclones form abortive colonies and are believed to consist mainly of terminally differentiated cells. Meroclones constitute a transitional class and are cultures formed by cells with limited growth potential, now thought to be TACs. Pellegrini et al. (1999) showed that cells cultured from human limbal tissue, but not central or paracentral corneal tissue, were able to produce holoclones, suggesting that stem cells are found in the human limbus but not in the cornea itself. As discussed in Sect. 19.8, care must be taken in the interpretation of ex vivo culture experiments, as it is possible to unmask proliferative potential that is not relevant to normal homeostasis.

Clinical data also suggest that the limbus contains a population of stem cells. If the corneal epithelium is completely destroyed along with the limbal zone, a situation that occurs in patients with chemical burns, standard cornea replacement is ineffective but transplantation of small pieces of healthy, human limbal epithelial tissue can reconstitute the entire corneal epithelium (Kenyon and Tseng 1989; Tseng 1989). However, grafts that include both limbal and corneal tissue seem to be more successful than limbal grafts alone and treatment with cultured limbal cells is also proving effective (Rama et al. 2010; Shortt et al. 2011).

19.3.3 The Quest for Markers of LESCs

Table 19.1 summarises a selection of the most promising positive and negative markers of the various human limbal/corneal epithelial compartments. Although a number of markers are specifically enriched in the basal limbal epithelium no single marker has been unambiguously confirmed as LESC specific. The most promising combination of markers currently appears to be a C/EBPδ-positive, Bmi1-positive, ΔNp63α-positive population identified in the human peripheral limbus (Barbaro et al. 2007).

The markers outlined in Table 19.1 are indicative of a tissue hierarchy that is initiated in the basal limbus and terminated in the suprabasal cornea, further supporting the LESC hypothesis. Markers of basal limbus are associated with SC-like properties including quiescence (C/EBPδ), self-renewal and proliferation (Bmi1 and ΔNp63α) and cell adhesion (N-cadherin and integrin α9) whereas markers of suprabasal cornea appear more associated with terminal differentiation (keratin 3/12, connexin 43). Five of the most promising and widely cited markers of LESCs/TACs are discussed in further detail below.

19.3.3.1 ΔNp63α

The p63 transcription factor has been identified as a marker of the most proliferative basal cells in the skin (Yang et al. 1998) and is required for the development of all stratified epithelia. Homozygous p63 null mouse embryos die postnatally and have severe limb defects and lack development of epidermal tissues (Mills et al. 1999). Six isoforms are generated from the p63 gene. Three isoforms (p63α, β and γ), having transactivating activity, are transcribed from the upstream promoter and have distinct C termini, whilst the three ΔN isoforms which lack the N-terminal coding exons are transcribed from a downstream intronic promoter (Parsa et al. 1999;

Table 19.1 Markers of human corneal and limbal epithelial compartments

	LESCs[a]	TACs	Differentiated			
Marker	bLE	bCE	sLE	sCE	Function	References
Positive limbal epithelial markers						
Bmi1	Yes (rLESC)				SC self-renewal	Barbaro et al. (2007)
C/EBPδ	Yes (rLESC)				Cell-cycle arrest	Barbaro et al. (2007)
ΔNp63	Yes	[b]			Proliferation	Pellegrini et al. (2001)
ΔNp63α	Yes				Proliferation	Di Iorio et al. 2005
ABCG2	Yes				Cell-surface transport	Chen et al. (2004)
Integrin α9	Yes				ECM binding	Chen et al. (2004)
N-cadherin	Yes				Cell adhesion	Hayashi et al. (2007)
Nonspecific markers						
GDNF	Yes		Yes		Cell survival	Qi et al. (2008)
GFR α1	Yes		Yes		Cell survival	Qi et al. (2008)
Keratin 15	Yes		Yes		Structural	Yoshida et al. (2006)
Keratin 5/ Keratin 14	Yes		Yes		Structural	Schlotzer-Schrehardt and Kruse (2005)
Nestin			Yes	Yes	Intermediate filaments	Chen et al. (2004)
α-enolase	Yes	Yes	Yes	Yes	Metabolic enzyme	Zieske et al. (1992)
Negative limbal epithelial markers						
NGF receptor (p75NTR)		Yes	Yes	Yes	Survival/ differentiation	Chen et al. (2004)
Involucrin		Yes	Yes	Yes	Structural	Chen et al. (2004)
Keratin 3/ Keratin 12		Yes	Yes	Yes	Structural	Chen et al. (2004)
Connexin 43		Yes	Yes	Yes	Structural	Chen et al. (2004)
E-cadherin		Yes	Yes	Yes	Cell adhesion	Chen et al. (2004)

Abbreviations: bCE basal corneal epithelium; *bLE* basal limbal epithelium; *sLE* suprabasal limbal epithelium; *sCE* suprabasal corneal epithelium; *aLESC* active limbal epithelial stem cell; *rLESC* resting limbal epithelial stem cell

[a]LESCs are located in the basal limbal epithelium. Some markers identify a sub-set of LESCs in the peripheral basal limbal epithelium that are thought to be resting LESCs (shown here as rLESC). Other LESC markers may be expressed in both active and resting LESCs or this information may be unknown

[b]ΔNp63 is not expressed in the human basal corneal epithelium (Pellegrini et al. 2001) but it is expressed in the mouse (Collinson et al. 2002; Moore et al. 2002; Ramaesh et al. 2005) and rat (Moore et al. 2002)

Yang and McKeon 2000). Di Iorio et al. (2005) showed that ΔNp63α is present in the human basal limbal epithelium and that ΔNp63γ may be expressed at low levels in superficial cornea and limbus. Neither ΔNp63α nor ΔNp63β is present in resting (non-wounded) human corneas but all three ΔNp63 isoforms are expressed widely in the corneal epithelium after wounding. ΔNp63α has, therefore, been proposed as a marker of both quiescent and active LESCs (Pellegrini et al. 2001; Barbaro et al. 2007) but its expression pattern suggests it is also likely to be expressed in early TACs. In the mouse, unspecified ΔNp63 isoforms are detectable in the basal layer of the corneal epithelium at all times (Collinson et al. 2002; Moore et al. 2002; Ramaesh et al. 2005). In agreement with a role for ΔNp63α in LESCs, a recent report shows that individuals who are heterozygous for some p63 mutations have LESC deficiency (Di Iorio et al. 2012).

19.3.3.2 ABCG2

The multi-drug resistance gene ATP-binding cassette sub-family G member 2 (ABCG2) was identified in a subpopulation of bone marrow cells (known as the side population) that efluxed Hoechst 33342 nuclear dye and exhibited SC-like properties (Goodell et al. 1996). ABCG2 has been proposed as a stem cell marker in other systems. It may protect stem cells from exogenous damage by exporting toxic compounds from the cell. It is detectable in the basal limbal epithelium by immunohistochemistry or RT-PCR of different species, including mice (Krulova et al. 2008). Cells from limbal explants that express ABCG2 show high clonogenic potential; however, its broad expression pattern in basal limbus suggests that it is a marker of early TACs as well as LESCs (Budak et al. 2005; Chen et al. 2004; de Paiva et al. 2005; Krulova et al. 2008).

19.3.3.3 C/EBPδ and Bmi1

Bmi1 is a proposed oncogene that has been shown to promote neural stem cell self-renewal by repressing the cell cycle inhibitors $p16^{Ink4a}$ and $p19^{Arf}$ (Molofsky et al. 2005). It is essential for the maintenance of neural stem cells (distinguishing their self-renewal from progenitor cell proliferation) and is responsible for the maintenance of adult self-renewing haematopoietic cells (Molofsky et al. 2003; Park et al. 2003). Knockout mice have defects in haematopoiesis, skeletal patterning and neurological development (Van der Lugt et al. 1994; Park et al. 2003). Bmi1 is expressed in limbal epithelium side-population cells (see earlier) but is expressed at extremely low levels if at all in the central cornea (Umemoto et al. 2006).

The CCAAT enhancer-binding protein δ (C/EBPδ) is a transcription factor that has been shown to regulate the cell cycle in an epithelial specific manner by inducing a G0/G1 arrest (O'Rourke et al. 1999; Hutt et al. 2000). Barbaro et al. (2007) demonstrated co-localisation of C/EBPδ and Bmi1 in human peripheral basal limbal epithelial cells constituting ~10% of the total population. They show that this expression overlaps with ΔNp63α expression but that after wounding there is a distinct population of cells that are ΔNp63α positive but Bmi1 and C/EBPδ negative. They suggest that ΔNp63α marks quiescent LESCs, activated SCs and early TACs but C/EBPδ and Bmi1 only mark a subpopulation of LESCs

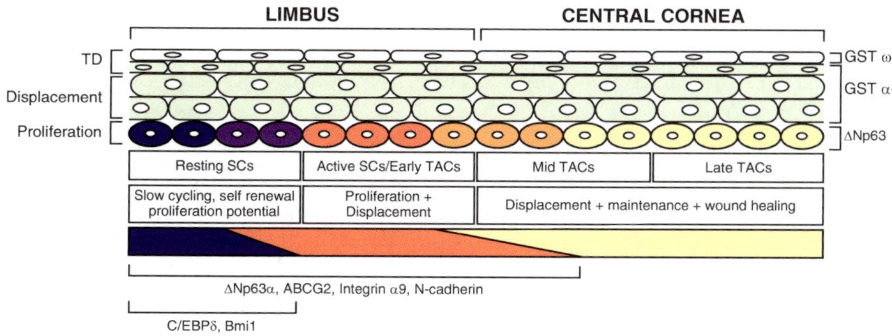

Fig. 19.2 A model of mammalian corneal regeneration and maintenance. Quiescent LESCs that express Bmi1, C/EBPδ (which sustain their quiescence and self-renewal) and ΔNp63α (which primes them for proliferation) reside at the periphery of the basal limbal epithelium (*dark shading*; Barbaro et al. 2007). They divide in response to injury or periodically during tissue maintenance to produce active LESCs and/or early TACs that down-regulate Bmi1 and C/EBPδ but continue to express ΔNp63α as they proliferate to produce TACs. Basal limbal cells also express integrin α9, N-cadherin and ABCG2 (Chen et al. 2004; Hayashi et al. 2007; Budak et al. 2005). In humans ΔNp63α is restricted to the basal limbal epithelium and although ΔNp63β and ΔNp63γ are expressed in the basal corneal epithelium during wound healing (not shown) no ΔNp63 isoforms are detectable in the resting basal corneal epithelium (Pellegrini et al. 2001; Di Iorio et al. 2005). However, in mice unspecified ΔNp63 isoforms are detectable in the resting basal corneal epithelium (Collinson et al 2002; Moore et al 2002; Ramaesh et al 2005). During normal maintenance TACs differentiate and migrate to the suprabasal layers. Here they become wing cells, down-regulate all p63 isoforms and begin to express GST-α4 in the intermediate layers (Norman et al. 2004). In the most superficial layer of the corneal epithelium they down-regulate GST-α4 and up-regulate GST-ω1 (Norman et al. 2004). It is not known whether GST-α4 and GST-ω1 are expressed in the limbal epithelium. The diagram is adapted from one published by Barbaro et al. (2007) but includes information from both human and mouse studies. Bmi1 and C/EBPδ have been identified in the human ocular surface but not yet in mouse whereas GST-α4 and GST-ω1 have been identified in mouse but not yet in human

that are quiescent during normal corneal maintenance (Fig. 19.2) but may be activated during wounding (also see Sect. 19.5). The balance of proliferative proteins (Bmi1 and ΔNp63α) and quiescence-associated factors (C/EBPδ) may allow slow cycling LESCs to exist in an intermediate state between quiescence and proliferation, poised to proliferate more rapidly if stimulated to do so by corneal injury.

19.3.3.4 Integrin α9

Integrins are membrane-bound glycoproteins important in cell–cell and cell–matrix adhesion. The corneo-limbal distribution of integrin α9 changes postnatally. At P7 in the mouse (a time when the cornea epithelium is still growing in size) integrin α9 positive cells are distributed throughout the epithelium but by P42 expression is restricted to the basal limbal epithelium (Pajoohesh-Ganji et al. 2004).

Table 19.2 Summary of differential expression studies

	Method	Species	Differentially regulated transcripts Group 1 (*n*)	Group 2 (*n*)	Group 3 (*n*)	References
1	PCR array	Human	LE only (21)	NA	CE only (24)	Nieto-Miguel et al. (2011)
2	Microarray	Human	LEC (95)	LE (169)	CE (1,237)	Kulkarni et al. (2010)
3	Microarray	Human	CE (93)	NA	CJE (211)	Turner et al. (2007)
4	Microarray	Human	FL (33)	LXE (22)	CE (11)	Figueira et al. (2007)
5	Microarray	Human	RA LE (499)	NA	N/SA LE (277)	Bian et al. (2010)
6	Microarray	Mouse	Basal LE (32[a])	NA	Basal CE (17[a])	Zhou et al. (2006)
7	SAGE	Mouse	P9 TC only (3,052)	Both (3,483)	Adult TC only (2,394)	Norman et al. (2004)
8	SAGE	Rat	LE only (759)	Both (2,292)	C only CE (844)	Adachi et al. (2006)
9	Microarray	Monkey	CJE (506)	LE (186)	CE (644)	Ding et al. (2008)
10	Microarray	Pig	LE SP (382)	NA	LE SP (296)	Akinci et al. (2009)
11	2D-Page/ MSpec	Human	LE (70)	NA	C CE (32)	Lyngholm et al. (2008b)

Abbreviations: LE limbal epithelium; *CE* corneal epithelium; *P9* postnatal day 9; *TC* total cornea; *C* central; *TL*, total limbus; *LEC* limbal epithelial crypt; *SP* side population; *CJE* conjunctival epithelium; *FL* foetal limbus; *FC* foetal cornea; *LXE* limbal explant epithelium; *LEP* limbal epithelial progenitors; *RA* rapidly adherent; *N/SA* none or slowly adherent. *2D-Page* 2-dimensional polyacrylamide gel electrophoresis; *Mspec* mass spectrometry

n = number of markers identified in each group

[a]Of the top 49 differentially expressed genes listed

The expression of integrin α9 is up-regulated during wound healing (see Sect. 19.5). This suggests that integrin α9 is a marker of LESCs and the early TAC populations (Stepp et al. 1995; Pal-Ghosh et al. 2004).

19.3.3.5 Identification of Markers Through mRNA and Proteomic Expression Studies

Table 19.2 summarises recent attempts to identify genes specific to corneo-limbal compartments using mRNA and proteomic expression techniques. A selection of the markers that have been independently validated from these studies is outlined in Table 19.3. There is a surprising lack of overlap with the markers described in Table 19.1 and also a lack of concordance between different studies. Crucially, the methods used to collect the tissue vary from laser capture dissection (Zhou et al. 2006) to selection of cells by their adhesive properties (Bian et al. 2010). Zhou et al. (2006) identified around 50 genes that are differentially expressed between basal limbal and basal corneal epithelium in the mouse. Further characterisation of randomly picked genes from the initial screen used semi-quantitative RT-PCR

Table 19.3 Top confirmed potential LESC markers from differential expression studies

Molecule	Species	Confirmed by	References
Epiregulin	Human	Microarray + IHC	Zhou et al. (2006)
Wnt-4	Human	Microarray + IHC	Figueira et al. (2007)
Keratin 14	Human	Microarray + IHC	Figueira et al. (2007)
P-Cadherin	Human	Microarray + IHC	Figueira et al. (2007)
SOD2	Human	Microarray + IHC + Mass Spec	Lyngholm et al. (2008a); Kulkarni et al. (2010)
Keratin 15	Human/Rat	Microarray + IHC + Mass Spec	Figueira et al. (2007); Akinci et al. (2009); Lyngholm et al. (2008a)
ID4	Human	2 independent expression studies	Takacs et al. (2009); Wolosin et al. (2000)
Spondin-1	Human	2 independent expression studies	Takacs et al. (2009); Zhou et al. (2006)
S100A8	Human	2 independent expression studies	Takacs et al. (2009); Lyngholm et al. (2008a)
Catenin-α2	Human	2 independent expression studies	Figueira et al. (2007); Zhou et al. (2006)
GPX2	Monkey/Pig	2 independent expression studies	Ding et al. (2008); Akinci et al. (2009)
Keratin 13	Rat/Monkey	2 independent expression studies	Adachi et al. (2006); Ding et al. (2008)

Abbreviations: IHC immunohistochemistry; *Mass Spec* mass spectrometry

and showed that *Ereg* (epiregulin), *Dach2* (dachshund 2) and *Sry* (sex determining region of chromosome Y) transcripts were found exclusively in limbal basal cells and not in corneal basal cells. Furthermore, diablo, cyclin M2 and multiple PDZ domain protein were expressed at significantly higher levels in corneal basal cells compared to limbal basal cells. Epiregulin-*LacZ* mice showed β-galactosidase-positive staining restricted to the basal limbal epithelium (Zhou et al. 2006). Because the basal limbal epithelium contains early TACs as well as LESCs it is not clear if these markers will be useful and further studies are required to determine if any are expressed specifically in LESCs.

19.3.4 The Limbal Niche

The microenvironment surrounding adult SCs plays a vital role in supporting, protecting and regulating the function of somatic and germ stem cells (Moore and Lemischka 2006). Evidence suggests that, in humans, LESCs reside in a specific anatomical location within the limbus known as the palisades of Vogt (Davanger and Evensen 1971; Townsend 1991; Dua and Azuara-Blanco 2000). These are subepithelial, vascularised papillae between which the epithelium projects downwards (Goldberg and Bron 1982). The limbus is a good candidate for a stem cell niche because it has blood vessels (which supply growth factors and nutrients), immune cells (such as Langerhans cells and T-lymphocytes which protect against pathogens) and, in some species, melanocytes which may protect SCs from ultraviolet light and potential DNA damage (Baum 1970) (Vantrappen et al. 1985; Boulton and Albon 2004; Li et al. 2007).

In humans three anatomical sites, associated with the palisades of Vogt, have been identified as probable LESC locations. These are known as the limbal epithelial crypts (LECs), limbal crypts (LCs) and focal stromal projections (FSPs). LECs were the first to be described (Dua et al. 2005) and are anatomical projections from the peripheral aspect of the limbal palisades that extend either radially into the conjunctival stroma or circumferentially along the limbus and contain epithelial cells that express K14 and the proposed LESC marker ABCG2 (see Sect. 19.3.3). Despite their similar names, LECs and LCs are anatomically distinct, with only the more peripheral LECs extending from the limbus into the conjunctival stroma. (It may be helpful to rename LECs and/or LCs because the similarity in their names is confusing and the original terminology has not always been used consistently.) Shortt et al. (2007b) describe LCs as downward focal projections of the basal limbal epithelium into the corneal limbal stroma and the FSPs as finger-like projections of stroma, containing a central blood vessel, extending upwards into the corneal limbal epithelium. LCs and FSPs are asymmetrically distributed, with the majority concentrated in the superior and inferior limbal quadrants (Shortt et al. 2007b). Other evidence also suggests the distribution of the limbal niche is asymmetrical in humans (Lauweryns et al. 1993; Wiley et al. 1991) and mice (Pajoohesh-Ganji et al. 2004). Epithelial cells in the LC/FSP regions are generally smaller, have a higher nuclear to cytoplasm ratio and express high levels of p63 and ABCG2. In addition, cells cultured from LC/FSP-rich regions of the limbus have a higher colony forming potential than those from non-LC/FSP-containing tissue (Shortt et al. 2007b). The palisades of Vogt are not present in most species and LCs have not been identified in mouse, rat or rabbit but have been identified in pig (Notara et al. 2011).

19.3.5 The Corneal Epithelial Stem Cell Hypothesis

Barrandon and colleagues propose that during normal homeostasis the corneal epithelium is maintained by stem cells residing throughout the basal layer of the corneal epithelium (Majo et al. 2008). Central to their argument was the observation that limbal and central corneal tissue, transplanted into immunocompromised mice, produced clones of centripetally migrating donor tissue only when the host corneal epithelium was removed. In these cases, both the corneal and limbal tissues behaved similarly. The authors proposed that LESCs were active during wound healing but played no role in normal maintenance of the unwounded corneal epithelium. Although Majo et al. (2008) do not address the question experimentally, they also suggest that, in contrast to the conventional LESC hypothesis, any movement in the

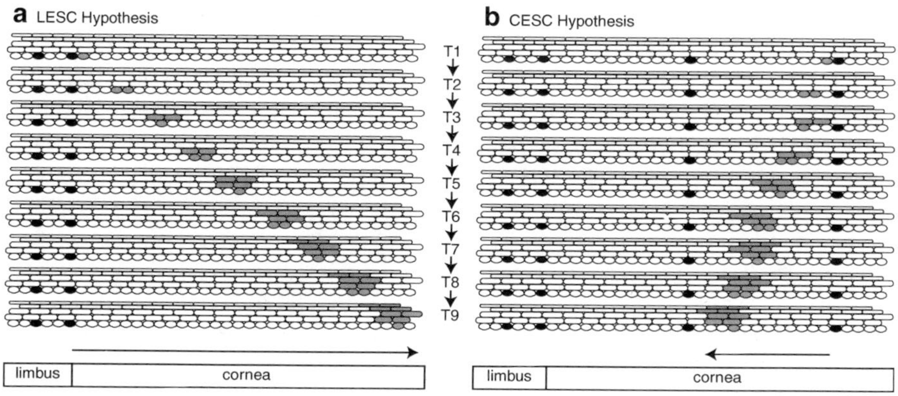

Fig. 19.3 Alternative hypotheses for maintaining the corneal epithelium. (**a**) Conventional Limbal Epithelial Stem Cell Hypothesis. Diagram representing the limbal epithelium and peripheral corneal epithelium with LESCs shown as *black* basal cells in the limbus. At time T1, one LESC produces a TAC (*shaded grey*). At later times (T2–T9) the daughter cells of this TAC are shown moving centripetally, dividing and, in some cases, leaving the basal layer, becoming non-mitotic and more differentiated and moving to the surface. For simplicity, only one clone of labelled cells is illustrated, only one TAC is shown dividing at each time point, basal and suprabasal cells move centripetally at the same rate and cells move vertically through the suprabasal layers to the surface relatively slowly. At T2, T4, T6 and T8 both daughter cells remain in the basal layer but at T3, T5, T7 and T9 both daughter cells move to the first suprabasal layer. (**b**) Alternative Corneal Epithelial Stem Cell Hypothesis (see Majo et al. 2008). Diagram representing the limbal epithelium and peripheral corneal epithelium with CESCs shown as *black* basal cells in the corneal epithelium and quiescent LESCs shown as *black* basal cells in the limbal epithelium. At time T1, one CESC produces a TAC (*shaded grey*). At times T2–T9 the daughter cells of this TAC are shown moving slowly centrifugally, dividing and, in some cases, leaving the basal layer. (TACs produced by CESCs in the corneal epithelium could either maintain the local area of corneal epithelium or move centrifugally (towards the limbus). For simplicity the clone of cells is shown moving centrifugally but relatively slowly)

corneal epithelium was more likely to be centrifugal than centripetal (Fig. 19.3). Using in vitro colony-forming assays, they also demonstrated that the replicative potential of the limbus and central cornea in culture was similar for many species (Majo et al. 2008). However, consistent with an earlier observation (Pellegrini et al. 1999), there was no evidence that the human central and intermediate cornea produced holoclones. For technical reasons, the holoclone assay currently does not work for the mouse cornea, so they were unable to investigate whether the mouse cornea produced holoclones.

The results reported by Majo et al. (2008) for the mouse are controversial and open to several interpretations but importantly they prompt re-evaluation of the conventional LESC hypothesis. Several authors have reported results that they consider support the new corneal epithelial stem cell (CESC) hypothesis for humans as well as mice but none is conclusive. A 12-h, ex vivo study of human corneal wound healing showed that the initial stage of wound closure was independent of the limbus (Chang et al. 2008). This experiment has been interpreted as supporting the CESC hypothesis (Sherwin 2009) but a 12-h study would be too short to evaluate the role of the limbus adequately. However, p63-positive, cultured clonogenic spheres can be isolated from the human central cornea as well as the limbus, albeit less efficiently, indicating that the human central cornea may contain some cells with progenitor potential (Chang et al. 2011). Other evidence suggests that the central cornea of both humans and rabbits has some ability to self-maintain. Thus, some patients with total LESC deficiency (with conjunctivalisation) retain central islands of normal corneal epithelium for several years (Dua et al. 2009). Similarly, the central cornea survives for months after surgical removal or isolation of the limbus in rabbits (Huang and Tseng 1991; Kawakita et al. 2011), but corneal integrity slowly degenerates and it is incapable of responding adequately when challenged by corneal wounding. These studies do not disprove the LESC hypothesis for humans or rabbits because although they imply that the central corneal epithelium has cells able to act as progenitors, either in culture or if LESCs are unable to maintain the corneal epithelium, they do not show these cells act as progenitors during normal homeostasis. Furthermore, autografts of ex vivo-labelled rabbit keratolimbal explants have shown that transplanted limbal tissue can colonise the corneal epithelium without wounding of the host tissue (Bradshaw et al. 1999). Thus, there is no reason to reject the LESC hypothesis for rabbits. If some progenitor TACs are able to assume the stem cell role in situations where LESCs are unable to maintain the corneal epithelium this would also explain the transplantation results obtained by Majo et al. (2008) with mice. This latent potential of some central corneal epithelial cells to proliferate in vitro and in transplantation experiments does not mean that they act as stem cells during normal tissue homeostasis in vivo. In younger individuals such cells could be equivalent to stem cells that persist from foetal stages (Tanifuji-Terai et al. 2006) and in older individuals these could be early TACs derived from LESCs.

The evidence for the existence of a LESC population in the mouse cornea is strong. In interpreting their results to formulate the new CESC hypothesis (Majo et al. 2008), the authors disregard convincing evidence that during normal maintenance of the unwounded cornea, corneal epithelial cells move centripetally not centrifugally (Kinoshita et al. 1981; Buck 1985; Nagasaki and Zhao 2003; Collinson et al. 2002; Endo et al. 2007; Mort et al. 2009), whereas conjunctival epithelial cells do not move significantly in either direction (Nagasaki and Zhao 2005). The distribution of individual β-galactosidase-positive radial stripes in the corneal epithelium of $KRT5^{LacZ/-}$ mosaic transgenic mice suggests the stripes extended centripetally from the limbus, so favouring the LESC hypothesis over the CESC hypothesis (Douvaras et al. 2012), but this experiment should be repeated with an inducible lineage marker to provide more conclusive evidence (see Sect. 19.8). Thus, while the transplantation experiment reported by Majo et al. (2008) has focused attention on the progenitor potential of the central corneal epithelium, current evidence favours the conventional LESC hypothesis over the new CESC hypothesis for the mouse as well as other species.

19.4 Maintaining Corneal Epithelial Homeostasis

19.4.1 Evolution of Ideas About Corneal Epithelial Maintenance (XYZ to LESC)

The corneal epithelium undergoes continuous cell renewal, which can replace the whole tissue every 2 weeks (Cenedella and Fleschner 1990) and corneal homeostasis must be carefully regulated to ensure a uniform epithelial thickness. Before the role of adult stem cells was understood, the corneal epithelium was thought to be a self-renewing tissue. Studies with ^{3}H-TdR labelling had shown that the basal corneal epithelial cells divided and moved vertically and were sloughed off within 3.5–7 days in mice and rats (Hanna and O'Brien 1960), so the corneal epithelium was thought to be maintained by proliferation of the basal epithelial cells. By 1982 evidence was beginning to emerge that corneal epithelial cells moved centripetally (Kinoshita et al. 1981; Buck 1982) and in 1983, Thoft and Friend proposed the 'X, Y, Z hypothesis' (Thoft and Friend 1983). This described the balance of proliferation, migration and cell loss by the simple formula $X + Y = Z$, where X was defined as 'the proliferation of basal epithelial cells', Y as 'the contribution to the cell mass by centripetal movement of peripheral cells' and Z as 'the epithelial cell loss from the surface'. It was not known whether the centripetal movement included movement of conjunctival cells across the limbus to the cornea as well as movement of cells from the peripheral cornea. It had been thought that conjunctival epithelium could transdifferentiate into corneal epithelium and was a source of new corneal epithelial cells during corneal wound healing (Friedenwald 1951). However, by 1986 it was becoming clear that this was not the case (Buck 1986) and the concept of centripetal movement was more firmly established (Buck 1985). Davanger and Evensen (1971) had already shown that the limbus produced cells that could colonise the corneal epithelium but the idea that the corneal epithelium was maintained by stem cells located in the limbus did not gain acceptance until later (Schermer et al. 1986; Cotsarelis et al. 1989).

Although the 'X, Y, Z' hypothesis of corneal epithelial maintenance has provided the framework for much of the subsequent research into corneal epithelial homeostasis leading to the LESC hypothesis, in its original form it does not take account of the role of stem cells in corneal maintenance. Nowadays the 'contribution to the cell mass by centripetal movement of peripheral cells' (Y) would be interpreted as production of basal corneal epithelial cells (TACs) by stem cells in the limbus (see Sect. 19.3.1). Production of cells by stem cells must precede proliferation of TACS, so if the X, Y, Z hypothesis was updated to allow for stem cells it could be redefined as: $Y_{SC} + X_{TAC} = Z_{L}$, where Y_{SC} is production of basal corneal epithelial cells by LESCs, X_{TAC} is proliferation of basal corneal epithelial TACs, as originally proposed (Thoft and Friend 1983), and Z_{L} is epithelial cell loss from the surface (Fig. 19.1). By including the qualification that the stem cells are located in the limbus, this revised formulation is equivalent to the LESC hypothesis.

Although the X, Y, Z hypothesis is still much cited it is now obsolete and has long been superseded by the LESC hypothesis, which currently provides the most likely explanation of how the corneal epithelium is maintained (as discussed in Sect. 19.3).

19.4.2 Centripetal Migration, Differentiation and Desquamation of Corneal Epithelial Cells

Centripetal migration in the mouse corneal epithelium has been demonstrated robustly both directly and indirectly by a number of experimental systems. Directly labelling superficial and wing cells with India ink allowed Buck (1985) to estimate a rate of centripetal migration of 17 μm/day. More recently Nagasaki and Zhao (2003) directly visualised centripetal migration using time-lapse imaging of GFP-bright clusters of corneal epithelial cells in ubiquitous-GFP reporter mice and estimated the rate of migration to be around 26 μm/day. Indirect methods using mouse genetic mosaics have also been very informative. When cells are randomly labelled with an X-linked *LacZ* transgene or a GFP-tagged lentiviral vector, a postnatal switch is observed from a randomly orientated mosaic pattern to a pattern of radial stripes in the corneal epithelium (Fig. 19.4). These patterns extend from the limbal region towards the central cornea and are evident from about 5 weeks postnatally. The stripes are thought to represent clones of centripetally migrating epithelial cells produced by LESCs that become active in the postnatal period prior to stripe formation (Collinson et al. 2002, 2004b; Mort et al. 2009; Endo et al. 2007). In most mosaic corneas, the stripes meet at a central clockwise or anticlockwise spiral (Collinson et al. 2002). The frequency of spirals increases with age, suggesting that they form stochastically but are stable once formed (Mort et al. 2009). Corneal stripes and spirals are also visualised in some human conditions (Bron 1973) and are unexplained. They may reflect failure of centripetally migrating cells to meet precisely at the centre of the tissue or could arise from small stochastic variations in movement of the epithelial sheet.

The causes of centripetal movement are still unknown and possibilities include (1) population pressure from the periphery due to production of new TACs by LESCs (Bron 1973; Sharma and Coles 1989; Wolosin et al. 2000); (2) preferential desquamation of epithelial cells from the central cornea drawing peripheral cells toward the centre (Lemp and Mathers 1989; Lavker et al. 1991); (3) chemotaxis involving a gradient resulting in either attraction to a central signal or repulsion from the periphery (e.g. limbal blood vessels) (Buck 1985); (4) stimulation by corneal nerves (Jones and Marfurt 1996); (5) response to endogenous electric currents due to ion flow in wounded and unwounded corneas (McCaig et al. 2005). Unlike the cornea, the conjunctival epithelium does not show any cell movement (Nagasaki and Zhao 2005) and mosaic patterns appear as patches rather than stripes (Collinson et al. 2002; Mort et al. 2009). This is consistent with the idea

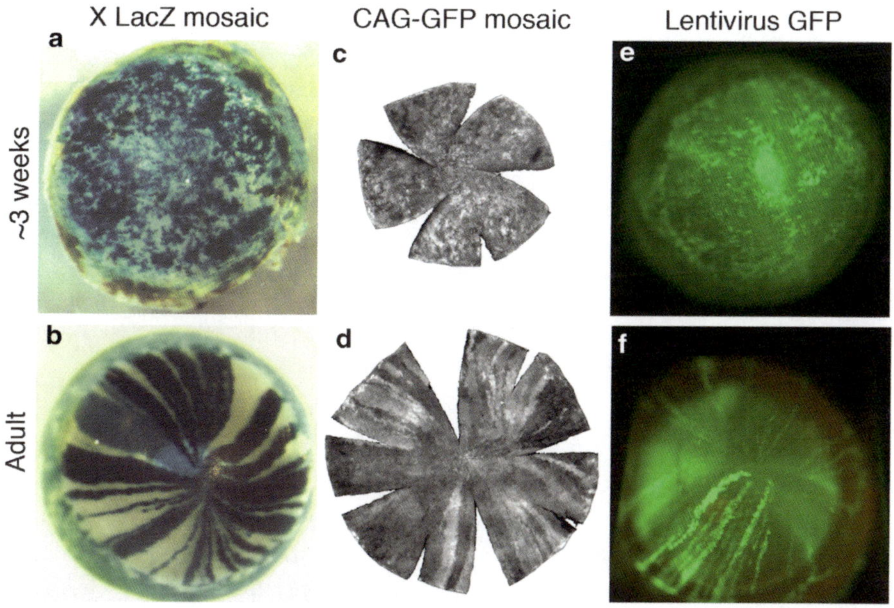

Fig. 19.4 Corneal epithelia of different types of mosaic mice show a transition from patches to radial stripes. (**a, b**) β-galactosidase staining in *XLacZ* X-inactivation mosaics (Collinson et al. 2002). (**c, d**) GFP fluorescence in CAG-GFP transgenic mosaics (Zhang et al. 2008). (**e, f**) GFP fluorescence in corneal epithelium after transfecting conceptuses with lentiviral vectors encoding green fluorescent protein (GFP) at embryonic day 9 or 10 (Endo et al. 2007). Images (**c, d**) are reproduced with permission of the authors and Molecular Vision. Images (**e, f**) are reproduced with permission of the authors and Molecular Therapy

that stem cells renewing this epithelium are distributed throughout the tissue (Nagasaki and Zhao 2005).

Differentiation of the corneal epithelium continues after birth. Evidence from expression of an ocular surface marker in the rat suggests that during development 'stem-like' cells reside throughout the basal layer of the corneal epithelium but become restricted to the limbus postnatally (Chung et al. 1992). This postnatal loss of stem cells from the central cornea is supported by analysis of mosaic mouse corneas which shows that transition to LESC-maintained corneal epithelium occurs between postnatal weeks 5 and 8 and the pattern is not fully mature until at least 10 weeks (Collinson et al. 2002; Mort et al. 2009). Changing postnatal expression of the keratins K12 and K14 suggests that the mouse corneal epithelium may not be fully mature until 3–6 months after birth (Tanifuji-Terai et al. 2006). Expression of integrin $\alpha9\beta1$ suggests that the limbus also matures progressively during the first 8 weeks after birth (Pajoohesh-Ganji et al. 2004). Care must be taken when extrapolating results from animal models to human because there are some notable species differences in corneal differentiation as well as more obvious differences of tissue size and life span. For example, mice have no *Krt3* (keratin 3) gene but keratin 5 is present in the mouse cornea (Byrne and Fuchs 1993; Lu et al. 2006; Ou et al. 2008). It has been suggested that K5 may pair with K12 in the cornea of mice and some other species with low K3 levels, including humans and dogs (Chaloin-Dufau et al. 1993; Hesse et al. 2004).

When TACs undergo their final division both daughter cells leave the basal layer together and begin the next stage of differentiation synchronously (Beebe and Masters 1996) and then move to the superficial layer within a few days (Hanna and O'Brien 1960). Few studies have been undertaken to measure the rate of epithelial shedding in mouse or human corneal epithelium. However, Ren and Wilson (1996), using both in vivo and in vitro measurements of corneal epithelial shedding rates in the rabbit, calculated that on average cells are shed at a rate of 5–15 cells/min from each cornea.

19.5 Corneal Epithelial Wound Healing and Homeostasis

Wound healing has been used as an experimental tool to challenge corneal homeostasis and, thus, has already been mentioned several times in this review. The nature of the corneal wound healing response depends on the type of injury and is regulated by molecules produced by the epithelium, stroma and lacrimal glands (Imanishi et al. 2000; Schultz et al. 1994; Wilson et al. 2001, 2003). Briefly, healing proceeds through three stages: an initial migratory stage to cover the wound, a proliferative stage to restore the epithelial thickness and a period of differentiation to restore the epithelium's complex structure (Suzuki et al. 2003). In essence this regenerative process mirrors normal corneal maintenance and has therefore been used as a tool to investigate this process. There are obvious caveats in the interpretation of results from wound healing studies, already alluded to elsewhere in this review (e.g. Sects. 19.3.3 and 19.3.5).

Several studies have shed light on the tissue hierarchy in the corneal epithelium by examining re-epithelialisation and have helped to identify compartments of putative resting SCs, activated SCs, TACs and differentiated cells. Cell-cycle double-labelling techniques have shown that the cornea uses three strategies to generate new epithelial cells in response to injury: (1) the activation of SCs to replenish the TAC population; (2) additional rounds of TAC proliferation and (3) the shortening of TAC cell cycle time to produce new tissue more quickly (Lehrer et al. 1998).

These cell cycle changes are accompanied by changes in the distribution of cells expressing proposed markers of LESCs and/or early TACs. Upon corneal wounding, early TACs (expressing ΔNp63α) migrate into the basal layer of the corneal epithelium from the limbus (Di Iorio et al. 2005). As outlined in Sect. 19.3.3, Barbaro and co-workers identified a subpopulation of basal limbal epithelial cells that are positive for C/EBPδ, Bmi1 and ΔNp63α. These are present during normal homeostasis and are thought to represent relatively quiescent LESCs. After wounding, some of these cells lose expression of C/EBPδ and Bmi1 but continue to express ΔNp63α as they proliferate, suggesting that some quiescent LESCs become active LESCs and early TACs, which move from the limbus to the cornea (Barbaro et al. 2007).

During healing of very large wounds, involving most of the cornea, integrin α9 expressing TACs move to repair the wound and this is accompanied by a depletion of integrin α9-expressing basal limbal epithelial cells (Pal-Ghosh et al. 2004). The loss of integrin α9 expression at the limbus after formation of large wounds seemed to correlate with the presence of goblet cells in the central cornea suggesting ingression of cells from the surrounding bulba-conjunctiva. Overall, this suggests that these very large corneal epithelial wounds place high demands on the stem cell reserve in the limbus, which may cause a depletion of stem cells and early TACs in the limbus and ingression of conjunctival cells.

19.6 Effects of Ageing on Corneal Epithelial Maintenance

Numerical analysis of striping patterns in corneas of adult X-inactivation mosaic mice (Fig. 19.4b) can give an indirect estimate of the number of coherent clones of LESCs maintaining the corneal epithelium (Collinson et al. 2002, 2004b; Mort et al. 2009). Although this is not equal to the number of active stem cells, it is useful for comparing dynamic patterns of SC maintenance in different experimental groups. The corrected number of radial stripes in the corneal epithelium declines from ~100 at 10 weeks postnatally to ~50 at 39 weeks, with no further decline up to 52 weeks (Collinson et al. 2002; Mort et al. 2009). This suggests that the number of coherent clones of LESCs maintaining the corneal epithelium decreases with age (Fig. 19.5), which could reflect either a loss of LESCs (Fig. 19.5b) or an increase in the proportion of quiescent LESCs (Fig. 19.5c). Alternatively, stochastic neutral drift in the distribution of LESCs may reduce the number of LESC clones but not the number of active LESCs (Fig. 19.5d). This possibility is suggested by lineage tracing studies which have identified similar coarsening of mosaic patterns derived from SC clones in the mouse testis (Nakagawa et al. 2007) and mouse intestine (Snippert et al. 2010; Lopez-Garcia et al. 2010; Klein and Simons 2011). Stochastic replacement of a SC clone by a neighbouring one could occur if SCs did not always divide asymmetrically to produce one SC and one TAC at each division, but sometimes divided symmetrically to produce two SCs or two TACs (Fig. 19.6). Similarly, the age-related decline in frequency of rare corneal epithelial stripes in mosaic $KRT5^{LacZ/-}$ transgenic mice (Douvaras et al. 2012) could reflect a decline in SC function with age or neutral drift in stem cell clones without an overall reduction in SC number. Other methods are required to investigate whether SC numbers decline with age.

Some observations on the human cornea and limbus support the notion that ageing affects stem cell numbers. The frequency of holoclones generated from skin of an infant was ten times greater than from an individual aged 64 (Barrandon and Green 1987) and a recent study showed that the efficiency of

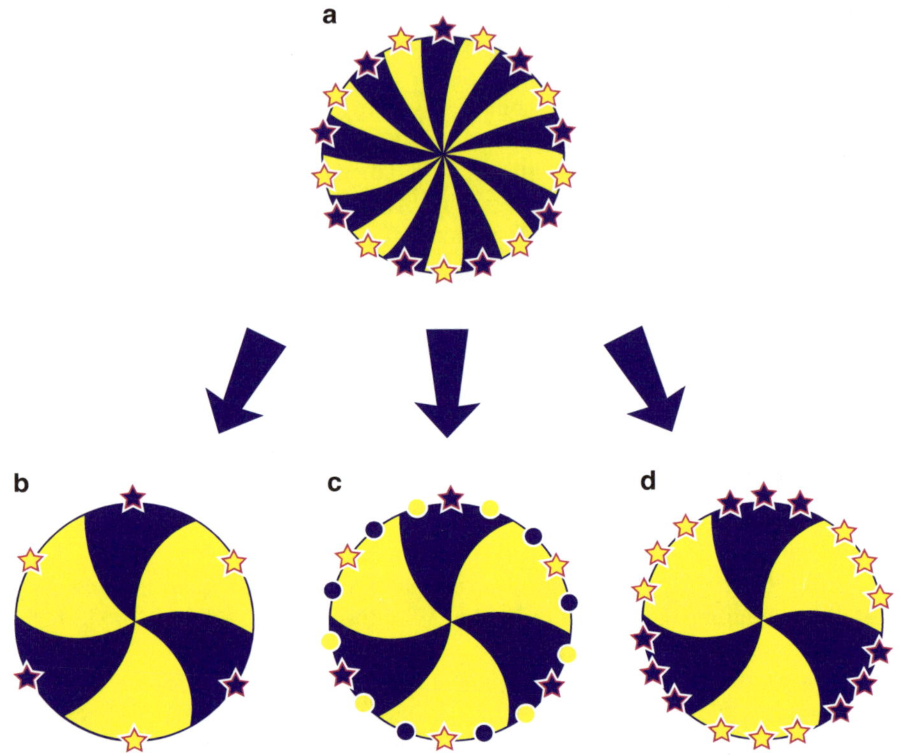

Fig. 19.5 Alternative explanations of age-related reduction in corrected stripe number in corneal epithelia of mouse X-inactivation mosaics. The mosaic corneal epithelium is shown as a disk with radial *dark* and *light* stripes produced by LESCs, represented by *dark* and *light stars*, at the edge of the cornea. (For purposes of illustration dark and light LESCs alternate around the circumference so all dark and light stripes are the same width.) *Dark* and *light circles* represent quiescent LESCs. (**a**) Full complement of active LESCs in young eye. (**b**) Reduced number of LESCs in older eye. (**c**) Increased proportion of quiescent LESCs in older eye. (**d**) Number of LESC clones declines (without reducing the number of LESCs) by stochastic neutral drift in the distribution of LESCs (also see Fig. 19.6). The corrected stripe number produced by quantitative analysis of X-inactivation mosaic patterns indicates the relative number of active LESC coherent clones rather than the actual or relative number of active LESCs (Collinson et al 2002; Mort et al 2009). Nevertheless, the possibilities illustrated in the figure still apply—i.e. stripe numbers can be reduced by stem cell loss, inactivation or redistribution. Adapted from Mort et al. (2009)

holoclone production from the human corneal limbus declined with age (Notara et al. 2012). Cells derived from younger donors (0–30 years old) had a significantly higher colony forming efficiency than cells from middle aged (30–60 years) and older donors (60–90 years). Other age-related changes noted for the human limbus include increased frequency of eyes without detectable palisades of Vogt (Zheng and Xu 2008), reduction in limbal crypts and focal stromal projections (see Sect. 19.3.4) in people over 60 years old (Notara et al. 2012)

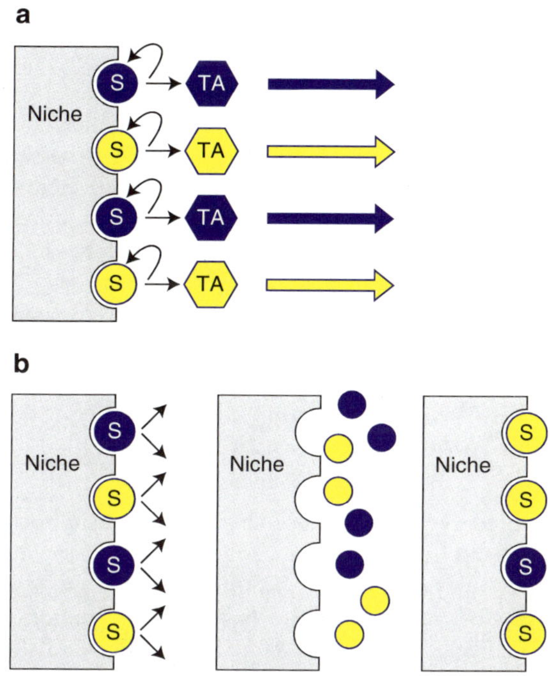

Fig. 19.6 Asymmetric and symmetric stem cell divisions. Stem cells (SCs) are represented by *dark* and *light circles* marked with the letter 'S' associated with a niche; transient (or transit) amplifying cells (TACs) are shown as *dark* and *light hexagons*, marked with the letters 'TA'. (**a**) Asymmetric division of stem cells always produces one stem cell and one TAC, which will divide a limited number of times to produce more differentiated cells. (**b**) Stem cells might also divide symmetrically to produce two equivalent daughter cells, which compete with neighbours to occupy limited space on the niche. Daughter cells securing contact with the niche would become stem cells and others would become TACs. This would provide scope for stochastic replacement of a SC by a daughter cell of a neighbouring SC. In the diagram, the *top dark* SC is lost and both daughter cells produced by the neighbouring light SC occupy the niche so the light SC clone expands

and an increase in average size of cells in the basal limbal epithelium (Zheng and Xu 2008). It has also been reported recently that the frequency of side-population cells (see discussion of ABCG2 marker in Sect. 19.3.3) declines with age and this could indicate an age-related decline in the number of SCs and early TACs (Chang et al. 2011).

Despite experimental evidence suggesting that stem cell function may decline with age, corneal epithelial maintenance appears to be remarkably robust and there is no evidence that ageing significantly compromises the integrity of the normal, healthy corneal epithelium. However, older mice of some strains are prone to corneal stromal mineralisation and some types of mouse cage bedding may trigger neovascularisation and keratitis in older mice (Smith et al. 2002).

19.7 Genetic Defects of Corneal Epithelial Maintenance

Corneal maintenance is impaired when LESCs are depleted by injury or disease (Daniels et al. 2001; Shortt et al. 2007a). When LESC deficiency only affects one eye, it can be treated with autografts of limbal tissue from the unaffected eye (Kenyon and Tseng 1989) or by transplanting cultured limbal epithelial cells (enriched for putative LESCs and early TACs) derived from the contralateral eye (Rama et al. 2010). LESC deficiency is also thought to underlie the corneal deterioration that occurs in people who are heterozygous for *PAX6* mutations that cause aniridia (discussed later) and those who are heterozygous for some p63 mutations (Di Iorio et al. 2012).

19.7.1 Aniridia-Related Keratopathy

Aniridia is an inherited eye disease caused by heterozygosity for a *PAX6* mutation ($PAX6^{+/-}$), which results in low levels of the PAX6 transcription factor. PAX6 expression is reduced in eyes of $PAX6^{+/-}$ aniridia patients and also in the abnormal corneal pannus tissue of patients with Stevens–Johnson syndrome (Li et al. 2008). Eyes of $PAX6^{+/-}$ aniridia patients develop abnormally and the cornea usually deteriorates progressively causing corneal opacity that sometimes leads to blindness (Mackman et al. 1979; Nelson et al. 1984; Nishida et al. 1995). This corneal deterioration is known as aniridic keratopathy or aniridia-related keratopathy (ARK) and features include irregular thickening of the peripheral corneal epithelium, in-growth of blood vessels from the limbus, associated with connective tissue (pannus), stromal scarring and accumulation of goblet cells within the corneal epithelium. Goblet cells are normally only found in the conjunctival epithelium (beyond the limbus) and their presence in the corneal epithelium has been interpreted as evidence for encroachment of conjunctival cells to compensate for poor LESC activity (Nishida et al. 1995). Limbal morphology is abnormal in $PAX6^{+/-}$ aniridia patients and the palisades of Vogt (putative LESC niche) are absent (Nishida et al. 1995). The conclusion that ARK is at least partly caused by LESC deficiency is supported by the clinical observation that transplantation treatment has much better success when transplanted tissue is healthy limbal and peripheral corneal tissue (kerato-limbal allograft) rather than central corneal tissue (penetrating keratoplasty) (Holland et al. 2003).

19.7.2 $Pax6^{+/-}$ *Mouse Aniridia Model*

Heterozygous $Pax6^{+/-}$ (e.g. $Pax6^{+/Sey\text{-}Neu}$) mice provide an excellent model of both the developmental eye defects seen in aniridia and the progressive corneal changes characteristic of ARK, although, unlike humans with aniridia, $Pax6^{+/-}$ mice also have small eyes (microphthalmia) (Ramaesh et al. 2003, 2005, 2006; Davis et al. 2003; Sivak et al. 2004; Leiper et al. 2006). It is clear that low levels of *Pax6* expression cause abnormal maintenance of the adult cornea in both mice and humans, but this may involve multiple mechanisms. The presence of goblet cells is consistent with LESC deficiency (Nishida et al. 1995) but this remains circumstantial because other explanations, such as abnormal differentiation, are possible (Ramaesh et al. 2005). Proliferation of TACs in the basal layer of the corneal epithelium is higher in *Pax6*$^{+/-}$ mice than in wild-type mice (Davis et al. 2003; Ramaesh et al. 2005) and this could be a secondary response to greater cell loss from a fragile epithelium (Davis et al. 2003) and/or reduced LESC function, rather than a direct effect of reduced Pax6 levels on cell proliferation.

Patterns visualised in the corneal epithelia of mosaic and chimeric mice provide clues about corneal epithelial maintenance. Wild-type (WT) *XLacZ*, X-inactivation mosaic mice exhibit radial striping patterns in the corneal epithelium, with stripes converging at a central spiral or midline and only very rarely do abnormal patterns occur (Fig. 19.7a–c). The pattern in *Pax6*$^{+/-}$ *XLacZ* mosaics is usually disorganised, implying that corneal epithelial cell movement is abnormal (Collinson et al. 2004a; Mort et al. 2011; Fig. 19.7d–f). However, it is unclear whether this is caused by some intrinsic movement defect or a response to chronic wounding of a thin and fragile corneal epithelium that diverts cells from their normal centripetal migration. Quantitative analysis implies that *Pax6*$^{+/-}$ mice have fewer active clones of LESCs maintaining the corneal epithelium than normal (Collinson et al. 2004a; Mort et al. 2011). However, it is unclear whether there are fewer active LESCs or if LESCs are simply clustered into larger clones in *Pax6*$^{+/-}$ eyes. Quantitative analysis of *Pax6*$^{+/-}$ $\leftrightarrow$ WT mouse chimeras showed that *Pax6*$^{+/-}$ cells were under-represented in the corneal epithelium of adults (Collinson et al. 2004a) but not in embryonic day (E) 16.5 foetuses (Collinson et al. 2001). This implies that *Pax6*$^{+/-}$ LESCs contribute less well to the adult corneal epithelium than WT cells, which indicates either that in these chimeras *Pax6*$^{+/-}$ LESCs are under-represented relative to WT LESCs or that they produce fewer TACs. Overall, the mosaic and chimera analyses are consistent with a mild LESC abnormality in the *Pax6*$^{+/-}$ mouse model of aniridia.

19.7.3 Other Examples of Abnormal Corneal Epithelial Maintenance in Mice

Although *Pax6*$^{+/-}$ *XLacZ* mosaic corneal patterns were mildly disrupted, other Pax6 defects are associated with more abnormal patterns. *Pax6*$^{+/Leca4}$ *XLacZ* mosaics (Mort et al. 2011) and *Pax6*$^{+/Sey\text{-}Neu}$ *Gli3*$^{+/XtJ}$ *XLacZ* double heterozygotes (Kucerova et al. 2012) both produced randomly orientated patterns of patches in the adults (Fig. 19.7g–j). This suggests that movement of cells from the limbus is severely reduced or absent and that the corneal epithelium may be maintaining itself, either because of LESC deficiency or a primary failure of centripetal cell

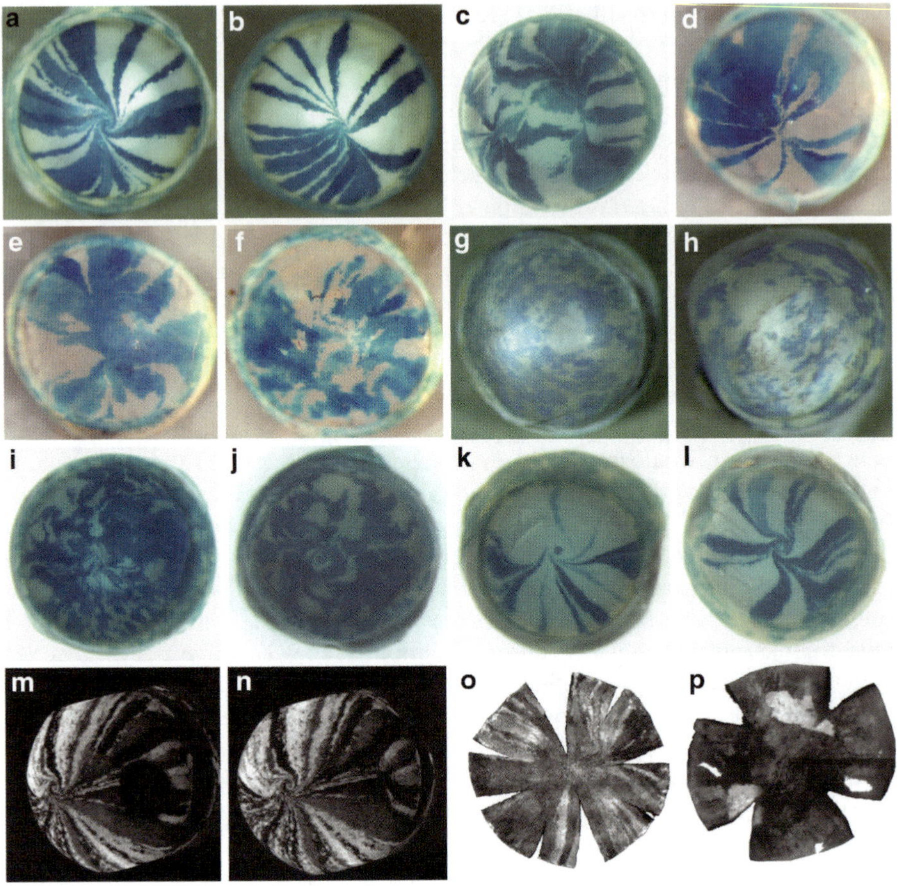

Fig. 19.7 Normal and abnormal mosaic patterns in the mouse corneal epithelium. (**a–l**) β-galactosidase staining in *XLacZ* X-inactivation mosaics; (**m–p**) GFP fluorescence in GFP transgenic mosaics. (**a**, **b**) Mosaic patterns in corneal epithelia of normal, wild-type *XLacZ* X-inactivation mosaic mice show radial stripes that meet at either a central spiral (**a**) or at the midline (**b**). (**c**) A rare wild-type *XLacZ* X-inactivation mosaic cornea with two points of stripe convergence. (**d–f**) *Pax6*$^{+/Sey\text{-}Neu}$ *XLacZ* mosaics may have normal striped patterns (**d**) but in most cases stripes are disrupted (**e**, **f**). (**g–j**) *Pax6*$^{+/Leca4}$ *XLacZ* mosaics (**g**, **h**) and double heterozygous *Pax6*$^{+/Sey\text{-}Neu}$ *Gli3*$^{+/XtJ}$ *XLacZ* mosaics (**i**, **j**) have a pattern of patches rather than stripes. Cornea (**j**) has a double spiral pattern. (**k**, **l**) *PAX77*$^{Tg/-}$ *XLacZ* mosaics over-express Pax6 but usually show a normal pattern of radial stripes. (**m**, **n**) Images from time-lapse confocal microscopy of healing of a 1 mm peripheral wound in a Y001deltaDRR (*PAX6-GFP*) mosaic corneal epithelium shows closure of the wound between 5.25 h. after wounding (**m**) and 18.75 h. (**n**) to form a second point of stripe convergence (Mort et al. 2009). (**o**, **p**) Flat whole mount *CAG-EGFP* corneas show radial stripes of GFP fluorescence (**o**) but adult homozygous *GFP-Dstn*$^{corn1/corn1}$ corneas (**p**) show globular and diffuse GFP patterns (Zhang et al. 2008). Images (**o**) and (**p**) are reproduced from Zhang et al. (2008) with permission of the authors and publishers of Molecular Vision. Others are from our own studies: **a**, **b**, **c**, **m**, **n** (Mort et al. 2009); **d** (Mort 2007); **e**, **f** (Collinson et al. 2004a); **i**, **j** (Kucerova et al. 2012); **g**, **h**, **k**, **l** unpublished

movement, as reported for *Dstn*$^{corn1/corn1}$ mosaics (Zhang et al. 2008), discussed below. Wild-type mice carrying the *PAX77* transgene (Schedl et al. 1996) over-express *Pax6* and show some corneal epithelial abnormalities characteristic of *Pax6*$^{+/-}$ heterozygotes (Dorà et al. 2008) but *PAX77*$^{Tg/-}$ *XLacZ* mosaics have normal radial stripes (Fig. 19.7k,l), implying centripetal movement is normal (Mort et al. 2011). Abnormal striped patterns also occur occasionally in wild-type *XLacZ* mosaics (Fig. 19.7c) and may be caused by healing of lateral wounds to generate a second point of stripe convergence during wound closure. Studies with GFP mosaics confirmed that lateral wounding can create a new convergence point and cell migration can reorientate after wounding (Fig. 19.7m,n; Mort et al. 2009).

In *Dstn*$^{corn1/corn1}$, CAG-EGFP mosaics, GFP radial stripes begin to emerge but their centripetal extension ceases at around 6 weeks of age and mosaic patterns become more globular (Zhang et al. 2008) (Fig. 19.7o,p). Cells in the corneal epithelium are stationary but proliferative (with regions of epithelial hyperplasia) and comprise a mixture of corneal epithelial and conjunctival epithelial phenotypes (including goblet cells in older mice), with BrdU label-retaining cells being distributed throughout the corneal epithelium instead of being confined to the limbus. The absence of cell movement suggests that the goblet cells arise by abnormal differentiation within the corneal epithelium, rather than by immigration from the conjunctiva. The authors proposed that the *Dstn*$^{corn1/corn1}$ corneal epithelium maintains itself with one or more populations of stem cells within the corneal epithelium that can produce K12-positive corneal epithelial cells, K8-positive conjunctiva-like epithelial cells and goblet cells (Zhang et al. 2008).

19.8 Research Directions for Identification of Stem Cells

As discussed in Sects. 19.2 and 19.3, the conventional view is that most adult tissues, including the corneal epithelium, are maintained by populations of long-lived, predominantly quiescent, SCs, residing in a specialised niche. These SCs show the hallmarks of being able to 'self-renew' (duplicate without losing developmental potential) and give rise to all differentiated cell types of the tissue or organ in which they reside (Potten and Loeffler 1990; He et al. 2009). In the case of the corneal epithelium, this SC population is unipotent generating only differentiated keratinocytes (and undifferentiated SCs). As discussed in the preceding sections the most plausible hypothesis to explain corneal epithelial maintenance at present is the LESC hypothesis.

A number of recent observations have forced stem cell biologists to re-appraise their interpretation of tissue stem cell hierarchies and may inform the current debate. For example, fast cycling populations of cells with stem cell or progenitor cell characteristics have been identified in mouse tail epidermis, the hair follicle epithelium and the small intestine. This evidence suggests either that quiescence is not a prerequisite for an adult SC population or that, in some tissues, TACs/progenitors are responsible for much of the normal maintenance (Clayton et al. 2007;

Barker et al. 2007; Jaks et al. 2008). Furthermore, evidence from the small intestine, spermatogonia and interfollicular epidermis suggest that some tissues may be maintained in a stochastic manner with competition between daughter cells to occupy the SC niche, rather than by rigid asymmetrical stem cell division (Lopez-Garcia et al. 2010; Klein et al. 2010; Li and Clevers 2010; Klein and Simons 2011); also see Fig. 19.6. In addition, SC-like potential can be 'unmasked' in hair follicle bulge cells that contribute to repair of the interfollicular epidermis despite the fact that they appear not to be involved during normal homoeostasis (Ito et al. 2005; Levy et al. 2007; Yu et al. 2008).

These studies suggest that mechanisms of stem cell maintenance may differ between tissues and that the interaction between the tissue 'environment' and the 'plasticity' of a particular cell population is key. Experimental approaches that modify the local environment (by wounding, transplantation or ex vivo expansion) may therefore induce SC-like behaviour in cells that do not normally contribute to homeostasis (Potten and Loeffler 1990). One current model suggests that in some tissues there may exist a rapidly cycling 'committed progenitor' (CP) responsible for the bulk of tissue maintenance as well as a population of quiescent or slow cycling SCs. These proposed CPs maintain the tissue and retain a degree of 'stemness' which enables them to act as an alternative source of stem cells in response to injury or physiological stress (Li and Clevers 2010; Klein and Simons 2011; Kaur and Potten 2011); however CPs have a lower proliferative potential than SCs. In tissues where this model fits the experimental data, identification of the relevant tissue-maintaining SC will require additional approaches. It is not yet known whether the corneal epithelium is maintained by such a combination of quiescent SCs and more active CPs. However it is intriguing that, for human cornea, Barbaro et al. (2007) identified a Bmi1-positive, C/EBPδ-positive and ΔNp63α-positive population (possibly a quiescent SC) that responds to injury and a separate Bmi1-negative, C/EBPδ-negative and ΔNp63α-positive population (possibly a CP) in uninjured limbus. It remains to be seen however, whether this potential quiescent SC population contributes to normal tissue maintenance or whether the potential CPs fulfil this role alone. Lineage tracing studies (see below) should resolve this.

What is clear is that, in order to identify the SC that is responsible for corneal epithelial maintenance, experimental approaches are required that seek to unpick the tissue hierarchy without altering the tissue 'environment' or unmasking latent 'plasticity'. To date the most useful and transparent techniques (in their execution if not in their interpretation) have been those that use genetic markers driven by the promoters of specific genes to demonstrate lineage, coupled with conditional diphtheria toxin-mediated cell ablation of those populations. For example, in the case of the small intestine, two potential SC populations have been identified. A slow cycling population that expresses *Bmi1* is located at the +4 position immediately above the base of the crypt (Sangiorgi and Capecchi 2008). The faster cycling crypt columnar cells (CBCs) are interspersed with the Paneth cells at the base of the crypt and expresses *Lgr5* (Barker et al. 2007). Lineage tracing has shown that both populations give rise to all mature intestinal epithelial cell lineages. However, whilst ablation of CBCs either genetically or by irradiation leads to expansion of the +4 cells to restore tissue maintenance (Tian et al. 2011; Yan et al. 2012), genetic ablation of the Bmi1-positive, +4 cell population results in widespread cell death and is not compatible with crypt maintenance (Sangiorgi and Capecchi 2008). The evidence suggests, therefore, that CBCs are responsible for the bulk of tissue maintenance, but that +4 cells sit at the top of the lineage as quiescent SCs able to divide in response to injury.

This demonstrates how the combination of cell lineage tracing and genetic cell ablation can be used as powerful tools to tease apart potential SC populations within a tissue. Combined with the identification of additional cell-specific markers and theoretical modelling of stem cell dynamics (Lopez-Garcia et al. 2010; Klein and Simons 2011), similar approaches are likely to be crucial in defining (or rejecting the existence of) SC/CP cell hierarchies in the corneal epithelium. The expression pattern of Bmi1 in the human cornea (Barbaro et al. 2007) and the availability of *Bmi1* transgenic mouse lines (Sangiorgi and Capecchi 2008) make the gene an obvious starting point for such an analysis. To unambiguously identify the corneal epithelium-maintaining stem cell population the following key points should be demonstrated: (1) The prospective SC population can be shown to populate the entire corneal epithelium though lineage analysis. (2) Ablation of the prospective SC population should result in corneal deterioration. Thus, although C/EBPδ and Bmi1 appear to be good candidates as markers of a subpopulation of quiescent LESCs in the human limbus, lineage tracing and ablation studies are required to determine whether these are the ultimate stem cells that maintain the corneal epithelium.

19.9 Conclusions

Ocular surface damage and disease are major causes of blindness and so are of great clinical significance. This has motivated the development of new clinical procedures, including transplantation of cells cultured from the corneal limbus to treat LESC deficiency (Rama et al. 2010; Shortt et al. 2011). In some respects clinical science is running ahead of the biological understanding and many fundamental basic research questions remain unresolved, not least the isolation and characterisation of the proposed LESC population.

The evidence reviewed here favours a model by which the corneal epithelium is maintained by LESCs. Evidence that describes a stem cell population within the corneal epithelium itself (CESCs) may represent the unmasking of latent ‘plasticity’ within the resident TAC population and does not support a fundamental difference between humans and the mouse. The balance of experimental evidence supports the conventional LESC hypothesis over the new CESC hypothesis for the mouse as well as other species. Nevertheless, there are significant differences among species, which need to be taken into account when extrapolating results from animal models to humans. Despite these differences the mouse has many advantages as an experimental animal and the availability of mouse genetic mutants, mosaics and various transgenic models provide powerful experimental resources with which to unravel the secrets of corneal epithelial maintenance and to model corneal epithelial defects.

There are currently some promising gene expression patterns that may open the door to reliable LESC biomarkers. Careful cell lineage and ablation studies combined with expression profiling will be required to validate these potential markers and unpick the tissue hierarchy responsible for corneal epithelial maintenance. With a better understanding of the basic biology, clinicians will be able to refine their existing therapies and achieve better clinical outcomes. Furthermore, they will gain better insight and a clearer understanding of the long-term efficacies of these still relatively new therapies.

Acknowledgements We thank Ronnie Grant for assistance with illustrations, Dr Alan W. Flake and Molecular Therapy for permission to reproduce images shown in Fig. 19.4e, f and Dr Takayuki Nagasaki and Molecular Vision for permission to reproduce images shown in Figs. 19.4c,d and 19.7o,p. We also thank Developmental Dynamics, BMC Developmental Biology, Investigative Ophthalmology & Vision Science and Molecular Vision for permission to reproduce our own previously published images included in Figs 19.4 and 19.7. We are grateful to the Wellcome Trust, Medical Research Council, Biotechnology and Biological Sciences Research Council, Fight for Sight (UK), the RS MacDonald Charitable Trust and the Royal College of Surgeons of Edinburgh for support for our own research.

References

Adachi W, Ulanovsky H, Li Y, Norman B, Davis J, Piatigorsky J (2006) Serial analysis of gene expression (SAGE) in the rat limbal and central corneal epithelium. Invest Ophthalmol Vis Sci 47(9):3801–3810. doi:10.1167/iovs.06-0216

Akinci MAM, Turner H, Taveras M, Wolosin JM (2009) Differential gene expression in the pig limbal side population: implications for stem cell cycling, replication, and survival. Invest Ophthalmol Vis Sci 50(12):5630–5638. doi:10.1167/iovs.09-3791

Barbaro V, Testa A, Di Iorio E, Mavilio F, Pellegrini G, De Luca M (2007) C/EBP delta regulates cell cycle and self-renewal of human limbal stem cells. J Cell Biol 177(6):1037–1049. doi:10.1083/jcb.200703003

Barker N, van Es JH, Kuipers J, Kujala P, van den Born M, Cozijnsen M, Haegebarth A, Korving J, Begthel H, Peters PJ, Clevers H (2007) Identification of stem cells in small intestine and colon by marker gene Lgr5. Nature 449(7165):1003–1007

Barrandon Y, Green H (1985) Cell size as a determinant of the clone forming ability of human keratinocytes. Proc Natl Acad Sci USA 82(16):5390–5394

Barrandon Y, Green H (1987) Three clonal types of keratinocyte with different capacities for multiplication. Proc Natl Acad Sci USA 84(8):2302–2306

Baum JL (1970) Melanocyte and Langerhans cell population of cornea and limbus in albino animal. Am J Ophthalmol 69(4):669–676

Beebe DC, Masters BR (1996) Cell lineage and the differentiation of corneal epithelial cells. Invest Ophthalmol Vis Sci 37(9):1815–1825

Bian F, Liu WB, Yoon KC, Lu R, Zhou N, Ma P, Pflugfelder SC, Li DQ (2010) Molecular signatures and biological pathway profiles of human corneal epithelial progenitor cells. Int J Biochem Cell Biol 42(7):1142–1153. doi:10.1016/j.biocel.2010.03.022

Boulton M, Albon J (2004) Stem cells in the eye. Int J Biochem Cell Biol 36(4):643–657. doi:10.1016/j.biocel.2003.10.013

Bradshaw JJ, Obritsch WF, Cho BJ, Gregerson DS, Holland EJ (1999) Ex vivo transduction of corneal epithelial progenitor cells using a retroviral vector. Invest Ophthalmol Vis Sci 40 (1):230–235

Braun KM, Watt FM (2004) Epidermal label-retaining cells: background and recent applications. J Investig Dermatol Symp Proc 9(3):196–201

Bron AJ (1973) Vortex patterns of corneal epithelium. Trans Ophthalmol Soc UK 93:455–472

Buck RC (1982) Hemidesmosomes of normal and regenerating mouse corneal epithelium. Virchows Archiv B 41(1–2):1–16

Buck RC (1985) Measurement of centripetal migration of normal corneal epithelial cells in the mouse. Invest Ophthalmol Vis Sci 26(9):1296–1299

Buck RC (1986) Ultrastructure of conjunctival epithelium replacing corneal epithelium. Curr Eye Res 5(2):149–159. doi:10.3109/02713688609015103

Budak MT, Alpdogan OS, Zhou MY, Lavker RM, Akinci MAM, Wolosin JM (2005) Ocular surface epithelia contain ABCG2-dependent side population cells exhibiting features associated with stem cells. J Cell Sci 118(8):1715–1724

Byrne C, Fuchs E (1993) Probing keratinocyte and differentiation specificity of the human K5 promoter in vitro and in transgenic mice. Mol Cell Biol 13(6):3176–3190

Cenedella RJ, Fleschner CR (1990) Kinetics of corneal epithelium turnover in vivo—studies of Lovastatin. Invest Ophthalmol Vis Sci 31(10):1957–1962

Chaloin-Dufau C, Pavitt I, Delorme P, Dhouailly D (1993) Identification of keratin-3 and keratin-12 in corneal epithelium of vertebrates. Epithelial Cell Biol 2(3):120–125

Chang CY, Green CR, McGhee CNJ, Sherwin T (2008) Acute wound healing in the human central corneal epithelium appears to be independent of limbal stem cell influence. Invest Ophthalmol Vis Sci 49(12):5279–5286. doi:10.1167/iovs.07-1260

Chang CYA, McGhee JJ, Green CR, Sherwin T (2011) Comparison of stem cell properties in cell populations isolated from human central and limbal corneal epithelium. Cornea 30 (10):1155–1162. doi:10.1097/ICO.0b013e318213796b

Chen Z, De Paiva CS, Luo LH, Kretzer FL, Pflugfelder SC, Li DQ (2004) Characterization of putative stem cell phenotype in human limbal epithelia. Stem Cells 22(3):355–366

Chung EH, Bukusoglu G, Zieske JD (1992) Localization of corneal epithelial stem-cells in the developing rat. Invest Ophthalmol Vis Sci 33(7):2199–2206

Clayton E, Doupe DP, Klein AM, Winton DJ, Simons BD, Jones PH (2007) A single type of progenitor cell maintains normal epidermis. Nature 446(7132):185–189

Collinson JM, Quinn JC, Buchanan MA, Kaufman MH, Wedden SE, West JD, Hill RE (2001) Primary defects in the lens underlie complex anterior segment abnormalities of the *Pax6* heterozygous eye. Proc Natl Acad Sci USA 98(17):9688–9693

Collinson JM, Morris L, Reid AI, Ramaesh T, Keighren MA, Flockhart JH, Hill RE, Tan SS, Ramaesh K, Dhillon B, West JD (2002) Clonal analysis of patterns of growth, stem cell activity, and cell movement during the development and maintenance of the murine corneal epithelium. Dev Dyn 224(4):432–440

Collinson JM, Chanas SA, Hill RE, West JD (2004a) Corneal development, limbal stem cell function, and corneal epithelial cell migration in the $Pax6^{+/-}$ mouse. Invest Ophthalmol Vis Sci 45(4):1101–1108

Collinson JM, Hill RE, West JD (2004b) Analysis of mouse eye development with chimeras and mosaics. Int J Dev Biol 48(8–9):793–804

Cotsarelis G, Cheng SZ, Dong G, Sun TT, Lavker RM (1989) Existence of slow-cycling limbal epithelial basal cells that can be preferentially stimulated to proliferate: implications on epithelial stem cells. Cell 57(2):201–209

Daniels JT, Dart JKG, Tuft SJ, Khaw PT (2001) Corneal stem cells in review. Wound Repair Regen 9(6):483–494

Davanger M, Evensen A (1971) Role of the pericorneal papillary structure in renewal of corneal epithelium. Nature 229:560–561

Davis J, Duncan MK, Robison WG, Piatigorsky J (2003) Requirement for Pax6 in corneal morphogenesis: a role in adhesion. J Cell Sci 116(11):2157–2167

de Paiva CS, Chen Z, Corrales RM, Pflugfelder SC, Li DQ (2005) ABCG2 transporter identifies a population of clonogenic human limbal epithelial cells. Stem Cells 23(1):63–73. doi:10.1634/stemcells.2004-0093

Di Iorio E, Barbaro V, Ruzza A, Ponzin D, Pellegrini G, De Luca M (2005) Isoforms of Delta Np63 and the migration of ocular limbal cells in human corneal regeneration. Proc Natl Acad Sci USA 102(27):9523–9528. doi:10.1073/pnas.0503437102

Di Iorio E, Kaye SB, Ponzin D, Barbaro V, Ferrari S, Böhm E, Nardiello P, Castaldo G, McGrath JA, Willoughby CE (2012) Limbal stem cell deficiency and ocular phenotype in ectrodactyly-ectodermal dysplasia-clefting syndrome caused by p63 mutations. Ophthalmology 119 (1):74–83. doi:10.1016/j.ophtha.2011.06.044

Ding ZH, Dong J, Liu J, Deng SX (2008) Preferential gene expression in the limbus of the vervet monkey. Mol Vis 14(240):2031–2041

Dorà N, Ou J, Kucerova R, Parisi I, West JD, Collinson JM (2008) PAX6 dosage effects on corneal development, growth and wound healing. Dev Dyn 237(5):1295–1306

Douvaras P, Webb S, Whitaker DA, Dorà N, Hill RE, Dorin JR, West JD (2012) Rare corneal clones in mice suggest an age-related decrease of stem cell activity and support the limbal epithelial stem cell hypothesis. Stem Cell Res 8:109–119

Du YD, Funderburgh ML, SundarRaj N, Funderburgh JL (2005) Multipotent stem cells in human corneal stroma. Stem Cells 23(9):1266–1275

Dua HS, Azuara-Blanco A (2000) Limbal stem cells of the corneal epithelium. Surv Ophthalmol 44(5):415–425. doi:10.1016/s0039-6257(00)00109-0

Dua HS, Shanmuganathan VA, Powell-Richards AO, Tighe PJ, Joseph A (2005) Limbal epithelial crypts: a novel anatomical structure and a putative limbal stem cell niche. Br J Ophthalmol 89(5):529–532

Dua HS, Miri A, Alomar T, Yeung AM, Said DG (2009) The role of limbal stem cells in corneal epithelial maintenance testing the dogma. Ophthalmology 116(5):856–863. doi:10.1016/j.ophtha.2008.12.017

Endo M, Zoltick PW, Chung DC, Bennett J, Radu A, Muvarak N, Flake AW (2007) Gene transfer to ocular stem cells by early gestational intraamniotic injection of lentiviral vector. Mol Ther 15(3):579–587

Figueira EC, Di Girolamo N, Coroneo MT, Wakefield D (2007) The phenotype of limbal epithelial stem cells. Invest Ophthalmol Vis Sci 48(1):144–156

Friedenwald JS (1951) Growth pressure and metaplasia of conjunctival and corneal epithelium. Doc Ophthalmol 5–6:184–192. doi:10.1007/bf00143661

Funderburgh ML, Du YQ, Mann MM, SundarRaj N, Funderburgh JL (2005) PAX6 expression identifies progenitor cells for corneal keratocytes. FASEB J 19(7):1371–1373

Gage PJ, Rhoades W, Prucka SK, Hjalt T (2005) Fate maps of neural crest and mesoderm in the mammalian eye. Invest Ophthalmol Vis Sci 46(11):4200–4208

Gaspar-Maia A, Alajem A, Polesso F, Sridharan R, Mason MJ, Heidersbach A, Ramalho-Santos J, McManus MT, Plath K, Meshorer E, Ramalho-Santos M (2009) Chd1 regulates open chromatin and pluripotency of embryonic stem cells. Nature 460(7257):863–868. doi:10.1038/nature08212

Goldberg MF, Bron AJ (1982) Limbal palisades of Vogt. Trans Am Ophthalmol Soc 80:155–171

Goodell MA, Brose K, Paradis G, Conner AS, Mulligan RC (1996) Isolation and functional properties of murine hematopoietic stem cells that are replicating in vivo. J Exp Med 183 (4):1797–1806. doi:10.1084/jem.183.4.1797

Hanna C, O'Brien JE (1960) Cell production and migration in the epithelial layer of the cornea. Arch Ophthal 64(4):536–539

Haustein J (1983) On the ultrastructure of the developing and adult mouse corneal stroma. Anat Embryol 168(2):291–305

Hayashi R, Yamato M, Sugiyama H, Sumide T, Yang J, Okano T, Tano Y, Nishida K (2007) N-cadherin is expressed by putative stem/progenitor cells and melanocytes in the human limbal epithelial stem cell niche. Stem Cells 25(2):289–296. doi:10.1634/stemcells.2006-0167

He SH, Nakada D, Morrison SJ (2009) Mechanisms of stem cell self-renewal. Ann Rev Cell Dev Biol 25:377–406. doi:10.1146/annurev.cellbio.042308.113248

Hesse M, Zimek A, Weber K, Magin TM (2004) Comprehensive analysis of keratin gene clusters in humans and rodents. Eur J Cell Biol 83(1):19–26

Holland EJ, Djalilian AR, Schwartz GS (2003) Management of aniridic keratopathy with keratolimbal allograft: a limbal stem cell transplantation technique. Ophthalmology 110 (1):125–130

Huang AJW, Tseng SCG (1991) Corneal epithelial wound healing in the absence of limbal epithelium. Invest Ophthalmol Vis Sci 32(1):96–105

Hutt JA, O'Rourke JP, DeWille J (2000) Signal transducer and activator of transcription 3 activates CCAAT enhancer-binding protein delta gene transcription in G(0) growth-arrested mouse mammary epithelial cells and in involuting mouse mammary gland. J Biol Chem 275 (37):29123–29131. doi:10.1074/jbc.M004476200

Imanishi J, Kamiyama K, Iguchi I, Kita M, Sotozono C, Kinoshita S (2000) Growth factors: importance in wound healing and maintenance of transparency of the cornea. Prog Retin Eye Res 19(1):113–129

Ito M, Liu YP, Yang ZX, Nguyen J, Liang F, Morris RJ, Cotsarelis G (2005) Stem cells in the hair follicle bulge contribute to wound repair but not to homeostasis of the epidermis. Nat Med 11(12):1351–1354. doi:10.1038/nm1328

Jaks V, Barker N, Kasper M, Van Es JH, Snippert HJ, Clevers H, Toftgard R (2008) Lgr5 marks cycling, yet long-lived, hair follicle stem cells. Nat Genet 40(11):1291–1299

Jones MA, Marfurt CF (1996) Sympathetic stimulation of corneal epithelial proliferation in wounded and nonwounded rat eyes. Invest Ophthalmol Vis Sci 37(13):2535–2547

Joyce NC (2003) Proliferative capacity of the corneal endothelium. Prog Retin Eye Res 22(3):359–389. doi:10.1016/s1350-9462(02)00065-4

Kaur P, Potten CS (2011) The interfollicular epidermal stem cell saga: sensationalism versus reality check. Exp Dermatol 20(9):697–702. doi:10.1111/j.1600-0625.2011.01338.x

Kawakita T, Higa K, Shimmura S, Tomita M, Tsubota K, Shimazaki J (2011) Fate of corneal epithelial cells separated from limbus in vivo. Invest Ophthalmol Vis Sci 52(11):8132–8137. doi:10.1167/iovs.11-7984

Kenyon KR, Tseng SCG (1989) Limbal autograft transplantation for ocular surface disorders. Ophthalmology 96(5):709–723

Kinoshita S, Friend J, Thoft RA (1981) Sex chromatin of donor corneal epithelium in rabbits. Invest Ophthalmol Vis Sci 21(3):434–441

Klein AM, Simons BD (2011) Universal patterns of stem cell fate in cycling adult tissues. Development 138(15):3103–3111. doi:10.1242/dev.060103

Klein AM, Nakagawa T, Ichikawa R, Yoshida S, Simons BD (2010) Mouse germ line stem cells undergo rapid and stochastic turnover. Cell Stem Cell 7(2):214–224. doi:10.1016/j.stem.2010.05.017

Krulova M, Pokorna K, Lencova A, Fric J, Zajicova A, Filipec M, Forrester JV, Holan V (2008) A rapid separation of two distinct populations of mouse corneal epithelial cells with limbal stem cell characteristics by centrifugation on Percoll gradient. Invest Ophthalmol Vis Sci 49 (9):3903–3908. doi:10.1167/iovs.08-1987

Kruse FE (1994) Stem cells and corneal epithelial regeneration. Eye 8:170–183

Kucerova R, Dorà N, Mort RL, Wallace K, Leiper LJ, Lowes C, Neves C, Walczysko P, Bruce F, Fowler PA, Rajnicek AM, McCaig CD, Zhao M, West JD, Collinson JM (2012) Interaction between hedgehog signalling and PAX6 dosage mediates maintenance and regeneration of the corneal epithelium. Mol Vis 18:139–150

Kulkarni BB, Tighe PJ, Mohammed I, Yeung AM, Powe DG, Hopkinson A, Shanmuganathan VA, Dua HS (2010) Comparative transcriptional profiling of the limbal epithelial crypt

demonstrates its putative stem cell niche characteristics. BMC Genomics 11:526. doi:10.1186/1471-2164-11-526
Lauweryns B, Vandenoord JJ, Devos R, Missotten L (1993) A new epithelial-cell type in the human cornea. Invest Ophthalmol Vis Sci 34(6):1983–1990
Lavker RM, Dong G, Cheng SZ, Kudoh K, Cotsarelis G, Sun TT (1991) Relative proliferative rates of limbal and corneal epithelia—implications of corneal epithelial migration, circadian rhythm, and suprabasally located DNA-synthesizing keratinocytes. Invest Ophthalmol Vis Sci 32(6):1864–1875
Lehrer MS, Sun TT, Lavker RM (1998) Strategies of epithelial repair: modulation of stem cell and transit amplifying cell proliferation. J Cell Sci 111:2867–2875
Leiper LJ, Walczysko P, Kucerova R, Ou JX, Shanley LJ, Lawson D, Forrester JV, McCaig CD, Zhao M, Collinson JM (2006) The roles of calcium signaling and ERK1/2 phosphorylation in a *Pax6*$^{+/-}$ mouse model of epithelial wound-healing delay. BMC Biol 4:27
Lemp MA, Mathers WD (1989) Corneal epithelial cell movement in humans. Eye 3:438–445
Levy V, Lindon C, Zheng Y, Harfe BD, Morgan BA (2007) Epidermal stem cells arise from the hair follicle after wounding. FASEB J 21(7):1358–1366. doi:10.1096/fj.06-6926com
Li LH, Clevers H (2010) Coexistence of quiescent and active adult stem cells in mammals. Science 327(5965):542–545. doi:10.1126/science.1180794
Li W, Hayashida Y, Chen YT, Tseng SCG (2007) Niche regulation of corneal epithelial stem cells at the limbus. Cell Res 17(1):26–36
Li W, Chen YT, Hayashida Y, Blanco G, Kheirkah A, He H, Chen SY, Liu CY, Tseng SCG (2008) Down-regulation of Pax6 is associated with abnormal differentiation of corneal epithelial cells in severe ocular surface diseases. J Pathol 214(1):114–122
Lopez-Garcia C, Klein AM, Simons BD, Winton DJ (2010) Intestinal stem cell replacement follows a pattern of neutral drift. Science 330(6005):822–825. doi:10.1126/science.1196236
Lu H, Zimek A, Chen J, Hesse M, Bussow H, Weber K, Magin TM (2006) Keratin 5 knockout mice reveal plasticity of keratin expression in the corneal epithelium. Eur J Cell Biol 85 (8):803–811
Lyngholm M, Hoyer PE, Vorum H, Nielsen K, Ehlers N, Mollgard K (2008a) Immunohistochemical markers for corneal stem cells in the early developing human eye. Exp Eye Res 87(2):115–121. doi:10.1016/j.exer.2008.05.004
Lyngholm M, Vorum H, Nielsen K, Ostergaard M, Honore B, Ehlers N (2008b) Differences in the protein expression in limbal versus central human corneal epithelium—a search for stem cell markers. Exp Eye Res 87(2):96–105. doi:10.1016/j.exer.2008.05.001
Mackman G, Brightbill F, Optiz J (1979) Corneal changes in aniridia. Am J Ophthalmol 8:497–502
Majo F, Rochat A, Nicolas M, Jaoude GA, Barrandon Y (2008) Oligopotent stem cells are distributed throughout the mammalian ocular surface. Nature 456(7219):250–255
McCaig CD, Rajnicek AM, Song B, Zhao M (2005) Controlling cell behavior electrically: current views and future potential. Physiol Rev 85(3):943–978. doi:10.1152/physrev.00020.2004
McGowan SL, Edelhauser HF, Pfister RR, Whikehart DR (2007) Stem cell markers in the human posterior limbus and corneal endothelium of unwounded and wounded corneas. Mol Vis 13(223–27):1984–2000
McKenna CC, Lwigale PY (2011) Innervation of the mouse cornea during development. Invest Ophthalmol Vis Sci 52(1):30–35. doi:10.1167/iovs.10-5902
Mills AA, Zheng BH, Wang XJ, Vogel H, Roop DR, Bradley A (1999) p63 is a p53 homologue required for limb and epidermal morphogenesis. Nature 398(6729):708–713
Molofsky AV, Pardal R, Iwashita T, Park IK, Clarke MF, Morrison SJ (2003) Bmi-1 dependence distinguishes neural stem cell self-renewal from progenitor proliferation. Nature 425 (6961):962–967. doi:10.1038/nature02060
Molofsky AV, He SH, Bydon M, Morrison SJ, Pardal R (2005) Bmi-1 promotes neural stem cell self-renewal and neural development but not mouse growth and survival by repressing the

p16^{Ink4a} and p19^{Arf} senescence pathways. Genes Dev 19(12):1432–1437. doi:10.1101/gad.1299505

Moore KA, Lemischka IR (2006) Stem cells and their niches. Science 311(5769):1880–1885

Moore JE, McMullen CBT, Mahon G, Adamis AP (2002) The corneal epithelial stem cell. DNA Cell Biol 21(5–6):443–451

Mort RA (2007) Stem cells function in the mouse corneal epithelium. PhD thesis, University of Edinburgh

Mort RL, Ramaesh T, Kleinjan DA, Morley SD, West JD (2009) Mosaic analysis of stem cell function and wound healing in the mouse corneal epithelium. BMC Dev Biol 9:4

Mort RL, Bentley AJ, Martin FL, Collinson JM, Douvaras P, Hill RE, Morley SD, Fullwood NJ, West JD (2011) Effects of aberrant Pax6 gene dosage on mouse corneal pathophysiology and corneal epithelial homeostasis. PLoS One 6(12):e28895. doi:doi:10.1371/journal.pone.0028895

Nagasaki T, Zhao J (2003) Centripetal movement of corneal epithelial cells in the normal adult mouse. Invest Ophthalmol Vis Sci 44(2):558–566

Nagasaki T, Zhao J (2005) Uniform distribution of epithelial stem cells in the bulbar conjunctiva. Invest Ophthalmol Vis Sci 46(1):126–132

Nakagawa T, Nabeshima YI, Yoshida S (2007) Functional identification of the actual and potential stem cell compartments in mouse spermatogenesis. Dev Cell 12(2):195–206

Nelson LB, Spaeth GL, Nowinski TS, Margo CE, Jackson L (1984) Aniridia—a review. Surv Ophthalmol 28(6):621–642. doi:10.1016/0039-6257(84)90184-x

Nieto-Miguel T, Calonge M, de la Mata A, Lopez-Paniagua M, Galindo S, de la Paz MF, Corrales RM (2011) A comparison of stem cell-related gene expression in the progenitor-rich limbal epithelium and the differentiating central corneal epithelium. Mol Vis 17(229–31):2102–2117

Nishida K, Kinoshita S, Ohashi Y, Kuwayama Y, Yamamoto S (1995) Ocular surface abnormalities in aniridia. Am J Ophthalmol 120(3):368–375

Norman B, Davis J, Piatigorsky J (2004) Postnatal gene expression in the normal mouse cornea by SAGE. Invest Ophthalmol Vis Sci 45(2):429–440. doi:10.1167/iovs.03-0449

Notara M, Schrader S, Daniels JT (2011) The porcine limbal epithelial stem cell niche as a new model for the study of transplanted tissue-engineered human limbal epithelial cells. Tissue Eng Part A 17(5–6):741–750. doi:10.1089/ten.TEA.2010.0343

Notara M, Shortt AJ, O'Callaghan AR, Daniels JT (2012) The impact of age on the physical and cellular properties of the human limbal stem cell niche. Age. Published online 18 January 2012. doi: 10.1007/s11357-011-9359-5

O'Rourke JP, Newbound GC, Hutt JA, DeWille J (1999) CCAAT/enhancer-binding protein delta regulates mammary epithelial cell G(0) growth arrest and apoptosis. J Biol Chem 274 (23):16582–16589. doi:10.1074/jbc.274.23.16582

Ou J, Walczysko P, Kucerova R, Rajnicek AM, McCaig CD, Zhao M, Collinson JM (2008) Chronic wound state exacerbated by oxidative stress in $Pax6^{+/-}$ aniridia-related keratopathy. J Pathol 215(4):421–430

Pajoohesh-Ganji A, Ghosh SP, Stepp MA (2004) Regional distribution of α9β1 integrin within the limbus of the mouse ocular surface. Dev Dynam 230(3):518–528

Pal-Ghosh S, Pajoohesh-Ganji A, Brown M, Stepp MA (2004) A mouse model for the study of recurrent corneal epithelial erosions: α9β1 integrin implicated in progression of the disease. Invest Ophthalmol Vis Sci 45(6):1775–1788

Park IK, Qian DL, Kiel M, Becker MW, Pihalja M, Weissman IL, Morrison SJ, Clarke MF (2003) Bmi-1 is required for maintenance of adult self-renewing haematopoietic stem cells. Nature 423(6937):302–305. doi:10.1038/nature01587

Parsa R, Yang A, McKeon F, Green H (1999) Association of p63 with proliferative potential in normal and neoplastic human keratinocytes. J Invest Dermatol 113(6):1099–1105. doi:10.1046/j.1523-1747.1999.00780.x

Pellegrini G, Golisano O, Paterna P, Lambiase A, Bonini S, Rama P, De Luca M (1999) Location and clonal analysis of stem cells and their differentiated progeny in the human ocular surface. J Cell Biol 145(4):769–782

Pellegrini G, Dellambra E, Golisano O, Martinelli E, Fantozzi I, Bondanza S, Ponzin D, McKeon F, De Luca M (2001) p63 identifies keratinocyte stem cells. Proc Natl Acad Sci USA 98 (6):3156–3161

Potten CS, Loeffler M (1990) Stem cells: attributes, cycles, spirals, pitfalls and uncertainties—lessons for and from the crypt. Development 110(4):1001–1020

Pratt T, Sharp L, Nichols T, Price DJ, Mason JO (2000) Embryonic stem cells and transgenic mice ubiquitously expressing a tau-tagged green fluorescent protein. Dev Biol 228(1):19–28

Qi H, Li DQ, Shine HD, Chen Z, Yoon KC, Jones DB, Pflugfelder SC (2008) Nerve growth factor and its receptor TrkA serve as potential markers for human corneal epithelial progenitor cells. Exp Eye Res 86(1):34–40. doi:10.1016/j.exer.2007.09.003

Rama P, Matuska S, Paganoni G, Spinelli A, De Luca M, Pellegrini G (2010) Limbal stem-cell therapy and long-term corneal regeneration. N Engl J Med 363(2):147–155. doi:10.1056/NEJMoa0905955

Ramaesh T, Collinson JM, Ramaesh K, Kaufman MH, West JD, Dhillon B (2003) Corneal abnormalities in *Pax6*$^{+/-}$ small eye mice mimic human aniridia-related keratopathy. Invest Ophthalmol Vis Sci 44(5):1871–1878

Ramaesh T, Ramaesh K, Collinson JM, Chanas SA, Dhillon B, West JD (2005) Developmental and cellular factors underlying corneal epithelial dysgenesis in the *Pax6*$^{+/-}$ mouse model of aniridia. Exp Eye Res 81(2):224–235

Ramaesh T, Ramaesh K, Leask R, Springbett A, Riley SC, Dhillon B, West JD (2006) Increased apoptosis and abnormal wound-healing responses in the heterozygous *Pax6*$^{+/-}$ mouse cornea. Invest Ophthalmol Vis Sci 47(5):1911–1917

Ren HW, Wilson G (1996) The cell shedding rate of the corneal epithelium—a comparison of collection methods. Curr Eye Res 15(10):1054–1059

Romano AC, Espana EM, Yoo SH, Budak MT, Wolosin JM, Tseng SCG (2003) Different cell sizes in human limbal and central corneal basal epithelia measured by confocal microscopy and flow cytometry. Invest Ophthalmol Vis Sci 44(12):5125–5129. doi:10.1167/iovs.03-0628

Sangiorgi E, Capecchi MR (2008) Bmi1 is expressed in vivo in intestinal stem cells. Nat Genet 40 (7):915–920

Schedl A, Ross A, Lee M, Engelkamp D, Rashbass P, van Heyningen V, Hastie ND (1996) Influence of *Pax6* gene dosage on development—overexpression causes severe eye abnormalities. Cell 86(1):71–82

Schermer A, Galvin S, Sun TT (1986) Differentiation-related expression of a major 64k corneal keratin in vivo and in culture suggests limbal location of corneal epithelial stem cells. J Cell Biol 103(1):49–62

Schlotzer-Schrehardt U, Kruse FE (2005) Identification and characterization of limbal stem cells. Exp Eye Res 81(3):247–264

Schultz G, Khaw PT, Oxford K, Macauley S, Vansetten G, Chegini N (1994) Growth factors and ocular wound healing. Eye 8:184–187

Sharma A, Coles WH (1989) Kinetics of corneal epithelial maintenance and graft loss—a population balance model. Invest Ophthalmol Vis Sci 30(9):1962–1971

Sherwin T (2009) A new niche for the corneal epithelial stem cell. Clin Exp Ophthalmol 37 (7):644–645. doi:10.1111/j.1442-9071.2009.02138.x

Shortt AJ, Secker GA, Limb GA, Khaw PT, Daniels JT (2007a) Transplantation of ex vivo cultured limbal epithelial stem cells: a review of techniques and clinical results. Surv Ophthalmol 52(5):483–502. doi:10.1016/j.survophthal.2007.06.013

Shortt AJ, Secker GA, Munro PM, Khaw PT, Tuft SJ, Daniels JT (2007b) Characterization of the limbal epithelial stem cell niche: novel imaging techniques permit in vivo observation and targeted biopsy of limbal epithelial stem cells. Stem Cells 25(6):1402–1409. doi:10.1634/stemcells.2006-0580

Shortt AJ, Tuft SJ, Daniels JT (2011) Corneal stem cells in the eye clinic. Br Med Bull 100(1): 209–225. doi:10.1093/bmb/ldr041

Sivak JM, West-Mays JA, Yee A, Williams T, Fini ME (2004) Transcription factors Pax6 and AP-2 alpha interact to coordinate corneal epithelial repair by controlling expression of matrix metalloproteinase gelatinase B. Mol Cell Biol 24(1):245–257

Smith RS, Sundberg JP, John SWM (2002) The anterior segment and ocular adnexae. In: Smith RS (ed) Systematic evaluation of the mouse eye. Anatomy, pathology and biomethods. CRC Press, Boca Raton, FL, pp 1–23

Snippert HJ, van der Flier LG, Sato T, van Es JH, van den Born M, Kroon-Veenboer C, Barker N, Klein AM, van Rheenen J, Simons BD, Clevers H (2010) Intestinal crypt homeostasis results from neutral competition between symmetrically dividing Lgr5 stem cells. Cell 143(1):134–144. doi:10.1016/j.cell.2010.09.016

Stepp MA, Zhu L, Sheppard D, Cranfill RL (1995) Localized distribution of α9 integrin in the cornea and changes in expression during corneal epithelial cell differentiation. J Histochem Cytochem 43(4):353–362

Suzuki K, Saito J, Yanai R, Yamada N, Chikama T, Seki K, Nishida T (2003) Cell-matrix, and cell-cell interactions during corneal epithelial wound healing. Prog Retin Eye Res 22 (2):113–133

Takacs L, Toth E, Berta A, Vereb G (2009) Stem cells of the adult cornea: from cytometric markers to therapeutic applications. Cytometry Part A 75A(1):54–66

Tani H, Morris RJ, Kaur P (2000) Enrichment for murine keratinocyte stem cells based on cell surface phenotype. Proc Natl Acad Sci USA 97(20):10960–10965. doi:10.1073/pnas.97.20.10960

Tanifuji-Terai N, Terai K, Hayashi Y, Chikama TI, Kao WWY (2006) Expression of keratin 12 and maturation of corneal epithelium during development and postnatal growth. Invest Ophthalmol Vis Sci 47(2):545–551

Thoft RA, Friend J (1983) The X, Y, Z hypothesis of corneal epithelial maintenance. Invest Ophthalmol Vis Sci 24(10):1442–1443

Tian H, Biehs B, Warming S, Leong K, Rangell L, Klein O, de Sauvage F (2011) A reserve stem cell population in small intestine renders Lgr5-positive cells dispensable. Nature 478 (7368):255–259. doi:10.1038/nature10408

Townsend W (1991) The limbal palisades of Vogt. Trans Am Ophthalmol Soc 89:721–756

Tseng SCG (1989) Concept and application of limbal stem cells. Eye 3:141–157

Tumbar T, Guasch G, Greco V, Blanpain C, Lowry WE, Rendl M, Fuchs E (2004) Defining the epithelial stem cell niche in skin. Science 303(5656):359–363

Turner HC, Budak MT, Akinci MAM, Wolosin JM (2007) Comparative analysis of human conjunctival and corneal epithelial gene expression with oligonucleotide microarrays. Invest Ophthalmol Vis Sci 48(5):2050–2061. doi:10.1167/iovs.06-0998

Umemoto T, Yamato M, Nishida K, Yang J, Tano Y, Okano T (2006) Limbal epithelial side-population cells have stem cell-like properties, including quiescent state. Stem Cells 24 (1):86–94. doi:10.1634/stemcells.2005-0064

Van der Lugt NMT, Domen J, Linders K, Vanroon M, Robanusmaandag E, Teriele H, Vandervalk M, Deschamps J, Sofroniew M, Vanlohuizen M, Berns A (1994) Posterior transformation, neurological abnormalities, and severe hematopoietic defects in mice with a targeted deletion of the Bmi-1 protooncogene. Genes Dev 8(7):757–769. doi:10.1101/gad.8.7.757

Vantrappen L, Geboes K, Missotten L, Maudgal PC, Desmet V (1985) Lymphocytes and Langerhans cells in the normal human cornea. Invest Ophthalmol Vis Sci 26(2):220–225

Wiley L, Sundarraj N, Sun TT, Thoft RA (1991) Regional heterogeneity in human corneal and limbal epithelia—an immunohistochemical evaluation. Invest Ophthalmol Vis Sci 32(3):594–602

Wilson SE, Mohan RR, Ambrosio R, Hong JW, Lee JS (2001) The corneal wound healing response: cytokine-mediated interaction of the epithelium, stroma, and inflammatory cells. Prog Retin Eye Res 20(5):625–637

Wilson SE, Netto M, Ambrosio R (2003) Corneal cells: chatty in development, homeostasis, wound healing, and disease. Am J Ophthalmol 136(3):530–536

Wolosin JM, Xiong XL, Schutte M, Stegman Z, Tieng A (2000) Stem cells and differentiation stages in the limbo-corneal epithelium. Prog Retin Eye Res 19(2):223–255. doi:10.1016/s1350-9462(99)00005-1

Yan KS, Chia LA, Li XN, Ootani A, Su J, Lee JY, Su N, Luo YL, Heilshorn SC, Amieva MR, Sangiorgi E, Capecchi MR, Kuo CJ (2012) The intestinal stem cell markers Bmi1 and Lgr5 identify two functionally distinct populations. Proc Natl Acad Sci USA 109(2):466–471. doi:10.1073/pnas.1118857109

Yang A, McKeon F (2000) P63 and p73: P53 mimics, menaces and more. Nat Rev Mol Cell Biol 1(3):199–207. doi:10.1038/35043127

Yang AN, Kaghad M, Wang YM, Gillett E, Fleming MD, Dotsch V, Andrews NC, Caput D, McKeon F (1998) p63, a p53 homolog at 3q27-29, encodes multiple products with transactivating, death-inducing, and dominant-negative activities. Mol Cell 2(3):305–316. doi:10.1016/s1097-2765(00)80275-0

Yoshida S, Shimmura S, Kawakita T, Miyashita H, Den S, Shimazaki J, Tsubota K (2006) Cytokeratin 15 can be used to identify the limbal phenotype in normal and diseased ocular surfaces. Invest Ophthalmol Vis Sci 47(11):4780–4786

Yu BD, Mukhopadhyay A, Wong C (2008) Skin and hair: models for exploring organ regeneration. Hum Mol Genet 17:R54–R59. doi:10.1093/hmg/ddn086

Zhang W, Zhao J, Chen L, Urbanowicz MM, Nagasaki T (2008) Abnormal epithelial homeostasis in the cornea of mice with a destrin deletion. Mol Vis 14:1929–1939

Zheng TY, Xu JJ (2008) Age-related changes of human limbus on in vivo confocal microscopy. Cornea 27(7):782–786

Zhou MY, Li XM, Lavker RM (2006) Transcriptional profiling of enriched populations of stem cells versus transient amplifying cells—a comparison of limbal and corneal epithelial basal cells. J Biol Chem 281(28):19600–19609. doi:10.1074/jbc.M600777200

Zieske JD (2004) Corneal development associated with eyelid opening. Int J Dev Biol 48 (8–9):903–911. doi:10.1387/ijdb.041860jz

Zieske JD, Bukusoglu G, Yankauckas MA, Wasson ME, Keutmann HT (1992) α-Enolase is restricted to basal cells of stratified squamous epithelium. Dev Biol 151(1):18–26. doi:10.1016/0012-1606(92)90209-y

Zieske JD, Francesconi CM, Guo XQ (2004) Cell cycle regulators at the ocular surface. Exp Eye Res 78(3):447–456. doi:10.1016/s0014-4835(03)00205-7

Chapter 20
Mouse-Induced Pluripotent Stem Cells

Hui Zhang and Ying Jin

20.1 Brief Introduction of Mouse-Induced Pluripotent Stem Cells

Pluripotent stem cells belong to a unique population of cells, which can self-renew indefinitely and have the potential to give rise to all cell types in an organism. Generally, pluripotent stem cells include embryonic stem cells (ESCs), embryonic germ cells (EGCs), and embryonal carcinoma cells (ECCs) (Boiani and Scholer 2005). In 2006, a new type of pluripotent stem cells, induced pluripotent stem cells (iPSCs), was established by direct reprogramming of differentiated somatic cells (Takahashi and Yamanaka 2006). The unique properties of iPSCs make them extremely attractive in regenerative medicine, drug screening, disease modeling, as well as study of developmental biology (Cyranoski 2008; Maherali and Hochedlinger 2008; Nishikawa et al. 2008).

Based on the hypothesis that factors playing important roles in the establishment and maintenance of pluripotency also play crucial roles in somatic cell reprograming, Yamanaka and Takahashi overexpressed 24 pluripotency-associated genes by a retroviral expression vector in mouse embryonic and adult fibroblasts. Interestingly, they found that a small fraction of fibroblasts in culture dishes acquired ESC-like properties and named the reprogrammed cells as iPSCs. After further narrowing down, they could generate iPSCs with genes encoding four transcription factors: *Oct4*, *Sox2*, *Klf4*, and *c-Myc* (Takahashi and Yamanaka 2006). This was the first time in history that differentiated somatic cells were reverted to an undifferentiated state directly by defined transcriptional factors in vitro and the technique carries huge application potential in regenerative medicine. It caused a tremendous sensation in both public and scientific circles. Although the first generation of iPSCs generated by

H. Zhang • Y. Jin (✉)
Key Laboratory of Stem Cell Biology, Institute of Health Sciences, Shanghai Jiao Tong University School of Medicine/Shanghai Institutes of Biological Sciences, CAS, Shanghai 200025, China
e-mail: yjin@sibs.ac.cn

J.Z. Kubiak (ed.), *Mouse Development*, Results and Problems in Cell Differentiation 55, DOI 10.1007/978-3-642-30406-4_20,

Yamanaka and colleagues resembled ESCs in many aspects, they differed from ESCs in that alive chimera were not born when iPSCs were injected into the blastocyst. In the following 2007, Jaenisch group further optimized the technique and generated mouse iPSCs that were epigenetically and developmentally indistinguishable from embryo-derived ESCs. The second generation of mouse iPSCs could produce chimera and were germline competent (Meissner et al. 2007). In 2008, human iPSCs were successfully generated by three groups independently through transduction of defined factors into human differentiated somatic cells, a step forward to the application of iPSC technology in regenerative medicine. Similar to mouse iPSCs, human iPSCs are similar to human ESCs in terms of the ability to self-renew and differentiate into all cell types of the three germ layers (Meissner et al. 2007; Takahashi et al. 2007; Yu et al. 2007). Finally, the debate of whether mouse iPSCs possess the same developmental potential as mouse ESCs do was settled by three reports in 2009 when viable mice were produced by iPSCs using the most stringent assay, i.e., tetraploid blastocyst complementation, to test the pluripotency of stem cells (Boland et al. 2009; Kang et al. 2009; Zhao et al. 2009).

This promising breakthrough provides an invaluable resource for regenerative medicine, enabling the generation of patient- or disease-specific cells of any lineages without the use of embryonic material involved in ethical controversies, as well as providing the potential solution to circumvent tissue rejection following transplantation. Furthermore, it is thought to herald the revolutionaries of regenerative medicine, pharmaceutical screening and disease models. In addition, the ability to revert somatic cells to an embryonic state provides a unique tool to dissect the molecular events that permit the conversion of one cell type to another. Thus, it is also of great significance to clarify vital principles of cell fate determination in basic researches (Ramalho-Santos 2009).

20.2 Considerations of Direct Reprogramming Technology

As mentioned above, the iPSCs were originally derived by retroviral over expression of certain transcriptional factors. The technology suffered from certain drawbacks: (1) some reprogramming factors are oncogenes such as c-Myc and Klf4, and their reactivation could lead to tumor formation in iPSC-produced chimeras; (2) the reprogramming process was slow and inefficient; (3) there exists insertional mutagenesis to the genome of starting cells by virus infection; and (4) incomplete reprogramming. It is critical to overcome these obstacles before the technology would be utilized in the clinical setting (Robinton and Daley 2012; Yamanaka 2009b; Zarzeczny et al. 2009).

While generation of iPSCs appears conceptually and technically simple, as shown in Fig. 20.1, several variables must be considered: (1) selection of starting cell types; (2) option of reprogramming factors; (3) approaches for factor delivery; (4) derivation conditions; (5) identification of iPSC colonies; and (6) expansion and characterization of iPSC lines. During the past several years, scientists have made

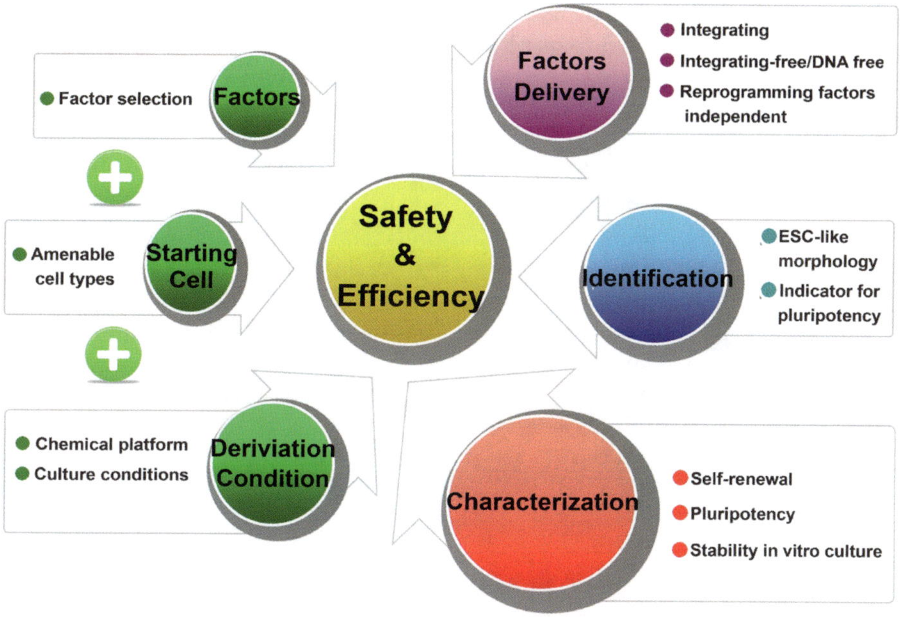

Fig. 20.1 Research directions and the current state of direct reprogramming technology: centering around the safety and efficiency for future clinic applications based on six variables of iPSC generation

extensive efforts to improve the iPSC technology, in order to realize somatic reprogramming in a safe and efficient manner. The exploration for new technologies of reprograming has advanced rapidly (Maherali and Hochedlinger 2008; Stadtfeld and Hochedlinger 2010).

20.2.1 Selection of Reprogramming Factors, Starting Cells, and Derivation Conditions

The original four-factor cocktail of *Oct4, Sox2, Klf4,* and *c-Myc* has been used to induce reprogramming successfully for multiple cell types of mouse and human as well as other species, including the rat, pig and rhesus monkey (Maherali and Hochedlinger 2008; Stadtfeld and Hochedlinger 2010). Another four-factor cocktail containing *OCT4, SOX2, NANOG*, and *LIN28* was also used to establish iPSC lines (Yu et al. 2007). Previous reports indicate that a variety of reprogramming cocktails, in certain cases only one factor, could convert somatic cells into a pluripotent state.

20.2.1.1 Factor Selection

To circumvent tumor risk caused by oncogenes in the pool of reprogramming factors, *Klf4* and *c-Myc* were first removed from the cocktail with a great sacrifice of the derivation efficiency (Huangfu et al. 2008b). In testing what was the minimal factor requirement for direct reprogramming, *Sox2* was found not essential in cell types with or without its endogenous expression (Eminli et al. 2008), while *Oct4* could play solo both in neural progenitor cells (NPCs) and fibroblast cells, demonstrating that *Oct4* is crucial in the reprogramming process (Kim et al. 2009b, c; Wang et al. 2011b). On the other hand, emphasizing on the efficiency of iPSC generation, some researchers induced reprogramming with additional factors (Liao et al. 2008). For example, *BAF* complex components could mediate reprogramming with higher efficiency by facilitating Oct4 binding to target promoters during reprogramming (Singhal et al. 2010). In addition, it was reported that the silencing of certain factors could also enhance the reprogramming efficiency. For instance, knockdown of DNA methyltransferase (*DNMT*) would permit full reprogramming in mouse fibroblast cells (Huangfu et al. 2008a) and the suppression of *p53* would convert up to 10% transduced MEF to iPSCs in the absence of the *Myc* retrovirus (Hong et al. 2009; Marion et al. 2009).

In the screening for the optimal reprogramming cocktail, *Sox1* and *Sox3* were found to be able to replace *Sox2*, albeit with a decrease in the reprogramming efficiency. Moreover, *Klf2* could replace *Klf4*. Importantly, less tumorigenic *L-Myc* or *N-Myc* could induce reprogramming in place of oncogenic *c-Myc* (Maherali and Hochedlinger 2008; Stadtfeld and Hochedlinger 2010). Recent studies demonstrated that *Oct4* could be replaced with nuclear receptor *Nr5a2* or *Esrrb*, or *E-cadherin* (Feng et al. 2009a; Heng et al. 2010; Redmer et al. 2011). These investigations have enhanced our understanding of the principles and mechanisms underlying the reprogramming process dramatically. Whereas avoiding oncogenes sacrificed the efficiency, small chemical compounds or specific cell types have been found to be capable of facilitating the conversion from somatic to pluripotent status.

20.2.1.2 The Chemical Platform

The initial attempts to use small compounds that might mimic the effect of transcription factors aimed at avoiding oncogene's involvement and eliminating the problem of genomic integration. Melton and collaborators reported that the reprogramming efficiency increased by 100-fold in the presence of valproic acid (VPA), a histone deacetylase (HDAC) inhibitor, and that pluripotency was established without the introduction of *Klf4* and *c-Myc*. It was further proposed that HDAC inhibitors enhance the histone acelylation level, which correlates with transcriptional activation, mimicking the effects of *c-Myc* (Huangfu et al. 2008a). Currently, VPA is commonly used in the induction of reprogramming to enhance the efficiency in various cell types. To date, small molecules used in the efficiency improvement mainly play roles in epigenetic modifications, such as 5-azacytidine (DNA methylating inhibitor) (Mikkelsen et al. 2008) and BIX01294 (histone methyltransferase *G9a* inhibitor) (Kubicek et al. 2007; Shi et al. 2008b), or in the regulation of certain signaling pathways such as BayK8644 (L-type calcium channel agonist) (Shi et al. 2008a) and Wnt3a (cell signaling molecule) (Feng et al. 2009b) in combination with CHIR99021 (a GSK3-β inhibitor) and PD0325901 (a MEK/ERK inhibitor) (Eminli et al. 2008). In addition, Wang and colleagues reported recently that histone demethylases Jhdm1a/1b enhance somatic cell reprogramming in a vitamin-C-dependent manne (Esteban et al. 2010; Wang et al. 2011a). With the evidence that BIX01294 could take the place of *Oct4* to induce pluripotency from NPCs and that a transforming growth factor TGF-β inhibitor (named 616452) could replace *Sox2* during iPSC generation, one of the ambitious goals in this field is ultimately to achieve complete chemical reprogramming by novel small molecules or novel cocktail of small molecules.

20.2.1.3 Cell Types Amenable to Reprogramming

Mouse iPSCs have been generated from multiple cell types, including fibroblast cells from embryo or adult tail tips, bone marrow cells, hepatocytes, gastric epithelial cells, pancreatic cells, NPCs, melanocyte, and mature B lymphocytes (Maherali and Hochedlinger 2008; Stadtfeld and Hochedlinger 2010). Among reported starting cell resources, several types of cells exhibited extremely high efficiency of induction of reprogramming. For example, the efficiency of iPSC generation from myeloid progenitors and hematopoietic stem cells could be 25% and 13%, respectively, which is a soaring increase compared with 0.02% from MEF (Maherali and Hochedlinger 2008; Raya et al. 2009). One explanation for this phenomenon is that progenitors possess the transcriptional and epigenetic features similar to stem cells. However, the detailed mechanisms underlying this phenomenon remain limited understanding. Taking advantage of endogenous expression of Sox2, NPCs could be reprogrammed by two factors (*Oct4* and *Klf4*) or even *Oct4* alone (Kim et al. 2009c; Shi et al. 2008a). Worthy of noting, a recent report indicated that *Oct4*-induced pluripotency could also be achieved in MEF in the presence of certain chemical compounds, which suggests the powerful and promising application of small-molecule compounds in somatic reprogramming (Shi et al. 2008a). Generally, the cell types which possess epigenetic and gene expression signatures closer to pluripotent cells tend to have a high efficiency in the establishment of iPSC lines.

20.2.1.4 Culture Conditions

Traditional conditions supporting the ESC maintenance are sufficient for both mouse and human iPSC generation. For instances, the substitution of knockout serum replacement (KSR) for fetal bovine serum that greatly enhances mouse ESC derivation also improves reprogramming process in mouse fibroblasts (Zhao et al. 2009). Moreover, conducting reprogramming under hypoxic conditions leads to an increased efficiency for both mouse and human cells, whereas a low O_2 tension prevents differentiation of human ESCs (Yoshida et al. 2009). A key goal for exploring favorable derivation conditions is to achieve defined and xeno-free cultures for the establishment of iPSCs suitable for clinical applications. By far, mouse iPSCs can be derived under feeder-free conditions (Silva et al. 2008). Basically, modification of iPSC derivation cultures relies on the experiences with ESC culture.

20.2.2 Approaches of Delivering Reprogramming Factor

In vitro reprogramming from somatic cells can serve as an appealing approach to produce patient-specific stem cells. This promising prediction has been convinced by reports that reprogrammed skin cells have alleviated the symptoms of Parkinson's disease (Wernig et al. 2008) and sickle cell anemia (Hanna et al. 2007) in mouse models. However, the original protocols employed retroviral vectors and constitutive/inducible lentiviruses to deliver reprogramming factors, which could lead to insertional mutagenesis and potential reactivation of the viral transgenes (Hanna et al. 2007; Meissner et al. 2007; Okita et al. 2007; Takahashi et al. 2007; Takahashi and Yamanaka 2006). The exploration for new approaches of pluripotency induction without ectopic integration has advanced significantly. So far integration-free iPSCs have been derived via excisable (transposon and floxed lentiviruses), nonintegrating vectors (plasmid, episomal, and adenovirus vectors), and DNA-free (recombination protein with a cell-penetrating peptide, Sendai virus, and synthetic modified mRNA) approaches.

20.2.2.1 Approaches with Genomic Integration

In vitro direct reprogramming methods were accomplished initially with the employment of Moloney-based retroviral vectors that are known to undergo silencing in the pluripotent state (Takahashi and Yamanaka 2006). This self-silencing property was considered as one of the criteria for full reprogramming (Stadtfeld et al. 2008a). However, besides genome integration, there are several drawbacks to preclude the utilization of retrovirus: (1) they usually target dividing cells, which restrict the resource of starting cells; (2) gradual silencing during cell conversion may result in a low efficiency of iPSC generation. Lentiviruses permit infection into nondividing cell types and several groups obtained lentivirus-mediated iPSCs. Nevertheless, they are not a preferable option due to their constitutive expression in cell conversion. Drug-inducible lentiviral vectors polished this technique by managing temporal control over factor expression (Markoulaki et al. 2009). Even though this method exhibits genome integration, it is a powerful model to conduct mechanistic analyses. Applying such system, Hockemeyer and colleagues established "secondary iPSCs" in which iPSC-derived differentiated cells displayed the same proviral integrations. Once induced by drug, the transgenes are homogeneously reactivated, resulting in a 100-fold increased efficiency in the secondary iPSC production (Hockemeyer et al. 2008).

20.2.2.2 Integration-Free Approaches

In 2008, Hochedlinger and his colleagues derived iPSCs by adenovirus-mediated factor induction from skin and liver cells of mice. The adenovirus excelled other viral vectors like lentivirus and retrovirus in non-incorporating its own genes into the host, thus avoiding potentially insertion-associated mutagenesis (Stadtfeld et al. 2008b). In the same year, Yamanaka group reprogrammed mouse cells by transfection with plasmids (Okita et al. 2008). Up to date, a number of integration-free methods for direct reprogramming have been reported, such as minicircles (Jia et al. 2010), transposons (Woltjen et al. 2009), episomal vectors (Yu et al. 2009), and DNA multicistronic construct targeting by homologous recombination (Wu et al. 2011). Several approaches involve sequential integration and excision of exogenous gene, for example, the piggyBac transposon system. In principle, these non-integration vectors are supposed to avoid transgene insertion, while in practice they could not exclude insertional mutagenesis entirely. Furthermore, non-integration approaches tend to possess dramatically lower efficiency compared with methods with integration.

20.2.2.3 DNA-Free Approaches

Suffering from low efficiency of iPSC generation with rare but possible risk of transgene insertion, the integrating-free approach did not satisfy the expectation for factor delivery. One attractive attempt is to induce iPSCs directly by transduction of proteins of reprogramming factors. Based on recombination proteins incorporating cell-penetrating peptide (11-arginine peptide), Ding's group demonstrated that protein-mediated transduction supported reprogramming murine somatic cells to pluripotency (Zhou et al. 2009); soon after, Kim and collaborators successfully applied a similar method on human fibroblast cells (Kim et al. 2009a). Although protein-based iPSC generation is the exclusive method to date without the use of genetic materials, it requires recombinant protein expression and purification expertise, and also suffers from extremely low efficiency as well as a long conversion process, which limits its further application. Nevertheless, protein based-iPSC generation deserves further research. In our lab, we have explored a simple, genetic material-free strategy for somatic cell reprogramming based on fusion proteins with protein transduction domain TAT (Schwarze and Dowdy 2000). We demonstrated that TAT fusion proteins of reprogramming factors could transduce multiple human and mouse cell lines and activate their corresponding reporter genes. However, we failed to establish iPSC lines using recombinant proteins alone. A huge step forward in the factor delivery field was taken in 2010. Warren and colleagues reported efficient induction of pluripotency by synthetic modified mRNA, which contained stabilizing modifications such as 5-methylguanosine capping or substituted ribonucleobases, e.g., pseudouracil, to alleviate the innate antiviral defence pathways such as interferon and NF-kB-dependent pathways triggered by single-stranded RNA biotypes (Fig. 20.2).

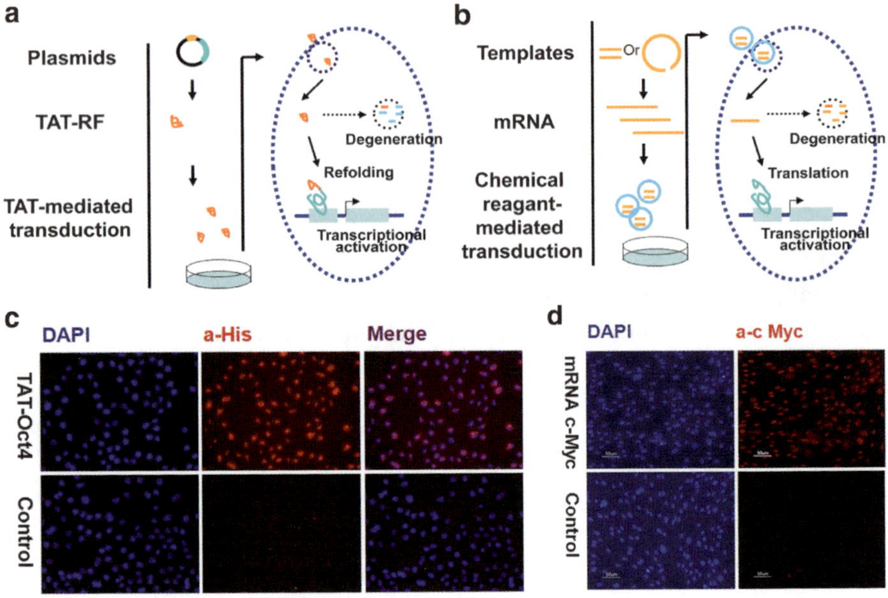

Fig. 20.2 Protein and mRNA based transduction. (**a**, **b**), the workflow of generation of reprogramming factors in TAT-fusion protein or mRNA forms and how they play their roles in vivo. (**c**, **d**) the transduced reprogramming factors (TAT-Oct4 and mRNA *c-MYC*) mediated by two approaches and localized primarily to the nucleus are readily detectable 24 h after transduction by immunofluorescen staining

This method satisfied both safety and efficiency of iPSC generation with a 35-fold higher efficiency than that of viral transduction (Warren et al. 2010). The repeatability of similar techniques in more research groups is critical for the eventual application in the clinic settings.

20.2.2.4 Reprogramming Factor Independent Approaches

Reprogramming approaches described above are dependent on overexpression of a set of transcription factors. Recently, Anokye-Danso and collaborators opened a novel way to direct reprogramming by the expression of the micro RNAs (miR302/367 cluster) in mouse and human somatic cells in the absence of exogenous transcription factors. In this report, the miRNA-based reprogramming process was surprisingly rapid and efficient with two orders of the magnitude more efficient than the standard four-factor mediated reprogramming. Nevertheless, suppression of HDAC2 is required. This work not only provides a powerful tool for reprogramming, but also challenges the conventional stereotype that miRANs regulate gene expression in a fine-turn-way (Anokye-Danso et al. 2011).

20.2.3 Identification, Expansion, and Characterization of iPSC Colonies

The first generation of mouse iPSCs was selected for the expression of ESC-specific, but nonessential, gene *Fbx15* (Takahashi and Yamanaka 2006). Subsequently, iPSCs were generated through selection for the expression of essential ESC-specific genes, *Nanog* or *Oct4* (Meissner et al. 2007; Okita et al. 2007; Wernig et al. 2007). Given that selection was enabled by the introduction of reporter alleles, the system appeared unfeasible in a clinical setting. Further investigation found the selection to be unnecessary and actually counterproductive. Currently, identification of iPSC colonies is solely based upon morphological observation, which relies on ESC culture experiences (Stadtfeld et al. 2008a). In general, mouse iPSC colonies can be distinguished by their ESC-like morphological properties. Aimed to monitor fully reprogrammed state, some groups established transgenic mice, harboring pluripotent gene (*Oct4* or *Nanog*) promoter driven EGFP. For example, Jaenisch's group generated transgenic mice that harbor an IRES-EGFP fusion cassette downstream of the stop codon of the *Oct4* (*Pou5f1*) gene. Using this system, EGFP is supposed to be turned on once somatic cells are converted into the pluripotent state and the expression of EGFP indicates the expression of endogenous *Oct4* (Lengner et al. 2007).

Characterization of iPSCs shares similar criteria with those of mouse ESCs and basic characteristics of pluripotent cells include unlimited self-renewal, the capability of differentiation into all cell lineages in vitro and in vivo. In addition, several examinations including pluripotency gene expression, genome stability, clonogenicity, karyotype, and extensive proliferation in vitro under certain culture conditions are also required. To determine the differentiation ability of iPSCs, in vitro assays such as embryoid body formation and in vivo assays such as teratoma formation, chimeric mice, and germline transmission are generally conducted. Under certain circumstances, the tetraploid complementation assay is performed to validate the full development potential. With the development of new techniques, more and more genome-scale profiles have been utilized to evaluate the status of iPSC lines in test, for instance, transcription and DNA methylation profiles (Meissner et al. 2007; Mikkelsen et al. 2007; Okita et al. 2007; Takahashi and Yamanaka 2006; Wernig et al. 2007).

20.3 Thoughts in Conflicts and Challenges in Current

Understanding mechanisms and principles underlying the reprogramming process is the guarantee of bringing all potentialities of iPSCs into full play. In fact, researches on this topic have turned out to be an international brainstorm spreading over almost all aspects in cell biology. Jaenisch has pondered and summarized that "these studies left important questions relevant to the basic mechanisms of epigenetic reprogramming unresolved: how does the reprogramming process progress over time and what happens to the majority of the cells that do not become reprogrammed upon continued cell growth and expression of the reprogramming factors? Why do some somatic cells that circumvent senescence or apoptosis induced by *Oct4, Sox2, Klf4*, and *c-Myc* convert into iPSCs earlier than others? Do all adult donor cells expressing *Oct4, Sox2, Klf4*, and *c-Myc* reprogramming factors eventually give rise to iPSCs or would this be achieved only upon additional genetic or small molecule manipulation? Is high reprogramming efficiency restricted to non-lineage committed or adult stem cells?" (Hanna et al. 2009) There are numerous questions by which hypotheses arise, and on which different thoughts in conflicts are argued and exchanged. Facts and evidences have accumulated and experts studying on this field have approximated the "truth" step by step. Unfortunately, limited to the space of this chapter, we can merely discuss several excellent while imperfect answers to these key questions in this rapidly moving field.

20.3.1 Cell Responding to Reprogramming: Elite or Stochastic?

With conventional viral expression system, the efficiency of initial germline-competent iPSC generation was only 0.05%, namely one cell is successfully converted into the ESC state among 2,000 originally plated starting cells (Maherali and Hochedlinger 2008; Takahashi and Yamanaka 2006). The low frequency implies that iPSCs might origin from rare tissue stem/progenitor cells that exist in the fibroblast cultures (studies have indicated that 0.067% of mouse skin cells are stem cells). Yamanaka's pioneering work attempted to exclude this possibility by showing that even from bone marrow stroma, which is more enriched in mesenchymal stem cells and other multipotent cells, the efficiency of iPSC generation was still comparably low (Takahashi and Yamanaka 2006). Therefore, reprogramming does not seem to favor multipotent tissue. In 2009, Yamanaka proposed "Elite and stochastic models for iPSC generation" to explain the low efficiency. In the elite model, only a small portion of cells, determined either before or after retroviral transduction, can be converted either partially or fully. While in the stochastic model, most (if not all) differentiated cells initiate the reprogramming process, but only a few can complete reprogramming (Yamanaka 2009a). From experimental observation, iPSC emergence seems to be a rare event with heterochrony. The quantified dissection of crucial events occurring during the in vitro reprogramming process has been problematic due to the cellular and genetic heterogeneity of de novo infected somatic cells. For instance, different infection efficiencies of four transcription factors and variable transcription factor expression in individual cells. It is also noteworthy that population-level measurements typically employed in reprogramming studies cannot distinguish between the above models (Hanna et al. 2009; Mikkelsen et al. 2008). Jaenisch's group answered this question in an ingenious and elegant way using the secondary generation iPSC system. They chose

B-cell lineage-committed cells from the secondary chimeras with a *Nanog*-GFP reporter and doxycycline-inducible lentiviral vectors encoding the four transcription factors. B-cell lineage-committed cells at the early pre-B-cell stage present a determined differentiated somatic cell type. They were efficiently cloned as single cells once isolated and served as a defined and homogenous starting cell population for reprogramming. With this powerful reprogramming system, they found that almost all donor cells eventually gave rise to iPSCs with continued growth and transcription factor expression. This evidence eloquently demonstrated that reprogramming by forced expression of transcription factors is a continuous stochastic process. In addition, they applied quantitative analyses to define distinct cell-division rate-dependent (inhibition of the p53/p21 pathway or over expression of Lin28) and independent (Nanog overexpression) modes for accelerating the dynamic process of reprogramming, and suggested that the number of cell divisions is a key parameter driving epigenetic reprogramming to pluripotency (Hanna et al. 2009).

20.3.2 iPSCs vs. ESCs: Are They Equivalent?

Fully reprogrammed iPSCs can form fertile mice by tetraploid blastocyst complementation (all iPSC mice), suggesting the equal developmental potential between mouse ESCs and iPSCs (Boland et al. 2009; Kang et al. 2009; Zhao et al. 2009). However, based on the observation of substantial variations among pluripotent cell lines, translational researches require better understanding of the differences between ESCs and iPSCs. This prime concern inspired flooded analyses of transcriptome and epigenome of iPSCs by reference to those of ESCs (Bock et al. 2011; De Carvalho et al. 2010; Gore et al. 2011; Guenther et al. 2010; Hussein et al. 2011; Laurent et al. 2011; Pasi et al. 2011; Stadtfeld et al. 2010). Inevitably, conclusions from these studies were highly dependent on the technique used and research aspects focused on. For instances, Young's group revealed that there is little difference between a panel of human ESCs and iPSCs (not genetic matched but with enlarged sample amounts) with respect to global chromatin structure (histone H3K4me3 and H3K27me3 modifications) and gene expression profiles (Guenther et al. 2010). Alexander Meissner's group reported the global similarity in DNA methylation and gene expression levels among a set of human ESCs and iPSCs line. Despite the deviation of a number of genes from the other cell lines, which is more prevalent among iPSC lines than among ESC lines, no epigenetic or transcriptional deviation was found unique and shared by all iPSC lines. Therefore, in their model, ESCs and iPSCs should not be considered as one or two defined points but rather as two partially overlapping point clouds (Bock et al. 2011).

In 2011, four independent studies came with unexpected haze: the genome itself of iPSCs suffered from mutations. Pasi and collaborators proposed that oncogenes (c-*Myc*) fuel the reprogramming-induced DNA damage by replicative stress (RS), which was supported by comparative genomic hybridization (cGH) analyses of iPSCs genomes (Pasi et al. 2011). Hussein and colleagues reinforced the link between RS and mutagenic-induced reprogramming by examining copy number variations (CNVs). This work suggested that de novo CNVs confer growth or survival disadvantages to the cells and that the genome of iPSCs would be dynamically influenced during culture and in vitro manipulation (Hussein et al. 2011). A much deeper study from Gore and colleagues revealed that iPSCs acquire epigenetic and genetic modifications as well. However, half of the mutations detected by sequencing the exons of 22 human iPSCs generated by five independent methods (including the most safe mRNA-iPSCs to date) existed in the starting cells before reprogramming. This finding suggested that the detected mutations (particularly for point coding mutations) might be selected for, rather than generated, during reprogramming (Gore et al. 2011). By performing a comprehensive high-resolution SNP analysis of 189 pluripotent (iPSCs and ESCs) and 119 non-pluripotent samples, Laurent and collaborators argued that intrinsic genomic instability is a general property of pluripotent genomes (both iPSCs and ESCs), but with different genomic aberration patterns (Laurent et al. 2011).

Stadtfeld and collaborators conducted genetically matched comparisons between mouse ESCs and iPSCs with respect to global mRNA and miRNA expression profiles (Stadtfeld et al. 2008a). They demonstrated that the overall messenger RNA and microRNA expression patterns are not distinguishable with the exception of a few transcripts encoded within the imprinted *Dlk1–Dio3* gene cluster. iPSCs with low levels of these transcripts contributed to chimeras poorly and failed to develop into all iPSC mice (Stadtfeld et al. 2010). A recent investigation showed that iPSCs with low expression of the *Dlk1-Dio3* cluster could support the development of all iPSC mice but with a lower frequency than those having normal expression of the cluster (Carey et al. 2011).

20.3.3 Patient-Specific iPSCs: Free From Auto-immunity?

Given that autologous cells are supposed to be immune tolerated by corresponding recipient, a granted promise for regenerative medicine is that applying patient-specific iPSCs into transplantation would avoid immune rejection response in recipient. Although this assumption is appealing and sound, astonishingly, Xu's group casted a timely warning that iPSCs were immunogenic even in the genetically matched recipients (Zhao et al. 2011). In this study, four pluripotent cell lines, namely, ESCs from B6 and 129/SvJ mice, and iPSCs by either retroviral (ViPSCs) or episomal approach (EiPSCs) derived from B6 MEF were injected into B6 mice. Subsequently, teratoma formation and immune responses were monitored. They found that B6 ESCs efficiently formed teratomas without any evident immune rejection, whereas allogeneic 129/SvJ ESCs failed to form teratomas due to rapid rejection by recipients. In contrast to B6 ESCs, teratomas formed by B6 EiPSCs were immunogenic in B6 mice with T-cell filtration. Moreover, they found multiple genes (e.g., *Zg16* and *Hormad1*) abnormally overexpressed in teratomas derived from EiPSCs, which possibly contributed to the immunogenicity of the B6 EiPSCs derived cells in B6 mice. These unexpected results added to a series of findings that iPSCs differ in subtle but potentially important ways from ESCs, and pointed out that some vital issues are required to be precisely evaluated before confidently clinical application of iPSCs: (1) whether differentiated cells from iPSCs would be immonogenic in autologous receipt? (2) To what extent the methods of iPSC generation will affect their immunogenicity; (3) whether the same response would happen in humans? In the meantime, the results underscore the importance to carry on researches of ESCs in order to fully understand the differences between iPSCs and ESCs.

20.4 Perspectives

Compared with human stem cells, mouse stem cells that benefit from being more amenable have served as the preferable research tools in stem cell study. With the same principle, mouse iPSCs also served as a rehearsal platform before iPSCs can be considered for human therapy (Robinton and Daley 2012). For example, Jaenisch's group established iPSC lines from fibroblasts of sickle anemia mice (Hanna et al. 2007). Importantly, they corrected the mutated gene in iPSCs and further differentiated iPSCs into hematopoietic stem cells, which rescued the disease phenotypes after transplanted back to sickle anemia mice. The study proves a principle that disease-specific iPSCs could be used for the cell therapy. In addition, mice models also are used as a reliable system to evaluate the long-term safety of iPSCs. Even with outwardly indistinguishable similarities between 4n-iPSCs (iPSC lines that have been tested by tetraploid complementation assay) and ESCs, evaluations and comparisons of the long-term safety and developmental process of animals produced from above two cell types are required. Recently, Man Tong and colleagues performed systematic comparison between mice derived from 4n-iPSCs and ESCs and found 4n-iPSC mice behaved normally in intelligence test and clinical hematological analysis (Tong et al. 2011). They suffered from developmental abnormities, including embryonic arrest, open-eyelid, and respiratory failure, at a comparable rate as that of ESC mice. However, tumorigenic cells were identified in 23% of iPSC mice, not in ESC mice, probably due to the reactivation of transgenic c-*Myc* (and *Klf4*). Furthermore, observations in human and mouse systems are complementary and should be discussed side by side to emphasize the similarities and the differences in the two systems. All these information would be combined to shed new light on understanding the epigenetic regulation of cellular programming and reprogramming processes.

Worthy of note, even with flooded and detailed dataset generated, several essential elements must be taken into consideration before any unimpeachable conclusions could be made to the critical questions (1) unavoidable contribution of genetic background and residual transgene expression in tested pluripotent cell lines; (2) imperfect reprogramming protocol: by far, even the most safe protocol for generation of mRNA-iPSCs still requires at least 18 rounds of RNA transfection through cationic lipid delivery vehicles. Moreover, single-stranded RNA biotypes trigger innate antiviral defence pathways, which would lead to apoptotic cell death (Warren et al. 2010). Therefore, further in-depth investigation is needed in order to fully understand the molecular basis of cellular reprogramming and develop safe cell therapeutic protocols. Undoubtedly, mouse iPSCs will serve as an essential player for realizing such goals.

References

Anokye-Danso F, Trivedi CM, Juhr D, Gupta M, Cui Z, Tian Y, Zhang Y, Yang W, Gruber PJ, Epstein JA et al (2011) Highly efficient miRNA-mediated reprogramming of mouse and human somatic cells to pluripotency. Cell Stem Cell 8:376–388

Bock C, Kiskinis E, Verstappen G, Gu H, Boulting G, Smith ZD, Ziller M, Croft GF, Amoroso MW, Oakley DH et al (2011) Reference maps of human ES and iPS cell variation enable high-throughput characterization of pluripotent cell lines. Cell 144:439–452

Boiani M, Scholer HR (2005) Regulatory networks in embryo-derived pluripotent stem cells. Nat Rev Mol Cell Biol 6:872–884

Boland MJ, Hazen JL, Nazor KL, Rodriguez AR, Gifford W, Martin G, Kupriyanov S, Baldwin KK (2009) Adult mice generated from induced pluripotent stem cells. Nature 461:91–94

Carey BW, Markoulaki S, Hanna JH, Faddah DA, Buganim Y, Kim J, Ganz K, Steine EJ, Cassady JP, Creyghton MP et al (2011) Reprogramming factor stoichiometry influences the epigenetic state and biological properties of induced pluripotent stem cells. Cell Stem Cell 9:588–598

Cyranoski D (2008) Stem cells: 5 things to know before jumping on the iPS bandwagon. Nature 452:406–408

De Carvalho DD, You JS, Jones PA (2010) DNA methylation and cellular reprogramming. Trends Cell Biol 20:609–617

Eminli S, Utikal J, Arnold K, Jaenisch R, Hochedlinger K (2008) Reprogramming of neural progenitor cells into induced pluripotent stem cells in the absence of exogenous Sox2 expression. Stem Cells 26:2467–2474

Esteban MA, Wang T, Qin B, Yang J, Qin D, Cai J, Li W, Weng Z, Chen J, Ni S et al (2010) Vitamin C enhances the generation of mouse and human induced pluripotent stem cells. Cell Stem Cell 6:71–79

Feng B, Jiang J, Kraus P, Ng JH, Heng JC, Chan YS, Yaw LP, Zhang W, Loh YH, Han J et al (2009a) Reprogramming of fibroblasts into induced pluripotent stem cells with orphan nuclear receptor Esrrb. Nat Cell Biol 11:197–203

Feng B, Ng JH, Heng JC, Ng HH (2009b) Molecules that promote or enhance reprogramming of somatic cells to induced pluripotent stem cells. Cell Stem Cell 4:301–312

Gore A, Li Z, Fung HL, Young JE, Agarwal S, Antosiewicz-Bourget J, Canto I, Giorgetti A, Israel MA, Kiskinis E et al (2011) Somatic coding mutations in human induced pluripotent stem cells. Nature 471:63–67

Guenther MG, Frampton GM, Soldner F, Hockemeyer D, Mitalipova M, Jaenisch R, Young RA (2010) Chromatin structure and gene expression programs of human embryonic and induced pluripotent stem cells. Cell Stem Cell 7:249–257

Hanna J, Wernig M, Markoulaki S, Sun CW, Meissner A, Cassady JP, Beard C, Brambrink T, Wu LC, Townes TM et al (2007) Treatment of sickle cell anemia mouse model with iPS cells generated from autologous skin. Science 318:1920–1923

Hanna J, Saha K, Pando B, van Zon J, Lengner CJ, Creyghton MP, van Oudenaarden A, Jaenisch R (2009) Direct cell reprogramming is a stochastic process amenable to acceleration. Nature 462:595–601

Heng JC, Feng B, Han J, Jiang J, Kraus P, Ng JH, Orlov YL, Huss M, Yang L, Lufkin T et al (2010) The nuclear receptor Nr5a2 can replace Oct4 in the reprogramming of murine somatic cells to pluripotent cells. Cell Stem Cell 6:167–174

Hockemeyer D, Soldner F, Cook EG, Gao Q, Mitalipova M, Jaenisch R (2008) A drug-inducible system for direct reprogramming of human somatic cells to pluripotency. Cell Stem Cell 3:346–353

Hong H, Takahashi K, Ichisaka T, Aoi T, Kanagawa O, Nakagawa M, Okita K, Yamanaka S (2009) Suppression of induced pluripotent stem cell generation by the p53-p21 pathway. Nature 460:1132–1135

Huangfu D, Maehr R, Guo W, Eijkelenboom A, Snitow M, Chen AE, Melton DA (2008a) Induction of pluripotent stem cells by defined factors is greatly improved by small-molecule compounds. Nat Biotechnol 26:795–797

Huangfu D, Osafune K, Maehr R, Guo W, Eijkelenboom A, Chen S, Muhlestein W, Melton DA (2008b) Induction of pluripotent stem cells from primary human fibroblasts with only Oct4 and Sox2. Nat Biotechnol 26:1269–1275

Hussein SM, Batada NN, Vuoristo S, Ching RW, Autio R, Narva E, Ng S, Sourour M, Hamalainen R, Olsson C et al (2011) Copy number variation and selection during reprogramming to pluripotency. Nature 471:58–62

Jia F, Wilson KD, Sun N, Gupta DM, Huang M, Li Z, Panetta NJ, Chen ZY, Robbins RC, Kay MA et al (2010) A nonviral minicircle vector for deriving human iPS cells. Nat Methods 7:197–199

Kang L, Wang J, Zhang Y, Kou Z, Gao S (2009) iPS cells can support full-term development of tetraploid blastocyst-complemented embryos. Cell Stem Cell 5:135–138

Kim D, Kim CH, Moon JI, Chung YG, Chang MY, Han BS, Ko S, Yang E, Cha KY, Lanza R et al (2009a) Generation of human induced pluripotent stem cells by direct delivery of reprogramming proteins. Cell Stem Cell 4:472–476

Kim JB, Greber B, Arauzo-Bravo MJ, Meyer J, Park KI, Zaehres H, Scholer HR (2009b) Direct reprogramming of human neural stem cells by OCT4. Nature 461:649–653

Kim JB, Sebastiano V, Wu G, Arauzo-Bravo MJ, Sasse P, Gentile L, Ko K, Ruau D, Ehrich M, van den Boom D et al (2009c) Oct4-induced pluripotency in adult neural stem cells. Cell 136:411–419

Kubicek S, O'Sullivan RJ, August EM, Hickey ER, Zhang Q, Teodoro ML, Rea S, Mechtler K, Kowalski JA, Homon CA et al (2007) Reversal of H3K9me2 by a small-molecule inhibitor for the G9a histone methyltransferase. Mol Cell 25:473–481

Laurent LC, Ulitsky I, Slavin I, Tran H, Schork A, Morey R, Lynch C, Harness JV, Lee S, Barrero MJ et al (2011) Dynamic changes in the copy number of pluripotency and cell proliferation genes in human ESCs and iPSCs during reprogramming and time in culture. Cell Stem Cell 8:106–118

Lengner CJ, Camargo FD, Hochedlinger K, Welstead GG, Zaidi S, Gokhale S, Scholer HR, Tomilin A, Jaenisch R (2007) Oct4 expression is not required for mouse somatic stem cell self-renewal. Cell Stem Cell 1:403–415

Liao J, Wu Z, Wang Y, Cheng L, Cui C, Gao Y, Chen T, Rao L, Chen S, Jia N et al (2008) Enhanced efficiency of generating induced pluripotent stem (iPS) cells from human somatic cells by a combination of six transcription factors. Cell Res 18:600–603

Maherali N, Hochedlinger K (2008) Guidelines and techniques for the generation of induced pluripotent stem cells. Cell Stem Cell 3:595–605

Marion RM, Strati K, Li H, Murga M, Blanco R, Ortega S, Fernandez-Capetillo O, Serrano M, Blasco MA (2009) A p53-mediated DNA damage response limits reprogramming to ensure iPS cell genomic integrity. Nature 460:1149–1153

Markoulaki S, Hanna J, Beard C, Carey BW, Cheng AW, Lengner CJ, Dausman JA, Fu D, Gao Q, Wu S et al (2009) Transgenic mice with defined combinations of drug-inducible reprogramming factors. Nat Biotechnol 27:169–171

Meissner A, Wernig M, Jaenisch R (2007) Direct reprogramming of genetically unmodified fibroblasts into pluripotent stem cells. Nat Biotechnol 25:1177–1181

Mikkelsen TS, Ku M, Jaffe DB, Issac B, Lieberman E, Giannoukos G, Alvarez P, Brockman W, Kim TK, Koche RP et al (2007) Genome-wide maps of chromatin state in pluripotent and lineage-committed cells. Nature 448:553–560

Mikkelsen TS, Hanna J, Zhang X, Ku M, Wernig M, Schorderet P, Bernstein BE, Jaenisch R, Lander ES, Meissner A (2008) Dissecting direct reprogramming through integrative genomic analysis. Nature 454:49–55

Nishikawa S, Goldstein RA, Nierras CR (2008) The promise of human induced pluripotent stem cells for research and therapy. Nat Rev Mol Cell Biol 9:725–729

Okita K, Ichisaka T, Yamanaka S (2007) Generation of germline-competent induced pluripotent stem cells. Nature 448:313–317

Okita K, Nakagawa M, Hyenjong H, Ichisaka T, Yamanaka S (2008) Generation of mouse induced pluripotent stem cells without viral vectors. Science 322:949–953

Pasi CE, Dereli-Oz A, Negrini S, Friedli M, Fragola G, Lombardo A, Van Houwe G, Naldini L, Casola S, Testa G et al (2011) Genomic instability in induced stem cells. Cell Death Differ 18:745–753

Ramalho-Santos M (2009) iPS cells: insights into basic biology. Cell 138:616–618

Raya A, Rodriguez-Piza I, Guenechea G, Vassena R, Navarro S, Barrero MJ, Consiglio A, Castella M, Rio P, Sleep E et al (2009) Disease-corrected haematopoietic progenitors from Fanconi anaemia induced pluripotent stem cells. Nature 460:53–59

Redmer T, Diecke S, Grigoryan T, Quiroga-Negreira A, Birchmeier W, Besser D (2011) E-cadherin is crucial for embryonic stem cell pluripotency and can replace OCT4 during somatic cell reprogramming. EMBO Rep 12:720–726

Robinton DA, Daley GQ (2012) The promise of induced pluripotent stem cells in research and therapy. Nature 481:295–305

Schwarze SR, Dowdy SF (2000) In vivo protein transduction: intracellular delivery of biologically active proteins, compounds and DNA. Trends Pharmacol Sci 21:45–48

Shi Y, Desponts C, Do JT, Hahm HS, Scholer HR, Ding S (2008a) Induction of pluripotent stem cells from mouse embryonic fibroblasts by Oct4 and Klf4 with small-molecule compounds. Cell Stem Cell 3:568–574

Shi Y, Do JT, Desponts C, Hahm HS, Scholer HR, Ding S (2008b) A combined chemical and genetic approach for the generation of induced pluripotent stem cells. Cell Stem Cell 2:525–528

Silva J, Barrandon O, Nichols J, Kawaguchi J, Theunissen TW, Smith A (2008) Promotion of reprogramming to ground state pluripotency by signal inhibition. PLoS Biol 6:e253

Singhal N, Graumann J, Wu G, Arauzo-Bravo MJ, Han DW, Greber B et al (2010) Chromatin-remodeling components of the BAF complex facilitate reprogramming. Cell 141:943–955

Stadtfeld M, Hochedlinger K (2010) Induced pluripotency: history, mechanisms, and applications. Genes Dev 24:2239–2263

Stadtfeld M, Maherali N, Breault DT, Hochedlinger K (2008a) Defining molecular cornerstones during fibroblast to iPS cell reprogramming in mouse. Cell Stem Cell 2:230–240

Stadtfeld M, Nagaya M, Utikal J, Weir G, Hochedlinger K (2008b) Induced pluripotent stem cells generated without viral integration. Science 322:945–949

Stadtfeld M, Apostolou E, Akutsu H, Fukuda A, Follett P, Natesan S, Kono T, Shioda T, Hochedlinger K (2010) Aberrant silencing of imprinted genes on chromosome 12qF1 in mouse induced pluripotent stem cells. Nature 465:175–181

Takahashi K, Yamanaka S (2006) Induction of pluripotent stem cells from mouse embryonic and adult fibroblast cultures by defined factors. Cell 126:663–676

Takahashi K, Tanabe K, Ohnuki M, Narita M, Ichisaka T, Tomoda K, Yamanaka S (2007) Induction of pluripotent stem cells from adult human fibroblasts by defined factors. Cell 131:861–872

Tong M, Lv Z, Liu L, Zhu H, Zheng QY, Zhao XY, Li W, Wu YB, Zhang HJ, Wu HJ et al (2011) Mice generated from tetraploid complementation competent iPS cells show similar developmental features as those from ES cells but are prone to tumorigenesis. Cell Res 21:1634–1637

Wang T, Chen K, Zeng X, Yang J, Wu Y, Shi X, Qin B, Zeng L, Esteban MA, Pan G et al (2011a) The histone demethylases Jhdm1a/1b enhance somatic cell reprogramming in a vitamin-C-dependent manner. Cell Stem Cell 9:575–587

Wang Y, Chen J, Hu JL, Wei XX, Qin D, Gao J, Zhang L, Jiang J, Li JS, Liu J et al (2011b) Reprogramming of mouse and human somatic cells by high-performance engineered factors. EMBO Rep 12:373–378

Warren L, Manos PD, Ahfeldt T, Loh YH, Li H, Lau F, Ebina W, Mandal PK, Smith ZD, Meissner A et al (2010) Highly efficient reprogramming to pluripotency and directed differentiation of human cells with synthetic modified mRNA. Cell Stem Cell 7:618–630

Wernig M, Meissner A, Foreman R, Brambrink T, Ku M, Hochedlinger K, Bernstein BE, Jaenisch R (2007) In vitro reprogramming of fibroblasts into a pluripotent ES-cell-like state. Nature 448:318–324

Wernig M, Zhao JP, Pruszak J, Hedlund E, Fu D, Soldner F, Broccoli V, Constantine-Paton M, Isacson O, Jaenisch R (2008) Neurons derived from reprogrammed fibroblasts functionally integrate into the fetal brain and improve symptoms of rats with Parkinson's disease. Proc Natl Acad Sci USA 105:5856–5861

Woltjen K, Michael IP, Mohseni P, Desai R, Mileikovsky M, Hamalainen R, Cowling R, Wang W, Liu P, Gertsenstein M et al (2009) piggyBac transposition reprograms fibroblasts to induced pluripotent stem cells. Nature 458:766–770

Wu G, Liu N, Rittelmeyer I, Sharma AD, Sgodda M, Zaehres H, Bleidissel M, Greber B, Gentile L, Han DW et al (2011) Generation of healthy mice from gene-corrected disease-specific induced pluripotent stem cells. PLoS Biol 9:e1001099

Yamanaka S (2009a) Elite and stochastic models for induced pluripotent stem cell generation. Nature 460:49–52

Yamanaka S (2009b) A fresh look at iPS cells. Cell 137:13–17

Yoshida Y, Takahashi K, Okita K, Ichisaka T, Yamanaka S (2009) Hypoxia enhances the generation of induced pluripotent stem cells. Cell Stem Cell 5:237–241

Yu J, Vodyanik MA, Smuga-Otto K, Antosiewicz-Bourget J, Frane JL, Tian S, Nie J, Jonsdottir GA, Ruotti V, Stewart R et al (2007) Induced pluripotent stem cell lines derived from human somatic cells. Science 318:1917–1920

Yu J, Hu K, Smuga-Otto K, Tian S, Stewart R, Slukvin II, Thomson JA (2009) Human induced pluripotent stem cells free of vector and transgene sequences. Science 324:797–801

Zarzeczny A, Scott C, Hyun I, Bennett J, Chandler J, Charge S, Heine H, Isasi R, Kato K, Lovell-Badge R et al (2009) iPS cells: mapping the policy issues. Cell 139:1032–1037

Zhao XY, Li W, Lv Z, Liu L, Tong M, Hai T, Hao J, Guo CL, Ma QW, Wang L et al (2009) iPS cells produce viable mice through tetraploid complementation. Nature 461:86–90

Zhao T, Zhang ZN, Rong Z, Xu Y (2011) Immunogenicity of induced pluripotent stem cells. Nature 474:212–215

Zhou H, Wu S, Joo JY, Zhu S, Han DW, Lin T, Trauger S, Bien G, Yao S, Zhu Y et al (2009) Generation of induced pluripotent stem cells using recombinant proteins. Cell Stem Cell 4:381–384

Chapter 21
Aging in the Mouse and Perspectives of Rejuvenation Through Induced Pluripotent Stem Cells (iPSCs)

Ken-ichi Isobe, Zhao Cheng, Sachiko Ito, and Naomi Nishio

Abstract The mouse is a perfect model to study aging in mammals. It has a relatively short life span and genetic manipulations in this species are well established. Most interestingly, the mouse is a fantastic tool to produce stem cells. Forced expression of only four transcription factors (Oct3/4, Sox2, Klf4, and c-Myc) in murine and human somatic cells resets the expression of genes that are characteristic of differentiated cells and consequently induces the formation of pluripotent stem cells (iPSCs). This technology opens new and exciting possibilities in medical research, especially personalized cell therapies for treating human disease. To treat damaged tissues or repair organs in elderly patients, it will be necessary to establish iPSCs from their tissues. To determine the feasibility of using this technology with elderly patients, we asked whether it was indeed possible to establish iPSCs from the tissues of aged mice and to differentiate them to tissue cells. We succeeded in establishing iPSC clones using bone marrow (BM) from 21-month-old EGFP-C57BL/6 mice, which had been cultured for 4 days in the presence of granulocyte macrophage-colony stimulating factor (GM-CSF). Our iPSCs from aged mice (aged iPSCs) and those from mouse embryonic fibroblasts (MEFs) strongly expressed SSEA-1 and Pou5f1, and showed strong alkaline phosphatase (AP) activity. Our aged iPSCs made teratomas when injected into the back skin of syngeneic mice, and differentiated to tissue cells of three germ lines in vitro. Further experiments to make chimeric mice and germ line cells will determine whether the aged iPSCs possess the properties of much younger cells and are capable of regenerating aged mice.

K.-i. Isobe (✉) • Z. Cheng • S. Ito • N. Nishio
Department of Immunology, Nagoya University Graduate School of Medicine, 65 Turumai-cho, Showa-ku, Nagoya, Aichi 466-8550, Japan
e-mail: kisobe@med.nagoya-u.ac.jp

J.Z. Kubiak (ed.), *Mouse Development*, Results and Problems in Cell Differentiation 55,
DOI 10.1007/978-3-642-30406-4_21, © Springer-Verlag Berlin Heidelberg 2012

21.1 Introduction

The mean life expectancy in developed countries has increased dramatically (Oeppen and Vaupel 2002). For instance, the mean life expectancy in Japan has increased from 50 to 80 years in no more than 6 decades. Socioeconomic factors, including the improvement of environmental conditions and medical care, may explain part of this increase in life span, but ample evidence suggests that genetic factors are also at play. Studies of twins and long-lived families have estimated that 20–30% of the variation in human life span is determined by genetic factors (Herskind et al. 1996; Mitchell et al. 2001; Hjelmborg et al. 2006).

The mouse is an ideal mammalian model for studying aging. It shares 99% of its genes with humans (Boguski 2002), has a short generation time, and its environment can be easily controlled (Yuan et al. 2011). Furthermore, the genetic resources for the mouse include hundreds of inbred strains and mutants and sophisticated genetic engineering technology for manipulating its genome (Paigen 1995; Peters et al. 2007).

Somatic cell nuclear transfer (SCNT) into oocytes and the cloning of entirely new animals from a single cell revolutionized the concepts of developmental biology. SCNT was first developed in *Xenopus* (Gurdon 1962), followed by "Dolly" the sheep (Wilmut et al. 1997) and then mice (Wakayama et al. 1998). This technique suggests that cellular factors affect chromatin structure and gene expression. The forced expression of only four transcription factors (Oct3/4, Sox2, Klf4, and c-Myc) in murine and human somatic cells can reset the genes of differentiated cells and induce pluripotent stem cells (iPSCs) (Takahashi and Yamanaka 2006; Takahashi et al. 2007; Wernig et al. 2007; Jaenisch and Young 2008; Yu, et al. 2007; Park et al. 2008; Lowry et al. 2008) (Fig. 21.1a). This technology opens exciting possibilities in medical research, especially personalized cell therapies for treating human diseases.

Because of the human leukocyte antigen (HLA) barrier, almost all organ transplantation therapies require immunosuppressive drugs which cause several side effects. If it were possible to generate iPSCs from a patient's own healthy cells and induce differentiation, we could avoid allogeneic organ transplantation. In developed countries, a large number of elderly people suffer from untreatable diseases. To treat damaged tissues or repair organs in elderly patients, it will be necessary to establish iPSCs from their tissues. To determine the feasibility of the application of this technology to elderly patients, we studied the possibilities to establish iPSCs from the tissues of aged mice and to redifferentiate them to a range of tissues. In the first section of this chapter, we will briefly summarize the molecular characteristics of iPSCs and then discuss the possibility of using this technique to rejuvenate aged damaged tissues.

21.2 Characteristics of Induced Pluripotent Stem Cells

Pluripotent stem cells exist in developing embryos and have been established in tissue culture. The first pluripotent cells were derived from a germ line tumor teratocarcinoma (Hogan 1976). Then, embryonic stem cells (ESCs) from the

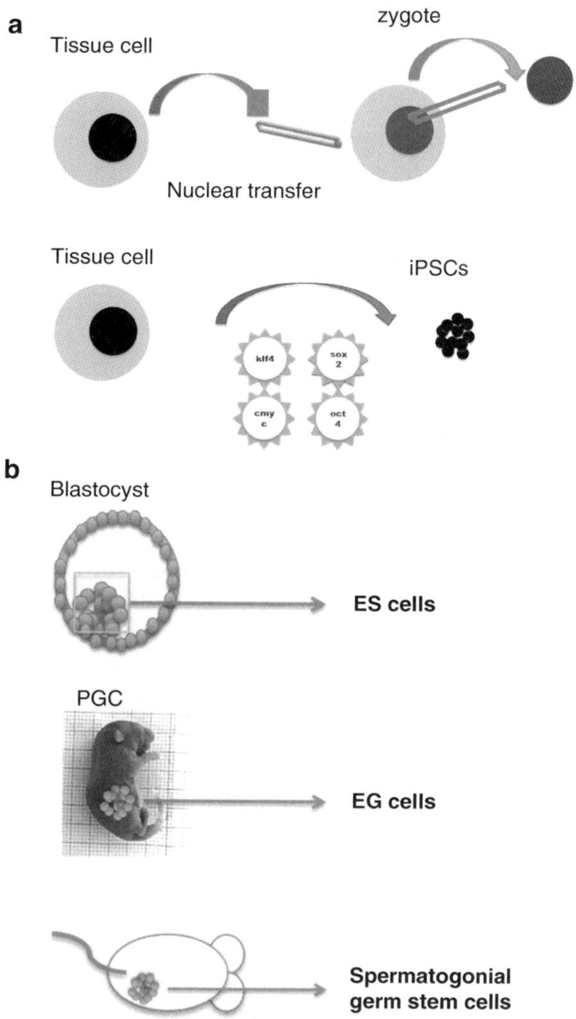

Fig. 21.1 Methods to make pluripotent cells. A. Adults somatic cells are reprogrammed by Somatic cell nuclear transfer (SCNT) into oocytes. The forced expression of transcription factors in somatic cells can reset the genes of differentiated cells and induce pluripotent stem cells (iPSCs). B. Pluripotent stem cells from fetal and adult stem cells

inner cell mass (ICM) of normal mouse embryos were developed, which made a great impact on biomedical research by permitting the creation of knockout mice (Evans and Kaufman 1981). Primordial germ cells isolated from embryonic day 8.5 embryos generated ES-like cells, termed embryonic germ cells (Surani 1999). From newborn or adult gonads, ES-like cells were generated and called spermatogonial stem cells, which are propagated in vitro in the presence of serum and leukemia inhibitory factor (LIF) (Kanatsu-Shinohara et al. 2004) (Fig. 21.1b). The pluripotent stem cells most commonly used in research are ESCs and iPSCs because they grow well in vitro. Many characteristics of ESCs and iPSCs are similar. ESCs or iPSCs are maintained on feeder fibroblasts in the presence of fetal bovine serum and LIF which support STAT3 (signal tranducer and activator of transcription 3)

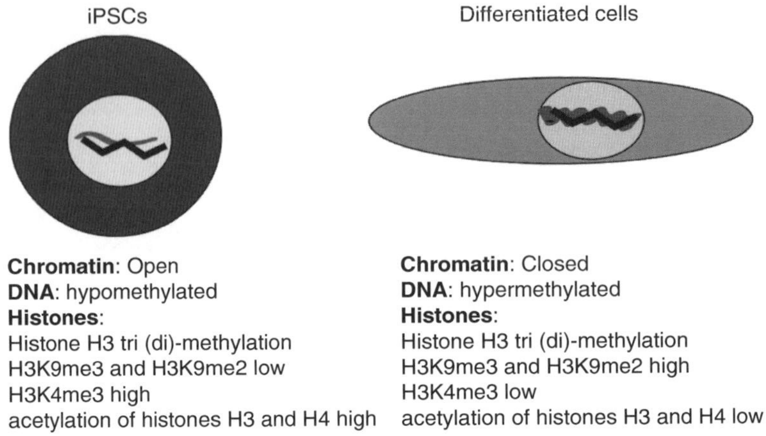

Fig. 21.2 Differences of chromatin of iPSCs and differentiated cells

signaling. Alternatively, feeder layers can be omitted by the inclusion of LIF, BMP-4 and other small molecules that stimulate Wnt signaling (Ying et al. 2008).

Chromatin—chromosomal DNA as packaged with histones—provides the cellular context for gene expression. On the basis of histological evidence, ESCs and iPSCs exhibit euchromatin (lightly stained and sparsely packed chromatin), which is mostly devoid of more densely stained, compacted areas of the heterochromatin. The chromatin of ESCs or iPSCs is in an "open" state (Gaspar-Maia et al. 2011). Cells in the ICM of the mouse blastocyst at day 3.5, which are the source of ES cells, share the same open chromatin conformation as ES cells. Upon differentiation, heterochromatin appears heterogeneous and clustered in distinct blocks. DNA of ESCs and iPSCs is hypomethylated. Especially, demethylation of Oct4 and Nanog is widely used to monitor successful reprogramming (Mikkelsen et al. 2008). Histone-modifying enzymes that repress genes encoding lineage-specific developmental regulators have the most profound impact on ESC or iPSC states. Histone H3 tri-methylation and di-methylation on Lys9 (H3K9me3 and H3K9me2) are indicators of heterochromatin. They are low in iPSCs and ESCs (Hawkins et al. 2010), but upon differentiation, H3K9 is hypermethylated. In contrast, active markers such as H3K4me3 and acetylation of histones H3 and H4 are upregulated in ESCs and iPSCs (Fig. 21.2). The pluripotency regulatory network is also directly linked to the Jumonji (Jmj) family H3K9 demethylases Jmjd1a/KDM2A and Jmjd2c/KDM4B, which act on H3K9me2 and H3K9me3, respectively (Loh et al. 2007). Both genes lie downstream of Oct4 and are regulated positively through its action. Recent discoveries have generated substantial excitement, as they show that cytosines in mammalian cells can be hydroxymethylated to 5hmC (5-hydroxymethylcytosine), which is especially abundant in tissues such as ESCs and iPSCs. 5-Methylcytosine (5mC) can be hydroxylated by the ten-eleven translocation (TET) family of enzymes. Koh and colleagues (2011) reported that TET levels are high in pluripotent cells and decline during differentiation. 5hmC was especially enriched at the start sites of genes whose promoters bear dual histone 3 lysine 27 trimethylation (H3K27me3) and histone 3 lysine 4 trimethylation (H3K4me3) marks (Williams et al. 2011; Wu et al. 2011). Tet1 is required for establishing a defined genomic pattern of 5hmC, and also for initiating an enzymatic cascade to maintain CpG-rich gene promoters in a DNA hypo-methylated state (Wu et al. 2011).

21.3 Is It Possible to Generate iPSCs from Aged Mice?

It has been shown that cellular senescence induced by reprogramming impairs successful reprogramming to iPSCs. Banito and collaborators (2009) showed that expression of four reprogramming factors mentioned above triggers senescence via upregulating p53, p16 (INK4a), and p21(CIP1). Upregulation of these senescence-related proteins blocks proliferation of target fibroblasts and reduces the generation of iPSCs. Genetic inhibition of the Ink4a/Arf locus has a profound positive effect on the efficiency of iPS cell generation, increasing both the kinetics of reprogramming and the number of emerging iPS cell colonies. In murine cells, Arf, rather than Ink4a, is the main barrier to reprogramming by activation of p53 (encoded by *Trp53*) and p21 (encoded by *Cdkn1a*), whereas, in human fibroblasts, INK4a is more important in reprogramming than ARF (Li et al. 2009). p53 is a stress-response protein, which suppresses tumor formation by promoting senescence (permanent cell-cycle arrest). Several papers have shown that inactivation of p53 markedly increases the efficiency of iPSC production (Hong et al. 2009; Kawamura et al. 2009; Utikal et al. 2009; Marión et al. 2009). These results indicate that during the course of induction of iPSCs, cellular aging occurs and reduces the capacity to generate iPSCs.

It was reported that fibroblasts from the lung of a human fetus had a fixed, replicative life span of about 50 population doublings before replication stopped (Hayflick 1965). This finding was reinforced by the report that the replicative potential of fibroblasts cultured from skin decreased with the increasing age of human donors (Martin et al. 1970). Additional support for the putative relationship between donor age and replicative life span of fibroblasts came from the report that the number of possible divisions in vitro was directly related to the average in vivo lifetime of the species (Röhme 1981). Later, the limited in vitro lifetime was attributed to telomere shortening (Harley et al. 1990), which also contributed to organismal aging by limiting the proliferative capacity of adult stem cells (Blasco 2007; Flores et al. 2006). These results indicated that the efficiency of iPSCs production might be lower in aged cells than in young cells.

We investigated the establishment of iPSCs from aged mice (Cheng et al. 2011). We used BM-derived myeloid (BM-M) cells that were actively proliferating in the presence of the granulocyte macrophage-colony stimulating factor (GM-CSF) (Fig. 21.3). First, we calculated the efficiency of iPSC generation from aged mice. We compared MEF cells and BM-M cells obtained from either 2-month-old C57BL/6 or 23-month-old C57BL/6 mice. Colonies appeared approximately 15 days after virus transduction from both MEF and 2-month (2M)-old BM-M cells, although the number of colonies produced by MEF cells increased more

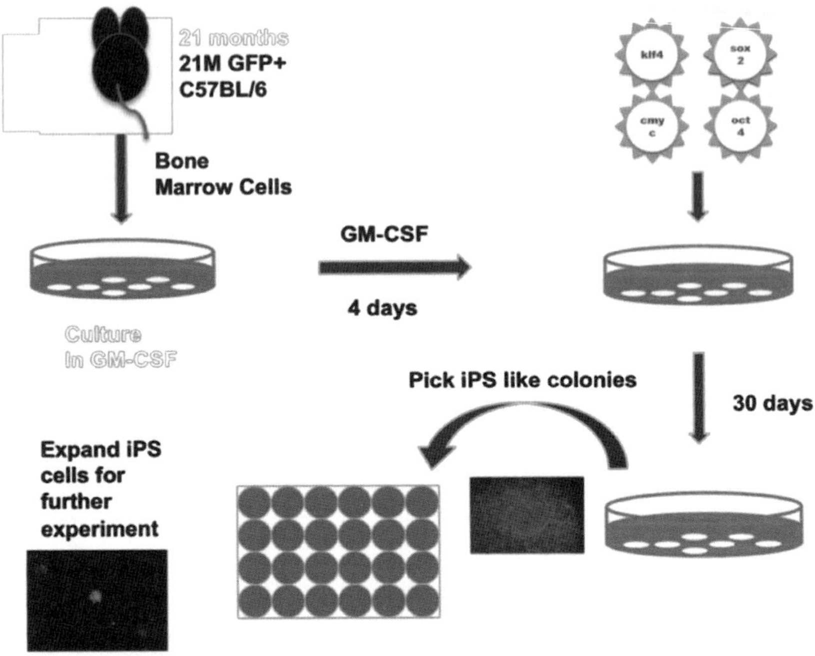

Fig. 21.3 Establishment of iPSCs from bone marrow (BM) derived macrophages (M) in GM-CSF of 21-month-old EGFP-positive C57BL/6 mice

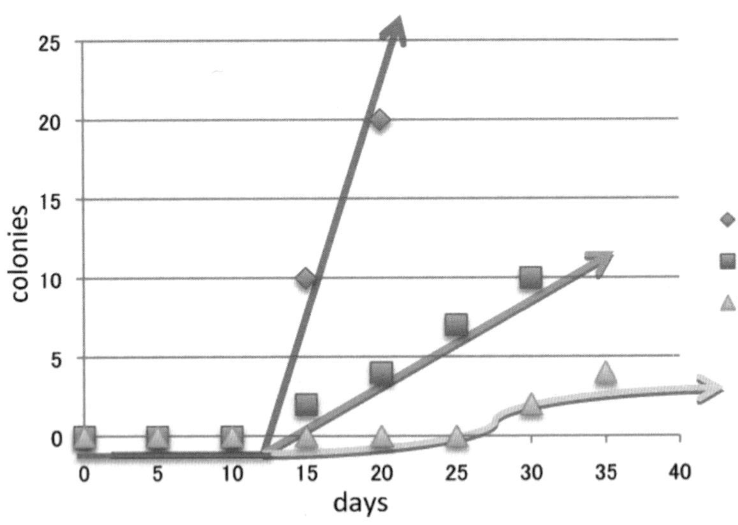

Fig. 21.4 Efficiency of the induction of MEF and BM-M cells. BM cells from 21-month-old EGFP-C57BL/6 mice. Graph shows the colony numbers at different time point

rapidly than those from 2M-old BM-M cells (10 colonies from MEF versus 2 colonies from 2M-old BM-M cells at 15 days). As expected, 23-month-old BM-M cells produced fewer colonies and these appeared later (10 colonies from 2M-old BM-M cells versus 2 colonies from 23M-old BM-M cells 30 days after transduction; Fig. 21.4). These results indicate that the efficiency of establishing iPSCs from aged mice is indeed lower than that from young mice (Cheng et al. 2011). Then we tried to verify whether iPSC cell lines from aged mice could grow and differentiate as well as iPSCs from young mice. To this end, we planned to transplant differentiated iPSCs to damaged tissues of syngeneic C57BL/6 mice. In order to discriminate transplanted iPSCs from recipient syngeneic mice, we selected GFP-positive C57BL/6 mice. We succeeded in establishing iPSCs clones using BM of 21-month-old EGFP-C57BL/6 mice, which had been cultured for 4 days in the presence of GM-CSF. We selected two clones (1 and 2) and expanded them (Fig. 21.3).

21.4 Are Aged Somatic Cells Rejuvenated By iPSC Production?

There is an overall decline in tissue regenerative potential with age (Rando 2006). For instance, wound healing becomes slower with age in skin. This age-related decline of wound healing is mainly caused by the decrease of regenerative potential of stem cells in the skin, although immune cells affect these phenomena (Nishio et al. 2008). During aging, the somatic cells accumulate DNA lesions, caused by a variety of stimuli, although the host has defense systems. These include single- and double-strand DNA breaks, chromosomal translocations and single base mutations. Telomere shortening occurs as a consequence of somatic cell division (Wang et al. 2009). Most of these DNA mutations that accumulate during aging are irreversible, except telomere shortening (described later). Unlike acquired DNA mutations, epigenomic changes (DNA methylation and histone methylations, acetylation, and ubiquitination) are potentially reversible. Loci in CpG islands gained methylation with age, while loci outside of CpG islands lost methylation with age (Christensen et al. 2009). Maegawa and collaborators (2010) demonstrated a surprisingly high rate of hyper- and hypomethylation as a function of age in normal mouse small intestine tissues and a strong tissue specificity to the process. During aging, histone methylation also occurs and may affect gene expression and function. It has been shown that regulators of histone H3K4 trimethylation complex associate during the life span in *C. elegans* (Greer et al. 2010). In mammals, histone methylation affects the aging of tissue cells. In young murine oocytes, dimethylation of lysines 4, 9, 36 and 79 in histone 3 (H3K4me2, H3K9me2, H3K36me2, H3K79me2) and methylation of lysine 20 in histone H4 (H4K20me2) and trimethylation of lysine 9 in histone 3 (H3K9me3) were observed. However, in old murine oocytes, H3K4me2 and H3K9me2 increased and demethylation of H3K9me3, H3K36me2, H3K79me2, and H4K20me2 occurred (Manosalva and González 2010) (Fig. 21.5).

Prior to the discovery of iPSCs, several approaches were taken to determine whether aged somatic cells could be rejuvenated and whether differentiation was reversible. Early cell fusion experiments between somatic cells and ESCs

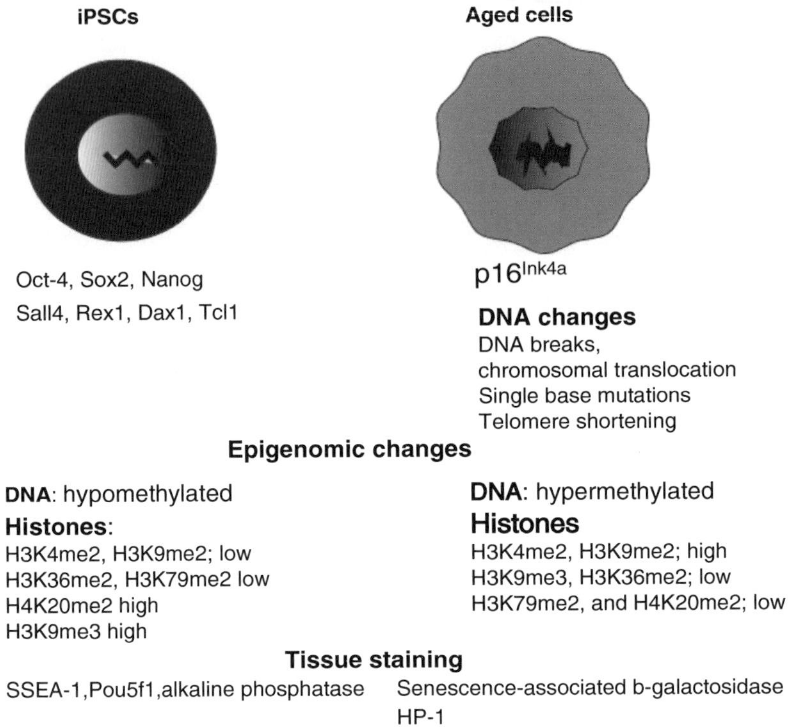

Fig. 21.5 DNA or epigenomic changes between iPSCs and aged cells

demonstrated the reversibility of differentiation (Blau et al. 1985). It was shown that the life span of mice cloned from somatic cells (immature Sertoli cells) was significantly shorter than that of genotype- and sex-matched controls, most likely due to severe pneumonia and hepatic failure (Ogonuki et al. 2002). Immune responses (antibody production and phagocytosis) in cloned mice were lower than in controls. However, they reported that two of the 12 cloned mice in their experiment appeared to have normal life spans.

An important question is whether telomeres of aged iPSCs become shorter than those of young iPSCs. Because telomerase activity is upregulated in both human and mouse iPSCs (Takahashi et al. 2007; Stadtfeld et al. 2008), it was suggested that telomeres might be elongated in iPSCs compared to parental cells. Consistent with this idea, telomeres were found to be elongated in iPSCs taken from an elderly person (Dimos et al. 2008).

If aged iPSCs were rejuvenated, their gene expression profile and epigenetic state might be similar to those of iPSCs from MEFs. Reprogramming of somatic cells to pluripotency is accompanied by extensive remodeling of epigenetic marks, including DNA demethylation of key pluripotency genes such as *Oct4* and *Nanog*. In somatic cells, the promoters of *Oct4* and *Nanog* are highly methylated, reflecting their transcriptionally repressed state. The formation of iPSCs involves activation of these genes, and their demethylation is widely used to monitor successful

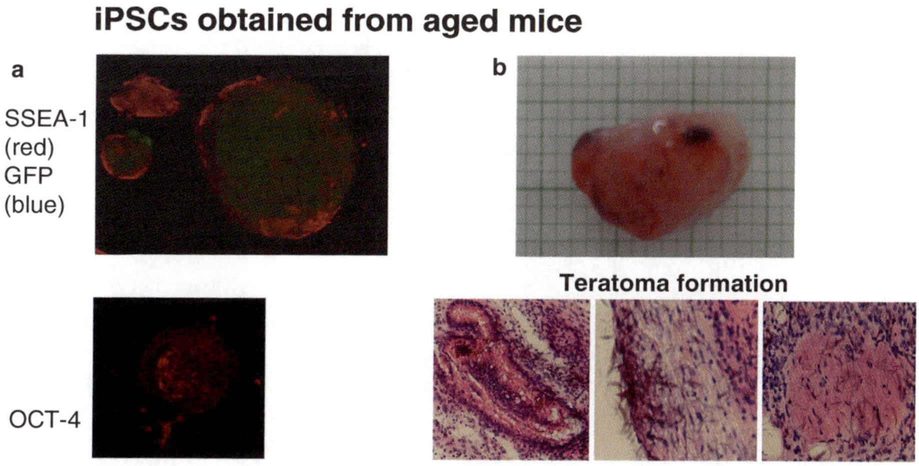

Fig. 21.6 Immunostaining of aged iPSCs and teratoma production on the back skin of syngeneic C57BL/6 mice

reprogramming (Mikkelsen et al. 2008). In principle, demethylation could occur by a passive mechanism, such as the inhibition of DNA methyltransferase 1 (Dnmt1) during DNA replication, or by an active mechanism in which the methylated base is removed from nonreplicating DNA.

Our iPSCs from aged mice and MEF iPSCs strongly expressed SSEA-1 and Oct-4 (Fig. 21.6a) and Pou5f1, and they showed strong alkaline phosphatase (AP) activity. Gene expression of *Nanog* and *Pou5f1* in aged iPSCs was as high as in MEF iPSCs and the promoters of *Nanog* and *Pou5f1* were hypomethylated in both aged iPSCs and MEF iPSCs by ChIP assays using anti-H3-K27 antibodies (Cheng et al. 2011). However, our examination of gene expression and the epigenetic state of aged iPSCs is preliminary. Whole genome sequencing and extensive Chip assays will likely reveal differences between aged iPSCs and MEF iPSCs and show unique characteristic features of age-derived iPSCs. Recently, it was shown that the reprogramming process and subsequent culture of iPSCs in vitro can induce genetic and epigenetic abnormalities in these cells (Hussein et al. 2011; Gore et al. 2011; Lister et al. 2011; Mayshar et al. 2010; Laurent et al. 2011). It has been shown that copy number variation and point mutations frequently occur during iPSC generation.

21.5 Do iPSCs from Aged Mice Differentiate Normally?

We examined whether iPSCs from aged mice (aged iPSCs) demonstrate pluripotency similar to iPSCs from MEF (MEF iPSCs). We transplanted aged iPSCs (1×10^7) or MEF iPSCs to the dorsal flank of syngeneic C57BL/6 mice. Teratomas appeared after the injection of aged iPSCs and MEF iPSCs. After 21 days, we observed distinct

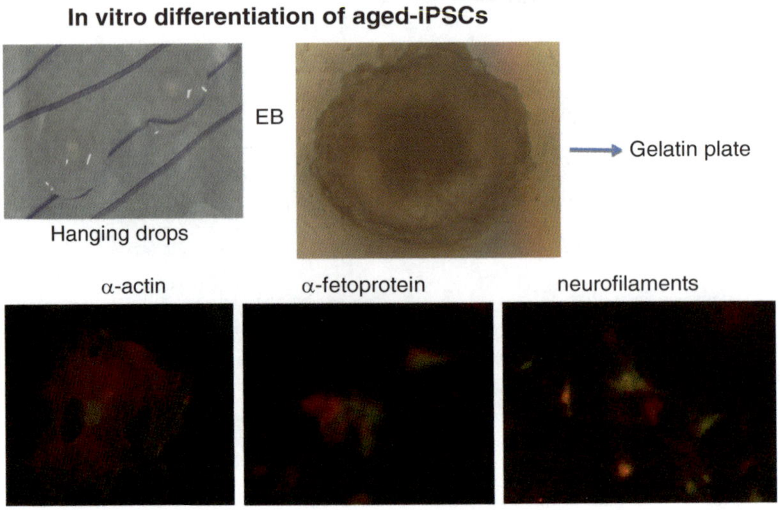

Fig. 21.7 In vitro differentiation of aged iPSCs by hanging drop methods

tumors that were collected and fixed in OCT (Optimal Cutting Temperature) Compound (Sakura Finetechnical Co., Ltd., Japan) for frozen 5-μm sections.

We observed various tissues belonging to three germ layers (endoderm, mesoderm, and ectoderm) (Fig. 21.6b). They were positively stained by antibodies against alpha-smooth muscle actin (mesoderm), α-fetoprotein (endoderm) and neurofilament H (ectoderm). We detected GFP-positive cells from the transplantation of aged iPSCs. Our observation indicates that iPSCs can differentiate to tissue cells and make tissues in vivo (Cheng et al. 2011).

However, another group showed that iPSCs made by a retroviral approach failed to form detectable teratomas or formed teratomas that were subsequently immune rejected by T-cell infiltration and massive necrosis when they were transplanted to syngeneic mice. Further, they showed that iPSCs generated by an episomal approach were also rejected by CD4 T cells in syngeneic mice (Zhao et al. 2011). The differences between our results and their results could be explained if aged iPSCs were not immunogenic. However, when we injected MEF-derived iPSCs to syngeneic mice, we observed teratoma formation in syngeneic mice. Another possible explanation can be attributed to the choice of the injection site. We injected iPSCs subcutaneously into the dorsal flank of syngeneic C57BL/6 mice. They were also injected subcutaneously into the hind leg region of C57BL/6 mice. They demonstrated that several genes were upregulated in differentiated teratoma, which were rejected by CD4 T cells. The immunogenicity of iPSCs presents major problems for personalized cell therapy and must be investigated in greater detail using our aged iPSCs.

Next, we attempted to differentiate aged iPSCs to three germ cell lines in vitro. We cultured aged iPSCs by the hanging drop method for 8 days to induce embryoid

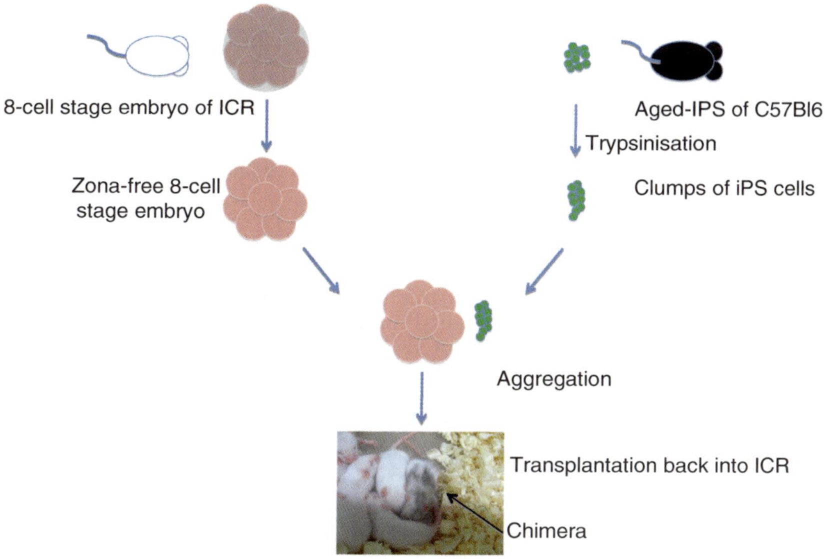

Fig. 21.8 Chimera production by aged iPSCs. Aged iPSCs are aggregated with 8-cell stage embryos derived from fertilized eggs of 2-month-old ICR mice

bodies. They were then transferred to 24-well trays. When these cultures were stained with tissue-specific antibodies, we detected tissue cells belonging to all the three germ layers (Fig. 21.7) (Cheng et al. 2011). Currently, we are attempting to differentiate aged iPSCs in vitro to various tissues.

21.6 Future Directions

In order to use iPSCs for regenerative medicine, especially iPSCs from elderly patients, the cells must be differentiated properly. We would like to use the murine model to determine whether aged iPSCs can be developed in vivo as are those from young mice. Namely, can iPS technology reverse aging and produce young mice, even if the original cells are derived from aged mice? We have set up chimera experiments (Fig. 21.8), which will ultimately induce newborns fully made by iPSCs (germ line transmission). When we performed chimeric experiments using 8-cell aggregation methods, we obtained some chimeric mice (Fig. 21.8), although we have not yet achieved germ line transmission.

Similar questions about aging have been raised by SCNT technology (Kishigami et al 2008). Using in vitro senescent cow cells, the birth of six healthy cloned calves was reported. Nuclear transfer extended the replicative life span of senescent cells (zero to four population doublings remaining) to greater than 90 population doublings (Lanza et al. 2000). Thus, SCNT can restore at least cellular senescence.

The first cloned mammal, Dolly, was found to have short telomeres that were comparable to the age of the cell donor (Shiels et al. 1999). However, later reports showed that murine clones grown over five generations did not have shorter telomeres (Wakayama et al. 2000). The first cloned mouse, Cumulina, died after 2 years and 7 months, which is slightly longer than the average life span of a mouse (Tamashiro et al. 2003). Although SCNT into oocytes and the cloning of animals from a single cell showed the possibility of the successful reprogramming of a somatic nucleus (Gurdon and Melton 2008), there are no reports of the application of SCNT using mouse cells that originated from old mice (e.g. 2 years old). Recently, Wakayama's group succeeded in establishing nuclear transfer embryonic stem (ntES) cell lines from aged mice with an establishment rate of 10–25%, irrespective of sex or strain (Mizutani et al. 2008). Our current trial to produce germ line transmission from aged iPSCs (21-month-old C57BL/6 mice) and examine their life span will hopefully answer fundamental questions of aging and determine the usefulness of iPSCs for future regenerative medicine.

References

Banito A, Rashid ST, Acosta JC, Li S, Pereira CF, Geti I et al (2009) Senescence impairs successful reprogramming to pluripotent stem cells. Genes Dev 23(18):2134–2139

Blasco MA (2007) The epigenetic regulation of mammalian telomeres. Nat Rev Genet 8(4):299–309

Blau HM, Pavlath GK, Hardeman EC, Chiu CP, Silberstein L, Webster SG et al (1985) Plasticity of the differentiated state. Science 230(4727):758–766

Boguski MS (2002) Comparative genomics: the mouse that roared. Nature 420(6915):515–516

Cheng Z, Ito S, Nishio N, Xiao H, Zhang R, Suzuki H et al (2011) Establishment of induced pluripotent stem cells from aged mice using bone marrow-derived myeloid cells. J Mol Cell Biol 3(2):91–98

Christensen BC, Houseman EA, Marsit CJ, Zheng S, Wrensch MR, Wiemels JL et al (2009) Aging and environmental exposures alter tissue-specific DNA methylation dependent upon CpG island context. PLoS Genet 5(8):e1000602

Dimos JT, Rodolfa KT, Niakan KK, Weisenthal LM, Mitsumoto H, Chung W et al (2008) Induced pluripotent stem cells generated from patients with ALS can be differentiated into motor neurons. Science 321(5893):1218–1221

Evans MJ, Kaufman MH (1981) Establishment in culture of pluripotential cells from mouse embryos. Nature 292(5819):154–156

Flores I, Benetti R, Blasco MA (2006) Telomerase regulation and stem cell behaviour. Curr Opin Cell Biol 18(3):254–260

Gaspar-Maia A, Alajem A, Meshorer E, Ramalho-Santos M (2011) Open chromatin in pluripotency and reprogramming. Nat Rev Mol Cell Biol 12(1):36–47

Gore A, Li Z, Fung HL, Young JE, Agarwal S, Antosiewicz-Bourget J et al (2011) Somatic coding mutations in human induced pluripotent stem cells. Nature 471(7336):63–67

Greer EL, Maures TJ, Hauswirth AG, Green EM, Leeman DS, Maro GS et al (2010) Members of the H3K4 trimethylation complex regulate lifespan in a germline-dependent manner in *C. elegans*. Nature 466(7304):383–387

Gurdon JB (1962) The transplantation of nuclei between two species of Xenopus. Dev Biol 5:68–83

Gurdon JB, Melton DA (2008) Nuclear reprogramming in cells. Science 322(5909):1811–1815

Harley CB, Futcher AB, Greider CW (1990) Telomeres shorten during ageing of human fibroblasts. Nature 345(6274):458–460

Hawkins RD, Hon GC, Lee LK, Ngo Q, Lister R, Pelizzola M et al (2010) Distinct epigenomic landscapes of pluripotent and lineage-committed human cells. Cell Stem Cell 6(5):479–491

Hayflick L (1965) The limited in vitro lifetime of human diploid cell strains. Exp Cell Res 37:614–636

Herskind AM, McGue M, Holm NV, Sørensen TI, Harvald B, Vaupel JW (1996) The heritability of human longevity: a population-based study of 2872 Danish twin pairs born 1870–1900. Hum Genet 97(3):319–323

Hogan BL (1976) Changes in the behaviour of teratocarcinoma cells cultivated in vitro. Nature 263(5573):136–137

Hong H, Takahashi K, Ichisaka T, Aoi T, Kanagawa O, Nakagawa M et al (2009) Suppression of induced pluripotent stem cell generation by the p53–p21 pathway. Nature 460(7259):1132–1135

Hussein SM, Batada NN, Vuoristo S, Ching RW, Autio R, Närvä E et al (2011) Copy number variation and selection during reprogramming to pluripotency. Nature 471(7336):58–62

Jaenisch R, Young R (2008) Stem cells, the molecular circuitry of pluripotency and nuclear reprogramming. Cell 132(4):567–582

Kanatsu-Shinohara M, Inoue K, Lee J, Yoshimoto M, Ogonuki N, Miki H et al (2004) Generation of pluripotent stem cells from neonatal mouse testis. Cell 119(7):1001–1012

Kawamura T, Suzuki J, Wang YV, Menendez S, Morera LB, Raya A et al (2009) Linking the p53 tumour suppressor pathway to somatic cell reprogramming. Nature 460(7259):1140–1144

Kishigami S, Wakayama S, Hosoi Y, Iritani A, Wakayama T (2008) Somatic cell nuclear transfer: infinite reproduction of a unique diploid genome. Exp Cell Res 314(9):1945–1950

Koh KP, Yabuuchi A, Rao S, Huang Y, Cunniff K, Nardone J et al (2011) Tet1 and Tet2 regulate 5-hydroxymethylcytosine production and cell lineage specification in mouse embryonic stem cells. Cell Stem Cell 8(2):200–213

Lanza RP, Cibelli JB, Blackwell C, Cristofalo VJ, Francis MK, Baerlocher GM et al (2000) Extension of cell life-span and telomere length in animals cloned from senescent somatic cells. Science 288(5466):665–669

Laurent LC, Ulitsky I, Slavin I, Tran H, Schork A, Morey R et al (2011) Dynamic changes in the copy number of pluripotency and cell proliferation genes in human ESCs and iPSCs during reprogramming and time in culture. Cell Stem Cell 8(1):106–118

Li H, Collado M, Villasante A, Strati K, Ortega S, Cañamero M et al (2009) The Ink4/Arf locus is a barrier for iPS cell reprogramming. Nature 460(7259):1136–1139

Lister R, Pelizzola M, Kida YS, Hawkins RD, Nery JR, Hon G et al (2011) Hotspots of aberrant epigenomic reprogramming in human induced pluripotent stem cells. Nature 471(7336):68–73

Loh YH, Zhang W, Chen X, George J, Ng HH (2007) Jmjd1a and Jmjd2c histone H3 Lys 9 demethylases regulate self-renewal in embryonic stem cells. Genes Dev 21(20):2545–2557

Lowry WE, Richter L, Yachechko R, Pyle AD, Tchieu J, Sridharan R et al (2008) Generation of human induced pluripotent stem cells from dermal fibroblasts. Proc Natl Acad Sci USA 105 (8):2883–2888

Maegawa S, Hinkal G, Kim HS, Shen L, Zhang L, Zhang J et al (2010) Widespread and tissue specific age-related DNA methylation changes in mice. Genome Res 20(3):332–340

Manosalva I, González A (2010) Aging changes the chromatin configuration and histone methylation of mouse oocytes at germinal vesicle stage. Theriogenology 74(9):1539–1547

Marión RM, Strati K, Li H, Murga M, Blanco R, Ortega S et al (2009) A p53-mediated DNA damage response limits reprogramming to ensure iPS cell genomic integrity. Nature 460 (7259):1149–1153

Martin GM, Sprague CA, Epstein CJ (1970) Replicative life-span of cultivated human cells. Effects of donor's age, tissue, and genotype. Lab Invest 23(1):86–92

Mayshar Y, Ben-David U, Lavon N, Biancotti JC, Yakir B, Clark AT et al (2010) Identification and classification of chromosomal aberrations in human induced pluripotent stem cells. Cell Stem Cell 7(4):521–531

Mikkelsen TS, Hanna J, Zhang X, Ku M, Wernig M, Schorderet P et al (2008) Dissecting direct reprogramming through integrative genomic analysis. Nature 454(7200):49–55
Mitchell BD, Hsueh WC, King TM, Pollin TI, Sorkin J, Agarwala R et al (2001) Heritability of life span in the Old Order Amish. Am J Med Genet 102(4):346–352
Mizutani E, Ono T, Li C, Maki-Suetsugu R, Wakayama T (2008) Propagation of senescent mice using nuclear transfer embryonic stem cell lines. Genesis 46(9):478–483
Nishio N, Okawa Y, Sakurai H, Isobe K (2008) Neutrophil depletion delays wound repair in aged mice. Age (Dordr) 30(1):11–19
Oeppen J, Vaupel JW (2002) Demography. Broken limits to life expectancy. Science 296 (5570):1029–1031
Ogonuki N, Inoue K, Yamamoto Y, Noguchi Y, Tanemura K, Suzuki O et al (2002) Early death of mice cloned from somatic cells. Nat Genet 30(3):253–254
Paigen K. (1995) A miracle enough the power of mice. Nat Med 1:215–220
Park IH, Zhao R, West JA, Yabuuchi A, Huo H, Ince TA et al (2008) Reprogramming of human somatic cells to pluripotency with defined factors. Nature 451(7175):141–146
Peters LL, Robledo RF, Bult CJ, Churchill GA, Paigen BJ, Svenson KL (2007) The mouse as a model for human biology: a resource guide for complex trait analysis. Nat Rev Genet 8(1):58–69
Rando TA (2006) Stem cells, ageing and the quest for immortality. Nature 441(7097):1080–1086
Röhme D (1981) Evidence for a relationship between longevity of mammalian species and life spans of normal fibroblasts in vitro and erythrocytes in vivo. Proc Natl Acad Sci USA 78(8):5009–5013
Shiels PG, Kind AJ, Campbell KH, Waddington D, Wilmut I, Colman A et al (1999) Analysis of telomere lengths in cloned sheep. Nature 399(6734):316–317
Stadtfeld M, Maherali N, Breault DT, Hochedlinger K (2008) Defining molecular cornerstones during fibroblast to iPS cell reprogramming in mouse. Cell Stem Cell 2(3):230–240
Surani MA (1999) Reprogramming a somatic nucleus by trans-modification activity in germ cells. Semin Cell Dev Biol 10(3):273–277
Takahashi K, Tanabe K, Ohnuki M, Narita M, Ichisaka T, Tomoda K et al (2007) Induction of pluripotent stem cells from adult human fibroblasts by defined factors. Cell 131(5):861–872
Takahashi K, Yamanaka S (2006) Induction of pluripotent stem cells from mouse embryonic and adult fibroblast cultures by defined factors. Cell 126(4):663–676
Tamashiro KL, Wakayama T, Yamazaki Y, Akutsu H, Woods SC, Kondo S et al (2003) Phenotype of cloned mice: development, behavior, and physiology. Exp Biol Med (Maywood) 228 (10):1193–1200
Utikal J, Polo JM, Stadtfeld M, Maherali N, Kulalert W, Walsh RM et al (2009) Immortalization eliminates a roadblock during cellular reprogramming into iPS cells. Nature 460(7259):1145–1148
vB Hjelmborg J, Iachine I, Skytthe A, Vaupel JW, McGue M, Koskenvuo M et al (2006) Genetic influence on human lifespan and longevity. Hum Genet 119(3):312–321
Wakayama T, Perry AC, Zuccotti M, Johnson KR, Yanagimachi R (1998) Full-term development of mice from enucleated oocytes injected with cumulus cell nuclei. Nature 394(6691):369–374
Wakayama T, Shinkai Y, Tamashiro KL, Niida H, Blanchard DC, Blanchard RJ et al (2000) Cloning of mice to six generations. Nature 407(6802):318–319
Wang C, Jurk D, Maddick M, Nelson G, Martin-Ruiz C, von Zglinicki T (2009) DNA damage response and cellular senescence in tissues of aging mice. Aging Cell 8(3):311–323
Wernig M, Meissner A, Foreman R, Brambrink T, Ku M, Hochedlinger K et al (2007) In vitro reprogramming of fibroblasts into a pluripotent ES-cell-like state. Nature 448(7151):318–324
Williams K, Christensen J, Pedersen MT, Johansen JV, Cloos PA, Rappsilber J et al (2011) TET1 and hydroxymethylcytosine in transcription and DNA methylation fidelity. Nature 473 (7347):343–348
Wilmut I, Schnieke AE, McWhir J, Kind AJ, Campbell KH (1997) Viable offspring derived from fetal and adult mammalian cells. Nature 385(6619):810–813
Wu H, D'Alessio AC, Ito S, Wang Z, Cui K, Zhao K et al (2011) Genome-wide analysis of 5-hydroxymethylcytosine distribution reveals its dual function in transcriptional regulation in mouse embryonic stem cells. Genes Dev 25(7):679–684

Ying QL, Wray J, Nichols J, Batlle-Morera L, Doble B, Woodgett J et al (2008) The ground state of embryonic stem cell self-renewal. Nature 453(7194):519–523
Yu J, Vodyanik MA, Smuga-Otto K, Antosiewicz-Bourget J, Frane JL, Tian S et al (2007) Induced pluripotent stem cell lines derived from human somatic cells. Science 318(5858):1917–1920
Yuan R, Peters LL, Paigen B (2011) Mice as a mammalian model for research on the genetics of aging. ILAR J 52(1):4–15
Zhao T, Zhang ZN, Rong Z, Xu Y (2011) Immunogenicity of induced pluripotent stem cells. Nature 474(7350):212–215

Index

J.Z. Kubiak (ed.), *Mouse Development*, Results and Problems in Cell Differentiation 55, DOI 10.1007/978-3-642-30406-4, © Springer-Verlag Berlin Heidelberg 2012

D

E

R

S

T

Printed by Printforce, the Netherlands